VOLUME THREE HUNDRED AND FORTY FIVE

INTERNATIONAL REVIEW OF CELL AND MOLECULAR BIOLOGY

Nucleic Acid Sensing and Immunity - Part B

INTERNATIONAL REVIEW OF CELL AND MOLECULAR BIOLOGY

VOLUME THREE HUNDRED AND FORTY FIVE

International Review of CELL AND MOLECULAR BIOLOGY

Nucleic Acid Sensing and Immunity - Part B

Edited by

CLAIRE VANPOUILLE-BOX
Weill Cornell Medical College, New York, NY, United States

LORENZO GALLUZZI
Weill Cornell Medical College, New York, NY, United States

ELSEVIER

ACADEMIC PRESS
An imprint of Elsevier

Academic Press is an imprint of Elsevier
50 Hampshire Street, 5th Floor, Cambridge, MA 02139, United States
525 B Street, Suite 1650, San Diego, CA 92101, United States
The Boulevard, Langford Lane, Kidlington, Oxford OX5 1GB, United Kingdom
125 London Wall, London, EC2Y 5AS, United Kingdom

First edition 2019

ISBN: 978-0-12-815981-1
ISSN: 1937-6448

Publisher: Zoe Kruze
Acquisition Editor: Ashlie M. Jackman
Editorial Project Manager: Teresa Pons-Ferrer
Production Project Manager: James Selvam
Cover Designer: Alan Studholme

Typeset by SPi Global, India

CONTENTS

CONTRIBUTORS

Julia Chronopoulos
Meakins Christie Laboratories, Research Institute of the McGill University Health Centre and McGill University, Montreal, QC, Canada

Hanne Dubois
Department of Internal Medicine; Ghent Gut Inflammation Group (GGIG), Ghent University; VIB-UGent Center for Inflammation Research, VIB, Ghent, Belgium

Soroor Farahnak
Meakins Christie Laboratories, Research Institute of the McGill University Health Centre and McGill University, Montreal, QC, Canada

Alan G. Goodman
School of Molecular Biosciences, College of Veterinary Medicine; Paul G. Allen School for Global Animal Health, College of Veterinary Medicine, Washington State University, Pullman, WA, United States

Jian-Guo He
MOE Key Laboratory of Aquatic Product Safety/State Key Laboratory of Biocontrol, School of Life Sciences; School of Marine Sciences, Sun Yat-Sen University, Guangzhou, China

Martin Heil
Departmento de Ingeniería Genética, CINVESTAV-Irapuato, Irapuato, Guanajuato, Mexico

Chaitali Khan
Department of Biological Sciences, Tata Institute of Fundamental Research, Mumbai, India

R. Marena Guzman
School of Molecular Biosciences, College of Veterinary Medicine, Washington State University, Pullman, WA, United States

James G. Martin
Meakins Christie Laboratories, Research Institute of the McGill University Health Centre and McGill University, Montreal, QC, Canada

Keesha M. Matz
School of Molecular Biosciences, College of Veterinary Medicine, Washington State University, Pullman, WA, United States

Sonia Muliyil
Department of Biological Sciences, Tata Institute of Fundamental Research, Mumbai, India

B.J. Rao
Department of Biological Sciences, Tata Institute of Fundamental Research, Mumbai, India

Geert van Loo
VIB-UGent Center for Inflammation Research, VIB; Department of Biomedical Molecular Biology, Ghent University, Ghent, Belgium

Isaac Vega-Muñoz
Departmento de Ingeniería Genética, CINVESTAV-Irapuato, Irapuato, Guanajuato, Mexico

Pei-Hui Wang
Advanced Medical Research Institute, Shandong University, Jinan, China; School of Biomedical Sciences, The University of Hong Kong, Pokfulam, Hong Kong

Andy Wullaert
Department of Internal Medicine; Ghent Gut Inflammation Group (GGIG), Ghent University; VIB-UGent Center for Inflammation Research, VIB, Ghent, Belgium

NUCLEIC ACID SENSING AT THE INTERFACE BETWEEN INNATE AND ADAPTIVE IMMUNITY

Innate immunity largely relies on the recognition of evolutionarily conserved structures on pathogens that are commonly known as microbe-associated molecular patterns (MAMPs) or pathogen-associated molecular patterns (PAMPs), through a limited number of pattern recognition receptors (PRRs) (Kawai and Akira, 2011). MAMPs are invariant among entire classes of microbes, are often (but not always) essential for their survival, and can be easily distinguished from "self" structures (Schlee and Hartmann, 2016). Importantly, PRRs can also sense endogenous factors, especially when these are present at ectopic locations (e.g., cytoplasmic nucleic acids and/or extracellular ATP) or in excessive amounts as a consequence of infection, inflammation, or other types of threats to cellular and organismal homeostasis (Galluzzi et al., 2018b; Vanpouille-Box et al., 2018). Endogenous factors that can activate PRRs are cumulatively known as damage-associated molecular patterns (DAMPs) (Krysko et al., 2012; Liston and Masters, 2017). Upon MAMP or DAMP recognition, PRRs deliver a danger signal to the host, generally resulting in the initiation of proinflammatory responses aiming at the preservation of organismal homeostasis (Cao, 2016). Ultimately, PRR engagement orchestrates the early response of the host to a threat (be it endogenous or exogenous in nature) and sets the stage for the activation of adaptive immunity (Iwasaki and Medzhitov, 2015).

The immunostimulatory properties of nucleic acids have been described as early as in 1963 by Warren and Isaac, who discovered that DNA and RNA molecules derived from microbes or host cells can induce interferon (IFN) production by fibroblasts (Jensen et al., 1963; Rotem et al., 1963). However, the existence of dedicated nucleic acid sensors was ignored at the time. Our current understanding of innate immunity began indeed to take shape only two decades later, following a hypothesis from Janeway (1989) that stimulated the discovery that Toll (a *Drosophila melanogaster* protein that was known for its role in embryonic development) can operate as a PRR capable to drive the secretion of antimicrobial peptides in response to fungal infection (Lemaitre et al., 1996). Shortly thereafter, the first human Toll-like receptor (TLR) was reported to stimulate the secretion of several proinflammatory cytokines on activation (Medzhitov et al., 1997), but it was only in the early 2000s that TLR3 and TLR9 were discovered to respond to viral double-stranded RNA and bacterial CpG-rich DNA, respectively (Alexopoulou et al., 2001; Hemmi et al., 2000). Since then, more than a dozen different human sensors for DNA and RNA have been discovered (Roers et al., 2016; Vanpouille-Box et al., 2018).

Given the central role of nucleic acid sensors in the control of invading pathogens and endogenous threats, it is not surprising that both microbes and malignant cells have evolved a large number of strategies to circumvent the activation of these PRRs (Chan and Gack, 2016; Chen et al., 2016; Galluzzi et al., 2017). As a stand-alone example, some cancer cells can evade the activation of cyclic GMP-AMP synthase (a sensor for cytosolic DNA) by upregulating three-prime repair exonuclease 1 (TREX1), resulting in limited type I IFN production in response to stress (be it endogenous or imposed by treatment) (Galluzzi et al., 2018a; Xia et al., 2016; Zitvogel et al., 2015), and hence a poorly inflammatory profile that supports immune evasion (Vanpouille-Box et al., 2017). Moreover, reflecting the potent proinflammatory responses driven by nucleic acid sensors, deregulated signaling via these PRRs has been associated with a plethora of pathologies, including rheumatic and autoimmune disorders (Crow et al., 2018; Muskardin and Niewold, 2018), neurological conditions (McGlasson et al., 2015), cardiac degeneration (King et al., 2017), and others (McNab et al., 2015). Thus, nucleic acid sensors and their downstream effectors have become attractive targets for the development of therapeutic agents with various applications (Junt and Barchet, 2015; Parker et al., 2016; Weiss et al., 2017).

In *Nucleic Acid Sensing and Immunity* (a special issue of the *International Review of Cell and Molecular Biology* series) world-class experts in the field offer an overview of (1) the major molecular and cellular aspects of nucleic acid sensing across the eukaryotic kingdom; (2) the complexity of innate and adaptive immune responses and their key role in the maintenance of cellular and organismal homeostasis; and (3) the intimate connections between deregulated nucleic acid sensing and human pathology. Each of the thirteen chapters of *Nucleic Acid Sensing and Immunity* focuses on one specific aspect of nucleic acid sensing, encompassing signaling, regulation, and pathophysiological relevance. The book (which consists of two parts, Part A and Part B) is equally addressed to expert investigators who may wish to extend their knowledge on nucleic acid sensing at the interface between innate and adaptive immunity, and to newcomers to this exciting and quickly progressing field of investigation.

ACKNOWLEDGMENTS

LG is supported by a Breakthrough Level 2 grant from the US Department of Defense (DoD), Breast Cancer Research Program (BRCP) (#BC180476P1), by a startup grant from the Dept. of Radiation Oncology at Weill Cornell Medicine (New York, US), by industrial collaborations with Lytix (Oslo, Norway) and Phosplatin (New York, US), and by donations from Phosplatin (New York, US), the Luke Heller TECPR2 Foundation (Boston, US) and Sotio a.s. (Prague, Czech Republic).

DISCLOSURES

LG provides remunerated consulting to OmniSEQ (Buffalo, NY, USA), Astra Zeneca (Gaithersburg, MD, USA), VL47 (New York, NY, USA) and the Luke Heller TECPR2 Foundation (Boston, MA, USA), and he is member of the Scientific Advisory Committee of OmniSEQ (Buffalo, NY, USA).

Claire Vanpouille-Box
Department of Radiation Oncology, Weill Cornell Medical College, New York, NY, United States
claire.vanpouille.box@gmail.com
Lorenzo Galluzzi
Department of Radiation Oncology, Weill Cornell Medical College, New York, NY, United States
Sandra and Edward Meyer Cancer Center, New York, NY, United States
Department of Dermatology, Yale School of Medicine, New Haven, CT, United States
Université Paris Descartes/Paris V, Paris, France
deadoc80@gmail.com

REFERENCES

Alexopoulou, L., Holt, A.C., Medzhitov, R., Flavell, R.A., 2001. Recognition of double-stranded RNA and activation of NF-kappaB by Toll-like receptor 3. Nature 413, 732–738.
Cao, X., 2016. Self-regulation and cross-regulation of pattern-recognition receptor signalling in health and disease. Nat. Rev. Immunol. 16, 35–50.
Chan, Y.K., Gack, M.U., 2016. Viral evasion of intracellular DNA and RNA sensing. Nat. Rev. Microbiol. 14, 360–373.
Chen, Q., Sun, L., Chen, Z.J., 2016. Regulation and function of the cGAS-STING pathway of cytosolic DNA sensing. Nat. Immunol. 17, 1142–1149.
Crow, M.K., Olferiev, M., Kirou, K.A., 2018. Type I interferons in autoimmune disease. Annu. Rev. Pathol. https://doi.org/10.1146/annurev-pathol-020117-043952.
Galluzzi, L., Buque, A., Kepp, O., Zitvogel, L., Kroemer, G., 2017. Immunogenic cell death in cancer and infectious disease. Nat. Rev. Immunol. 17, 97–111.
Galluzzi, L., Chan, T.A., Kroemer, G., Wolchok, J.D., Lopez-Soto, A., 2018a. The hallmarks of successful anticancer immunotherapy. Sci. Transl. Med. 10 (459), pii: eaat7807. https://doi.org/10.1126/scitranslmed.aat7807.

Galluzzi, L., Yamazaki, T., Kroemer, G., 2018b. Linking cellular stress responses to systemic homeostasis. Nat. Rev. Mol. Cell Biol. 19, 731–745.

Hemmi, H., Takeuchi, O., Kawai, T., Kaisho, T., Sato, S., Sanjo, H., Matsumoto, M., Hoshino, K., Wagner, H., Takeda, K., Akira, S., 2000. A Toll-like receptor recognizes bacterial DNA. Nature 408, 740–745.

Iwasaki, A., Medzhitov, R., 2015. Control of adaptive immunity by the innate immune system. Nat. Immunol. 16, 343–353.

Janeway Jr., C.A., 1989. Approaching the asymptote? Evolution and revolution in immunology. Cold Spring Harb. Symp. Quant. Biol. 54 (Pt. 1), 1–13.

Jensen, K.E., Neal, A.L., Owens, R.E., Warren, J., 1963. Interferon responses of chick embryo fibroblasts to nucleic acids and related compounds. Nature 200, 433–434.

Junt, T., Barchet, W., 2015. Translating nucleic acid-sensing pathways into therapies. Nat. Rev. Immunol. 15, 529–544.

Kawai, T., Akira, S., 2011. Toll-like receptors and their crosstalk with other innate receptors in infection and immunity. Immunity 34, 637–650.

King, K.R., Aguirre, A.D., Ye, Y.X., Sun, Y., Roh, J.D., Ng Jr., R.P., Kohler, R.H., Arlauckas, S.P., Iwamoto, Y., Savol, A., Sadreyev, R.I., Kelly, M., Fitzgibbons, T.P., Fitzgerald, K.A., Mitchison, T., Libby, P., Nahrendorf, M., Weissleder, R., 2017. IRF3 and type I interferons fuel a fatal response to myocardial infarction. Nat. Med. 23, 1481–1487.

Krysko, D.V., Garg, A.D., Kaczmarek, A., Krysko, O., Agostinis, P., Vandenabeele, P., 2012. Immunogenic cell death and DAMPs in cancer therapy. Nat. Rev. Cancer 12, 860–875.

Lemaitre, B., Nicolas, E., Michaut, L., Reichhart, J.M., Hoffmann, J.A., 1996. The dorsoventral regulatory gene cassette spätzle/Toll/cactus controls the potent antifungal response in Drosophila adults. Cell 86, 973–983.

Liston, A., Masters, S.L., 2017. Homeostasis-altering molecular processes as mechanisms of inflammasome activation. Nat. Rev. Immunol. 17, 208–214.

McGlasson, S., Jury, A., Jackson, A., Hunt, D., 2015. Type I interferon dysregulation and neurological disease. Nat. Rev. Neurol. 11, 515–523.

McNab, F., Mayer-Barber, K., Sher, A., Wack, A., O'Garra, A., 2015. Type I interferons in infectious disease. Nat. Rev. Immunol. 15, 87–103.

Medzhitov, R., Preston-Hurlburt, P., Janeway Jr., C.A., 1997. A human homologue of the Drosophila Toll protein signals activation of adaptive immunity. Nature 388, 394–397.

Muskardin, T.L.W., Niewold, T.B., 2018. Type I interferon in rheumatic diseases. Nat. Rev. Rheumatol. 14, 214–228.

Parker, B.S., Rautela, J., Hertzog, P.J., 2016. Antitumour actions of interferons: implications for cancer therapy. Nat. Rev. Cancer 16, 131–144.

Roers, A., Hiller, B., Hornung, V., 2016. Recognition of endogenous nucleic acids by the innate immune system. Immunity 44, 739–754.

Rotem, Z., Cox, R.A., Isaacs, A., 1963. Inhibition of virus multiplication by foreign nucleic acid. Nature 197, 564–566.

Schlee, M., Hartmann, G., 2016. Discriminating self from non-self in nucleic acid sensing. Nat. Rev. Immunol. 16, 566–580.

Vanpouille-Box, C., Alard, A., Aryankalayil, M.J., Sarfraz, Y., Diamond, J.M., Schneider, R.J., Inghirami, G., Coleman, C.N., Formenti, S.C., Demaria, S., 2017. DNA exonuclease Trex1 regulates radiotherapy-induced tumour immunogenicity. Nat. Commun. 8, 15618.

Vanpouille-Box, C., Demaria, S., Formenti, S.C., Galluzzi, L., 2018. Cytosolic DNA sensing in organismal tumor control. Cancer Cell 34, 361–378.

Weiss, J.M., Guerin, M.V., Regnier, F., Renault, G., Galy-Fauroux, I., Vimeux, L., Feuillet, V., Peranzoni, E., Thoreau, M., Trautmann, A., Bercovici, N., 2017. The STING agonist DMXAA triggers a cooperation between T lymphocytes and myeloid cells that leads to tumor regression. Oncoimmunology 6, e1346765.

Xia, T., Konno, H., Ahn, J., Barber, G.N., 2016. Deregulation of STING signaling in colorectal carcinoma constrains DNA damage responses and correlates with tumorigenesis. Cell Rep. 14, 282–297.

Zitvogel, L., Galluzzi, L., Kepp, O., Smyth, M.J., Kroemer, G., 2015. Type I interferons in anticancer immunity. Nat. Rev. Immunol. 15, 405–414.

CHAPTER ONE

Nucleic Acid Sensing in Allergic Disorders

Soroor Farahnak, Julia Chronopoulos, James G. Martin[1]
Meakins Christie Laboratories, Research Institute of the McGill University Health Centre and McGill University, Montreal, QC, Canada
[1]Corresponding author: e-mail address: james.martin@mcgill.ca

Contents

Abstract

Recent advances indicate that there is crosstalk between allergic disorders and nucleic acid sensing. Triggers that activate inflammatory mechanisms via nucleic acid sensors affect both allergic phenotypes and anti-viral responses, depending on the timing and the order of exposure. Viral respiratory infections, such as those caused by the rhinovirus, influenza, and respiratory syncytial virus, are the most frequent cause of significant asthma exacerbations through effects mediated predominantly by TLR3. However, agonists of other nucleic acid sensors, such as TLR7/8 and TLR9 agonists, may inhibit allergic inflammation and reduce clinical manifestations of disease. The allergic state can predispose the immune system to both exaggerated responses to viral infections or protection from anti-viral inflammatory responses. T_H2 cytokines appear to alter the epithelium, leading to defective viral clearance or exaggerated responses to viral infections.

International Review of Cell and Molecular Biology, Volume 345
ISSN 1937-6448
https://doi.org/10.1016/bs.ircmb.2018.11.001

However, a T_H2 skewed allergic response may be protective against a T_H1-dependent inflammatory anti-viral response. This review briefly introduces the receptors involved in nucleic acid sensing, addresses mechanisms by which nucleic acid sensing and allergic responses can counteract one another, and discusses the strategies in experimental settings, both in animal and human studies, to harness the nucleic acid sensing machinery for the intervention of allergic disorders.

1. INTRODUCTION

Eczema, allergic rhinitis, and the majority of cases of asthma are associated with an adaptive immune response related to T cell expression of so-called T_H2 cytokines. Although T cells may differentiate into a wide range of subsets, T_H2 cells and their signature cytokines, including interleukins (IL) 4, 5, 9, 10, 13, and 25 are the key players in allergic diseases (Caza and Landas, 2015; Hirahara et al., 2013; Paul and Zhu, 2010). These cytokines drive the synthesis of immunoglobulin E (IgE), tissue eosinophilia, and in the case of allergic asthma, hyperresponsiveness of the airways to inhaled bronchoconstrictors (Cockcroft et al., 1977; Galli et al., 2008). T_H2 differentiation is highly influenced by the cytokine milieu from innate immune cells as well as naïve $CD4^+$ T cells (Noben-Trauth et al., 2000). There is emerging evidence that nucleic acid sensing by innate and adaptive immune cells and structural cells condition the tissue milieu to modulate the predisposition to develop a T_H2 inflammatory response.

Nucleic acid sensing and self/non-self recognition is an immune mechanism that has been well preserved during evolution. In invertebrates, the main mechanism to sense and defend against non-self nucleic acids is RNA interference (RNAi) and direct degradation of viral RNA (Svoboda, 2014). However, present in vertebrates, RNAi is not the only mechanism to sense foreign nucleic acids. Pathogenic and self-nucleic acids as well as their metabolites can also be recognized as pathogen associated molecular patterns (PAMPs) and danger associated molecular patterns (DAMPs), respectively, via pathogen recognition receptors (PRRs) which sense nucleic acids based on their localization, availability, sequence, molecular structures, and modifications. These PRRs activate and shape the adaptive immune response by modulating cytokine and chemokine milieus (Iwasaki and Medzhitov, 2010). Signaling through PRRs is necessary to trigger an immune response and assure survival following infection, but receptors and downstream pathways that are activated and the by-products generated can sometimes have adverse effects on the host.

Responses of structural cells such as the epithelium, the primary target of foreign insults and pathogens, generate key host defense molecules, such as so-called alarmins including IL-33, thymic stromal lymphopoietin (TSLP) and IL-25, which may contribute to the magnitude of the characteristic eosinophilic inflammation of allergic airway disease (Lee et al., 2012; Mitchell and O'Byrne, 2017). Apart from innate immune cell types, the potential of adaptive immune cells, such as B and T lymphocytes, to respond to self and non-self molecular patterns via PRRs may contribute to the adaptive immune response to allergens (Imanishi et al., 2014; Sharma et al., 2015).

Not surprisingly, there has been enormous interest in understanding the mechanisms behind the rise of diseases that are associated with T_H2 driven inflammation. Strachan et al. described lower rates of allergic disease in children among families with multiple siblings. These findings gave rise to the concept of the hygiene hypothesis—that an excessively clean environment and lack of exposure to pathogens may be responsible for the rise of these allergic conditions, possibly by failing to promote the development of T_H1 and regulatory T cell (Treg) oriented immune response (Romagnani, 2004; Strachan, 1989). Alternative causes of allergic disease need consideration and among those are viral illnesses that may alter the maturation of the immune system in early life. Respiratory viruses such as rhinoviruses, influenza, and respiratory syncytial virus (RSV) that activate nucleotide sensing receptors, such as toll-like receptors (TLR) 3 and 7, are among the major culprits causing asthma attacks necessitating hospital visits (Ritchie et al., 2015). RSV infections in infants lead to the development of bronchiolitis and increased risk of asthma in later childhood (Stensballe et al., 2009). Although it is tempting to attribute the pathogenesis of asthma to these viral illnesses, the causal role of the viral infections in inducing asthma *de novo* is still uncertain.

This review focuses on the role of nucleic acid sensing receptors in modulating allergic inflammation and the consequences of disease prevalence. The discussion is focused on asthma as the most frequently explored pathology in human and in model systems. In addition to the pathogenic role of nucleotide sensing receptors, this review will also discuss the modulation of these receptors for therapeutic purposes. Based on the rationale that T_H2 inflammation is the basis for most cases of asthma, attempts to bias the response to environmental triggers toward alternative patterns of inflammation have been made using TLR agonists. In particular, agonists of TLR 7/8 and 9 have been tested for their potential therapeutic effects on models of allergic asthma and in human trials of allergic rhinitis (Camateros et al., 2007; Kline et al., 1998; Moisan et al., 2006).

2. CATEGORIZING NUCLEIC ACID SENSING RECEPTORS BASED ON MECHANISMS OF ACTION (TABLE 1)

2.1 Group I Nucleic Acid Sensing Receptors

The first group of receptors includes TLR 3, 7, 8 and 9, nucleotide-binding oligomerization domain-like receptors (NLRs), absent in melanoma 2 (AIM2) and cyclic GMP–AMP synthetase (cGAS). This group of immune sensing pattern recognition receptors triggers an immune response by activating transcription factors such as nuclear factor-κB (NF-κB) and interferon (IFN)-regulatory factors (IRFs) to upregulate the expression of anti-viral cytokines, mainly type I IFNs (IFN-α and IFN-β) and IFN-stimulated genes (ISGs) (Franchi et al., 2009; Kawasaki et al., 2011; Ori et al., 2017; Pichlmair and Reis e Sousa, 2007; Poltorak et al., 2016).

2.1.1 TLRs, Membrane-Bound Pattern Recognition Receptors

TLRs have been more extensively explored than other receptors in the context of allergic inflammation and airway disease. Intracellular TLRs (3, 7, 8, 9) are located in the endoplasmic reticulum (ER), endosomes, and lysosomes (Kawai and Akira, 2011). TLRs are composed of three major domains: the ectodomain or leucine-rich repeat (LRR) which mediates recognition of PAMPs and DAMPs, a transmembrane domain, and an intracellular domain, homologous to IL-1R, known as Toll/IL-1R (TIR) domain, which initiates downstream signaling pathways (Akira et al., 2006). Upon ligand binding, TLRs undergo dimerization and conformational changes leading to the recruitment of Toll/interleukin-1 receptor (TIR) domain containing adaptor molecules, including myeloid differentiation factor 88 (MyD88), TIR-associated protein (TIRAP), TIR-domain-containing adaptor protein-inducing IFN-β (TRIF) and TRIF-related adaptor molecule (TRAM). These adaptors are essential for the activation of NF-κB and IRFs which induce the expression of proinflammatory cytokines, such as type I IFNs, IL-6, and IL-1β (Honda et al., 2006; Liu et al., 2017).

TLR3 mediates responses to double stranded RNA (dsRNA) and is, therefore activated by the common respiratory viruses, rhinovirus (positive sense RNA virus) and RSV (negative-sense RNA virus). TLR3 is expressed by conventional dendritic cells (cDCs), hematopoietic cells, as well as stromal cells, including epithelial cells and fibroblasts (Schulz et al., 2005). TLR7 senses single stranded (ss) RNA and siRNA, and can also be activated by the chemical imiquimod which has anti-viral properties (Hemmi et al., 2002).

Table 1 Nucleic Acid Sensing Receptors and Their Ligands, Sites of Localization and Functions

Nucleic Acid Sensing Receptors	Receptors	Major Ligands	Localization	Function
Group I	TLR3	dsRNA	Membrane-bound in endosome, lysosome, ER	Activation of NF-κB, IRF3 and MAPKs
	TLR7	ssRNA, siRNA	Membrane-bound in endosome, lysosome, ER	Activation of NF-κB, IRF7 and MAPKs
	TLR8	ssRNA	Membrane-bound in endosome, lysosome, ER	Activation of NF-κB, IRF7, IRF5 and MAPKs
	TLR9	Unmethylated DNA	Membrane-bound in endosome, lysosome, ER	Activation of NF-κB, IRF7 and MAPKs
	NLRX1	dsRNA	Cytoplasm	Inflammasome mediated caspase-1 activation, Mitochondrial mediated ROS production and activation of NF-κB signaling pathway
	NLRP3	Pathogenic RNA and DNA, extracellular ATP	Cytoplasm	Inflammasome mediated caspase-1 activation, activation of NF-κB signaling pathway
	AIM2	dsDNA	Cytoplasm	Inflammasome mediated caspase-1 activation
	cGAS	dsDNA	Cytoplasm	STING mediated NF-κB and IRF3 activation

Continued

Table 1 Nucleic Acid Sensing Receptors and Their Ligands, Sites of Localization and Functions—cont'd

Nucleic Acid Sensing Receptors	Receptors	Major Ligands	Localization	Function
Group II	PKR	dsRNA	Cytoplasm	Suppressing translation initiation
	OAS	dsRNA	Cytoplasm	RNase L mediated RNA degradation, RIG-1 and MDA5 mediated type I IFN production
	RNA pol III	Cytosolic DNA	Cytoplasm	Synthesizing RNA ligands for RIG-1 from intracellular DNA
	ADAR1	dsRNA	Cytoplasm	Deamination of adenosine in dsRNA
Group III (RLRs)	RIG-1	ss/dsRNA	Cytoplasm	RNA helicase, MAVS mediated IRF3 and NF-κB activation
	MDA5	dsRNA	Cytoplasm	RNA helicase, MAVS mediated IRF3 and NF-κB activation
	LGP2	dsRNA	Cytoplasm	RNA helicase, regulating RIG-1 and MDA5 functions

TLR7 is the principal receptor expressed by plasmacytoid DCs (pDC), B cells, and monocytes (Gantier et al., 2008; Hornung et al., 2002, 2005; Kim et al., 2008). Eosinophils, natural killer (NK) cells, $CD4^+$ T cells, $CD8^+$ T cells and airway epithelial cells have been reported to express TLR7 as well (Dominguez-Villar et al., 2015; Hart et al., 2005; Nagase et al., 2003; Shikhagaie et al., 2014; Song et al., 2009). TLR8 is another ssRNA sensing receptor that is phylogenetically similar to TLR7 and is mainly expressed by myeloid DCs (mDC), pDCs, and monocytes (Alexopoulou et al., 2012; Bao and Liu, 2013; Hornung et al., 2002). Compared to males, a greater production of IFN-α by peripheral blood mononuclear cells of female subjects stimulated with the TLR7 and TLR8 agonists, resiquimod and imiquimod, has been reported (Berghofer et al., 2006). While both

TLR7 and 8 are X chromosome linked, whether or not they are responsible for a sex difference in anti-viral immune response has yet to be determined. TLR9 was the first discovered nucleic acid sensing PRR which recognizes non methylated 2′-(cytidine-phosphate-guanosine) DNA and is expressed predominantly by pDCs and B cells (Blasius and Beutler, 2010; Roers et al., 2016).

2.1.2 NLRs and Nucleic Acid Sensing

Although primarily expressed in immune cells, NLRs can also be expressed in nonimmune cells, such as epithelial cells. NLRs are cytoplasmic PRRs and are categorized into five main subfamilies based on their N-terminal structures: NLRA, NLRB, NLRC, and NLRX (Koonin and Aravind, 2000; Ting et al., 2008). The major signaling pathways that are induced by NLR include NF-κB, mitogen activated protein kinases (MAPKs), and caspase-1. Ligand binding to NLRs initiates the oligomerization, recruitment, and approximation of effector proteins in a large molecule complex, known as the inflammasome, that recruits and activates caspase-1. This interaction is required for the cleavage of pro-IL-1β, pro-IL-18, and pro-IL-33 into their biologically active forms. Nucleic acid sensing NLRs include NLRP3 and NLRX1. NLRP3 detects bacterial and viral RNA, as well as viral DNA and has been found to respond to infection with the Sendai and influenza viruses (Kanneganti et al., 2006). Moreover, extracellular ATP, a well-known endogenous DAMP, activates the NLRP3 inflammasome (Mariathasan et al., 2006). NLRX1 is bound to the mitochondrial outer membrane and detects cytoplasmic viral dsRNA. It regulates anti-viral signaling by inducing the mitochondrial respiratory chain to produce reactive oxygen species (ROS), which in turn activate NF-κB dependent signaling pathways (De Nardo et al., 2014; Geddes et al., 2009; Kanneganti et al., 2006; Leiva-Juarez et al., 2018; Muruve et al., 2008; Tattoli et al., 2008).

2.1.3 AIM2 and cGAS

AIM2 and cGAS are cytosolic PRRs that sense dsDNA (Kanneganti et al., 2006; Luecke and Paludan, 2017). AIM2 activates caspase-1 via the AIM2 inflammasome, while cGAS belongs to the nucleotidyl transferase family and synthesizes the dinucleotide second messenger cyclic guanosine monophosphate-adenosine monophosphate (cGAMP) to activate the STING dependent signaling pathway (Li et al., 2013). STING then induces NF-κB and IRF3 activation and type I IFN production (Ishikawa and Barber, 2008).

2.2 Group II Nucleic Acid Sensing Receptors

The second group of receptors includes dsRNA-activated protein kinase R (PKR; also known as eukaryotic translation initiation factor 2-alpha kinase 2: eIF2AK2), 2′-5′-oligoadenylate synthetases (OASs), RNA pol III, and adenosine deaminase acting on RNA 1 (ADAR1). This group of cytosolic receptors demonstrates direct anti-viral activity by degradation, chemical modification, and inhibition of translation (Choi et al., 2015; Ezelle et al., 2016; Ori et al., 2017; Zlotorynski, 2018). PKR is a serine/threonine kinase, activated by binding to dsRNA and suppressing translation initiation (Williams, 2001). OAS activation occurs by binding to dsRNA, which catalyzes the conversion of ATP to 2′–5′-linked oligoadenylates, and in turn, bind to and activate ribonuclease L (RNase L). RNase L degrades ssRNAs of viral and cellular origin (Silverman, 2007) and generates cleaved RNA products which induce type I IFN production by activating the cytosolic nucleic acid receptors, the retinoic acid inducible gene I (RIG-I; also known as DDX58) and melanoma differentiation-associated protein 5 (MDA5; also known as IFIH1) (Malathi et al., 2007). Pol III synthesizes RNA from intracellular DNA which is recognized by RIG-I (Ablasser et al., 2009; Chiu et al., 2009).

2.3 Group III Nucleic Acid Sensing Receptors: RLRs

Collectively called RIG-1 like receptors (RLR), RIG-I, MDA5 and laboratory of genetics and physiology 2 (LGP2) have the characteristics of both groups I and II nucleic acid sensing receptors, functioning as both helicases and inducers of an anti-viral immune response (Liu and Gu, 2011). In contrast to nucleic acid sensing TLRs that are transmembrane receptors, RLRs are localized in the cytosol, suggesting their role in sensing replicating viral genomes in the cytoplasm. RLRs detect both ssRNA and short dsRNA and activate the inflammasome. The downstream signaling pathway is initiated by the interaction between the mitochondrial outer membrane associated IPS-1 (also called MAVS, Cardif, and VISA), inducing the activation of transcription factors IRF3, IRF7 and NF-κB, and expression of type I IFN genes (Kawai and Akira, 2010, 2011; Kawai et al., 2005). Both RIG-1 and MDA5 consist of three distinct domains, including an N-terminal region with tandem caspase activation and recruitment domains (CARD), a C-terminal repressor domain (RD) and a central DExD/H box RNA helicase domain which is capable of binding to and unwinding RNA.

RIG-I detects 5′triphosphate blunted ends in the presence of short ss/dsRNA; in contrast, MDA5 binds to long dsRNA without sequence specificity. LGP2 contains the same helicase and C-terminal domains as RIG-I and MDA5 but lacks the CARD domains. However, LGP2 regulates RIG-I and MDA5 activity in response to certain viruses (Kell and Gale, 2015; Satoh et al., 2010; Schlee, 2013). Despite their redundant structures and functions, certain viruses will preferentially signal through one pathway more strongly than the other. For example, the success of recovery from human metapneumovirus infection, an antisense ssRNA virus that accounts for 5–15% of child hospital visits for respiratory illness, depends on signaling through MDA5, but not RIG-I, and the subsequent production of type I IFNs (Banos-Lara Mdel et al., 2013).

3. IMMUNE SYSTEM AND NUCLEIC ACID SENSING

As the first line of defense, activation of the innate immune system by PAMPs and DAMPs is essential for acute inflammation and shaping the adaptive immune response (Medzhitov, 2007; Schaefer, 2014). The innate immune response is initiated and largely mediated by epithelial cells as well as phagocytes.

Type I IFNs, produced by innate cells, modulate T cell and B cell response, especially during viral infections. $CD4^+$ T cells from mice lacking the IFN-1 receptor exhibit reduced clonal expansion under lymphocytic choriomeningitis virus infection (Havenar-Daughton et al., 2006). IFNs affect B-cell survival, trafficking, antibody production, expression of surface markers as well as TLR7 (Badr et al., 2010; Braun et al., 2002; Evans et al., 1993; Green et al., 2009). Moreover, type I IFNs, specifically IFN-α, upregulate the transcription factor T-bet in B cells to promote IFN-γ production (de Goer de Herve et al., 2011). Cytokines produced in response to the inflammasome (IL-1β, IL-18 and IL-33) in addition to activating innate immune cells, such as monocytes and macrophages, promote the differentiation of $CD4^+$ T cells into T_H17, T_H1 or T_H2 effector cells, respectively (Chung et al., 2009; Dinarello, 1999; Schmitz et al., 2005). However, the role of the NLRP3 inflammasome in allergic airway disease studied in mice is controversial. In an allergic model of asthma induced by house dust mite (HDM) and ovalbumin (OVA), NLR3 knockout mice had unchanged airway function and inflammation; however, in the latter model, the level of IL-13 and IL-33 was lower (Allen et al., 2012). In contrast, it has been shown that mice deficient in *NLRP3 or caspase-1 had an elevated allergic response to HDM as reflected in* eosinophilic inflammation and T_H2 signature cytokines in the lung (Madouri et al., 2015). Differences may reflect the particular allergen used to trigger airway responses.

3.1 Nucleic Acid Sensing by Dendritic Cells

DCs participate in the innate immune response and initiate adaptive immunity through the regulation of T cell activation and differentiation. The DC population is heterogeneous and consists of various subsets expressing different patterns of nucleic acid sensing receptors (Mildner and Jung, 2014). pDCs reside in the blood, lymphoid organs, thymus, tonsils, and lung tissue. While functioning as antigen presenting cells, the major role of pDCs is a rapid response against viral infections (Colonna et al., 2004). pDCs are equipped to recognize viral structures efficiently by expressing nucleic acid sensors TLR7, TLR8, TLR9, and RLRs, resulting in potent type I IFN production, up to 100-fold greater than other cell types. pDCs from subjects with allergic rhinitis, non-atopic asthma, and atopic dermatitis are defective in type I IFN production, a factor that may contribute to their inability to initiate an effective anti-viral immune response (Gill et al., 2010; Pritchard et al., 2012, 2014). In a skin model, IFN-I from pDCs was crucial for epidermal regeneration, raising the possibility of a potentially protective role of viral infection against epithelial dysfunction induced by the proteolytic activity associated with many allergens (Gregorio et al., 2010; Lynch et al., 2014). Hence, targeting pathways which cause a defective anti-viral response of pDC in asthmatics may be a potential therapeutic approach to induce tolerance to harmless antigens.

cDCs are known for their superior potential to cross-present antigens to $CD8^+$ T cells, essential for induction of virus-specific cytotoxic T cells (Geurtsvankessel et al., 2008). In allergic asthma and dermatitis however, cDCs have been shown to induce a T_H2 response and may therefore have the potential to support an atopic immune response (Deckers et al., 2017; Gao et al., 2013).

3.2 Nucleic Acid Sensing by CD4 T Cells

Nucleic acid sensing by $CD4^+$ T cells may contribute to the initiation and exacerbation of allergic inflammatory responses. CpG DNA increases survival of activated $CD4^+$ T cells which is dependent on MyD88 and activation of NF-kB (Gelman et al., 2006). CpG can reverse the suppressive function of Tregs through the TLR8-MyD88-IRAK4 signaling pathway (Peng et al., 2005). TLR7/8 agonists stimulate the proliferation and cytokine synthesis of IFN-γ, IL-8, and IL-10 but not IL-4 by human $CD4^+$ T cells, thereby polarizing the immune system toward T_H1 inflammation (Caron et al., 2005). Interestingly, complexes of self-DNA with antimicrobial peptides or core histones induce inhibition of T-bet expression and upregulate GATA3 and T_H2 cytokine expression *in vitro*, independent of known nucleic acid sensors including TLRs, RLRs, the inflammasome and STING (Imanishi et al., 2014). This finding suggests that release of self-DNA from inefficiently cleared apoptotic airway epithelial cells induced by intranasal allergen challenge may be responsible for elevated T_H2 cytokine production and AHR (Juncadella et al., 2013).

3.3 Nucleic Acid Sensing by B Cells

B cells play a prominent role in the development of responses to self and foreign antigens by cytokine and immunoglobulin synthesis. B cell receptor (BCR) stimulation with concomitant T cell help results in the proliferation and differentiation of naïve B cells into plasma and memory B cells. Memory and activated B cells upregulate the expression of nucleic acid sensing TLRs, namely TLR3, 7 and 9 in mouse, and TLR7 and 9 in human, upon activation through the BCR and stimulation by cytokines (Browne, 2012). Stimulating human B cells with TLR7 and TLR9 agonists results in proliferation as well as expansion of memory B cells and plasma cells (Simchoni and Cunningham-Rundles, 2015).

4. NUCLEIC ACID SENSING BY EPITHELIAL CELLS

As the first line of defense in innate immune system, epithelial cells at the mucosal surfaces of the gastrointestinal tract and airways have dynamic interactions with allergens and pathogens, acting as both a physical barrier and initiator of innate and subsequent adaptive immune responses. Epithelial cells are armed with nucleic acid sensing receptors, such as TLRs and RLRs, which are actively involved in sensing pathogenic nucleic acids and production of inflammatory cytokines and chemokines (Bulek et al., 2010; Kato and Schleimer, 2007; McClure and Massari, 2014; Tengroth et al., 2014). Human nasal epithelial cells (HNEC) express TLR3, TLR7, TLR9, RIG-I, and MDA5. Stimulation of TLR3 and 7 in HNEC induces the expression of IL-6 and GM-CSF; while TLR9 and RIG-I/MDA5 stimulation causes IL-8 and IFN-β production, respectively (Tengroth et al., 2014). TLRs are not restricted to mucosal surfaces but are also expressed on skin. For instance, TLR3 has been implicated in noninfectious inflammatory diseases of the skin and contact dermatitis; TLR3-deficient mice are protected against these conditions (Nakamura et al., 2015). TLR3 mediates the secretion of TSLP from keratinocytes (Kubo et al., 2014), and may prime DCs and type 2 innate-like lymphoid cells (ILC2s). The activation of ILC2s by respiratory viruses not only favors the establishment of a pro-allergic environment through type 2 cytokine secretion but may also preserve the epithelial barrier function (Monticelli et al., 2011), a property essential to exclude aeroallergens from interaction with underlying immune effector cells.

Respiratory epithelial cells are the first target of respiratory viruses, initiating inflammatory anti-viral innate immune responses. RSV and pneumonia virus of mice (PVM) trigger IL-25 expression, while influenza virus and PMV induce IL-33 expression by airway epithelial cells (Kumar et al., 2014). Human lung epithelial cell lines respond to both influenza virus infection and dsRNA by producing inflammatory cytokines via the TLR3 and TRIF pathway (Guillot et al., 2005). Moreover, TLR3 stimulation of human airway epithelial cells *in vitro* by dsRNA induces TSLP production (Kato et al., 2007), while, influenza A virus induces IL-1β production in a RIG-I/TLR3/NLRP3 dependent manner (Pothlichet et al., 2013).

By instructing the cytokine milieu in the tissue, the epithelial layer can potentially influence the activation and recruitment of antigen presenting cells and the subsequent recruitment and differentiation of adaptive immune cells. For instance, while TSLP, IL-25, IL-33, IL-6, and GM-CSF contribute to the promotion of T_H2 responses, type I IFNs contribute to a T_H1 biased response (Huber and Farrar, 2011; Rincon et al., 1997; Saenz et al., 2008; Sheih et al., 2017). Moreover, IL-1β induces polarization of T_H1/T_H17 cells; however its effect on T_H2 differentiation is controversial (Santarlasci et al., 2013).

5. VIRAL ILLNESS AND ACUTE ASTHMA EXACERBATIONS

While exacerbations of asthma are largely attributable to viral respiratory tract infections, allergens can potentially attenuate an anti-viral immune response via impaired nucleic acid sensing machinery (Deschildre et al., 2017; Ritchie et al., 2015). Given the primordial role of the airway epithelium in asthma, current evidence suggests that the interaction of viruses with the epithelium may be altered by prior exposure of the epithelium to T_H2 cytokines. The T_H2 environment and allergic sensitization contribute to the increased susceptibility to viral infections. Among the respiratory viruses, the rhinovirus has been thoroughly studied and documented to be the most frequent cause of asthma attacks (Johnston et al., 1996). Deficient type I and III IFN synthesis and the inability to prevent uncontrolled viral replication have been implicated in the increased susceptibility to rhinoviral infection in asthmatics (Baraldo et al., 2012; Hammond et al., 2015; Wark et al., 2005).

Asthmatic airway epithelial cells in culture have attenuated anti-viral responses to infection with RV-16 that are associated with a failure of apoptosis and enhanced viral replication. There is lower type I IFN synthesis by the asthmatic epithelial cells with RV infection than in control cells but the deficient response to viral infection can be normalized by administering exogenous IFN-β (Wark et al., 2005). Subsequent work by the same group showed a deficiency of IFN-λ production by asthmatic airway epithelial cells and macrophages (Contoli et al., 2006). Exposure of human bronchial epithelial cells to HDM, prior to stimulation of the cells with polyinosinic: polycytidylic acid (poly I:C), which mimics dsRNA viruses, led to reduced synthesis of IFN-β and -λ (Akbarshahi et al., 2018). Additionally, the upregulation of TLR3, MDA5, and RIG-I that follows poly I:C treatment was markedly attenuated by prior exposure to HDM. The results of cell culture work are supported by experimental infection of human asthmatic subjects; human subjects infected with rhinovirus upregulate TLR3, MDA5, and RIG-I in their airway tissues. Asthmatic subjects show a deficiency of IFN-α/β expression that is not attributable to expression patterns of the aforementioned receptors that are also upregulated in the asthmatic subjects *in vivo* (Zhu et al., 2018).

Similar findings have been reported for children. Specifically, levels of IFN-β and IFN-λ were lower in response to rhinovirus in bronchial epithelial cells of asthmatic children, atopic non-asthmatic children, and atopic asthmatic children compared to non-asthmatic children, correlating with an increase in viral RNA, IL-4, eosinophilia, epithelial damage, and serum IgE levels (Baraldo et al., 2012). Other investigators have not found that these observations extend to RSV and influenza (Patel et al., 2014). The exacerbation in asthmatics that are infected with influenza may be due to upregulation of VCAM-1 on the endothelial surface, resulting in the pulmonary influx of mast cell progenitors expressing the α4-integrin (Zarnegar et al., 2017). This process has been linked to the recognition of viral ssRNA through RIG-I in mast cells *in vitro* (Zarnegar et al., 2017). The relationship between viral infections of epithelial cells and the subsequent neutrophilic inflammatory response to the development of bronchospasm in asthma is still unclear. Interestingly, the production of cysteinyl leukotrienes, characteristic mediators of allergic responses, can be increased during respiratory viral infection as has been shown in RSV bronchiolitis in children, suggesting an additional contributory mechanism of viral infection to asthma exacerbation (Kim et al., 2006). The cellular origin of these and other bronchoconstrictive mediators and their relationship to nucleic acid sensing remain to be elucidated.

5.1 Rhinovirus

Infection with rhinovirus affects primarily the epithelial cells of the upper and lower airways (Fig. 1). TLR3 in conjunction with MDA5 and RIG-I are important in signaling the production of type I IFNs during rhinoviral infection and in the establishment of an overall anti-viral environment (Unger et al., 2014). Rhinovirus or poly I:C, activates nucleotide-binding oligomerization domain, leucine-rich repeat containing X1 (NLRX-1) in immortalized and primary airway epithelial cells and generates mitochondrial reactive oxygen species which, although necessary for viral clearance, can impair barrier integrity by disrupting tight junctions and compromise host tolerance (Unger et al., 2014). These findings have potential implications for asthma, as a disrupted epithelial barrier increases mucosal permeability allowing for inhaled allergens to trigger an immune response. Additionally, the adaptive response may be affected. TLR3 stimulation has been shown to promote DC activation, migration, and antigen processing by mDCs, much as does TLR4, and promotes allergic sensitization in mice treated concurrently with poly I:C and OVA (Reuter et al., 2012).

Allergen exposure prior to viral infection also has consequences for the host response. Treatment of mice with HDM instillations intra-nasally prior to administration of poly I:C to mimic TLR3 activation reduced IFN-β and -λ mRNA expression in lung tissue and reduced IFN-β protein in bronchoalveolar lavage fluid (BALF) (Akbarshahi et al., 2018). The downstream signaling molecules, IRF-3 and IRF-7, were also reduced in expression by exposure to HDM prior to poly I:C treatments. It was suggested that enzymatic activity in HDM altered the glycosylation state of TLR3 (Mahmutovic Persson et al., 2016). In mice exposed to dsRNA, HDM pretreatment to induce the allergic phenotype caused a marked upregulation of

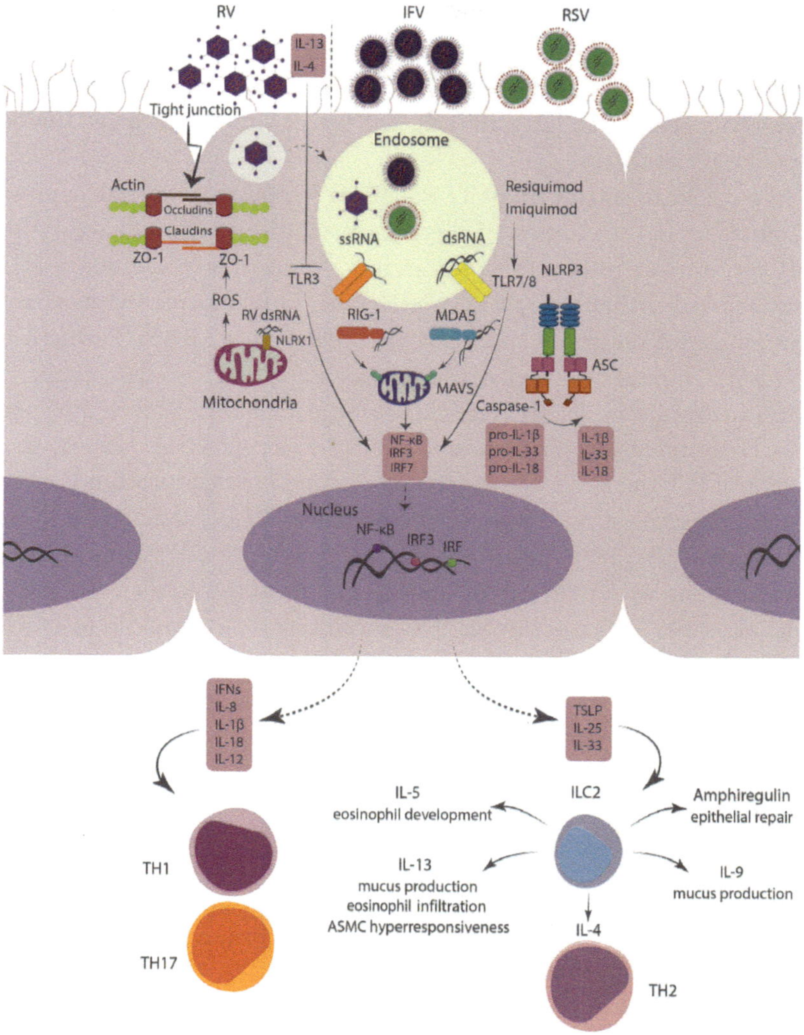

Fig. 1 Modulation of immune responses by airway epithelial cells via nucleic acid sensors. Airway viral infections trigger anti-viral immune responses by airway epithelial cells which are the initiators of activation and modulation of innate and adaptive responses. ssRNA viruses such as RV, IFV and RSV enter the epithelial cells endosomes via receptor mediated endocytosis and subsequent release of viral RNA into cytoplasm (Chen et al., 2018; Griffiths et al., 2017; Jacobs et al., 2013). In the endosomes viral ssRNA as well as intermediate dsRNA, generated as a result of viral replication, serve as ligands for nucleic acid sensors TLR3 and TLR7/8 respectively. Cytosolic viral RNA activates RIG-1 and MDA5 as well as NLRs, NLRX-1 and NLRP3. TLR3 and 7/8 as well as RIG-1 and MDA5 which

(Continued)

the alarmins TSLP, IL-25, and IL-33 (Mahmutovic-Persson et al., 2014). No surrogates for asthma, such as AHR and mucus secretion, were reported, although eosinophilic inflammation was present after HDM exposure with and without poly I:C and an additional neutrophilic inflammation was induced by poly I:C treatments.

In contrast to activation of TLR3, TLR7 appears to exercise a protective function in the context of allergic airway disease. The specific role of TLR7 in murine responses to rhinovirus infection was assessed in HDM sensitized and challenged mice. Mice lacking TLR7 developed augmented eosinophilic inflammation and AHR (Hatchwell et al., 2015). Furthermore, IFN-α/β synthesis was reduced and was implicated in the exaggerated asthmatic phenotype. Transfer of TLR7-competent DCs rescued the phenotype. These results support a role for TLR7 as a protective pathway in the context of RV infection by suppressing eosinophil inflammation. Repeated IL-5 administration recapitulated the effects of TLR7 deficiency by suppression of TLR7 expression and signaling. These results are particularly interesting since they shed light on the benefits of reduced airway eosinophilia in asthma, whether by corticosteroid treatment or biologics such as mepolizumab, that reduce the frequency of exacerbations. Perhaps the interaction of IL-5 with TLR7 accounts for the frequency of RV infection induced attacks in persistently eosinophilic asthma and the reduction in exacerbations seen following neutralization of IL-5 in exacerbation-prone asthmatics (Haldar et al., 2009).

Fig. 1—Cont'd interact with mitochondrial outer membrane protein MAVS induce activation and nuclear translocation of transcription factors NF-κB, IRF3 and IRF7, resulting in gene expression of proinflammatory cytokines, including type-1 IFNs, IL-1β, IL-18, IL-12 and IL-8, which can induce T_H1 and T_H17 differentiation. Treating airway epithelial cells by T_H2 cytokines, IL-13 and IL-4, reduces anti-viral immune response in RV infection by inhibiting TLR3 gene expression. Activation of NLRP3 via both ssRNA and dsRNA induces caspase-1 recruitment and activation which in turn activates pro-IL-33, pro-IL-1β and pro-IL-18 by proteolytic cleavage. RV dsRNA is shown to activate mitochondrial bound NLRX-1, inducing ROSs production which in turn can both activate NF-κB signaling pathway and disrupt epithelial tight junctions. Lack of epithelial barrier integrity in RV infection which increases the susceptibility to IFV infection has been postulated to enhance immune responses to allergens. Viral infection induces activation of IL-33 and expression of TSLP and IL-25 which activate ILC2 in the lung tissues. Amphiregulin expression by ILC2 induces epithelial repair which may protect against both viral infection and allergen entry. Cytokines produced by ILC2 cells can exacerbate allergic asthma by increasing mucus production (IL-9 and IL-13), eosinophilia (IL-5 and IL-13), T_H2 differentiation (IL-4) and ASM hyperresponsiveness (IL-13).

Since the exacerbating phenotype of asthma is of the T_H2 type but exacerbations are most often viral in origin, then it is reasonable to postulate that epithelial viral responses may be altered by T_H2 cytokines. Several studies have addressed this possibility with support for the hypothesis. Treatment of bronchial epithelial cells with T_H2 cytokines (IL-4 and IL-13) impaired RV-16-induced IFN-β and IFN-λ production, increased viral replication, and reduced TLR3 expression (Contoli et al., 2015). These investigators also confirmed the T_H2 environment in the nasal mucosa of atopic subjects with increased IL-4 mRNA levels in nasal epithelial cells and reduced TLR3 expression. Findings were in direct relation with RV-16 replication in nasal brushings of atopic rhinitis patients. IL-4 and IL-13 impaired viral activation of IRF3 in bronchial epithelial cell cultures (Contoli et al., 2015). Not all studies have linked asthmatic reactions after RV infection to deficient IFN responses. For example, among children with episodes of RV induced wheezing, elevated levels of IFN-λ in nasal washes are strongly associated with severity of symptoms (Miller et al., 2012).

5.2 Influenza Virus

Influenza is a frequent trigger for asthma exacerbations (Fig. 1). The relationship of viral responses to the asthmatic condition has been modeled in mice. In healthy mice inoculated with influenza, innate immune cell recruitment to the lungs consists mostly of neutrophils, bone marrow-derived macrophages, and natural killer cells (Iwasaki and Pillai, 2014). In contrast, in allergen-sensitized and challenged mice when infected with influenza at the peak of the inflammatory response the preferentially recruited cell type is the eosinophil (Samarasinghe et al., 2014). This may account for asthma exacerbations in subjects that have an eosinophilic response to infection. Interestingly, these mice experience less immunopathology and clear the virus more rapidly. During periods of heightened inflammation and eosinophilia, mimicking acute asthma, mice infected with influenza A virus have shown a substantial reduction in lung damage attributed to the upregulation of costimulatory molecules on eosinophils which primed $CD8^+$ T cells to enhance anti-viral immunity (Samarasinghe et al., 2017). Despite lower type I and III IFNs in allergen challenged mice, these animals had a reduced viral burden and morbidity compared to the influenza infected controls due to the dual role of type I IFNs in inducing both anti-viral immunity and contributing to host immunopathology (Samarasinghe et al., 2014). Influenza infection preceding either

OVA or HDM challenge provided protection against the simulated asthma exacerbation. The protection was attributed to cross-reactivity between influenza-derived peptides and allergens, which prime the immune system (Furuya et al., 2014; Skevaki et al., 2017). These results find support from studies on human. Although asthmatics accounted for a large proportion of hospitalizations during the 2009 H1N1 pandemic, they were least likely to die of infection. This phenomenon was studied in multiple laboratory settings and seemed to be in part related to the maintenance and structure of epithelial integrity following influenza infection during different stages of inflammation in the allergic airways (Samarasinghe et al., 2014). Collectively, these data, in part, explain the discrepancies in responses to influenza infection within asthmatics and when compared to non-asthmatics. The epithelium in a T_H2 immune setting seems to be beneficial for the host in reducing airway pathology during influenza infection.

The synergistic role of $CD4^+$ T cells and ILCs in driving asthma exacerbations post-inoculation with influenza virus is of interest. A protective function of ILC2s has been shown during influenza infection by inducing epithelial repair through the secretion of the epidermal growth factor, amphiregulin (Califano et al., 2018; Monticelli et al., 2011). However, when activated with IL-15 and IL-33, ILC2s produce IL-5 and IL-13 causing AHR, mucus production, and eosinophilia, mimicking an allergic phenotype which can predispose to exacerbations in asthmatics (Califano et al., 2018). Alveolar macrophages, which are the first immune cells to encounter the virus, were shown to be potent producers of IL-33 (Li et al., 2018) in addition to epithelial cells. ILC2s help to establish a predominant T_H2 environment through the secretion of IL-5 and IL-13 in late and early infection, respectively (Li et al., 2018). The delayed contribution of ILC2s in asthma exacerbations during influenza infection is thought to be due to the abundance of certain cytokines, such as IFN-γ, in early infection and their ability to suppress ILC2 activation (Califano et al., 2018; Li et al., 2018).

5.3 Respiratory Syncytial Virus (RSV)

Infections by negative-sense RNA viruses, such as RSV, have long been linked with the risk of asthma (Fig. 1). Early life severe bronchiolitis is a strong predictor of the subsequent development of asthma. The Avon Longitudinal Study of Parents and Children by Henderson et al. showed that RSV infection before the age of one lead to the sensitization of common inhaled or ingested allergens by the age of three and increased the risk of developing asthma by the age of seven (Henderson et al., 2005). Exaggerated bronchial epithelial cell synthesis of type I and III IFNs when infected with RSV *in vitro* has been associated with impaired lung function in asthmatic children (Altman et al., 2018), supporting a causal link between the development of persistent airway obstruction and early innate immune responses to RSV. However, the cause and effect relationships of this comorbidity require further exploration.

Pre-clinical studies of RSV infection show complex interactions with allergen-driven airway dysfunction and inflammation that depend on the timing of infection in relation to allergen sensitization and challenge. RSV infection prior to allergen challenge diminishes the impact of the allergen challenge (Peebles et al., 2001a). In contrast, infection at the time of sensitization or following sensitization prolongs AHR and inflammation in the mouse. The augmented responses are not associated with enhanced T_H2 cytokine synthesis (Peebles et al., 2001b) but rather, with an increase in IL-17A (Newcomb et al., 2013). However, IL-17A appears to confer protection in this setting. Thus, the AHR is unexplained. RSV infection breaks tolerance of neonatal mice previously protected from OVA via passive transfer of the allergen through the mother's breast milk. This break in tolerance was associated with the differentiation of Tregs toward T_H2 cytokine expressing cells (Krishnamoorthy et al., 2012). Another study supported this finding by showing that an impaired Treg response correlated with an increase in costimulatory molecules, CD40 and CD86, on the surface of DCs increased the likelihood of sustaining chronic inflammation to allergens, potentially leading to the development of asthma (Knudson and Varga, 2015). A commonly held theory relates to the predisposition of a T_H2-biased inflammatory milieu during infection which drives allergic manifestations rather than a proper anti-viral response.

Other studies have shown an increase in IL-17 in children infected with RSV which resulted in neutrophilia, mucus secretion, and AHR, potentially contributing to the likelihood of the subsequent development of childhood asthma (Knudson and Varga, 2015). A crucial determinant in disease severity following infection depends on the age of the host at the time of the infection and the developmental stage of the lungs. Priming neonatal mice with RSV led to a predominantly T_H2 response during reinfection, whereas delayed priming was associated with less severe disease and increased IFN-γ following reinfection (Culley et al., 2002). Therefore, the time of infection is important in biasing the immune profile to either reduce or enhance disease scoring in adulthood. Similar to infection with influenza, RSV infection induced an upregulation of IL-13 producing ILC2s associated with increased AHR, mucus secretion, higher viral load, and morbidity (Stier et al., 2016). This phenotype was dependent on TSLP which has the capacity to drive ILC2 activation and proliferation in early infection and to enhance $CD4^+$ T_H2 cells during late infection (Stier et al., 2016). A phase I randomized clinical trial in which patients received the monoclonal antibody against TSLP (AMG 157) showed reduced airway obstruction and inflammation compared to placebo. In line with this observation, TSLP has been shown to be upregulated in airway epithelial cells isolated from asthmatic children upon *in vitro* stimulation with RSV (Stier et al., 2016). These data suggest an important role for asthmatic airways in their ability to drive more TSLP production in a feed-forward loop through the secretion of T_H2 cytokines and shifting the immune response towards a T_H2 phenotype associated with an exaggerated ILC2 response.

6. NUCLEIC ACID-BASED THERAPIES FOR ALLERGIC DISORDERS

There are several potential therapeutic applications which utilize nucleic acid sensing pathways. Strong pre-clinical data support a therapeutic role for agonists of TLR7/8 in the mitigation of allergic disease. Repeated challenge of the OVA-sensitized BN rat leads to extensive airway remodeling characterized by an increase in airway smooth muscle mass from hyperplasia and goblet cell differentiation that is T cell dependent (Ramos-Barbon et al., 2005) and mediated by cysteinyl leukotrienes (Wang et al., 1993) and the epidermal growth factor receptor (Tamaoka et al., 2008). Resiquimod, a TLR7/8 agonist, is markedly inhibitory of allergen-induced inflammation and the remodeling process (Camateros et al., 2007). The anti-allergic mechanism of action is not clear, but is not attributable to a biasing toward a T_H1 response. Similar studies in the non-human primate naturally sensitized to *Ascaris suum* and challenged via the airways and pre-treated with resiquimod conferred protection against inflammation and airway dysfunction (Camateros et al., 2018). Stimulation of DCs by imiquimod triggers IFN synthesis, reduces airway hyperresponsiveness (Hornung et al., 2005) and inflammation in the airways and downregulates the T_H2 response associated with asthma (Gibson et al., 2002). Despite the sex-linked relationship to expression of IFN-α in females with TLR7 stimulation (Berghofer et al., 2006), there appear to be no studies comparing male to female responses to therapeutic interventions.

Unmethylated oligodeoxynucleotides containing CpG with specific base sequence motifs (CpG ODNs) can induce immunomodulatory effects of bacterial DNA via TLR9. Therapeutic application of TLR9 modulation for the treatment of atopic disorders has been studied and exploited in clinical trials in recent years. CpG ODNs as adjuvants can be applied for immunotherapy to treat such disorders. The immune therapeutic role of CpG ODNs is believed to be partly attributed to their potential in initiating a T_H1 biased adaptive immune response and releasing cytokines such as IL-12 and IFN-γ; hence, reversing and preventing the hallmark T_H2 immune response of allergic diseases. In murine models of acute atopic asthma CpG ODN administration with antigen inhibits the synthesis of IL-4 and IL-5, as well as eosinophilia and AHR (Hussain et al., 2002; Kline et al., 1998, 1999; Sur et al., 1999). In an allergic mouse model of asthma induced by ragweed, administration of CpG 48 h before allergen challenge conferred protection from the T_H2 allergic response by inducing a T_H1 response thereby reducing the number of the cells producing ragweed allergen-specific IgE (Sur et al., 1999). In a cockroach induced asthma model, administration of CpG ODN prior to allergen exposure reduced airway eosinophilia, goblet cell hyperplasia, AHR, IL-13, IL-5, serum cockroach specific IgE, and IgG1/IgG2a ratio. This inhibitory function was accompanied by an increase in IFN-γ in the lung (Kim et al., 2016). In established murine asthma induced by OVA, treating mice with allergen and CpG ODN reduced IL-5, eosinophilic bronchial inflammation, and AHR while increasing the T_H1 cytokine IFN-γ (Kline et al., 2002). There is supporting evidence showing nucleic acid sensing by B cells having anti-allergic immune responses by inducing T_H1 cytokines with reciprocal suppression of the T_H2 response. CpG induces T-bet mrnA expression by B cells and inhibition of immunoglobulin isotype class switching by IL-4 and CD40 to IgG1 and IgE (Liu et al., 2003). CpG DNA has been shown to induce the production of T_H1 immunoglobulin isotypes, IgG2a, IgG2b and IgG3, and to suppress the T_H2 isotypes, IgG1 and IgE, in a TLR9 and MYD88 dependent manner in mice (Lin et al., 2004). Given the presence of a distinct airway microbiota in asthmatic subjects (Hilty et al., 2010), it is perhaps surprising that TLR9 signaling is not more effective in reducing T_H2 mediated inflammation. However early life exposure to bacterial products may indeed be protective (von Mutius, 2016), although this protection is more likely TLR4-mediated.

The therapeutic effect of TLR9 agonists has also been studied in skin allergic disorders in the context of atopic dermatitis (AD) in mice, although no clinical trial has been conducted in human as of yet. AD is a chronic inflammatory skin disease characterized by increased IgE level, inflammatory cell infiltration, and the elevated expression of T_H2 cytokines (Brandt and Sivaprasad, 2011). Topical application of CpG ODNs suppresses the signs of atopic dermatitis in skin lesions of NC/Nga mice which is accompanied by a reduction in IL-4 expressing cells, serum IgE and IgG1, as well as increased IL-12p40 and serum IgG2a (Kim et al., 2007).

In a clinical trial, a TLR9 agonist, called CYT003-QbG10, was administered to subjects with HDM allergy. Compared to subjects on placebo, CYT003-QbG10 reduced rhinoconjunctivitis, increased quality of life score, and caused a median 10-fold increase in allergen tolerance measured by a conjunctival provocation test (Klimek et al., 2011). Treatment of individuals allergic to ragweed with a ragweed-pollen antigen conjugated to CpG ODN (AIC) reduced the nasal inflammatory response. AIC treatment reduced eosinophils and IL-4 mRNA-positive cells while increasing the number of IFN-γ mRNA-positive cells in nasal biopsies. Furthermore, the IgE levels were attenuated, while antibody production was shifted toward IgG (Creticos et al., 2006; Tulic et al., 2004). While 1018 ISS, a form of CpG ODN, application successfully reduced T_H2 mediated lung inflammation and AHR in mice and primates, human subjects with allergic rhinitis CpG ODN coupled to ragweed allergen showed efficacy in increased expression of IFNs and ISGs but did not affect the airway response to allergen nor the T_H2 response that accompanied the airway narrowing (Broide et al., 1998; Fanucchi et al., 2004; Gauvreau et al., 2006).

TLR7 targeted therapy has been clinically tested on patients with allergic rhinitis. Intranasal administration of a selective TLR7 agonist, AZD8848, prior to allergen challenge reduced nasal symptoms (Greiff et al., 2012). Additionally, stimulation of dendritic cells by imiquimod, a TLR7 agonist that has been explored in the treatment of asthma using model systems, triggers IFN synthesis as well as reduces AHR and inflammation in the airways and downregulated the T_H2 response associated with asthma (Gibson et al., 2002).

7. CONCLUSIONS

The outcome of nucleic acid sensing depends on the nature and intensity of the stimuli. The interaction between nucleic acid sensing and allergic responses is complex and in model systems may depend on the order of the exposure; whether allergic sensitization and exposure precedes or follows infection. Whether viral infections at early ages are causally related to the development of allergic responses is also not clear. Unlike lung associated allergic diseases, nucleic acid sensing has not been studied in other types of allergic disorders extensively. There are candidate agonists for nucleic acid receptors to attenuate allergic responses, mainly asthma and allergic rhinitis, that have the potential to be used as concomitant treatments with well established current medications. Animal and cell culture models show lower anti-viral responses in the allergic milieu perhaps accounting for asthma exacerbations but may also be associated with less immunopathology in some circumstances. The notion that allergic responses can lead to the inability of the immune system to mount an efficient anti-viral response to the rhinovirus brings up the issue of devising the right strategies to combat this common cause of asthma attacks in patients.

ACKNOWLEDGMENT

We would like to thank Dr. Toby K. McGovern, McGill University, Montreal, for reading and providing us with an invaluable critique. The study was supported by the Richard and Edith Strauss Foundation.

REFERENCES

Ablasser, A., Bauernfeind, F., Hartmann, G., Latz, E., Fitzgerald, K.A., Hornung, V., 2009. RIG-I-dependent sensing of poly(dA:dT) through the induction of an RNA polymerase III-transcribed RNA intermediate. Nat. Immunol. 10, 1065–1072.

Akbarshahi, H., Menzel, M., Ramu, S., Mahmutovic Persson, I., Bjermer, L., Uller, L., 2018. House dust mite impairs antiviral response in asthma exacerbation models through its effects on TLR3. Allergy 73, 1053–1063.

Akira, S., Uematsu, S., Takeuchi, O., 2006. Pathogen recognition and innate immunity. Cell 124, 783–801.

Alexopoulou, L., Desnues, B., Demaria, O., 2012. Toll-like receptor 8: the awkward TLR. Med. Sci. (Paris) 28, 96–102.

Allen, I.C., Jania, C.M., Wilson, J.E., Tekeppe, E.M., Hua, X., Brickey, W.J., Kwan, M., Koller, B.H., Tilley, S.L., Ting, J.P., 2012. Analysis of NLRP3 in the development of allergic airway disease in mice. J. Immunol. 188, 2884–2893.

Altman, M.C., Reeves, S.R., Parker, A.R., Whalen, E., Misura, K.M., Barrow, K.A., James, R.G., Hallstrand, T.S., Ziegler, S.F., Debley, J.S., 2018. Interferon response to respiratory syncytial virus by bronchial epithelium from children with asthma is inversely correlated with pulmonary function. J. Allergy Clin. Immunol. 142, 451–459.

Badr, G., Saad, H., Waly, H., Hassan, K., Abdel-Tawab, H., Alhazza, I.M., Ahmed, E.A., 2010. Type I interferon (IFN-alpha/beta) rescues B-lymphocytes from apoptosis via PI3Kdelta/Akt, Rho-A, NFkappaB and Bcl-2/Bcl(XL). Cell. Immunol. 263, 31–40.

Banos-Lara Mdel, R., Ghosh, A., Guerrero-Plata, A., 2013. Critical role of MDA5 in the interferon response induced by human metapneumovirus infection in dendritic cells and in vivo. J. Virol. 87, 1242–1251.

Bao, M., Liu, Y.J., 2013. Regulation of TLR7/9 signaling in plasmacytoid dendritic cells. Protein Cell 4, 40–52.

Baraldo, S., Contoli, M., Bazzan, E., Turato, G., Padovani, A., Marku, B., Calabrese, F., Caramori, G., Ballarin, A., Snijders, D., Barbato, A., Saetta, M., Papi, A., 2012. Deficient antiviral immune responses in childhood: distinct roles of atopy and asthma. J. Allergy Clin. Immunol. 130, 1307–1314.

Berghofer, B., Frommer, T., Haley, G., Fink, L., Bein, G., Hackstein, H., 2006. TLR7 ligands induce higher IFN-alpha production in females. J. Immunol. 177, 2088–2096.

Blasius, A.L., Beutler, B., 2010. Intracellular toll-like receptors. Immunity 32, 305–315.

Brandt, E.B., Sivaprasad, U., 2011. Th2 cytokines and atopic dermatitis. J. Clin. Cell. Immunol. 2, 1–25.

Braun, D., Caramalho, I., Demengeot, J., 2002. IFN-alpha/beta enhances BCR-dependent B cell responses. Int. Immunol. 14, 411–419.

Broide, D., Schwarze, J., Tighe, H., Gifford, T., Nguyen, M.D., Malek, S., Van Uden, J., Martin-Orozco, E., Gelfand, E.W., Raz, E., 1998. Immunostimulatory DNA sequences inhibit IL-5, eosinophilic inflammation, and airway hyperresponsiveness in mice. J. Immunol. 161, 7054–7062.

Browne, E.P., 2012. Regulation of B-cell responses by Toll-like receptors. Immunology 136, 370–379.

Bulek, K., Swaidani, S., Aronica, M., Li, X., 2010. Epithelium: the interplay between innate and Th2 immunity. Immunol. Cell Biol. 88, 257–268.

Califano, D., Furuya, Y., Roberts, S., Avram, D., McKenzie, A.N.J., Metzger, D.W., 2018. IFN-gamma increases susceptibility to influenza A infection through suppression of group II innate lymphoid cells. Mucosal Immunol. 11, 209–219.

Camateros, P., Tamaoka, M., Hassan, M., Marino, R., Moisan, J., Marion, D., Guiot, M.C., Martin, J.G., Radzioch, D., 2007. Chronic asthma-induced airway remodeling is prevented by toll-like receptor-7/8 ligand S28463. Am. J. Respir. Crit Care Med. 175, 1241–1249.

Camateros, P., Kanagaratham, C., Najdekr, L., Holub, D., Vrbkova, J., Cote, L., Fournier, J., Gourdon, J., Creery, D., Olivenstein, R., Kopriva, F., Adam, T., Friedecky, D., Dzubak, P., Hajduch, M., Radzioch, D., 2018. Toll-like receptor 7/8 ligand, S28463, suppresses Ascaris suum-induced allergic asthma in nonhuman primates. Am. J. Respir. Cell Mol. Biol. 58, 55–65.

Caron, G., Duluc, D., Fremaux, I., Jeannin, P., David, C., Gascan, H., Delneste, Y., 2005. Direct stimulation of human T cells via TLR5 and TLR7/8: flagellin and R-848 up-regulate proliferation and IFN-gamma production by memory CD4+ T cells. J. Immunol. 175, 1551–1557.

Caza, T., Landas, S., 2015. Functional and phenotypic plasticity of CD4(+) T cell subsets. Biomed. Res. Int. 2015, 521957.

Chen, X., Liu, S., Goraya, M.U., Maarouf, M., Huang, S., Chen, J.L., 2018. Host immune response to influenza A virus infection. Front. Immunol. 9, 320.

Chiu, Y.H., Macmillan, J.B., Chen, Z.J., 2009. RNA polymerase III detects cytosolic DNA and induces type I interferons through the RIG-I pathway. Cell 138, 576–591.

Choi, U.Y., Kang, J.S., Hwang, Y.S., Kim, Y.J., 2015. Oligoadenylate synthase-like (OASL) proteins: dual functions and associations with diseases. Exp. Mol. Med. 47, e144.

Chung, Y., Chang, S.H., Martinez, G.J., Yang, X.O., Nurieva, R., Kang, H.S., Ma, L., Watowich, S.S., Jetten, A.M., Tian, Q., Dong, C., 2009. Critical regulation of early Th17 cell differentiation by interleukin-1 signaling. Immunity 30, 576–587.

Cockcroft, D.W., Ruffin, R.E., Dolovich, J., Hargreave, F.E., 1977. Allergen-induced increase in non-allergic bronchial reactivity. Clin. Allergy 7, 503–513.

Colonna, M., Trinchieri, G., Liu, Y.J., 2004. Plasmacytoid dendritic cells in immunity. Nat. Immunol. 5, 1219–1226.

Contoli, M., Message, S.D., Laza-Stanca, V., Edwards, M.R., Wark, P.A., Bartlett, N.W., Kebadze, T., Mallia, P., Stanciu, L.A., Parker, H.L., Slater, L., Lewis-Antes, A., Kon, O.M., Holgate, S.T., Davies, D.E., Kotenko, S.V., Papi, A., Johnston, S.L., 2006. Role of deficient type III interferon-lambda production in asthma exacerbations. Nat. Med. 12, 1023–1026.

Contoli, M., Ito, K., Padovani, A., Poletti, D., Marku, B., Edwards, M.R., Stanciu, L.A., Gnesini, G., Pastore, A., Spanevello, A., Morelli, P., Johnston, S.L., Caramori, G., Papi, A., 2015. Th2 cytokines impair innate immune responses to rhinovirus in respiratory epithelial cells. Allergy 70, 910–920.

Creticos, P.S., Schroeder, J.T., Hamilton, R.G., Balcer-Whaley, S.L., Khattignavong, A.P., Lindblad, R., Li, H., Coffman, R., Seyfert, V., Eiden, J.J., Broide, D., 2006. Immunotherapy with a ragweed-toll-like receptor 9 agonist vaccine for allergic rhinitis. N. Engl. J. Med. 355, 1445–1455.

Culley, F.J., Pollott, J., Openshaw, P.J., 2002. Age at first viral infection determines the pattern of T cell-mediated disease during reinfection in adulthood. J. Exp. Med. 196, 1381–1386.

De Nardo, D., De Nardo, C.M., Latz, E., 2014. New insights into mechanisms controlling the NLRP3 inflammasome and its role in lung disease. Am. J. Pathol. 184, 42–54.

Deckers, J., Sichien, D., Plantinga, M., Van Moorleghem, J., Vanheerswynghels, M., Hoste, E., Malissen, B., Dombrowicz, D., Guilliams, M., De Bosscher, K., Lambrecht, B.N., Hammad, H., 2017. Epicutaneous sensitization to house dust mite allergen requires interferon regulatory factor 4-dependent dermal dendritic cells. J. Allergy Clin. Immunol. 140, 1364–1377.e2.

Deschildre, A., Pichavant, M., Engelmann, I., Langlois, C., Drumez, E., Pouessel, G., Boileau, S., Romero-Cubero, D., Decleyre-Badiu, I., Dewilde, A., Hober, D., Neve, V., Thumerelle, C., Lejeune, S., Mordacq, C., Gosset, P., 2017. Virus-triggered exacerbation in allergic asthmatic children: neutrophilic airway inflammation and alteration of virus sensors characterize a subgroup of patients. Respir. Res. 18, 191.

Dinarello, C.A., 1999. IL-18: a TH1-inducing, proinflammatory cytokine and new member of the IL-1 family. J. Allergy Clin. Immunol. 103, 11–24.

Dominguez-Villar, M., Gautron, A.S., De Marcken, M., Keller, M.J., Hafler, D.A., 2015. TLR7 induces anergy in human CD4(+) T cells. Nat. Immunol. 16, 118–128.

Evans, S.S., Collea, R.P., Appenheimer, M.M., Gollnick, S.O., 1993. Interferon-alpha induces the expression of the L-selectin homing receptor in human B lymphoid cells. J. Cell Biol. 123, 1889–1898.

Ezelle, H.J., Malathi, K., Hassel, B.A., 2016. The roles of RNase-L in antimicrobial immunity and the cytoskeleton-associated innate response. Int. J. Mol. Sci. 17, 1–24.

Fanucchi, M.V., Schelegle, E.S., Baker, G.L., Evans, M.J., McDonald, R.J., Gershwin, L.J., Raz, E., Hyde, D.M., Plopper, C.G., Miller, L.A., 2004. Immunostimulatory oligonucleotides attenuate airways remodeling in allergic monkeys. Am. J. Respir. Crit. Care Med. 170, 1153–1157.

Franchi, L., Eigenbrod, T., Munoz-Planillo, R., Nunez, G., 2009. The inflammasome: a caspase-1-activation platform that regulates immune responses and disease pathogenesis. Nat. Immunol. 10, 241–247.

Furuya, Y., Roberts, S., Hurteau, G.J., Sanfilippo, A.M., Racine, R., Metzger, D.W., 2014. Asthma increases susceptibility to heterologous but not homologous secondary influenza. J. Virol. 88, 9166–9181.

Galli, S.J., Tsai, M., Piliponsky, A.M., 2008. The development of allergic inflammation. Nature 454, 445–454.

Gantier, M.P., Tong, S., Behlke, M.A., Xu, D., Phipps, S., Foster, P.S., Williams, B.R., 2008. TLR7 is involved in sequence-specific sensing of single-stranded RNAs in human macrophages. J. Immunol. 180, 2117–2124.

Gao, Y., Nish, S.A., Jiang, R., Hou, L., Licona-Limon, P., Weinstein, J.S., Zhao, H., Medzhitov, R., 2013. Control of T helper 2 responses by transcription factor IRF4-dependent dendritic cells. Immunity 39, 722–732.

Gauvreau, G.M., Hessel, E.M., Boulet, L.P., Coffman, R.L., O'byrne, P.M., 2006. Immunostimulatory sequences regulate interferon-inducible genes but not allergic airway responses. Am. J. Respir. Crit. Care Med. 174, 15–20.

Geddes, K., Magalhaes, J.G., Girardin, S.E., 2009. Unleashing the therapeutic potential of NOD-like receptors. Nat. Rev. Drug Discov. 8, 465–479.

Gelman, A.E., Larosa, D.F., Zhang, J., Walsh, P.T., Choi, Y., Sunyer, J.O., Turka, L.A., 2006. The adaptor molecule MyD88 activates PI-3 kinase signaling in CD4+ T cells and enables CpG oligodeoxynucleotide-mediated costimulation. Immunity 25, 783–793.

Geurtsvankessel, C.H., Willart, M.A., Van Rijt, L.S., Muskens, F., Kool, M., Baas, C., Thielemans, K., Bennett, C., Clausen, B.E., Hoogsteden, H.C., Osterhaus, A.D., Rimmelzwaan, G.F., Lambrecht, B.N., 2008. Clearance of influenza virus from the lung depends on migratory langerin+CD11b- but not plasmacytoid dendritic cells. J. Exp. Med. 205, 1621–1634.

Gibson, S.J., Lindh, J.M., Riter, T.R., Gleason, R.M., Rogers, L.M., Fuller, A.E., Oesterich, J.L., Gorden, K.B., Qiu, X., Mckane, S.W., Noelle, R.J., Miller, R.L., Kedl, R.M., Fitzgerald-Bocarsly, P., Tomai, M.A., Vasilakos, J.P., 2002. Plasmacytoid dendritic cells produce cytokines and mature in response to the TLR7 agonists, imiquimod and resiquimod. Cell. Immunol. 218, 74–86.

Gill, M.A., Bajwa, G., George, T.A., Dong, C.C., Dougherty, I.I., Jiang, N., Gan, V.N., Gruchalla, R.S., 2010. Counterregulation between the FcepsilonRI pathway and antiviral responses in human plasmacytoid dendritic cells. J. Immunol. 184, 5999–6006.

de Goer de Herve, M.G., Durali, D., Dembele, B., Giuliani, M., Tran, T.A., Azzarone, B., Eid, P., Tardieu, M., Delfraissy, J.F., Taoufik, Y., 2011. Interferon-alpha triggers B cell effector 1 (Be1) commitment. PLoS One 6, e19366.

Green, N.M., Laws, A., Kiefer, K., Busconi, L., Kim, Y.M., Brinkmann, M.M., Trail, E.H., Yasuda, K., Christensen, S.R., Shlomchik, M.J., Vogel, S., Connor, J.H., Ploegh, H., Eilat, D., Rifkin, I.R., van Seventer, J.M., Marshak-Rothstein, A., 2009. Murine B cell response to TLR7 ligands depends on an IFN-beta feedback loop. J. Immunol. 183, 1569–1576.

Gregorio, J., Meller, S., Conrad, C., Di Nardo, A., Homey, B., Lauerma, A., Arai, N., Gallo, R.L., Digiovanni, J., Gilliet, M., 2010. Plasmacytoid dendritic cells sense skin injury and promote wound healing through type I interferons. J. Exp. Med. 207, 2921–2930.

Greiff, L., Cervin, A., Ahlstrom-Emanuelsson, C., Almqvist, G., Andersson, M., Dolata, J., Eriksson, L., Hogestatt, E., Kallen, A., Norlen, P., Sjolin, I.L., Widegren, H., 2012. Repeated intranasal TLR7 stimulation reduces allergen responsiveness in allergic rhinitis. Respir. Res. 13, 53.

Griffiths, C., Drews, S.J., Marchant, D.J., 2017. Respiratory syncytial virus: infection, detection, and new options for prevention and treatment. Clin. Microbiol. Rev. 30, 277–319.

Guillot, L., Le Goffic, R., Bloch, S., Escriou, N., Akira, S., Chignard, M., Si-Tahar, M., 2005. Involvement of toll-like receptor 3 in the immune response of lung epithelial cells to double-stranded RNA and influenza A virus. J. Biol. Chem. 280, 5571–5580.

Haldar, P., Brightling, C.E., Hargadon, B., Gupta, S., Monteiro, W., Sousa, A., Marshall, R.P., Bradding, P., Green, R.H., Wardlaw, A.J., Pavord, I.D., 2009. Mepolizumab and exacerbations of refractory eosinophilic asthma. N. Engl. J. Med. 360, 973–984.

Hammond, C., Kurten, M., Kennedy, J.L., 2015. Rhinovirus and asthma: a storied history of incompatibility. Curr. Allergy Asthma Rep. 15, 502.

Hart, O.M., Athie-Morales, V., O'connor, G.M., Gardiner, C.M., 2005. TLR7/8-mediated activation of human NK cells results in accessory cell-dependent IFN-gamma production. J. Immunol. 175, 1636–1642.
Hatchwell, L., Collison, A., Girkin, J., Parsons, K., Li, J., Zhang, J., Phipps, S., Knight, D., Bartlett, N.W., Johnston, S.L., Foster, P.S., Wark, P.A., Mattes, J., 2015. Toll-like receptor 7 governs interferon and inflammatory responses to rhinovirus and is suppressed by IL-5-induced lung eosinophilia. Thorax 70, 854–861.
Havenar-Daughton, C., Kolumam, G.A., Murali-Krishna, K., 2006. Cutting edge: the direct action of type I IFN on CD4 T cells is critical for sustaining clonal expansion in response to a viral but not a bacterial infection. J. Immunol. 176, 3315–3319.
Hemmi, H., Kaisho, T., Takeuchi, O., Sato, S., Sanjo, H., Hoshino, K., Horiuchi, T., Tomizawa, H., Takeda, K., Akira, S., 2002. Small anti-viral compounds activate immune cells via the TLR7 MyD88-dependent signaling pathway. Nat. Immunol. 3, 196–200.
Henderson, J., Hilliard, T.N., Sherriff, A., Stalker, D., Al Shammari, N., Thomas, H.M., 2005. Hospitalization for RSV bronchiolitis before 12 months of age and subsequent asthma, atopy and wheeze: a longitudinal birth cohort study. Pediatr. Allergy Immunol. 16, 386–392.
Hilty, M., Burke, C., Pedro, H., Cardenas, P., Bush, A., Bossley, C., Davies, J., Ervine, A., Poulter, L., Pachter, L., Moffatt, M.F., Cookson, W.O., 2010. Disordered microbial communities in asthmatic airways. PLoS One 5, e8578.
Hirahara, K., Poholek, A., Vahedi, G., Laurence, A., Kanno, Y., Milner, J.D., O'shea, J.J., 2013. Mechanisms underlying helper T-cell plasticity: implications for immune-mediated disease. J. Allergy Clin. Immunol. 131, 1276–1287.
Honda, K., Takaoka, A., Taniguchi, T., 2006. Type I interferon [corrected] gene induction by the interferon regulatory factor family of transcription factors. Immunity 25, 349–360.
Hornung, V., Rothenfusser, S., Britsch, S., Krug, A., Jahrsdorfer, B., Giese, T., Endres, S., Hartmann, G., 2002. Quantitative expression of toll-like receptor 1-10 mRNA in cellular subsets of human peripheral blood mononuclear cells and sensitivity to CpG oligodeoxynucleotides. J. Immunol. 168, 4531–4537.
Hornung, V., Guenthner-Biller, M., Bourquin, C., Ablasser, A., Schlee, M., Uematsu, S., Noronha, A., Manoharan, M., Akira, S., de Fougerolles, A., Endres, S., Hartmann, G., 2005. Sequence-specific potent induction of IFN-alpha by short interfering RNA in plasmacytoid dendritic cells through TLR7. Nat. Med. 11, 263–270.
Huber, J.P., Farrar, J.D., 2011. Regulation of effector and memory T-cell functions by type I interferon. Immunology 132, 466–474.
Hussain, I., Jain, V.V., Kitagaki, K., Businga, T.R., O'shaughnessy, P., Kline, J.N., 2002. Modulation of murine allergic rhinosinusitis by CpG oligodeoxynucleotides. Laryngoscope 112, 1819–1826.
Imanishi, T., Ishihara, C., Badr Mel, S., Hashimoto-Tane, A., Kimura, Y., Kawai, T., Takeuchi, O., Ishii, K.J., Taniguchi, S., Noda, T., Hirano, H., Brombacher, F., Barber, G.N., Akira, S., Saito, T., 2014. Nucleic acid sensing by T cells initiates Th2 cell differentiation. Nat. Commun. 5, 3566.
Ishikawa, H., Barber, G.N., 2008. STING is an endoplasmic reticulum adaptor that facilitates innate immune signalling. Nature 455, 674–678.
Iwasaki, A., Medzhitov, R., 2010. Regulation of adaptive immunity by the innate immune system. Science 327, 291–295.
Iwasaki, A., Pillai, P.S., 2014. Innate immunity to influenza virus infection. Nat. Rev. Immunol. 14, 315–328.
Jacobs, S.E., Lamson, D.M., St George, K., Walsh, T.J., 2013. Human rhinoviruses. Clin. Microbiol. Rev. 26, 135–162.

Johnston, S.L., Pattemore, P.K., Sanderson, G., Smith, S., Campbell, M.J., Josephs, L.K., Cunningham, A., Robinson, B.S., Myint, S.H., Ward, M.E., Tyrrell, D.A., Holgate, S.T., 1996. The relationship between upper respiratory infections and hospital admissions for asthma: a time-trend analysis. Am. J. Respir. Crit. Care Med. 154, 654–660.

Juncadella, I.J., Kadl, A., Sharma, A.K., Shim, Y.M., Hochreiter-Hufford, A., Borish, L., Ravichandran, K.S., 2013. Apoptotic cell clearance by bronchial epithelial cells critically influences airway inflammation. Nature 493, 547–551.

Kanneganti, T.D., Body-Malapel, M., Amer, A., Park, J.H., Whitfield, J., Franchi, L., Taraporewala, Z.F., Miller, D., Patton, J.T., Inohara, N., Nunez, G., 2006. Critical role for Cryopyrin/Nalp3 in activation of caspase-1 in response to viral infection and double-stranded RNA. J. Biol. Chem. 281, 36560–36568.

Kato, A., Schleimer, R.P., 2007. Beyond inflammation: airway epithelial cells are at the interface of innate and adaptive immunity. Curr. Opin. Immunol. 19, 711–720.

Kato, A., Favoreto Jr., S., Avila, P.C., Schleimer, R.P., 2007. TLR3- and Th2 cytokine-dependent production of thymic stromal lymphopoietin in human airway epithelial cells. J. Immunol. 179, 1080–1087.

Kawai, T., Akira, S., 2010. The role of pattern-recognition receptors in innate immunity: update on Toll-like receptors. Nat. Immunol. 11, 373–384.

Kawai, T., Akira, S., 2011. Toll-like receptors and their crosstalk with other innate receptors in infection and immunity. Immunity 34, 637–650.

Kawai, T., Takahashi, K., Sato, S., Coban, C., Kumar, H., Kato, H., Ishii, K.J., Takeuchi, O., Akira, S., 2005. IPS-1, an adaptor triggering RIG-I- and Mda5-mediated type I interferon induction. Nat. Immunol. 6, 981–988.

Kawasaki, T., Kawai, T., Akira, S., 2011. Recognition of nucleic acids by pattern-recognition receptors and its relevance in autoimmunity. Immunol. Rev. 243, 61–73.

Kell, A.M., Gale Jr., M., 2015. RIG-I in RNA virus recognition. Virology 479–480, 110–121.

Kim, C.K., Koh, J.Y., Han, T.H., Kim, D.K., Kim, B.I., Koh, Y.Y., 2006. Increased levels of BAL cysteinyl leukotrienesinacute [corrected] RSV bronchiolitis. Acta Paediatr. 95, 479–485.

Kim, Y.S., Kim, Y., Lee, K.J., Kwon, H.J., Kim, D.S., Kim, T.Y., 2007. Improvement of atopic dermatitis in NC/Nga mice by topical application of CpG phosphodiester-ODN. Int. Arch. Allergy Immunol. 144, 315–324.

Kim, Y.M., Brinkmann, M.M., Paquet, M.E., Ploegh, H.L., 2008. UNC93B1 delivers nucleotide-sensing toll-like receptors to endolysosomes. Nature 452, 234–238.

Kim, D.H., Sohn, J.H., Park, H.J., Lee, J.H., Park, J.W., Choi, J.M., 2016. CpG oligodeoxynucleotide inhibits cockroach-induced asthma via induction of IFN-gamma(+) Th1 cells or Foxp3(+) regulatory T cells in the lung. Allergy Asthma Immunol. Res. 8, 264–275.

Klimek, L., Willers, J., Hammann-Haenni, A., Pfaar, O., Stocker, H., Mueller, P., Renner, W.A., Bachmann, M.F., 2011. Assessment of clinical efficacy of CYT003-QbG10 in patients with allergic rhinoconjunctivitis: a phase IIb study. Clin. Exp. Allergy 41, 1305–1312.

Kline, J.N., Waldschmidt, T.J., Businga, T.R., Lemish, J.E., Weinstock, J.V., Thorne, P.S., Krieg, A.M., 1998. Modulation of airway inflammation by CpG oligodeoxynucleotides in a murine model of asthma. J. Immunol. 160, 2555–2559.

Kline, J.N., Krieg, A.M., Waldschmidt, T.J., Ballas, Z.K., Jain, V., Businga, T.R., 1999. CpG oligodeoxynucleotides do not require TH1 cytokines to prevent eosinophilic airway inflammation in a murine model of asthma. J. Allergy Clin. Immunol. 104, 1258–1264.

Kline, J.N., Kitagaki, K., Businga, T.R., Jain, V.V., 2002. Treatment of established asthma in a murine model using CpG oligodeoxynucleotides. Am. J. Physiol. Lung Cell. Mol. Physiol. 283, L170–L179.

Knudson, C.J., Varga, S.M., 2015. The relationship between respiratory syncytial virus and asthma. Vet. Pathol. 52, 97–106.

Koonin, E.V., Aravind, L., 2000. The NACHT family - a new group of predicted NTPases implicated in apoptosis and MHC transcription activation. Trends Biochem. Sci. 25, 223–224.

Krishnamoorthy, N., Khare, A., Oriss, T.B., Raundhal, M., Morse, C., Yarlagadda, M., Wenzel, S.E., Moore, M.L., Peebles Jr., R.S., Ray, A., Ray, P., 2012. Early infection with respiratory syncytial virus impairs regulatory T cell function and increases susceptibility to allergic asthma. Nat. Med. 18, 1525–1530.

Kubo, T., Kamekura, R., Kumagai, A., Kawata, K., Yamashita, K., Mitsuhashi, Y., Kojima, T., Sugimoto, K., Yoneta, A., Sumikawa, Y., Yamashita, T., Sato, N., Himi, T., Ichimiya, S., 2014. DeltaNp63 controls a TLR3-mediated mechanism that abundantly provides thymic stromal lymphopoietin in atopic dermatitis. PLoS One 9, e105498.

Kumar, R.K., Foster, P.S., Rosenberg, H.F., 2014. Respiratory viral infection, epithelial cytokines, and innate lymphoid cells in asthma exacerbations. J. Leukoc. Biol. 96, 391–396.

Lee, H.C., Headley, M.B., Loo, Y.M., Berlin, A., Gale Jr., M., Debley, J.S., Lukacs, N.W., Ziegler, S.F., 2012. Thymic stromal lymphopoietin is induced by respiratory syncytial virus-infected airway epithelial cells and promotes a type 2 response to infection. J. Allergy Clin. Immunol. 130, 1187–1196.e5.

Leiva-Juarez, M.M., Kolls, J.K., Evans, S.E., 2018. Lung epithelial cells: therapeutically inducible effectors of antimicrobial defense. Mucosal Immunol. 11, 21–34.

Li, X.D., Wu, J., Gao, D., Wang, H., Sun, L., Chen, Z.J., 2013. Pivotal roles of cGAS-cGAMP signaling in antiviral defense and immune adjuvant effects. Science 341, 1390–1394.

Li, B.W.S., de Bruijn, M.J.W., Lukkes, M., Van Nimwegen, M., Bergen, I.M., Kleinjan, A., Geurtsvankessel, C.H., Andeweg, A., Rimmelzwaan, G.F., Hendriks, R.W., 2018. T cells and ILC2s are major effector cells in influenza-induced exacerbation of allergic airway inflammation in mice. Eur. J. Immunol. 1–13.

Lin, L., Gerth, A.J., Peng, S.L., 2004. CpG DNA redirects class-switching towards "Th1-like" Ig isotype production via TLR9 and MyD88. Eur. J. Immunol. 34, 1483–1487.

Liu, F., Gu, J., 2011. Retinoic acid inducible gene-I, more than a virus sensor. Protein Cell 2, 351–357.

Liu, N., Ohnishi, N., Ni, L., Akira, S., Bacon, K.B., 2003. CpG directly induces T-bet expression and inhibits IgG1 and IgE switching in B cells. Nat. Immunol. 4, 687–693.

Liu, T., Zhang, L., Joo, D., Sun, S.C., 2017. NF-kappaB signaling in inflammation. Signal Transduct. Target. Ther. 2, 1–9.

Luecke, S., Paludan, S.R., 2017. Molecular requirements for sensing of intracellular microbial nucleic acids by the innate immune system. Cytokine 98, 4–14.

Lynch, J.P., Mazzone, S.B., Rogers, M.J., Arikkatt, J.J., Loh, Z., Pritchard, A.L., Upham, J.W., Phipps, S., 2014. The plasmacytoid dendritic cell: at the cross-roads in asthma. Eur. Respir. J. 43, 264–275.

Madouri, F., Guillou, N., Fauconnier, L., Marchiol, T., Rouxel, N., Chenuet, P., Ledru, A., Apetoh, L., Ghiringhelli, F., Chamaillard, M., Zheng, S.G., Trovero, F., Quesniaux, V.F., Ryffel, B., Togbe, D., 2015. Caspase-1 activation by NLRP3 inflammasome dampens IL-33-dependent house dust mite-induced allergic lung inflammation. J. Mol. Cell Biol. 7, 351–365.

Mahmutovic Persson, I., Akbarshahi, H., Menzel, M., Brandelius, A., Uller, L., 2016. Increased expression of upstream TH2-cytokines in a mouse model of viral-induced asthma exacerbation. J. Transl. Med. 14, 52.

Mahmutovic-Persson, I., Akbarshahi, H., Bartlett, N.W., Glanville, N., Johnston, S.L., Brandelius, A., Uller, L., 2014. Inhaled dsRNA and rhinovirus evoke neutrophilic exacerbation and lung expression of thymic stromal lymphopoietin in allergic mice with established experimental asthma. Allergy 69, 348–358.

Malathi, K., Dong, B., Gale Jr., M., Silverman, R.H., 2007. Small self-RNA generated by RNase L amplifies antiviral innate immunity. Nature 448, 816–819.

Mariathasan, S., Weiss, D.S., Newton, K., McBride, J., O'rourke, K., Roose-Girma, M., Lee, W.P., Weinrauch, Y., Monack, D.M., Dixit, V.M., 2006. Cryopyrin activates the inflammasome in response to toxins and ATP. Nature 440, 228–232.

McClure, R., Massari, P., 2014. TLR-dependent human mucosal epithelial cell responses to microbial pathogens. Front. Immunol. 5, 386.

Medzhitov, R., 2007. Recognition of microorganisms and activation of the immune response. Nature 449, 819–826.

Mildner, A., Jung, S., 2014. Development and function of dendritic cell subsets. Immunity 40, 642–656.

Miller, E.K., Hernandez, J.Z., Wimmenauer, V., Shepherd, B.E., Hijano, D., Libster, R., Serra, M.E., Bhat, N., Batalle, J.P., Mohamed, Y., Reynaldi, A., Rodriguez, A., Otello, M., Pisapia, N., Bugna, J., Bellabarba, M., Kraft, D., Coviello, S., Ferolla, F.M., Chen, A., London, S.J., Siberry, G.K., Williams, J.V., Polack, F.P., 2012. A mechanistic role for type III IFN-lambda1 in asthma exacerbations mediated by human rhinoviruses. Am. J. Respir. Crit. Care Med. 185, 508–516.

Mitchell, P.D., O'Byrne, P.M., 2017. Biologics and the lung: TSLP and other epithelial cell-derived cytokines in asthma. Pharmacol. Ther. 169, 104–112.

Moisan, J., Camateros, P., Thuraisingam, T., Marion, D., Koohsari, H., Martin, P., Boghdady, M.L., Ding, A., Gaestel, M., Guiot, M.C., Martin, J.G., Radzioch, D., 2006. TLR7 ligand prevents allergen-induced airway hyperresponsiveness and eosinophilia in allergic asthma by a MYD88-dependent and MK2-independent pathway. Am. J. Physiol. Lung Cell. Mol. Physiol. 290, L987–L995.

Monticelli, L.A., Sonnenberg, G.F., Abt, M.C., Alenghat, T., Ziegler, C.G., Doering, T.A., Angelosanto, J.M., Laidlaw, B.J., Yang, C.Y., Sathaliyawala, T., Kubota, M., Turner, D., Diamond, J.M., Goldrath, A.W., Farber, D.L., Collman, R.G., Wherry, E.J., Artis, D., 2011. Innate lymphoid cells promote lung-tissue homeostasis after infection with influenza virus. Nat. Immunol. 12, 1045–1054.

Muruve, D.A., Petrilli, V., Zaiss, A.K., White, L.R., Clark, S.A., Ross, P.J., Parks, R.J., Tschopp, J., 2008. The inflammasome recognizes cytosolic microbial and host DNA and triggers an innate immune response. Nature 452, 103–107.

Nagase, H., Okugawa, S., Ota, Y., Yamaguchi, M., Tomizawa, H., Matsushima, K., Ohta, K., Yamamoto, K., Hirai, K., 2003. Expression and function of Toll-like receptors in eosinophils: activation by Toll-like receptor 7 ligand. J. Immunol. 171, 3977–3982.

Nakamura, N., Tamagawa-Mineoka, R., Ueta, M., Kinoshita, S., Katoh, N., 2015. Toll-like receptor 3 increases allergic and irritant contact dermatitis. J. Invest. Dermatol. 135, 411–417.

Newcomb, D.C., Boswell, M.G., Reiss, S., Zhou, W., Goleniewska, K., Toki, S., Harintho, M.T., Lukacs, N.W., Kolls, J.K., Peebles Jr., R.S., 2013. IL-17A inhibits airway reactivity induced by respiratory syncytial virus infection during allergic airway inflammation. Thorax 68, 717–723.

Noben-Trauth, N., Hu-Li, J., Paul, W.E., 2000. Conventional, naive CD4 + T cells provide an initial source of IL-4 during Th2 differentiation. J. Immunol. 165, 3620–3625.

Ori, D., Murase, M., Kawai, T., 2017. Cytosolic nucleic acid sensors and innate immune regulation. Int. Rev. Immunol. 36, 74–88.

Patel, D.A., You, Y., Huang, G., Byers, D.E., Kim, H.J., Agapov, E., Moore, M.L., Peebles Jr., R.S., Castro, M., Sumino, K., Shifren, A., Brody, S.L., Holtzman, M.J., 2014. Interferon response and respiratory virus control are preserved in bronchial epithelial cells in asthma. J. Allergy Clin. Immunol. 134, 1402–1412.e7.

Paul, W.E., Zhu, J., 2010. How are T(H)2-type immune responses initiated and amplified? Nat. Rev. Immunol. 10, 225–235.

Peebles Jr., R.S., Hashimoto, K., Collins, R.D., Jarzecka, K., Furlong, J., Mitchell, D.B., Sheller, J.R., Graham, B.S., 2001a. Immune interaction between respiratory syncytial virus infection and allergen sensitization critically depends on timing of challenges. J. Infect. Dis. 184, 1374–1379.
Peebles Jr., R.S., Sheller, J.R., Collins, R.D., Jarzecka, A.K., Mitchell, D.B., Parker, R.A., Graham, B.S., 2001b. Respiratory syncytial virus infection does not increase allergen-induced type 2 cytokine production, yet increases airway hyperresponsiveness in mice. J. Med. Virol. 63, 178–188.
Peng, G., Guo, Z., Kiniwa, Y., Voo, K.S., Peng, W., Fu, T., Wang, D.Y., Li, Y., Wang, H.Y., Wang, R.F., 2005. Toll-like receptor 8-mediated reversal of CD4+ regulatory T cell function. Science 309, 1380–1384.
Pichlmair, A., Reis e Sousa, C., 2007. Innate recognition of viruses. Immunity 27, 370–383.
Poltorak, A., Kurmyshkina, O., Volkova, T., 2016. Stimulator of interferon genes (STING): a "new chapter" in virus-associated cancer research. Lessons from wild-derived mouse models of innate immunity. Cytokine Growth Factor Rev. 29, 83–91.
Pothlichet, J., Meunier, I., Davis, B.K., Ting, J.P., Skamene, E., von Messling, V., Vidal, S.M., 2013. Type I IFN triggers RIG-I/TLR3/NLRP3-dependent inflammasome activation in influenza A virus infected cells. PLoS Pathog. 9, e1003256.
Pritchard, A.L., Carroll, M.L., Burel, J.G., White, O.J., Phipps, S., Upham, J.W., 2012. Innate IFNs and plasmacytoid dendritic cells constrain Th2 cytokine responses to rhinovirus: a regulatory mechanism with relevance to asthma. J. Immunol. 188, 5898–5905.
Pritchard, A.L., White, O.J., Burel, J.G., Carroll, M.L., Phipps, S., Upham, J.W., 2014. Asthma is associated with multiple alterations in anti-viral innate signalling pathways. PLoS One 9, e106501.
Ramos-Barbon, D., Presley, J.F., Hamid, Q.A., Fixman, E.D., Martin, J.G., 2005. Antigen-specific CD4+ T cells drive airway smooth muscle remodeling in experimental asthma. J. Clin. Invest. 115, 1580–1589.
Reuter, S., Dehzad, N., Martin, H., Bohm, L., Becker, M., Buhl, R., Stassen, M., Taube, C., 2012. TLR3 but not TLR7/8 ligand induces allergic sensitization to inhaled allergen. J. Immunol. 188, 5123–5131.
Rincon, M., Anguita, J., Nakamura, T., Fikrig, E., Flavell, R.A., 1997. Interleukin (IL)-6 directs the differentiation of IL-4-producing CD4+ T cells. J. Exp. Med. 185, 461–469.
Ritchie, A.I., Farne, H.A., Singanayagam, A., Jackson, D.J., Mallia, P., Johnston, S.L., 2015. Pathogenesis of viral infection in exacerbations of airway disease. Ann. Am. Thorac. Soc. 12 (Suppl. 2), S115–S132.
Roers, A., Hiller, B., Hornung, V., 2016. Recognition of endogenous nucleic acids by the innate immune system. Immunity 44, 739–754.
Romagnani, S., 2004. The increased prevalence of allergy and the hygiene hypothesis: missing immune deviation, reduced immune suppression, or both? Immunology 112, 352–363.
Saenz, S.A., Taylor, B.C., Artis, D., 2008. Welcome to the neighborhood: epithelial cell-derived cytokines license innate and adaptive immune responses at mucosal sites. Immunol. Rev. 226, 172–190.
Samarasinghe, A.E., Woolard, S.N., Boyd, K.L., Hoselton, S.A., Schuh, J.M., McCullers, J.A., 2014. The immune profile associated with acute allergic asthma accelerates clearance of influenza virus. Immunol. Cell Biol. 92, 449–459.
Samarasinghe, A.E., Melo, R.C., Duan, S., Lemessurier, K.S., Liedmann, S., Surman, S.L., Lee, J.J., Hurwitz, J.L., Thomas, P.G., McCullers, J.A., 2017. Eosinophils promote antiviral immunity in mice infected with influenza A virus. J. Immunol. 198, 3214–3226.
Santarlasci, V., Cosmi, L., Maggi, L., Liotta, F., Annunziato, F., 2013. IL-1 and T helper immune responses. Front. Immunol. 4, 182.

Satoh, T., Kato, H., Kumagai, Y., Yoneyama, M., Sato, S., Matsushita, K., Tsujimura, T., Fujita, T., Akira, S., Takeuchi, O., 2010. LGP2 is a positive regulator of RIG-I- and MDA5-mediated antiviral responses. Proc. Natl. Acad. Sci. U. S. A. 107, 1512–1517.

Schaefer, L., 2014. Complexity of danger: the diverse nature of damage-associated molecular patterns. J. Biol. Chem. 289, 35237–35245.

Schlee, M., 2013. Master sensors of pathogenic RNA—RIG-I like receptors. Immunobiology 218, 1322–1335.

Schmitz, J., Owyang, A., Oldham, E., Song, Y., Murphy, E., McClanahan, T.K., Zurawski, G., Moshrefi, M., Qin, J., Li, X., Gorman, D.M., Bazan, J.F., Kastelein, R.A., 2005. IL-33, an interleukin-1-like cytokine that signals via the IL-1 receptor-related protein ST2 and induces T helper type 2-associated cytokines. Immunity 23, 479–490.

Schulz, O., Diebold, S.S., Chen, M., Naslund, T.I., Nolte, M.A., Alexopoulou, L., Azuma, Y.T., Flavell, R.A., Liljestrom, P., Reis e Sousa, C., 2005. Toll-like receptor 3 promotes cross-priming to virus-infected cells. Nature 433, 887–892.

Sharma, S., Fitzgerald, K.A., Cancro, M.P., Marshak-Rothstein, A., 2015. Nucleic acid-sensing receptors: rheostats of autoimmunity and autoinflammation. J. Immunol. 195, 3507–3512.

Sheih, A., Parks, W.C., Ziegler, S.F., 2017. GM-CSF produced by the airway epithelium is required for sensitization to cockroach allergen. Mucosal Immunol. 10, 705–715.

Shikhagaie, M.M., Andersson, C.K., Mori, M., Kortekaas Krohn, I., Bergqvist, A., Dahl, R., Ekblad, E., Hoffmann, H.J., Bjermer, L., Erjefalt, J.S., 2014. Mapping of TLR5 and TLR7 in central and distal human airways and identification of reduced TLR expression in severe asthma. Clin. Exp. Allergy 44, 184–196.

Silverman, R.H., 2007. Viral encounters with 2',5'-oligoadenylate synthetase and RNase L during the interferon antiviral response. J. Virol. 81, 12720–12729.

Simchoni, N., Cunningham-Rundles, C., 2015. TLR7- and TLR9-responsive human B cells share phenotypic and genetic characteristics. J. Immunol. 194, 3035–3044.

Skevaki, C., Hudemann, C., Matrosovich, M., Mobs, C., Paul, S., Wachtendorf, A., Alashkar Alhamwe, B., Potaczek, D.P., Hagner, S., Gemsa, D., Garn, H., Sette, A., Renz, H., 2017. Influenza-derived peptides cross-react with allergens and provide asthma protection. J. Allergy Clin. Immunol. 142, 804–814.

Song, Y., Zhuang, Y., Zhai, S., Huang, D., Zhang, Y., Kang, W., Li, X., Liu, Q., Yu, Q., Sun, Y., 2009. Increased expression of TLR7 in CD8(+) T cells leads to TLR7-mediated activation and accessory cell-dependent IFN-gamma production in HIV type 1 infection. AIDS Res. Hum. Retroviruses 25, 1287–1295.

Stensballe, L.G., Simonsen, J.B., Thomsen, S.F., Larsen, A.M., Lysdal, S.H., Aaby, P., Kyvik, K.O., Skytthe, A., Backer, V., Bisgaard, H., 2009. The causal direction in the association between respiratory syncytial virus hospitalization and asthma. J. Allergy Clin. Immunol. 123, 131–137.e1.

Stier, M.T., Bloodworth, M.H., Toki, S., Newcomb, D.C., Goleniewska, K., Boyd, K.L., Quitalig, M., Hotard, A.L., Moore, M.L., Hartert, T.V., Zhou, B., McKenzie, A.N., Peebles Jr., R.S., 2016. Respiratory syncytial virus infection activates IL-13-producing group 2 innate lymphoid cells through thymic stromal lymphopoietin. J. Allergy Clin. Immunol. 138, 814–824.e11.

Strachan, D.P., 1989. Hay fever, hygiene, and household size. BMJ 299, 1259–1260.

Sur, S., Wild, J.S., Choudhury, B.K., Sur, N., Alam, R., Klinman, D.M., 1999. Long term prevention of allergic lung inflammation in a mouse model of asthma by CpG oligodeoxynucleotides. J. Immunol. 162, 6284–6293.

Svoboda, P., 2014. Renaissance of mammalian endogenous RNAi. FEBS Lett. 588, 2550–2556.

Tamaoka, M., Hassan, M., McGovern, T., Ramos-Barbon, D., Jo, T., Yoshizawa, Y., Tolloczko, B., Hamid, Q., Martin, J.G., 2008. The epidermal growth factor receptor mediates allergic airway remodelling in the rat. Eur. Respir. J 32, 1213–1223.

Tattoli, I., Carneiro, L.A., Jehanno, M., Magalhaes, J.G., Shu, Y., Philpott, D.J., Arnoult, D., Girardin, S.E., 2008. NLRX1 is a mitochondrial NOD-like receptor that amplifies NF-kappaB and JNK pathways by inducing reactive oxygen species production. EMBO Rep. 9, 293–300.

Tengroth, L., Millrud, C.R., Kvarnhammar, A.M., Kumlien Georen, S., Latif, L., Cardell, L.O., 2014. Functional effects of toll-like receptor (TLR)3, 7, 9, RIG-I and MDA-5 stimulation in nasal epithelial cells. PLoS One 9, e98239.

Ting, J.P., Lovering, R.C., Alnemri, E.S., Bertin, J., Boss, J.M., Davis, B.K., Flavell, R.A., Girardin, S.E., Godzik, A., Harton, J.A., Hoffman, H.M., Hugot, J.P., Inohara, N., Mackenzie, A., Maltais, L.J., Nunez, G., Ogura, Y., Otten, L.A., Philpott, D., Reed, J.C., Reith, W., Schreiber, S., Steimle, V., Ward, P.A., 2008. The NLR gene family: a standard nomenclature. Immunity 28, 285–287.

Tulic, M.K., Fiset, P.O., Christodoulopoulos, P., Vaillancourt, P., Desrosiers, M., Lavigne, F., Eiden, J., Hamid, Q., 2004. Amb a 1-immunostimulatory oligodeoxynucleotide conjugate immunotherapy decreases the nasal inflammatory response. J. Allergy Clin. Immunol. 113, 235–241.

Unger, B.L., Ganesan, S., Comstock, A.T., Faris, A.N., Hershenson, M.B., Sajjan, U.S., 2014. Nod-like receptor X-1 is required for rhinovirus-induced barrier dysfunction in airway epithelial cells. J. Virol. 88, 3705–3718.

von Mutius, E., 2016. The microbial environment and its influence on asthma prevention in early life. J. Allergy Clin. Immunol 137, 680–689.

Wang, C.G., Du, T., Xu, L.J., Martin, J.G., 1993. Role of leukotriene D4 in allergen-induced increases in airway smooth muscle in the rat. Am. Rev. Respir. Dis. 148, 413–417.

Wark, P.A., Johnston, S.L., Bucchieri, F., Powell, R., Puddicombe, S., Laza-Stanca, V., Holgate, S.T., Davies, D.E., 2005. Asthmatic bronchial epithelial cells have a deficient innate immune response to infection with rhinovirus. J. Exp. Med. 201, 937–947.

Williams, B.R., 2001. Signal integration via PKR. Sci. STKE 2001, re2.

Zarnegar, B., Mendez-Enriquez, E., Westin, A., Soderberg, C., Dahlin, J.S., Gronvik, K.O., Hallgren, J., 2017. Influenza infection in mice induces accumulation of lung mast cells through the recruitment and maturation of mast cell progenitors. Front. Immunol. 8, 310.

Zhu, J., Message, S.D., Mallia, P., Kebadze, T., Contoli, M., Ward, C.K., Barnathan, E.S., Mascelli, M.A., Kon, O.M., Papi, A., Stanciu, L.A., Edwards, M.R., Jeffery, P.K., Johnston, S.L., 2018. Bronchial mucosal IFN-alpha/beta and pattern recognition receptor expression in patients with experimental rhinovirus-induced asthma exacerbations. J. Allergy Clin. Immunol. 30610-9.

Zlotorynski, E., 2018. RNA metabolism: interfering with interferon by RNA editing. Nat. Rev. Mol. Cell Biol. 19, 140–141.

CHAPTER TWO

The Role of Nucleic Acid Sensing in Controlling Microbial and Autoimmune Disorders

Keesha M. Matz*[,2], R. Marena Guzman*[,2], Alan G. Goodman*[,†,1]
*School of Molecular Biosciences, College of Veterinary Medicine, Washington State University, Pullman, WA, United States
†Paul G. Allen School for Global Animal Health, College of Veterinary Medicine, Washington State University, Pullman, WA, United States
[1]Corresponding author: e-mail address: alan.goodman@wsu.edu

Contents

Abstract

Innate immunity, the first line of defense against invading pathogens, is an ancient form of host defense found in all animals, from sponges to humans. During infection, innate immune receptors recognize conserved molecular patterns, such as microbial surface molecules, metabolites produces during infection, or nucleic acids of the microbe's

[2] These authors contributed equally to this work.

International Review of Cell and Molecular Biology, Volume 345
ISSN 1937-6448
https://doi.org/10.1016/bs.ircmb.2018.08.002

genome. When initiated, the innate immune response activates a host defense program that leads to the synthesis proteins capable of pathogen killing. In mammals, the induction of cytokines during the innate immune response leads to the recruitment of professional immune cells to the site of infection, leading to an adaptive immune response. While a fully functional innate immune response is crucial for a proper host response and curbing microbial infection, if the innate immune response is dysfunctional and is activated in the absence of infection, autoinflammation and autoimmune disorders can develop. Therefore, it follows that the innate immune response must be tightly controlled to avoid an autoimmune response from host-derived molecules, yet still unencumbered to respond to infection. In this review, we will focus on the innate immune response activated from cytosolic nucleic acids, derived from the microbe or host itself. We will depict how viruses and bacteria activate these nucleic acid sensing pathways and their mechanisms to inhibit the pathways. We will also describe the autoinflammatory and autoimmune disorders that develop when these pathways are hyperactive. Finally, we will discuss gaps in knowledge with regard to innate immune response failure and identify where further research is needed.

1. INTRODUCTION

Several nucleic sensing pathways have been identified over the last few decades that have increased our understanding of diseases that occur when these pathways are dysfunctional. The last decade has experienced an upsurge of new discoveries regarding nucleic acid sensing, many due to advances in technology. Nevertheless, intricacies of how the failure of nucleic acid-sensing mechanisms leads to autoinflammation and autoimmunity remain unsolved. In this review, we explore what is known about how the failure of nucleic acid-sensing mechanisms leads to autoimmunity and autoinflammation as well as highlight questions that remain to be solved.

Autoinflammation and autoimmunity are often mistakenly used interchangeably but refer to the origin of disease. Namely, while autoinflammatory diseases are driven primarily by dysregulation of the innate immune system and do not rely on T cells or B cells for disease to progress, autoimmune diseases are driven primarily by T-cell and B-cell mediation (Arakelyan et al., 2017). Several autoimmune diseases originate from the innate immune system but then require lymphocytes for progression, indicating that disease is often a result of more than one component of immunity. Details that differentiate autoimmunity and autoinflammation remain controversial, but our current understanding of these pathologies is based on our ample understanding of innate immune recognition.

Central to the autoinflammatory and autoimmune disorders discussed in this review are the sensors that detect cytosolic nucleic acids to stimulate an innate immune response. The primary role of the innate immune system is to be the first line of defense against foreign microbes. The recognition of microbes and nucleic acid depends on pattern recognition receptors (PRRs) that recognize a variety of pathogen-derived molecules called pathogen-associated molecular patterns (PAMPs). While many types of PAMPs are encountered only during microbial infection, nucleic acid-mediated PRR activation can be microbe- or host-derived. That is, the innate immune response can be activated by the presence of self-nucleic acids that escape the nucleus, a major etiological cause of the autoinflammatory and autoimmune disorders discussed here. As such, in this review, we will describe the receptors of RNA and DNA nucleic acids and the signaling pathways that each stimulate. Then, for both the RNA and DNA sensing pathways, we will identify the specific types of microbial infections that activate innate immune responses and how each microbe has evolved mechanisms to inhibit these pathways. Finally, we will discuss the autoinflammatory and autoimmune disorders that develop when these nucleic acid signaling pathways are dysfunctional.

2. RNA NUCLEIC ACID SENSING IN VIRAL IMMUNOLOGY AND AUTOIMMUNITY

2.1 Detection of Intracellular RNA

2.1.1 Protein Kinase R

Protein kinase R (PKR) is encoded by the *EIF2AK2* gene, and it is an interferon (IFN)-induced, double-stranded RNA (dsRNA)-dependent protein kinase that phosphorylates the alpha subunit of eukaryotic initiation factor 2 (eIF2α), resulting in the inhibition of mRNA translation initiation (Kitajewski et al., 1986). PKR has also been shown to be a component of IFN activation to facilitate a robust innate immune response (Balachandran et al., 2000). PKR itself is induced by IFN, resulting in a positive feed-forward loop that further amplifies the innate immune response (Li et al., 2011). Double-stranded RNA activation of PKR results in its dimerization and autophosphorylation (Dever, 2002). Downstream IFN gene induction is induced primarily by the NF-κB transcription factor following PKR-mediated phosphorylation of IκB (Kumar et al., 1994), but not IRF3 activation (Smith et al., 2001). A number of viruses have encoded mechanisms to block PKR activation to allow for enhanced virus replication, such as adenovirus, reovirus, influenza virus, and hepatitis C virus (Gale et al., 1997; Katze et al., 1987; Lloyd and Shatkin, 1992; Lu et al., 1995).

The activation of PKR is also inhibited by other cellular factors, such as the gene encoded by *DNAJC3*, namely $P58^{IPK}$ (Lee et al., 1992). Specifically, $P58^{IPK}$ interacts with PKR at the site that promotes its dimerization and autophosphorylation (Gale et al., 1996). Importantly, $P58^{IPK}$ is activated during influenza virus infection and $P58^{IPK}$ dysfunction results in late onset type 1 diabetes, which will be discussed in a following section (Ladiges et al., 2005; Melville et al., 1999).

2.1.2 Toll-Like Receptors

Toll-like receptors (TLRs) are integral membrane glycoproteins, have a trimodular structure, and contain 16–28 leucine-rich repeats (LRRs), which are necessary for interaction and recruitment of several adaptor proteins (Kawasaki et al., 2011; Matsushima et al., 2007). TLRs are a subset of PRRs expressed on the cell membrane of professional immune cells like monocytes, macrophages, dendritic cells, B cells, and non-immune cells like keratinocytes and epithelial cells (Kawasaki et al., 2011; Novak et al., 2010). TLRs are classified by their ectodomain for ligand binding (Kawai and Akira, 2009; Kumar et al., 2009). Thirteen TLRs have been identified in mammals but only TLR3, TLR7/8, and TLR9 recognize microbial nucleic acids in endolysosomal compartments while the others bind to bacterial or parasitic PAMPs such as triacyl lipopeptides, peptidoglycan, or lipopolysaccharide on the cell surface (Kawai and Akira, 2009; Kawasaki et al., 2011).

TLR3 recognizes dsRNA, a product of RNA virus replication lifecycles, in endolysosomal compartments. Recognition of dsRNA by TLR3 leads to signaling through NF-κB and subsequent activation of IFNs (Alexopoulou et al., 2001). This IFN-β promoter activation is uniquely mediated by the adaptor protein, Toll-interleukin 1 receptor domain (TIR)-containing adaptor inducing IFN-β (TRIF or TICAM-1) (Oshiumi et al., 2003). This recognition process by TLR3 must be tightly regulated to ensure IFN activation only in the presence of non-self nucleic acid. The structural composition of the dsRNA is important in efficient recognition by TLR3. For example, the 2′-OH group in cytidylic acid is necessary for the dsRNA to be recognized by TLR3 (Okahira et al., 2005). More recently, work has been done to examine the functional structure of TLR3 that has the ability to recognize dsRNA. Glycosylation and cathepsin cleavage of TLR3 occur as it is transported from the endoplasmic reticulum through the Golgi apparatus and into an endolysosome (Toscano et al., 2013). The TLR3 C terminal and N terminal cleavage product complex that results is necessary for the recognition of dsRNA (Garcia-Cattaneo et al., 2012; Murakami et al., 2014). This mature TLR3 is fully functional in the endolysosome and ready to accurately recognize non-self dsRNA (Toscano et al., 2013). An interesting exception has also been discovered whereby poliovirus-derived single-stranded RNA segments that have loop structures resembling dsRNA can also activate TLR3 when they come from damaged or inflamed cells (Tatematsu et al., 2013). As we will discuss below, TLR3 plays a key role in controlling viral infections that involve dsRNA structures and it is also involved in the manifestation of autoimmune disorders such as type 1 diabetes.

TLR7/8 resides in endolysosomal compartments and recognizes ssRNA as the ligand of activation. Signaling occurs through the adaptor molecule MyD88 and IRF7 (Diebold et al., 2004; Heil et al., 2004; Kawai et al., 2004). For example, TLR7 and TLR8 have been shown to aid in host defense against the paramyxovirus, Sendai virus (Melchjorsen et al., 2005). HIV is also an antagonist for TLR7/8 mediated antiviral responses robust sensing by these TLRs leads to recruitment of effector cells to the site of viral infection (Schlaepfer and Speck, 2008). Mechanistically, TLR7/8 sensing of ssRNA occurs in a sequence-independent manner. The uridine and ribose molecules of RNA are known antagonists of TLR7 (Diebold et al., 2006). TLR7/8 activation induces a robust IFN response alongside the production of other cytokines such as: interleukin (IL)-1β, IL-6 and IL-12. IL-1β in particular is produced when non-self ssRNA and TLR7/8 activation results in activation caspase-1 (Nicholas et al., 2011). More recently, researchers also determined that the RNA editing phenomenon of adenosine-to-inosine conversion enhances TLR7/8 activation. In this experiment, TLR7 sensing of ssRNA was enhanced in inosine-modified viral RNA (Sarvestani et al., 2014). The cooperative role between TLR7 and TLR8 is also an important part of the mechanism that results in an IFN response to ssRNA. Influenza virus infection was effectively controlled in a rat study where the dual administration of TLR7/8 was administered and effectively suppressed viral load (Hammerbeck et al., 2007). Additionally, during Japanese encephalitis virus infection, TLR8 can compensate for a lack of TLR7 by activating an effective response alone (Awais et al., 2017). TLR7/8 is a necessary component of the antiviral response that responds to foreign ssRNA in endolysosomal compartments and activates an IFN response.

2.1.3 RIG-I-Like Receptors

RIG-I-like receptors (RLRs) recognize viral RNA and initiate innate immune response signaling (Loo and Gale, 2011). RLRs are characterized by their central DExD/H box RNA helicase domain that senses cytosolic dsRNA (Gack, 2014). There are three known RLRs with unique functions: retinoic acid-inducible gene I (RIG-I) (Yoneyama et al., 2004), melanoma differentiation-associated gene 5 (MDA5) (Kang et al., 2004), and laboratory of genetics and physiology 2 (LGP2) (Cui et al., 2001). RLRs are expressed in most human tissues to allow for widespread type I IFN induction upon viral infection. RLRs are expressed at low levels in resting cells, and then the expression is increased in response to viral infection. RIG-I and MDA5 have similar functions such that they initiate antiviral signals to induce IFN gene activation (Kang et al., 2004; Yoneyama et al., 2004). LGP2 functions as a regulator of RIG-I and MDA5 (Yoneyama et al., 2005). Therefore, the antiviral response is a balanced system where the outcome of a viral infection is determined by the level of viral replication as compared to the level of antiviral response activation.

RIG-I and MDA5 are structurally similar with two caspase-recruitment domains (CARDs) at the N-terminus region, a central DExD/H domain, and C-terminal domain (CTD) (Kang et al., 2004, Yoneyama et al., 2004). Once the RIG-I and MDA5 central DExD/H domain and CTD bind viral RNA, the CARD domains interact with the mitochondrial antiviral signaling (MAVS) adaptor protein. MAVS is composed of an N-terminal CARD-like domain and C-terminal transmembrane domain, both of which are necessary for protein function and signaling (Seth et al., 2005). The activation signal is then transmitted through Fas-associated protein with the death domain (FADD)/receptor-interacting protein 1 (RIP1) that leads to the translocation of the NF-κB transcription factor into the nucleus (Honda et al., 2006). Signaling through MAVS can also activate IKKε and TANK-binding kinase 1 (TBK1) that phosphorylate IRF3 and allows for translocation to the nucleus (Sharma et al., 2003). NF-κB and IRF3 are transcription factors that lead to the production of proinflammatory cytokines, namely IFN-β for type I IFN response to viral infection (Yoneyama et al., 2005).

Mitochondrial antiviral-signaling protein (MAVS, also known as IPS-1, VISA, and Cardif) is an adaptor molecule that also induces IFN from RIG-I and MDA5 signaling (Kawai et al., 2005). The MAVS molecule has a CARD domain that interacts with the CARD domains of RIG-I and MDA5. MAVS signaling requires TBK1 and IKKε protein kinases and activates transcription factors NF-κB and IRF3, leading to IFN induction (Kawai et al., 2005). Autoamplification of IFN signaling ensues because RIG-I and MDA5 are IFN-inducible (Honda et al., 2006). The abundance of IFNs induces up-regulation of IFN-stimulated genes (ISGs) (Pine et al., 1990). This can occur through the Jak/STAT (signal transducer and activator of transcription) upon binding by IFN-β (Darnell et al., 1994). The antiviral signals are then spread to surrounding infected and uninfected cells. Cells enter into antiviral states that control the infection by resisting viral replication. ISGs are responsible for amplifying the antiviral response and encoding proteins that have direct antiviral activity (Yoneyama et al., 1996). The production of innate immune cytokines and chemokines recruits professional immune cells to the site of infection and initiate the adaptive immune response (Kadowaki et al., 2000). Activation of RIG-I and MDA5 by dsRNA leads to activation of a signaling cascade and subsequent IFN induction during antiviral innate immune responses (Fig. 1).

Since an effective IFN response requires equilibrium within the system, regulators such as LGP2 are key in modulating activation signals. LGP2 has a different structure as compared to RIG-I and MDA5 because it lacks the two CARD domains (Murali et al., 2008). The ATPase domain of LGP2 functions upstream of RIG-I and MDA5 to facilitate recognition of cytosolic viral RNA (Satoh et al., 2010). Studies have shown that LGP2 binds dsRNA and negatively regulates RIG-I and MDA5 activation. These studies observed negative feedback regulation during Sendai virus and Newcastle disease virus infection (Rothenfusser et al., 2005; Yoneyama et al., 2005). However, a recent study has suggested that LGP2 might also have a positive enhancement regulation mechanism for MDA5 and RIG-I. Structural analysis of LGP2 revealed that RNA-dependent binding of dsRNA by LGP2 enhanced MDA5 antiviral signaling (Uchikawa et al., 2016). LGP2 has also been described as a necessary component of an effective IFN response during positive-sense picornavirus infection (Satoh et al., 2010). However, other studies have shown that LGP2 negatively regulates RIG-I function by sequestering viral RNA or competing with IKKε for interaction with MAVS (Komuro and Horvath, 2006; Rothenfusser et al., 2005; Saito et al., 2007). To clarify these seemingly disparate roles for LGP2, it was shown that in mice lacking LGP2, there was increased resistance during negative-sense vesicular stomatitis virus (VSV) infection while IFN signaling was defective during Cardiovirus A, a member of the *Picornaviridae* family, infection, causing the mice to be more susceptible to infection (Venkataraman et al., 2007). Taken together, LGP2 has an important

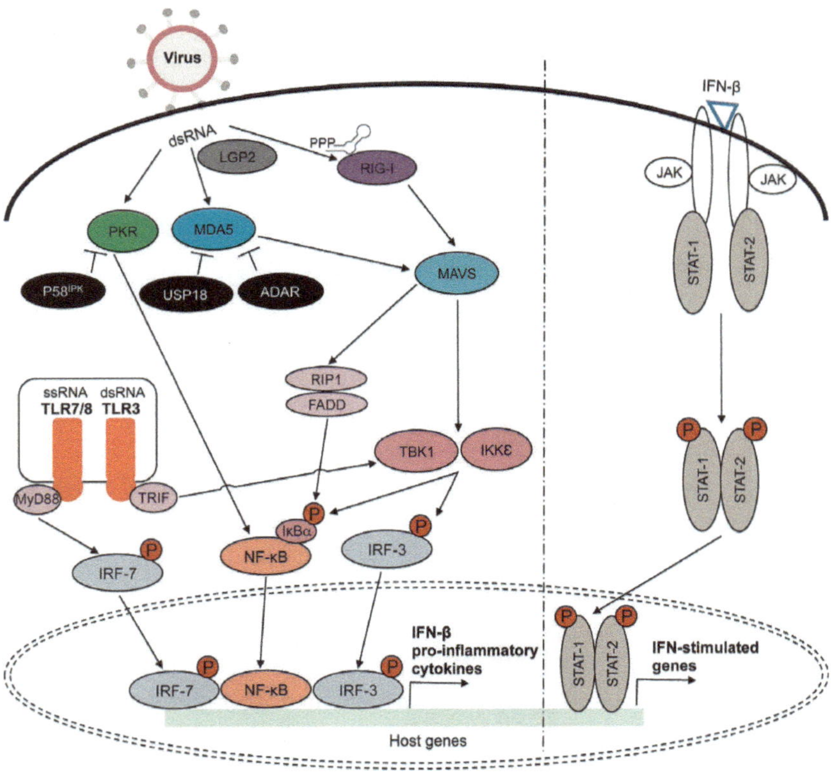

Fig. 1 RIG-I, MDA5, LGP2, PKR, and TLRs3/7/8 are activated during viral infection in the presence of non-self dsRNA. RIG-I recognizes short dsRNA and non-self ssRNA with 5′-triphosphate or 5′-diphosphate, while MDA5 recognizes long dsRNA. LGP2 acts upstream to regulate RIG-I and MDA5 activation in the presence of dsRNA. RIG-I and MDA5 interact with MAVS adaptor protein located on the mitochondrial membrane. MAVS then activates FADD/RIP1 and TBK1/IKKε protein kinases that induce nuclear translocation of NF-κB and IRF3 following phosphorylation of IκBα or IRF3. PKR is also activated in the presence of dsRNA and it transmits activation signals through NF-κB. Endolysosomal TLRs3 and 7/8 lead to TBK1 and IRF7 activation via adaptors TRIF and MyD88, respectively. NF-κB, IRF3, and IRF7 are transcription factors that induce pro-inflammatory cytokine production, including IFN-β. Autoamplification of IFN signaling occurs through activation of the Jak/STAT pathway upon IFN-β binding. While IFN-stimulated genes spread antiviral signals to surrounding cells, viral proteins inhibit multiple steps of these pathways (see Table 1). Inhibitors of these signaling pathway also work to control activation signals and dysfunctions in these molecules can contribute to various autoimmune disorders. For example, P58^{IPK} inhibits PKR while USP18 and ADAR suppress MDA5 activity (see Table 2).

regulatory role in IFN signaling, but the response is variable among different RNA viruses.

RIG-I and MDA5 have variable recognition abilities for foreign cytosolic RNA (Kato et al., 2006). The varying abilities for RIG-I and MDA5 to discriminate between self and non-self allows for specification in the antiviral innate immune response. RIG-I and MDA5 both have the ability to recognize dsRNA. However, RIG-I is known to recognize *Orthomyxoviridae*, *Flaviviridae*, and *Paramyxoviridae* family viruses while MDA5 specifically recognizes picornavirus infections (Kato et al., 2006; Loo et al., 2008). RIG-I and MDA5 activation together can also be essential to induce strong IFN responses during viral infections, such as West Nile virus (Errett et al., 2013). The length of the dsRNA plays a key role in activation of RIG-I or MDA5. RIG-I recognizes short dsRNA (up to 1 kb) while MDA5 recognizes long dsRNA molecules (more than 2 kb) (Kato et al., 2008). For example, the replicative form of the picornavirus genome is a 7.5 kb dsRNA intermediate that has been shown to robustly activate MDA5 in infected cells (Feng et al., 2012). RIG-I has additional abilities to discriminate between self and non-self by recognition of single-stranded RNA with a 5′-triphosphate (5′-PPP) or with a 5′-diphosphate (5′-PP) (Goubau et al., 2014; Hornung et al., 2006; Pichlmair et al., 2006). This was determined because influenza virus infection activates RIG-I in the absence of a dsRNA intermediate during the replication lifecycle (Pichlmair et al., 2006). Host ssRNA exists in the cytosol with a 5′-guanosine cap because it has undergone post-transcriptional modifications. Therefore, the 5′-PPP is an important feature in discrimination because self and non-self because it is unique to viral cytosolic ssRNA. The 5′-PPP is a component of the RNA molecule after viral polymerase replication has taken place (Hornung et al., 2006). It was also described that RIG-I can recognize ssRNA with 5′-PP, a special feature of mammalian reoviruses (Goubau et al., 2014). Recently, it was shown that a conserved residue (H830) of RIG-I is essential to prevent sensing of self RNA that bears a N_1-2′-*O*-methyl group. Additionally, yellow fever virus encodes a methyl transferase to allow escape of viral RNA recognition (Schuberth-Wagner et al., 2015). Together, RIG-I and MDA5 recognize different features of foreign RNA molecules that lead to unique activation of these RLRs during different viral infections. The length of the viral dsRNA along with 5′-PPP, 5′-PP, or cap methylation features on ssRNA dictates the activation of either RIG-I or MDA5 in antiviral innate immune responses.

RNA binding to RIG-I and MDA5 requires specific molecular mechanisms to facilitate controlled and effective activation of the RLRs. Positive and negative regulatory mechanisms of RIG-I and MDA5 are necessary to tightly control IFN signaling. It has been shown that RIG-I retains an inactive configuration until RNA binding occurs and the ATPase activity increases to induce signal transduction of the antiviral response (Gee et al., 2008). Exposure of the RIG-I and MDA5 CARD domains leads to structural changes, such as ubiquitination and phosphorylation, that vary levels of activation. The RNF125 ubiquitin E3 ligase suppresses RIG-I and MDA5 activity by causing ubiquitination of their CARD domains and subsequent proteasomal degradation (Arimoto et al., 2007). Further, RNF125 enhancement by IFN creates a negative feedback loop that controls RIG-I and MDA5 activation during infection (Arimoto et al., 2007). In contrast, strong activation of RIG-I is facilitated by K63-polyubiquitin chains binding to RIG-I at the CARD domains (Zeng et al., 2010). RIG-I and MDA5 are also regulated by phosphorylation and dephosphorylation of its CARD domains through PP1α and PP1γ phosphatases. The function of PP1α and PP1γ to dephosphorylate the CARD domains of RIG-I and MDA5 is needed to facilitate strong induction of an IFN response (Wies et al., 2013). This positive and negative regulation of RIG-I and MDA5 is important in maintaining appropriate control of IFN induction and serving as a first line of defense against viral infections. There are many regulatory mechanisms for RIG-I- and MDA5-mediated IFN induction that help in effective viral control. However, dysfunctions in recognition and signaling can lead to increased viral susceptibility and even autoimmunity.

2.2 RNA Sensing During Viral Infections

Mutations within the genes coding for RNA sensors have been shown to increase susceptibility to a variety of viral infections. Additionally, viruses have mechanisms to antagonize specific aspects of the antiviral innate immune response. Decreasing IFN induction during viral infection is advantageous for the virus because it increases its infectivity potential. Several RNA viruses that exhibit these characteristics include: hepatitis C virus, Dengue virus, West Nile virus, Zika virus, respiratory syncytial virus, Nipah virus, Ebola virus, and Rotavirus (Table 1).

2.2.1 Hepatitis C Virus

Hepatitis C virus (HCV) is a positive-sense ssRNA virus within the *Flaviviridae* family that activates RIG-I during infection (Sumpter et al., 2005).

Table 1 Summary of Host Proteins That Sense RNA Virus Infection, Viral Proteins That Inhibit Host Responses, and Polymorphisms That Affect Pathogenesis

Virus	Family	RLR Detection	Innate Immune Inhibition by Viral Components	Genetic Polymorphisms That Increase Susceptibility
Hepatitis C virus (HCV)	Flaviviridae (+) ssRNA	• RIG-I (Sumpter et al., 2005) • TLR7 (Stone et al., 2014)	• NS3/4A viral protease (Ding et al., 2012; Li et al., 2005) • NS3 (Kasama et al., 2012) • NS4B (Ding et al., 2013)	• Polymorphisms in TLRs3/7/8 result in increased HCV susceptibility (El-Bendary et al., 2018)
Dengue virus (DENV)	Flaviviridae (+) ssRNA	• RIG-I and MDA5 (Nasirudeen et al., 2011) • TLR3 (Rodriguez-Madoz et al., 2010)	• Virus induced double-membrane vesicles (Junjhon et al., 2014; Uchida et al., 2014) • NS4B, NS2A, NS4A (Muñoz-Jordán et al., 2003) • NS2B/NS3 (Angleró-Rodríguez et al., 2014; Rodriguez-Madoz et al., 2010) • NS2A/NS4B (Dalrymple et al., 2015) • NS5 (Ashour et al., 2009)	• Type 2 diabetes increases susceptibility to severe viral infection (Lee et al., 2013)
West Nile virus (WNV)	Flaviviridae (+) ssRNA	• RIG-I and MDA5 (Fredericksen et al., 2008) • TLR7 (Xie et al., 2013) • TLR3 (Szretter et al., 2010)	• NS1 (Zhang et al., 2017b)	• Lack of MAVS in myeloid cells (Pinto et al., 2014) • *RFC1*-DNA polymerase activator (Loeb et al., 2011) • Lack of MAVS (Suthar et al., 2010) • Type 2 diabetes (Kumar et al., 2012)

Continued

Table 1 Summary of Host Proteins That Sense RNA Virus Infection, Viral Proteins That Inhibit Host Responses, and Polymorphisms That Affect Pathogenesis—cont'd

Virus	Family	RLR Detection	Innate Immune Inhibition by Viral Components	Genetic Polymorphisms That Increase Susceptibility
Zika virus (ZIKA)	Flaviviridae (+) ssRNA	• RIG-I, MDA5, and TLR3 (Hamel et al., 2015) • TLRs3/8 (Luo et al., 2018)	• NS5 (Grant et al., 2016; Kumar et al., 2016) • NS1 (Xia et al., 2018)	• MAVS (Piret et al., 2018) • IRF3, IRF5, IRF7 (Lazear et al., 2016)
Respiratory syncytial virus (RSV)	Paramyxoviridae (−)ssRNA	• RIG-I (Loo et al., 2008) • TLRs3/7 (Qi et al., 2015)	• NS2 (Ramaswamy et al., 2004, 2006) • NS1 (Xu et al., 2014b)	• *FokI*-vitamin D receptor (Hansdottir et al., 2010; Janssen et al., 2007; Stoppelenburg et al., 2014) • Plasmacytoid dendritic cells (Marr et al., 2014)
Nipah virus (NiV)	Paramyxoviridae (−)ssRNA	• RIG-I (Habjan et al., 2008)	• Phosphoprotein (Ciancanelli et al., 2009; Rodriguez et al., 2004; Shaw et al., 2004) • Nucleoprotein (Sugai et al., 2017) • Matrix protein (Bharaj et al., 2016) • Nonstructural C protein (Yamaguchi et al., 2014)	
Ebola virus (EBOV)	Filoviridae (−) ssRNA	• RIG-I (Habjan et al., 2008)	• VP35 (Caballero et al., 2016; Hartman et al., 2008; Luthra et al., 2013; Yen et al., 2014)	
Rotavirus (RV)	Reoviridae dsRNA-RT	• RIG-I and MDA5 (Broquet et al., 2011) • TLRs3/7 (Yang et al., 2016)	• NSP1 (Barro and Patton, 2005; Graff et al., 2009; Qin et al., 2011)	• Variable RNA transcripts (Uzri and Greenberg, 2013)

HCV uses viral proteases and viral proteins to target specific components of IFN signaling and decrease antiviral responses. The HCV-NS3/4A viral protease is involved in many mechanisms of antiviral suppression. HCV-NS3/4A is known to cleave MAVS and decrease IFN signaling. A point mutation in MAVS at Cys-508 renders it resistant to NS3/4A cleavage (Li et al., 2005). HCV-NS3/4A are also known to induce expression of the translocase of outer mitochondrial membrane 70 (TOM70) and induce IFN signaling in hepatocytes (Kasama et al., 2012). However, the HCV nonstructural (NS) protein 3 (HCV-NS3) then suppresses TOM70 induction of IRF3 mediated immunity by cleaving MAVS upstream of TOM70. Also, HCV-NS3/4A viral protease has also been shown to inhibit IL-28 induction. IL-28 contains NF-κB and IRF3 binding sites, meaning it can be induced by these transcription factors (Ding et al., 2012). Thus, the actions of HCV-NS3/4A aid in viral persistence within the host. The HCV-NS4B viral protein is also involved in antiviral suppression. HCV-NS4B interferes with TBK1 interactions, inhibiting these proteins from properly relaying IFN signaling (Ding et al., 2013). Finally, the HCV core protein was shown to inhibit TLR7-mediated IFN induction as well as IRF7 and STAT1 expression in plasmacytoid dendritic cells (Stone et al., 2014), and polymorphisms in genes encoding TLR3/7/8 result in increased susceptibility to HCV infection (El-Bendary et al., 2018). Together, HCV-NS3/4A, HCV-NS4B, and HCV-Core are viral components that suppress IFN induction and evade innate immune responses.

2.2.2 Dengue Virus and West Nile Virus

Dengue virus (DENV) and West Nile virus (WNV) are both mosquito-borne viruses within the *Flaviviridae* family (Ahlers and Goodman, 2018; Fredericksen et al., 2008; Loo et al., 2008; Nasirudeen et al., 2011). DENV and WNV both activate the IFN-mediated innate immune response through RIG-I and MDA5. Knockdown of RIG-I and MDA5 resulted in increased susceptibility to DENV infection (Nasirudeen et al., 2011). This highlights the importance of RIG-I and MDA5 in controlling DENV infection. Innate immune responses to WNV infection occur in two major phases, with RIG-I and MDA5 being important sensors at both phases. The initial response to WNV leads to IRF3 activation and ISG induction. Later stages in WNV infection are dominated by IFN-dependent antiviral gene expression. In response to WNV infection, RIG-I and MDA5 act through MAVS as they work together to stimulate a strong IFN response and subsequent signal amplification (Fredericksen et al., 2008). DENV and WNV both have antiviral evasion mechanisms that antagonize IFN induction after RIG-I and MDA5 sensing.

DENV has many mechanisms of innate immune suppression at different stages of infection. During early antiviral responses, before IFN induction, DENV induces the production of autophagic proteins, and autophagy activation suppresses the antiviral response (Huang et al., 2016). As the infection progresses, DENV also suppresses IFN induction by keeping viral dsRNA within intracellular membranes to hide the dsRNA from recognition by RLR receptors (Uchida et al., 2014). Virus induced double-membrane vesicles form early in infection and contain the dsRNA along with viral nonstructural proteins and replication machinery (Junjhon et al., 2014; Mackenzie et al., 1996). Therefore, viral replication can be occur within these vesicles in the first 48 h of infection with little dsRNA exposure in the cytosol (Uchida et al., 2014). However, other studies have shown that DENV NS4A is part of the membrane-bound replication complex that is also associated with viral RNA (Miller et al., 2007), also similar to the flavivirus Kunjin virus (Roosendaal et al., 2006). Additionally, DENV viral RNA was shown to associate with the rough ER (Grief et al., 1997). Therefore, it remains to be fully clarified if DENV viral RNA is fully incorporated within virus-induced membrane structures to avoid RLR recognition or if they are on the surface of cytosolic membrane structures. To evade host immune responses, DENV also uses nonstructural viral proteins direct suppression IFN signaling. Three nonstructural proteins of DENV are known to down-regulate IFN-β expression. The presence of DENV-NS4B caused the most significant decrease in IFN activity while DENV-NS2A and DENV-NS4A also antagonized IFN, but to a lesser extent than DENV-NS4B (Muñoz-Jordán et al., 2003). Additional research following this study described the mechanism by which the nonstructural proteins decreased innate immune responses during DENV infection. A viral nonstructural protein complex (DENV-NS2B/NS3) interacts with IKKε to inhibit kinase activity and ultimately decrease IFN induction (Angleró-Rodríguez et al., 2014). Also, DENV-NS2A/NS4B inhibits TBK1 phosphorylation in a dose dependent manner (Dalrymple et al., 2015). Experiments also indicated that DENV infection caused a reduction in STAT2 expression suggesting the virus antagonized this innate immune response gene (Ashour et al., 2009). DENV-NS5 is the viral polymerase and it has a role in inhibiting STAT2 function. Mature DENV-NS5 is the product of a polyprotein and this was important for STAT2 binding and subsequent inhibition. Precursor DENV-NS5 in the form of a polyprotein had a single role in inducing degradation of STAT2 while the proteolytically processed, mature DENV-NS5 protein had a single role in binding STAT2 (Ashour et al., 2009). DENV-NS5-mediated degradation of STAT2 is facilitated by the host E3 ubiquitin ligase, UBR4 (Morrison et al., 2013). Finally, the DENV protease complex (NS2B/NS3) has been shown to inhibit TLR3-mediated IFN induction (Rodriguez-Madoz et al., 2010). Together these studies highlight that many DENV nonstructural proteins function in suppressing IFN induction during innate immune antiviral responses.

WNV also encodes mechanisms of innate immune suppression. IFN responses in myeloid cells are necessary for control of WNV infection. Mice lacking MAVS and the type I IFN receptor had extremely low cytokine production coupled with high WNV replication. These findings highlight the importance of IFN signaling through MAVS to control WNV infection in myeloid cells (Pinto et al., 2014). Due to the neuroinvasive nature of WNV, IFN responses to control WNV infection in the central nervous system have been studied (Ramos et al., 2012). *In vivo* experiments showed a significant upregulation of IL-1β during acute WNV infection. IL-1β showed a synergistic role with IFN signaling to control WNV infection in neurons (Pinto et al., 2014). WNV susceptibility has also been linked to gene expression changes in peripheral blood cells. Individuals with resistance to WNV infection had higher IL-4 levels in serum as compared to individuals who developed severe symptoms. These IL-4 levels altered expression of many genes and correlated with disease outcome (Qian et al., 2014). Neuroinvasive human cases from WNV infection were also linked to mutations in *RFC1* gene through a screen of common genetic polymorphisms. *RFC1* aids in proper activation of DNA polymerase (Loeb et al., 2011). Additionally, experiments in mice revealed that during WNV infection, MAVS is essential for RLR signaling. Mice that lacked MAVS exhibited uncontrolled infection and a lack of regulatory T-cell expansions that is normally a characteristic of acute WNV infection. These findings highlight the importance of MAVS mediated RLR signaling to control WNV infection (Suthar et al., 2010). WNV NS1 inhibited K63-linked ubiquitination of RIG-I and blocked IRF3 signaling upon WNV recognition. It was specifically determined that WNV-NS1 interaction with MDA5 and RIG-I induced proteasome degradation of these intracellular receptors (Zhang et al., 2017b). Lastly, an attenuated WNV strain containing a mutant NS4B protein (P38G), exhibited increased T-cell priming via a TLR7-mediated mechanism (Xie et al., 2013), and TLR3 is important in blocking WNV replication and spread into the brain (Szretter et al., 2010; Wang et al., 2004). Together, there are a variety of antiviral mechanisms in different cell types employed by WNV to antagonize the IFN response.

Interestingly, type 2 diabetes has been shown to increase susceptibility of severe DENV and WNV infection. One study showed that DENV infected patients with type 2 diabetes were at a higher risk of developing dengue hemorrhagic fever as compared to DENV infected patients without type 2 diabetes. This was quantified by an increase in IL-4 and IL-10 cytokine production in patients with DENV and type 2 diabetes because these cytokines are an important immunopathogenesis marker for dengue hemorrhagic fever (Lee et al., 2013). Another study described an increase in WNV titer in serum, peripheral tissue, and the brain of the diabetic mouse model. Type 2 diabetes caused a non-specific WNV response that increased susceptibility of neuroinvasive WNV infection (Kumar et al., 2012). These studies highlight the important balance in innate immune signaling that must be present to efficiently clear a viral infection.

2.2.3 Zika Virus

Similar to DENV and WNV, Zika virus (ZIKV) is part of the *Flaviviridae* family and is transmitted by mosquitoes. A robust IFN response is also needed to control ZIKV infection. ZIKV infection induced RLR signaling in human skin fibroblasts to control infection. Human skin fibroblasts infected with ZIKV showed upregulation of TLR3, RIG-I, MDA5, and ISGs. There was also sequential activation observed where TLR3 was activated 6 h post-infection and RIG-I and MDA5 were activated later in infection (Hamel et al., 2015). Other mosquito-borne flaviviruses have been known to infect dendritic cells, and this was consistent with ZIKV. Researchers observed that RIG-I response to ZIKV in human dendritic cells was activated during infection. However, there was observable inhibition of IFN protein translation. Mechanistically, ZIKV was also able to control IFN signaling by blocking STAT1 and STAT2 phosphorylation (Bowen et al., 2017). These studies show that ZIKV infection leads to the activation of RIG-I and MDA5 but that ZIKV also has mechanisms to antagonize downstream IFN signaling.

There are specific IFN adaptor molecules that increase ZIKV susceptibility and some nonstructural proteins of the virus antagonize IFN signaling, as shown for DENV and WNV. The MAVS adaptor protein within the IFN signaling cascade is important in the early IFN response because mice lacking MAVS had higher viremia than control mice (Piret et al., 2018). During the development of a ZIKV mouse model, experiments revealed that a triple knockout of IRF3, IRF5, and IRF7 increased susceptibility to ZIKV infection through decreased IFN signaling. Also, mice lacking the IFN receptor had higher viral load in the brain and spinal cord that could not be controlled, which correlates with the neuroinvasive nature of the virus (Lazear et al., 2016). On the other hand, increased IFN induction and placental inflammation can lead to brain damage in newborns, and inhibition of TLR3 and TLR8 led to decreased pro-inflammatory cytokine responses in trophoblasts (Luo et al., 2018). Many ZIKV nonstructural proteins have been shown to facilitate the inhibition of IFN signaling observed during infection. For example, ZIKV-NS5 binds STAT2 and its expression correlates with subsequent proteasome degradation of STAT2 (Kumar et al., 2016). The IFN inhibitory role of ZIKV-NS5 in other flaviviruses is consistent for that of ZIKV. DENV-NS5 acts on STAT2 but the mechanism is slightly different than that of ZIKV because it involves the E3 ubiquitin ligase UBR4 to induce degradation (Grant et al., 2016). ZIKV-NS1 has also been shown to inhibit IFN signaling induction through binding of TBK1. This inhibitory role of the ZIKV-NS1 was observed only in ZIKV strains that caused epidemics after 2012. This highlights the importance of a fixed mutation in viral evolution because the ZIKV-NS1 fixed mutation helped the virus increase infectivity by decreasing IFN-β induction (Xia et al., 2018). Therefore, adaptor molecules are important for regulating an efficient antiviral response and without them IFN signaling may be uncontrolled. Additionally, ZIKV-NS5 and ZIKV-NS1 have IFN inhibitory roles during viral infection.

2.2.4 Respiratory Syncytial Virus

Respiratory syncytial virus (RSV) disproportionately causes more severe symptoms in infants as compared to older populations. RSV has a negative-sense non-segmented RNA genome and it is part of the *Pneumovirus* genus of the *Paramyxoviridae* family. RSV dsRNA is sensed by RLRs and TLRs (reviewed in Klein Klouwenberg et al., 2009; Mukherjee and Lukacs, 2013), and host immune responses play a major role in the differences in viral susceptibility (Loo et al., 2008; van Drunen Littel-van den Hurk and Watkiss, 2012). RSV infection is known to specifically cause severe lower respiratory tract infection in newborn infants (Marr et al., 2014). Differences in the innate immune response to RSV determine the disease severity in a child. One study found polymorphisms in immune-related genes of pre-term babies that correlated with increased disease susceptibility as compared to babies carried to term (Siezen et al., 2009). Another study found a consistent association in 22 single-nucleotide polymorphisms (SNPs) within 21 innate immune genes and the development of severe RSV bronchiolitis. These SNPs were identified in a cohort of children hospitalized for severe RSV bronchiolitis as compared to the control population. One SNP identified in this study with a highly significant association with bronchiolitis was the vitamin D receptor (Janssen et al., 2007). Vitamin D is known to mediate NF-κB and STAT1 expression (Stoppelenburg et al., 2014). Vitamin D deficiencies have been correlated with increased risk and severity of RSV (Hansdottir et al., 2010). Specifically, the *FokI* vitamin D receptor polymorphism abrogated vitamin D's control of the STAT1-mediated antiviral response (Stoppelenburg et al., 2014). Together, these findings illustrate the importance of vitamin D in controlling the induction of STAT1 during antiviral responses and the importance of this regulation in RSV infection. Many SNPs have been linked to RSV susceptibility and loss-of-function SNPs in the vitamin D receptor have a strong association with the development of severe RSV bronchiolitis.

Increased susceptibility to RSV has also been linked to dysfunctions in IFN signaling from loss-of-function in plasmacytoid dendritic cells (pDCs) and direct inhibition of IFN signaling by RSV nonstructural proteins. Researchers have determined that pDCs are important in activating IFN responses in lungs during RSV infection. This epidemiological study found that newborn infants had increased susceptibility to RSV because they did not have fully functioning pDCs and therefore there was low IFN induction by RIG-I (Marr et al., 2014). Respiratory macrophages and pDCs are critical in combating RSV infection, not only through RLR signaling, but also through TLR3- and TLR7-mediated mechanisms (Qi et al., 2015). Using mouse pneumonia virus, the rodent-specific form of RSV, it was shown that TLR7 was critical for the host defense response and IFN induction during infection (Davidson et al., 2011). These results were corroborated through analysis of RSV-infected human pDCs (Schijf et al., 2013). RSV is also known to decrease IFN signaling in airway epithelial cells through degradation of STAT2 (Ramaswamy et al., 2004) and through an RSV-NS1-dependent but TLR3-independent mechanism (Xu et al., 2014b). Additionally, RSV-NS2 was identified as inducing these inhibitory effects on the IFN response (Ramaswamy et al., 2006). This innate immune response inhibition by RSV aids in viral replication specifically within airway epithelial cells. The control of RSV is complex, and it requires fully functioning proteins within the IFN pathway. As seen for other viruses, the RSV nonstructural proteins have individual roles in antagonizing the antiviral response.

2.2.5 Nipah Virus

Nipah virus (NiV) is another virus in the *Paramyxoviridae* family, but unlike RSV, it is part of the *Henipavirus* genus (Ciancanelli et al., 2009). NiV is a deadly zoonotic virus, and there is a 40–90% case mortality rate in human infected with NiV in Southeast Asia (Bharaj et al., 2016). NiV dsRNA is sensed by RIG-I, but not MDA5, to initiate IFN induction (Habjan et al., 2008). Research has shown that many of the NiV proteins contribute the deadly symptoms of NiV by antagonizing IFN stimulated antiviral responses (Bharaj et al., 2016). First, the NiV phosphoprotein (P) gene encodes the NiV-P, NiV-V, and NiV-W proteins that all antagonize IFN signaling (Ciancanelli et al., 2009). The NiV-V protein, mainly found in the cytoplasm, binds STAT1 and STAT2 proteins to prevent dimerization and nuclear transport (Rodriguez et al., 2004). The NiV-W protein has a very similar role to the NiV-V protein except it sequesters STAT1 in the nucleus to inhibit subsequent ISG activation. Therefore, the NiV-V and NiV-W proteins have dual roles but are located in either the nucleus or cytoplasm to block STAT1 and induce antiviral functions. The NiV-P protein can also bind STAT1 to sequester the inactive protein in the nucleus, but to a lesser extent than NiV-V and NiV-W (Ciancanelli et al., 2009; Shaw et al., 2004). In comparison, Cedar virus is a henipavirus that is not pathogenic to humans and the Cedar virus P gene products do not antagonize STAT proteins as seen in NiV infection. This suggests that the antagonizing properties of NiV-P gene protein products greatly contribute to the highly lethal nature of the virus (Lieu et al., 2015). Second, the NiV nucleoprotein also decreased IFN responses in a dose dependent manner. The specific mechanism of inhibition is through targeting of STAT1 and STAT2 complex formation in the cytoplasm. NiV nucleoprotein decreases STAT1/STAT2 nuclear transport and subsequently down-regulates ISG expression (Sugai et al., 2017). Third, the NiV matrix protein has been shown to inhibit IKKε kinase activity that is involved in IFN signaling. The NiV matrix protein acts by degrading TRIM6, an E3-ubiquitin ligase that generates unanchored polyubiquitin chains for IKKε activation. Therefore, this degradation by the NiV matrix protein results in decreased activity of IKKε activity and decreases IFN signaling (Bharaj et al., 2016). Finally, the nonstructural C protein of paramyxoviruses inhibits IFN signaling (Gotoh et al., 2003; Mathieu et al., 2012), via inhibition of TLR7 in pDCs (Yamaguchi et al., 2014). Together, the gene products from NiV phosphoprotein, nucleoprotein, and matrix protein have been shown to antagonize IFN induction at different steps in its signaling pathway. This likely contributes to the high virulence of NiV in humans.

2.2.6 Ebola Virus

Ebola virus is negative-sense RNA virus with a non-segmented genome similar to that of RSV and NiV. Ebola virus is a hemorrhagic fever virus in the *Filoviridae* family. *Zaire ebolavirus* (EBOV) has a fatality rate of up to 90% in humans and the viral dsRNA is sensed by RIG-I to activate innate immune responses (Habjan et al., 2008; Kash et al., 2006). Early studies on EBOV were performed in human liver cells. In this study, researchers determined that IFN signaling was suppressed during infection. Through genomic analysis of gene expression, many antiviral genes were suppressed during infection. Researchers hypothesized a global IFN suppression model and also identified specific antiviral genes, such as IRF3, as being suppressed during EBOV infection (Kash et al., 2006). Further research revealed that the EBOV viral protein 35 (EBOV-VP35) had a specific inhibitor mechanism for IRF3. This was a highly specific suppression model because a point mutation in EBOV-VP35 altered its inhibitor function. EBOV-VP35 also had a role in enhancing viral replication (Hartman et al., 2008). Furthermore, EBOV-VP35 binds dsRNA, inhibiting RIG-I-mediated detection of viral dsRNA (Cardenas et al., 2006), and EBOV-VP35 is an important cofactor in the viral polymerase complex (Prins et al., 2010). Together, EBOV-VP35 plays an important role in EBOV pathogenesis both for spread of the virus within the host as well as IFN specific suppression. Further research on EBOV-VP35 enhanced the understanding of its inhibition mechanism. PACT (PKR activator) is a dsRNA binding protein that is known to induce activation of RIG-I. Experiments showed that EBOV-VP35 was able to inhibit PACT from activating RIG-I through direct binding. This was observed in a dose-dependent manner where increased presence of EBOV-VP35 increased PACT inhibition. EBOV-VP35 can also bind viral dsRNA and decrease activation of RIG-I during EBOV infection (Luthra et al., 2013). Finally, EBOV-VP35 can inhibit TLR3-mediated signaling, but this occurs in a dsRNA-independent manner suggesting that the TLRs can circumvent VP35-mediated IFN inhibition (Leung et al., 2011; Yen et al., 2014). Taken together, EBOV-VP35 is a key antagonist of IFN induction during EBOV infection.

Further *in vivo* experiments have shown a more complex picture of EBOV infection than just the antagonizing properties of EBOV-VP35 (Caballero et al., 2016). Transcriptomics analysis of peripheral blood mononuclear cells from cynomolgus macaques infected with EBOV showed strong innate immune activation during viral infection. Many ISGs were upregulated in response to viral infection via an intramuscular injection. These findings were contrasting to previous studies that had shown the EBOV-VP35 protein targeted the IFN response for suppression. The authors of this study proposed a model of infection where EBOV-VP35 inhibits IRF3 activation in the single infected cell. Then, in neighboring cells, EBOV-VP35 might be inducing IRF3 nuclear translocation to facilitate induction of IFN signaling through an unknown mechanism. This hypothesis aims to explain why in isolated EBOV-VP35 experiments, the protein appears to decrease IFN signaling but then during *in vivo* EBOV infection, ISGs are upregulated (Caballero et al., 2016). This elevated pro-inflammatory response to EBOV has been described in other studies. Rhesus macaques were infected with EBOV and monitored daily throughout the infection (Ebihara et al., 2011). Blood samples described the cytokine and chemokine profile throughout the course of infection. IL-1β and IL-6 pro-inflammatory cytokines were upregulated during infection. Interestingly, anti-inflammatory cytokines such as IL-10 and IL-13 were also increased in fatal EBOV cases. This shows that the cytokine balance during infection might be an important indicator of disease outcome. This study also showed anti-coagulation induced by the virus and ultimately contributed to overall pathogenesis (Ebihara et al., 2011). These findings were also evident in samples for fatal human cases of Ebola virus disease, symptoms of which include hemorrhagic fever. Many cytokines and chemokines were upregulated in fatal human cases. When looking at the disease progression *in vivo*, uncontrolled pro-inflammatory responses are a characteristic of fatal EBOV cases. Biomarkers of infection were identified as IL-1β and IL-6 along with IL-8 at late stages of infection (McElroy et al., 2014). These EBOV studies highlight the important balance of the innate immune response in determining disease outcome. EBOV viral proteins have been identified to inhibit IFN signaling during infection.

2.2.7 Rotavirus

Rotavirus (RV) is a segmented dsRNA virus within the *Reoviridae* family. It causes severe diarrhea in young children and is known to infect epithelial cells of the small intestine (Barro and Patton, 2005). RV dsRNA has been shown to activate RIG-I like receptors within intestinal epithelial cells. This virus is unique in that it is sensed by both RIG-I and MDA5 either together or separately. When MAVS, RIG-I, or MDA5 were silenced, RV titer increased and IFN-β production decreased. These findings suggested that RV induced IFN-β production through MAVS signaling after RIG-I and MDA5 activation (Broquet et al., 2011). Researchers have looked further into the specific RV RNA transcripts that activate IFN signaling. Nascent single-stranded RNA transcripts produced during viral replication were a strong inducer of IFN signaling (Uzri and Greenberg, 2013). Also, large RNA transcripts produced 6 h after cells were infected with RV activated IFN signaling. This finding was only observed 6 h after infection but not 1 h after infection. Both the single-stranded RNA transcripts and large RNA produced 6 h after infection had uncapped RNA at the 5′ end. RNA lacking 2′-O-methylated 5′ cap was detected in the large RNA as well. The viral protein 3 enzyme of RV is known to encode guanylyltransferase and methyltransferase. Therefore, inaccuracy in this enzyme would result in a lack of a 5′ cap or 2′-O-methyl group on the 5′ cap structure of viral RNA. Then, RNA lacking these structural components would activate RIG-I-like receptors and subsequent IFN-β production (Uzri and Greenberg, 2013). In conclusion, RNA sensing by RIG-I or MDA5 is important for controlled IFN induction during RV infection. Variations in RNA transcripts and processing are also an important indicator of IFN-β production levels. Regarding TLR signaling and RV, interestingly, TLR3 and TLR7 were required to protect the gut from inflammatory-inducing conditions through proper recognition of the gut virome, which includes RV (Yang et al., 2016).

RV nonstructural proteins are also known to antagonize innate immune responses. Specifically, RV-NS1 (or NSP1) is an antagonist of IRF3 through a common mechanism found in other viruses where it induces proteasome-dependent degradation of IRF3 to decrease IFN signaling (Barro and Patton, 2005). RV-NS1 also has a novel mechanism of decreasing NF-κB activation. This viral protein induces proteasome-dependent degradation of β-transducin repeat containing protein, a cellular protein that functions within the multi-subunit complex responsible for NF-κB activation (Graff et al., 2009). Another study also demonstrated that RV-NS1 decreased IFN-β production through a mechanism independent of IRF3 degradation as previously reported. RV-NS1 directly bound to RIG-I to decrease activation and IFN-β production. This direct interaction was confirmed with immunoprecipitation and RV-NS1 interacted with RIG-I outside the IRF3 binding domain (Qin et al., 2011). Therefore, RV is sensed by RLRs and this sensing is enhanced with inefficiency of viral enzymes that make RNA products of replication more noticeable to the RLR sensing domain. However, the virus also has ways to combat this innate immune activation and RV-NS1 is a main driver that decreased IFN-β production through antagonizing of RIG-I, NF-κB, and IRF3 activation.

2.3 Defects in RNA Sensing and Autoimmunity

Hyperactive or non-functional mutations within the genes coding for RNA sensors or their interacting proteins have been show to lead to autoimmune-like phenotypes, such as type 1 diabetes, Aicardi–Goutières syndrome, Crohn's disease, Singleton–Merten syndrome, and systemic lupus erythematosus. As described below, many of these phenotypes are the result of imbalanced IFN or insulin signaling, autophagy, or responses to pathogens (Table 2).

2.3.1 The Role of MDA5 in Type 1 Diabetes and Aicardi–Goutières Syndrome

Type 1 diabetes (T1D) is characterized by the development of autoantibodies that target beta cells of the human pancreas. These autoantibodies and auto-reactive immune cells then trigger the destruction of insulin-producing pancreatic beta cells and lead to T1D. Researchers have found that there is a genetic component to disease susceptibility along with viral infection and environmental variables. Environmental factors are known to be associated with disease manifestation, and one example is vitamin D deficiency. Also, Caucasians living in Europe have a high disease incidence that highlights the exogenous factors of T1D. Cohort studies linked enterovirus or coxsackie B virus infection with increased susceptibility to the development of T1D (Knip et al., 2005). Enterovirus is known to target beta cells in the human pancreas. Autopsied pancreases of T1D patients revealed a tropism of enterovirus for human pancreatic islet cells (Ylipaasto et al., 2004). A Finnish Diabetes Prediction and Prevention Study indicated a positive association between the development of autoantibodies for human pancreatic beta cells and enterovirus infection. Autoantibodies were more likely to be present in children infected with enterovirus as compared to the control groups (Lönnrot et al., 2000). In Europe there is a high prevalence of T1D and a low prevalence of enterovirus as compared to other regions of the world, such as Cuba. Therefore, another study observed the relationship between enterovirus and T1D in Cuba, where T1D incidence is low and enterovirus prevalence is high. The researchers found a direct correlation between enterovirus infection and the development of preclinical and clinical stages of T1D (Sarmiento et al., 2013). Molecular mimicry is a proposed mechanism for this correlation because antibodies made against enterovirus could also target beta cells (Jang et al., 2015). Acute cytolytic damage from enterovirus was also a proposed mechanism for the

Table 2 Summary of Polymorphisms in RNA Sensors and Their Associated Syndromes

RNA Sensor	RNA Sensor Associated Protein	Genetic Polymorphisms That Causes Dysfunction	Associated Syndrome/ Pathology
MDA5		1. Four loss-of-function variants *IFIH1*-gene that encodes MDA5 (Lincez et al., 2015; Liu et al., 2009; Nejentsev et al., 2009; Shigemoto et al., 2009) 2. Gain-of-function *IFIH1* mutations (Crow et al., 2015) 3. Missense mutation in *IFIH1* (Funabiki et al., 2014) 4. Missense mutation in *IFIH1* leading to overactive MDA5 (Rutsch et al., 2015)	1. Type 1 diabetes 2. Aicardi–Goutières syndrome 3. Lupus-like symptoms 4. Singleton–Merten Syndrome
	USP18	1. Loss-of-function in USP18 induces overactivation of MDA5 (Santin et al., 2012)	1. Type 1 diabetes
	ADAR	1. Loss-of-function in *ADAR* induces overactivation of MDA5 (Crow et al., 2015; Pestal et al., 2015) 2. Gain-of-function mutation in *IFIH1* led to over expressed MDA5 (Crow et al., 2015; Schmelzer et al., 2018)	1-2. Aicardi–Goutières syndrome
PKR			
	$P58^{IPK}$ interaction with PERK	1. Knockout of $P58^{IPK}$ (Ladiges et al., 2005) 2. Loss-of-function mutations in *Dnajc3* (Synofzik et al., 2014) 3. Homozygous stop mutation (Synofzik et al., 2014)	1. Symptoms associated with type 1 and late stage type 2 diabetes 2. Juvenile-onset diabetes and multisystemic neurodegenerative disorders 3. Monogenic recessive diabetes mellitus

Table 2 Summary of Polymorphisms in RNA Sensors and Their Associated Syndromes—cont'd

RNA Sensor	RNA Sensor Associated Protein	Genetic Polymorphisms That Causes Dysfunction	Associated Syndrome/ Pathology
RIG-I		1. Knockout of RIG-I (Wang et al., 2007) 2. Downregulation of G protein αi2 subunit (Hampe et al., 2001b) 3. Decreased expression of RIG-I (Funke et al., 2011) 4. Variants of *DDX58* causing overactive RIG-I (Jang et al., 2015)	1. Colitis-like phenotype associated with inflammatory bowel disease 2. Human inflammatory bowel disease 3. Crohn's disease 4. Singleton–Merten Syndrome
	NOD2	1. NOD2 mutations that imbalance IFN signaling through RIG-I (Morosky et al., 2011)	1. Crohn's disease
TLR3		1. Polymorphisms in *TLR3* (Assmann et al., 2014)	1. Type 1 diabetes
TLR7		1. Polymorphisms in *TLR7* promoter (Skonieczna et al., 2018)	1. Systemic lupus erythematosus

association of enterovirus infection with T1D (Sarmiento et al., 2013). A population-based cohort study in Taiwan of age- and sex-matched children also showed that T1D incidence was higher in the enterovirus-infected cohort as compared to non-enterovirus-infected cohort (Jang et al., 2015). A similar correlation is also evident for coxsackie B virus. Data collected throughout many countries in Europe concluded that diabetic children were more likely to have antibodies against coxsackie B virus 1 than the control group of children without diabetes (Oikarinen et al., 2014). Therefore, infections with enterovirus or coxsackie B virus increase susceptibility to T1D. This highlights the connection between innate immune signaling dysfunction in response to viral infection and the development of autoimmunity.

Dysfunctions in MDA5 have also been linked to the development of T1D. Genetic analysis through a genome-wide association study (GWAS) revealed genetic components of T1D susceptibility. SNPs have been associated with the development of T1D. Four SNPs with a strong association to disease susceptibility were located within the *IFIH1* gene that codes for MDA5. High *IFIH1* gene expression in peripheral blood mononuclear cells also correlated with susceptible genotypes. This study highlights an interesting genetic association and gene expression importance for *IFIH1* in the development of T1D (Liu et al., 2009). Another GWAS further linked T1D and *IFIH1*. It identified four rare variants of *IFIH1* that strongly correlated with a decreased risk of developing T1D (Nejentsev et al., 2009). Two of these *IFIH1* variants were correlated with a loss-of-function to MDA5 (Shigemoto et al., 2009). This resistance to T1D was evident when mice heterozygous for *IFIH1* ($MDA5^{+/-}$) on a NOD/Ltj background, a known mouse model of T1D, had decreased levels of MDA5 protein as compared to wild-type mice. Mice heterozygous for MDA5 drove a regulatory T-cell response that was protective against T1D during coxsackievirus infection. Mice homozygous for MDA5 on the NOD/Ltj background had high MDA5 protein levels that resulted in a strong effector T-cells response and beta cell destruction during T1D during coxsackievirus infection (Lincez et al., 2015). These findings show how MDA5 function correlates with T1D susceptibility. Reduced MDA5 expression has also been linked to increased susceptibility to encephalomyocarditis virus strain D (EMCV-D) and diabetes. EMCV-D infects insulin-producing beta cells in the pancreas similar to enterovirus or coxsackie B virus. Therefore, a lack of viral infection control is likely to lead to the production of beta cell antigens that can cause destruction of the pancreas and increase susceptibility of T1D. Mouse experiments revealed that MDA5 along with TLR3 were essential in initiating a controlled IFN response against EMCV-D. Also, mice with knockouts in MDA5 and/or TLR3 had increased susceptibility to T1D (McCartney et al., 2011). Taken together, loss-of-function in MDA5 due to genetic polymorphisms or viral infection can lead to autoimmune disorders, namely T1D.

Over activation of MDA5 can also lead to autoimmune disorders. Researchers found that mice with a missense mutation in *IFIH1* quickly developed lupus-like autoimmune symptoms without the presence of a viral infection (Funabiki et al., 2014). This means that MDA5 not only has a role in controlling viral infections, but can also control autoimmune disorders. The *IFIH1* missense mutation leads to a gain-of-function in MDA5 and IFN induction spread throughout multiple organs. The cytokine production was directly correlative to MAVS activity, and without MAVS, the mice with the *IFIH1* missense mutation did not develop autoimmune symptoms. This clarified the importance of MDA5 signaling through MAVS in the development of autoimmune disorders (Funabiki et al., 2014). A regulator of IFN signaling, ubiquitin-specific peptidase (USP18), has an important cross talk function with MDA5. USP18 controls IFN-stimulated gene 15 protease during IFN signaling through MAVS. A lack of USP18 resulted in over activation of MDA5 and an uncontrolled pro-inflammatory response in pancreatic beta cells. The authors suggested that USP18 was a major regulator of IFN responses and clarified a mechanism by which MDA5 expression might influence the development of T1D (Santin et al., 2012). Therefore, MDA5 gain-of-function also causes unregulated IFN signaling that can lead uncontrolled beta cell destruction and the development of T1D.

Dysfunctions in the tight control of MDA5-mediated signaling have also been implicated in another autoimmune disease, Aicardi–Goutières syndrome (AGS) (Crow et al., 2015). AGS is an autoinflammatory, neurodevelopmental disorder that affects the brain, skin, and immune system. Characteristics range from chilblains (skin lesions) to cerebral calcifications and atrophy that may be present before birth but ultimately result in slow neurological decline due to the excess production of IFN-α (Rice et al., 2007). Analysis of AGS patient genomes revealed seven gene mutations linked associated with the disease. With regards to RNA sensing pathways, candidate genes of particular interest were *ADAR1*, encoding for an RNA-editing enzyme, and *IFIH1*, encoding for MDA5. Mutations in these genes led to an increase in IFN signaling in cerebrospinal fluid and serum that correlated with disease symptoms (Crow et al., 2015). *ADAR1* is a RNA-editing enzyme known to inhibit MDA5 signaling through MAVS (Pestal et al., 2015). Therefore, a loss-of-function mutation in *ADAR1* contributed to uncontrolled MDA5 antiviral response (Mannion et al., 2014; Pestal et al., 2015). More specifically, mice with a knock-in mutation in *ADAR1* that inhibited its RNA-editing ability exhibited embryonic lethality at embryonic day 13.5 due to MDA5 hyperactivity (Liddicoat et al., 2015). A study has also shown that the loss of *ADAR1* in human cells results in PKR hyperactivation and translational shutdown (Chung et al., 2018). A gain-of-function mutation in *IFIH1* led to over expression of MDA5 and contributed to an over-reactive IFN response (Crow et al., 2015). It is important to note that clinical studies have shown that these gene mutations do not always lead to the same clinical symptoms in patients. One clinical study observed two siblings both with the loss-of-function mutation in *ADAR1*. Even though they both had the same mutation, their symptoms presented at different ages in childhood and symptoms were also variable. This is important information because early diagnosis and treatment of AGS can help with early administration of IFN suppression therapy to minimize brain damage induced by inflammation (Schmelzer et al., 2018). These disease phenotypes in relationship to uncontrolled IFN signaling show the functional importance of MDA5 and other genes involved in facilitating a controlled IFN response (Crow et al., 2015). Moreover, mutations in *IFIH1* in AGS patients reduce the tolerance of MDA5 to Alu retroelements, and that these retroelements can also activate MDA5 during the loss of *ADAR1* function. A gain-of-function mutation in MDA5 (G495R) also retains the minimal Alu duplex length to 30–40 bp (Ahmad et al., 2018). In conclusion, mutations in genes related to MDA5 signaling have been linked to the development of AGS.

2.3.2 P58^{IPK}, PERK, TLRs, and Type 1 Diabetes

The dynamic balance between the eIF2α kinases, such as PKR, with P58^{IPK} is important in determining the outcome of viral infection and the development of diabetes. As discussed previously, PKR is involved in amplifying the IFN response, and P58^{IPK} is a known inhibitor of PKR. P58^{IPK} is important for innate immune responses to viral infection along with the development of autoimmune disorders. P58^{IPK} was originally discovered and purified from influenza virus-infected cells (Lee et al., 1990). During influenza virus infection, P58^{IPK} inhibits PKR's ability to phosphorylate eIF2α, thus facilitating efficient translation of influenza virus mRNAs. In influenza virus infected cells lacking P58^{IPK}, there was a notable decrease in influenza mRNA translation and increase in eIF2α phosphorylation. In cells lacking the inhibitory target of P58^{IPK}, PKR, the results were reversed, and influenza virus mRNA translation increased while eIF2α phosphorylation decreased. This suggests efficient influenza virus replication if facilitated by P58^{IPK} inhibition of PKR and this has downstream ramifications on eIF2α phosphorylation. These results were also observed during VSV infection (Goodman et al., 2007). *In vivo* experiments enhanced our understanding of P58^{IPK} and identified it as a cellular inhibitor of the host defense (CIHD). Observation of P58$^{IPK-/-}$ mice infected with influenza virus showed the P58^{IPK} helps in host survival. P58$^{IPK-/-}$ mice had increased mortality but similar viral load as compared to the wild-type control mice. Additionally, in P58$^{IPK-/-}$ mice, there was an observed increase in eIF2α and PKR phosphorylation during influenza virus infection. These findings suggested that P58^{IPK} activation during viral infection actually enhanced host survival while also prolonging viral replication (Goodman et al., 2009). Through computational modeling, it was shown that during influenza virus infection there was a significant increase in $P58^{IPK}$ while PKR and eIF2α phosphorylation were decreased. The infectious dose of virus also influenced this relationship. These findings were confirmed with vaccinia virus infection where rapid activation of $P58^{IPK}$ was also observed (Goodman et al., 2011). Proviral roles of $P58^{IPK}$ were further demonstrated during coxsackievirus B3 infection where $P58^{IPK}$ suppressed virally induced apoptosis (Zhang Huifang et al., 2013). $P58^{IPK}$ is an important regulatory in influenza virus and vesicular stomatitis virus infection within the antiviral innate immune response signaling cascade. The tight regulation of PKR by $P58^{IPK}$ is an important indicator of viral infection outcome.

$P58^{IPK}$ also inhibits PERK, pancreatic ER-localized eIF2α kinase, which is encoded by *EIF2AK3*. Similar to that of PKR, PERK has a kinase binding domain very similar to that found in PKR where $P58^{IPK}$ interacts to inhibit function. During ER stress, defined as the continued presence of unfolded proteins, $P58^{IPK}$ is activated. $P58^{IPK}$ then interacts with PERK and inhibits its activity through decreased phosphorylation. PERK functions during ER stress to decrease protein production in the ER. However, it is important to have regulatory proteins, such as $P58^{IPK}$, to prevent excess protein loss during stress. These findings show that $P58^{IPK}$ is an important regulator of PERK in maintaining a controlled ER-stress response (Yan et al., 2002). $P58^{IPK}$ and PERK are known to be expressed in pancreatic cells, hinting at an important connection between these proteins and the development of diabetes (Shi et al., 1999). In mice lacking PERK, progressive diabetes mellitus and exocrine pancreatic insufficiency develop (Harding et al., 2001). PERK has a high importance in controlling protein synthesis during the ER stress response. The important inhibitory role of $P58^{IPK}$ on PERK was confirmed in *in vivo* experiments where adult mice lacking $P58^{IPK}$ developed glucosuria, hyperglycemia, and hypoinsulinemia (Ladiges et al., 2005). These mice had greater levels of insulin producing beta cell destruction in the pancreas. The gene expression profiles in these mice were significantly altered to favor apoptosis of pancreatic islets. These findings suggest an important regulatory role of $P58^{IPK}$ in preventing uncontrolled cell destruction during stress. In the absence of $P58^{IPK}$, these mice developed insulin deficiency phenotypes that are very similar to what is seen in type 1 and late stage type 2 diabetes (Ladiges et al., 2005). When PERK or $P58^{IPK}$ were dysfunctional, ER stress homeostasis was disrupted leading to uncontrolled pancreatic beta cell apoptosis and severe diabetic phenotypes. The importance of the *DNAJC3* gene has also been shown recently in humans to be linked to autoimmunity. A large screen of individuals with diabetes revealed a loss-of-function mutation in *DNAJC3* that correlated with juvenile-onset diabetes and multisystemic neurodegenerative disorders. The identified homozygous stop mutation in *Dnajc3* in humans leads to a monogenic recessive form of diabetes mellitus (Synofzik et al., 2014). Taken together, $P58^{IPK}$ is a key regulator in viral infection and autoimmune diseases through its interactions with PKR and PERK.

In addition to the importance of translational control pathways, TLR signaling has also been implicated in regulating T1D. Pancreatic beta cells respond to dsRNA treatment via a TLR3 and TRIF dependent manner, leading to type I IFN induction. Additional treatment with IFN-γ leads to beta cell apoptosis (Rasschaert et al., 2005). Furthermore, beta cells that were knocked out for TLR3 or the type I IFN receptor were protected from apoptosis during dsRNA treatment (Dogusan et al., 2008). The presence of reactive oxygen species was also required for TLR3-mediated NF-κB activation and the induction of IFN-β and TNF-α (Seleme et al., 2012). In patients that died from fulminant T1D, a subtype of diabetes mellitus, TLR3 expression was detected in 85% of T cells that had infiltrated the pancreas and in 63% of infiltrated macrophages, leading to beta cell death (Shibasaki et al., 2010). Considering the important role of TLR3 in T1D pathogenesis, it follows that polymorphisms in *TLR3* were shown to be associated with increased risk for T1D (Assmann et al., 2014). Finally, in insulitic islets isolated by laser capture microdissection from patients with recent onset T1D, there was increased ISG expression. Specifically, *TLR3* and *EIF2AK2* were of the significantly overexpressed ISGs, which further bolsters the link between translational control and innate immune signaling in T1D (Lundberg et al., 2016). In addition to TLR3, TLR7 also plays an important role in T1D autoimmune diabetes, since treating non-obese diabetic mice with TLR7 agonists accelerated the onset of autoimmune diabetes (Lee et al., 2011). In patients that are genetically susceptible to T1D who display increased levels of autoantibodies, there were increased levels of IL-1β, and treating peripheral blood mononuclear cells with TLR3 or TLR7 agonists led to increased percentages of IL-1β dendritic cells (Alkanani et al., 2012). As such, it has been shown that in TLR7 deficient mice, there was attenuated diabetic retinopathy (Liao et al., 2017). Taken together, targeting the TLR pathway may provide an opportunity for therapy in patients that may be susceptible to the onset of T1D.

2.3.3 Implication of RIG-I in Crohn's Disease

RIG-I has also been linked to autoimmune diseases such as inflammatory bowel disease (IBD) that encompasses Crohn's disease (CD) and ulcerative colitis (UC). Dysfunctions in innate immune response signaling, specifically through RIG-I, have been associated with the development of IBD symptoms. In RIG-I knockout mice, colitis-like phenotypes developed. In these mice, there was a noticeable decrease in the size of Peyer's patches that are an essential component of immunity as they defend against pathogens in the intestine. Therefore, the observed cellular apoptosis and decreased size of Peyer's patches in RIG-I deficient mice may increase susceptibility to the colitis-like phenotypes observed. These deficient mice also showed down-regulation of G protein αi2 subunit (Gαi2) in many tissues and negatively regulated T-cell responses (Wang et al., 2007). Gαi2 is necessary for many cellular processes and is a candidate gene associated with the development of human IBD (Hampe et al., 2001b). IBD developed in Gαi2$^{-/-}$ mice and increases in lymphocyte apoptosis lead to a decrease in the size of Peyer's patches (Ohman et al., 2002). The importance of RIG-I in the development of CD was confirmed in a global gene expression analysis. Gut tissue samples of CD and UC diagnosed patients were obtained. A decrease in transcription of RIG-I in epithelial layer of the ileum was associated with CD patients specifically. These results show that RIG-I is not only important in controlling viral infection, but it is also important in controlling the development of CD (Funke et al., 2011). Lack of tight regulation in RIG-I can lead to changes in Peyer's patches that increase susceptibility to autoimmune diseases, such as IBD.

An emerging correlation between autophagy regulation and RLR signaling in innate immune responses is important in autoimmune disease pathology of CD. Autophagy and IFN-mediated immunity are meticulously balanced in healthy individuals. Dysfunction in this balance can lead to autoimmune disorders, such as CD (reviewed in Deretic, 2016; Plantinga et al., 2012; Takahama et al., 2018). Autophagy is a degradation process that has been tied to antiviral innate immune responses. Atg5–Atg12 conjugation is a known regulator of autophagy and directly interacted with the CARD domain of RIG-I and MAVS individually. This binding decreases IFN pathway signaling and production. In this study, VSV replication was enhanced by Atg5–Atg12 activity because IFN production was suppressed. Therefore, autophagic regulation is an important indicator for infection outcome. Importantly, Atg5–Atg12 can interact with MAVS even in the absence of viral infection. This means that Atg5–Atg12 has an important role in cellular homeostasis that negatively regulates the IFN response through RIG-I and MAVS signaling in healthy individuals (Jounai et al., 2007). In autophagic deficient cells lacking Atg5, RLR signaling and IFN secretion increase to resist VSV replication. This deficiency also results in increased dysfunctional mitochondrial function and the mitochondrial associated protein, MAVS. Dysfunctional mitochondria enhance reactive oxygen species present in the cell and this further enhances RLR signaling to further amplify IFN signals. Therefore, autophagic signaling is an important regulator of RLR signaling and these data show the important balance between autophagy and IFN-mediated immunity (Tal et al., 2009). NOD2, which is a member of the NOD-like receptor (NLR) family, is a PRR for bacterial lipopolysaccharides (Inohara et al., 2001; Ogura et al., 2001b), and mutations in NOD2 have been associated with the onset of CD (Hampe et al., 2001a; Hugot et al., 2001; Ogura et al., 2001a). NOD2-mediated autophagy has also been proposed as a component of CD onset. NOD2 directly interacts with RIG-I to negatively regulate IFN induction. Three specific NOD2 mutations associated with CD enhance the capacity of NOD2 and RIG-I to negatively signal IFN as compared to wild-type NOD2, while mutations in NOD2 exhibit impaired autophagosome formation. Together, RIG-I signaling with NOD2 not only functions in response to pathogens, but it also influences the onset of CD (Morosky et al., 2011). Negative regulation of IFN via autophagy through RIG-I with MAVS or NOD2 protects healthy individuals from developing autoimmune disorders such as CD, highlighting the interactions among NLRs, RLRs, autophagy and CD (further reviewed in Coutermarsh-Ott et al., 2016; de Bruyn and Vermeire, 2017).

In CD individuals, the balance of autophagy negative regulation and IFN signaling is disrupted. The loss of autophagy regulation has been proposed as a key pathogenesis mechanism in CD. One study found that an important autophagic response pathway is suppressed in CD inflamed epithelial tissue. The EIF2AK4-EIF2A-ATF4 pathway was identified as being important in controlling intracellular replication of adherent-invasive *Escherichia coli* in leading to robust autophagic gene expression. *EIF2AK4* encodes the GCN2 eIF2α kinase, which is activated during amino acid starvation. GCN2 then activates ATF4 and subsequently autophagy (B'Chir et al., 2013). This autophagic pathway is suppressed in inflamed CD tissue and there is not sufficient autophagic induction to control the intracellular *E. coli* replication. The adherent-invasive *E. coli* that colonize intestinal mucosa are in high abundance because autophagy is not properly activated thereby leading to increased inflammation in previously inflamed CD epithelial cells. This study highlights the relationships among translational control, autophagy, and inflammatory autoimmunity (Bretin et al., 2016). Autophagy can also be increased when the NF-κB transcription factor is inhibited. Porcine follicle development experiments showed that the follicle stimulating hormone inhibits NF-κB and subsequent IFN induction. This inhibition of NF-κB then led to enhanced autophagic activity through Jun N-terminal kinase signaling (Gao et al., 2016). Meta-analysis of individuals with and without CD linked two gene polymorphisms with disease susceptibility. Polymorphisms in *ATG16L1* that limit interaction with Atg12p–Atg5p to induce autophagy signaling were correlated with an increase in an individual's susceptibility to CD. Mutations in *IRGM* were also strongly associated with the development of CD and this gene encodes the GTP-binding protein that induces autophagy signaling through RIG-I. Polymorphisms in *ATG16L1* and *IRGM* were also documented as being risk factors for UC disease (Palomino-Morales et al., 2009). One of the CD-risk mutations in *ATG16L1*, T300A, was shown to improve overall survival in colorectal cancer patients. While this mutation was not associated with a change in autophagy, there was increased type I IFN production and sensitivity to dsRNA treatment via MAVS (Grimm et al., 2016). The mechanism behind the control that IRGM has on autophagy was shown to be through the interaction between IRGM and NOD2. In fact, IRGM, NOD2, and ATG16L1 form a complex that regulates the autophagic response to microbes (Chauhan et al., 2015). Together, IRGM confers antimicrobial and anti-inflammatory states that are important in regulating CD.

2.3.4 Singleton–Merten Syndrome, Systemic Lupus Erythematosus, and Prolidase Disorder

Dysfunction in IFN induction through MDA5 and RIG-I has also been linked to Singleton–Merten Syndrome (SMS) (Jang et al., 2015; Rutsch et al., 2015). SMS is a rare autoimmune disorder and classical SMS manifests in many ways such as dental dysplasmia, aortic calcification, glaucoma, and osteopenia. The disorder is known to have autosomal-dominant inheritance but clues about the genetic profile contributing to disease susceptibility have only recently been reported. Mutations in *IFIH1* and *DDX58*, the gene that encodes RIG-I, led to increased induction of IFN-β production and contributed to disease symptoms. The missense mutation in *IFIH1* enhanced MDA5 function and contributed to the enhanced IFN-β production in blood and dental tissue (Rutsch et al., 2015). Researchers also identified an atypical SMS phenotype in patients that exhibited variations in the clinical manifestations listed above. Two different variants of *DDX58* led to constitutively active functionality of RIG-I that led to uncontrolled IFN induction. The Glu373Ala variant of *DDX58* correlated with classical SMS symptoms except there were no dental abnormalities observed in these patients. The patients with the Cys268Phe variant of *DDX58* had glaucoma and skeletal abnormalities without dental dysplasmia or aortic calcification (Jang et al., 2015). These genetic analyses show the importance of regulated RIG-I and MDA5 induction because overproduction of IFN-β can lead to the characteristic clinical manifestations of SMS. Another rare autosomal disorder, prolidase disorder (PD), is associated with dysfunction in regulated in IFN induction. Mutations in the gene that encodes prolidase, *PEPD*, cause PD in newborns that present with dermatological symptoms, that range from rashes to lower extremity ulcers, due to increased immunoglobulin levels and decreased complement factor C1q. (Falik-Zaccai et al., 2010; Kurien et al., 2013; Pandit et al., 2013). Prolidase deficiency is also associated with systemic lupus erythematosus (SLE), an autoinflammatory disorder characterized by the production of antibodies against self-DNA and RNA (Arbuckle et al., 2003; Dubois and Tuffanelli, 1964; Lisnevskaia et al., 2014; Tan et al., 1966), which helps explain the phenotypic similarities between PD and SLE (Butbul Aviel et al., 2012; Kurien et al., 2013). Interestingly, during flaviviral infection, viral NS5 binds PEPD, resulting in decreased expression of the type I IFN receptor. Additionally, PD patients also exhibit decreased type I IFN receptor expression, further linking PD with defects in innate immunity (Lubick et al., 2015).

TLR-mediated detection of nucleic acids has been shown to exacerbate pathogenesis of SLE (Deane et al., 2007). Furthermore, lupus-prone TLR7-deficient mice do not generate antibodies to RNA-containing antigens, thus displaying less severe symptoms and prolonged survival (Christensen et al., 2006). Since the production of autoantibodies against self-nucleic acids is a major driving force behind SLE pathogenesis, it follows that B cells play a major role in the development of SLE (Fan et al., 2018). Indeed, galectin-9, an s-type lectin that induces apoptosis in activated T_H1 and T_H17 cells while promoting T_{reg} cell differentiation (Wu et al., 2014), inhibits maturation of pDCs and B cells, thereby decreasing cytokine and antibody production in response to TLR7 ligands (Panda et al., 2018). A recent study also showed that polymorphisms in the promoter of *TLR7* were associated with SLE (Skonieczna et al., 2018). Finally, while pDCs secrete increased levels of type I IFN in a TLR7-mediated manner in SLE patients (Murayama et al., 2017), lupus nephritis was recently shown to be independent of type I IFN yet remained dependent on TLR7 signaling (Wolf et al., 2018).

Taken together, diabetes, Aicardi–Goutières syndrome, inflammatory bowel diseases, lupus, and Singleton–Merten syndrome have all been linked to dysfunctions in RLR- and TLR-mediated initiation of IFN signaling. These autoimmune disorders demonstrate the important balance of the innate immune response needed to prevent the development of disorders.

3. INTRACELLULAR RECOGNITION OF DNA

Since 2000, multiple research groups have identified at least 15 different DNA binding proteins that induce an innate immune response, referred to as DNA sensors. For example, the DNA-dependent activator of IFN-regulatory factors (DAI) was the first cytosolic DNA sensor identified to bind directly to the Z and B forms of DNA as well as detect herpes simplex virus infection (Furr et al., 2011; Takaoka et al., 2007). When activated, DAI, also known as Z-DNA binding protein 1 (ZBP1), induces receptor-interacting protein kinase 3 (RIPK3)-mediated necroptosis that is inhibited by RIPK1 (Lin et al., 2016; Newton et al., 2016). However, in addition to DAI signaling, two discoveries in the last 10 years have exposed the complexity of DNA-sensing mechanisms as well as highlight the conserved function of different sensors to protect the host from exogenous DNA. The first discovery was the identification of STING (also known as MYPS/ERIS/MITA), encoded by the *TMEM173* gene, as a central adaptor protein for immunity to cytosolic nucleic acids and a PRR for cyclic dinucleotides (CDN) (Burdette et al., 2011; Ishikawa and Barber, 2008; Jin et al., 2008; Sun et al., 2009; Zhong et al., 2008). The second discovery came when cGAS was shown to be a direct DNA sensor that synthesizes cGAMP for STING activation (Gao et al., 2013b; Sun et al., 2013; Wu et al., 2013). Several other DNA sensors have been shown to interact with STING signaling such as DDX41, LSm14A, and MRE11 (Kondo et al., 2013; Li et al., 2012b; Zhang et al., 2011c). Other DNA sensors induce an innate immune response independent of STING, such as RNA polymerase III, LRRFIP1, Sox2, Rad50, DHX9/36, and AIM2 (Bürckstümmer et al., 2009; Chiu et al., 2009; Fernandes-Alnemri et al., 2009; Kim et al., 2010b; Roth et al., 2014; Xia et al., 2015; Yang et al., 2010). Additionally, the DNA sensors Ku70 and IFI16 have been shown to activate STING in cell- and pathogen-specific manners (Sui et al., 2017; Unterholzner et al., 2010). It is possible that other

DNA sensors will be identified or that previously identified proteins will be shown to serve as DNA sensors whether dependent or independent of STING. Importantly, the existence of diverse DNA sensors, including synthases, DNA repair proteins, helicases, and inflammasomes, indicates that recognition of endogenous and exogenous DNA is a critical component of the host immune response. Below we describe the centrality of the cGAS/STING signaling pathway, the pathways induced by STING-dependent sensors, and lastly the pathways induced by STING-independent sensors (Fig. 2).

3.1 A Central DNA-Sensing Pathway via cGAS and STING

The presence of host- or pathogen-derived dsDNA in the cytosol can be recognized by cyclic guanosine monophosphate (GMP)-adenosine monophosphate (AMP) synthase (cGAS), an enzyme that was recently identified as a critical DNA sensor (Gao et al., 2013b; Li et al., 2013). cGAS is encoded by the *MB21D1* gene and single-molecule assays have shown that the N terminus of human cGAS plays an important role in its own activation upon binding nonspecific DNA inducing downstream signaling through STING (Tao et al., 2017). Unlike some DNA sensors that recognize dsDNA in a sequence-specific manner, cGAS activation is independent of the DNA sequence because binding depends on electrostatic and hydrogen bonding interactions between the negative sugar-phosphate backbone of DNA and positive-charged surfaces of cGAS (Civril et al., 2013). Indeed, the crystallization of human cGAS revealed its unique zinc-ribbon motif insertion which provides DNA-binding specificity (Kranzusch et al., 2013). In this manner, cGAS is a general DNA sensor for its unique ability to bind sequence-independent B form dsDNA. In addition, cGAS is unique for its synthase function which catalyzes the production of 2′-3′-cyclic GMP-AMP (cGAMP), the noncanonical CDN after binding DNA (Gao et al., 2013b). cGAMP has been shown to bind to STING to induce an innate immune response that results in the production of type I IFN (Ablasser et al., 2013; Diner et al., 2013; Shang et al., 2012; Sun et al., 2013). Several biochemical studies have also shown that cGAMP contains mixed phosphodiester linkages, a unique distinction from bacterial CDNs, and that the strength of IFN response depends on the phosphodiester linkage. That is, 2′-3′-cGAMP has the strongest affinity for human STING. Nonetheless, the versatility of STING in recognizing CDNs continues to be an area of intense investigation and suggests that some regulation of this

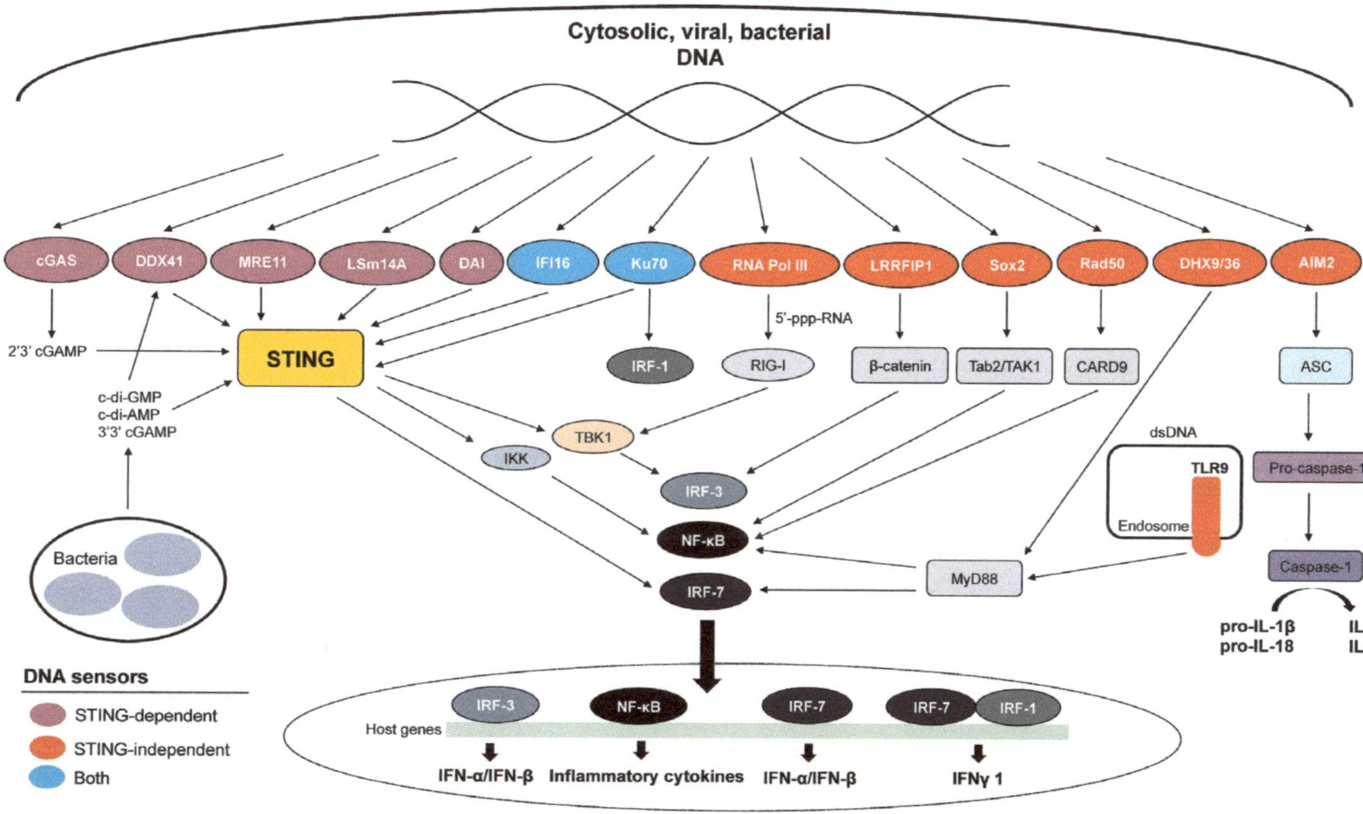

Fig. 2 Sensing of cytosolic DNA or cyclic dinucleotides via multiple sensors leads through the induction of interferon and proinflammatory cytokines. Cytosolic DNA derived from viral or bacterial infection (see Table 3), as well as self-DNA, is sensed by receptors that signal through STING (pink), receptors that are independent of STING (red), and receptors that signal via STING-dependent or -independent mechanisms (blue). STING binds to cyclic dinucleotides, by-products of some bacterial infection and also metabolized by cGAS upon binding to dsDNA. While AIM2 leads to caspase and inflammasome activation, DNA-mediated activation of the other receptors leads to IRF or NF-κB transcription factor activation and the induction of IFN and proinflammatory cytokines. Dysfunction or hyperactivation of these DNA-sensing pathways can lead to autoimmunity (see Table 4).

pathway depends on binding affinity (Ablasser et al., 2013; Gao et al., 2013c; Opoku-Temeng et al., 2016; Sun et al., 2013; Wu et al., 2013).

Upon binding of cGAMP, STING dimerizes and translocates in autophagosomes to a perinuclear region which is necessary for downstream signaling (Barker et al., 2013; Diner et al., 2013; Ishikawa et al., 2009; Moretti et al., 2017; Sun et al., 2009, 2013). STING is an ER-resident host protein that contains four transmembrane domains and globular carboxy-terminal domain (CTD) that enable the binding of STING to TBK1 (Bhat and Fitzgerald, 2014; Burdette and Vance, 2013). The binding of STING and TBK1 causes autophosphorylation of TBK1 at Serine-172, which then facilitates the direct phosphorylation of STING at its Serine-366 and Leucine-374 residues by TBK1. Consequently, phosphorylation of STING by TBK1 recruits IRF3 to also be phosphorylated by TBK1 (Li et al., 2017c; Liu et al., 2015a; Tanaka and Chen, 2012). The association of TBK1 with STING facilitates the dsDNA-mediated activation of the NF-κB and IRF3 transcription factors. Specifically, TBK1 controls the activation of STING-mediated NF-κB signaling through its IKKαβ activation loop while the kinase domain of the ribosomal protein S6 kinase 1 (S6K1) binds to STING to facilitate the formation of a S6K1-STING-TBK1 complex that is necessary for the phosphorylation of IRF3 (Abe and Barber, 2014; Wang et al., 2016). Additionally, the activation of NF-κB by IKK depends on the phosphorylation of IκBα, which induces its degradation and activates NF-κB (Baeuerle and Baltimore, 1988; Beg and Baldwin, 1993; Israel, 2010; Mathes et al., 2008). Once activated in a cGAS/STING-dependent manner, IRF3 and NF-κB translocate to the nucleus and induce the expression of type I IFN, IL-6, IL-1β, and the production of proinflammatory cytokines like TNF-α (Paludan and Bowie, 2013). Additionally, STING translocation from the ER to an ER-Golgi intermediate compartment and the Golgi apparatus is an important rate-limiting event in signal transduction even in the absence of cGAMP (Chen et al., 2016; Dobbs et al., 2015). A recent study has demonstrated that STING senses bacterial c-di-AMP as a *vita*-PAMP, a sub-class of PAMPs derived only form living microbes, to induce ER stress and produce IFN (Moretti et al., 2017). This study and others indicate that ER-phagy and autophagy pathways play an important role in STING regulation during bacterial infections (Watson et al., 2015).

The cGAS/STING signaling pathway is induced by a variety of pathogens including DNA viruses, RNA viruses, retroviruses, and bacteria, all of which have evolved mechanisms to avoid detection by this pathway. Through a variety of mechanisms, several DNA viruses such as adenovirus, herpes simplex virus (HSV)-1, and human papilloma virus (HPV) have been shown to induce or actively inhibit a STING-dependent type I IFN response (Anghelina et al., 2016; Ishikawa et al., 2009; Lam and Falck-Pedersen, 2014; Lam et al., 2014; Liang et al., 2015; Sunthamala et al., 2014). For example, over a dozen proteins of HSV-1 have been found to actively suppress cytosolic-DNA recognition by the cGAS/STING pathway (Christensen et al., 2016; Horan et al., 2013; Ishikawa et al., 2009; Kalamvoki and Roizman, 2014; Su and Zheng, 2017; Xu et al., 2017; Ye et al., 2017; Zhang et al., 2016; Zheng, 2018). Similarly, the E2 proteins of HPV16 inhibit the transcription of different ISGs by targeting STING (Sunthamala et al., 2014). Another study has shown that both human and mouse cytomegalovirus (CMV) induce a cGAS/STING-dependent type I IFN response or actively inhibit STING through its UL82 tegument protein or US9 glycoprotein (Choi et al., 2018; Fu et al., 2017; Lio et al., 2016). Several studies have shown that some retroviruses including HIV, murine leukemia virus, and simian immunodeficiency virus can induce a type I IFN response due to recognition of reverse-transcribed DNA by cGAS and subsequent production of cGAMP (Gao et al., 2013a; Lahaye et al., 2013; Rasaiyaah et al., 2013). DENV can antagonize STING signaling by utilizing viral NS2B3 proteases to bind and cleave human STING although this cleavage does not occur in mouse or nonhuman primate STING (Aguirre and Fernandez-Sesma, 2017; Aguirre et al., 2012; Stabell et al., 2018; Yu et al., 2012). HCV has also been shown to antagonize STING signaling, highlighting the multifaceted role of STING in pathogen evasion beyond direct DNA sensing through cGAS (Ding et al., 2013; Maringer and Fernandez-Sesma, 2014; Moriyama et al., 2007; Nitta et al., 2013).

The fact that STING potentiates a type I IFN response in response to bacterial CDNs introduced a role for STING in regulating bacterial infection. First, it was found that synthetic c-di-GMP could induce a type I IFN response (McWhirter et al., 2009). Then, it was found that *Listeria monocytogenes* secretes c-di-AMP that also induced a type I IFN response (Woodward et al., 2010). Further, a mutant mouse strain called Goldenticket (Gt) had a null mutation in its STING allele (*Sting*$^{Gt/Gt}$) that rendered it unable to detect c-di-GMP and c-di-AMP from *Listeria* infection (Sauer et al., 2011). Finally, STING was shown to be a direct sensor of c-di-GMP (Burdette et al., 2011). Since then, studies have shown that STING mediates a type I IFN response specifically in response to CDNs such as those produced by *Staphylococcus aureus* and *Chlamydia trachomatis* (Barker et al., 2013; Gries et al., 2016; Zhang et al., 2014). Undoubtedly, the role of STING in detecting CDN-mediated type I IFN response was authenticated by the discovery that cGAS produces cGAMP in a DNA-dependent manner (Sun et al., 2013). Further studies since the discovery of cGAS and STING have further connected the CDN and DNA-mediated activation of this central pathway. In 2013, Chen's group showed that $cGAS^{-/-}$ macrophages, fibroblasts, and dendritic cells as well as $cGAS^{-/-}$ mice could not produce type I IFNs or cytokines in response to DNA treatment or infection with HSV-1 or vaccinia virus, but infection with Sendai virus, an RNA virus, did induce type I IFN (Li et al., 2013). These infections were further tested in $Sting^{Gt/Gt}$ mice and cells. Notably, the production of IFN-β in $cGAS^{-/-}$ cells was rescued by the delivery of cGAMP, but not in $Sting^{Gt/Gt}$ cells, further supporting the role of STING in binding to CDNs. Importantly, this was the first study to use cGAMP as a potential adjuvant in the context of STING. They found that injection of the protein antigen ovalbumin (OVA) in the presence or absence of cGAMP in wild-type or $Sting^{Gt/Gt}$ mice could boost the development of OVA-specific antibodies in wild-type mice only (Desmet and Ishii, 2012; Li et al., 2013). Clinical trials using STING agonists have followed. For example, a current Phase I clinical study is using the synthetic, STING-activating CDN agonist MIW815 (ADU-S100) to study its safety and efficacy in treating patients with advanced/metastatic solid tumors or lymphomas via intratumoral injections (ClinicalTrials.gov Identifier: NCT02675439). Another study has shown that c-di-GMP can be complexed with simple cell-penetrating peptides that enhance cellular delivery and biological activity in murine splenocytes but also that the bacterial 3′-3′-cGAMP is a superior stimulator of IFN genes ligand than c-di-GMP in human peripheral blood mononuclear cells (Yildiz et al., 2015). A more recent study has shown that expression of an inducible c-di-GMP-producing diguanylate cyclase in *Klebsiella pneumonia* increased endogenous concentrations of c-di-GMP which attenuated its virulence in the lung of mice (Rosen et al., 2017). Importantly, the attenuation of virulence was independent of STING, supporting the idea that CDNs can be used to influence pathogen virulence (Karaolis et al., 2007a,b; Rosen et al., 2017). Indeed, the use of CDNs as adjuvants has increased in the last few years and will likely continue to expand as more pathogens and diseases are studied under this framework (Dubensky et al., 2013; Junkins et al., 2018; Karaolis et al., 2007a; Miyabe et al., 2014; Škrnjug et al., 2014).

As seen by the diversity of pathogens that cGAS and STING signaling are able to regulate, the centrality of this pathway is evident. Many DNA sensors had been identified before cGAS, but our understanding of host defense mechanisms has greatly increased due to the unique synthase function of cGAS in producing cGAMP in response to the presence of viral DNA, thus acting as an amplifier of cGAS-mediated DNA sensing. In addition, the centrality of STING is highlighted by its ability to use host-derived CDN, 2′-3′-cGAMP, or bacterial CDNs like 3′-3′-cGAMP, c-di-GMP, and c-di-AMP to an innate immune response. Indeed, several groups are investigating the potential use of CDNs as adjuvants in treating human diseases but also in controlling bacterial factors like virulence. Other groups are investigating novels ways, such as ER-phagy, that STING induces an innate immune response. All of these ongoing studies reflect the centrality of STING in the innate immune response to cytosolic nucleic acids (Table 3).

Table 3 Summary of Host Proteins That Sense DNA Virus or Bacterial Infections

DNA Sensor	DNA Sensor Associated Protein	Pathogens That Stimulate Sensor
cGAS	STING	• Adenovirus (AdV) (Anghelina et al., 2016; Ishikawa et al., 2009; Lam and Falck-Pedersen, 2014; Lam et al., 2014) • Herpes simplex virus 1 (HSV-1) (Christensen et al., 2016; Horan et al., 2013; Ishikawa et al., 2009; Kalamvoki and Roizman, 2014; Li et al., 2013; Su and Zheng, 2017; Xu et al., 2017; Ye et al., 2017; Zhang et al., 2016; Zheng, 2018) • Human papillomavirus (HPV) (Sunthamala et al., 2014) • Human and mouse cytomegalovirus (HCMV or MCMV) (Choi et al., 2018; Fu et al., 2017; Lio et al., 2016) • Human immunodeficiency (HIV) (Gao et al., 2013a; Lahaye et al., 2013; Rasaiyaah et al., 2013) • Dengue virus (DENV) (Aguirre and Fernandez-Sesma, 2017; Aguirre et al., 2012; Stabell et al., 2018; Yu et al., 2012) • Hepatitis C virus (HCV) (Ding et al., 2013; Maringer and Fernandez-Sesma, 2014; Moriyama et al., 2007; Nitta et al., 2013) • *Listeria monocytogenes* (Sauer et al., 2011; Woodward et al., 2010) • *Staphylococcus aureus* (Gries et al., 2016; Zhang et al., 2014) • *Chlamydia trachomatis* (Barker et al., 2013)

Continued

Table 3 Summary of Host Proteins That Sense DNA Virus or Bacterial Infections—cont'd

DNA Sensor	DNA Sensor Associated Protein	Pathogens That Stimulate Sensor
DDX41	STING	• HSV-1 (Zhang et al., 2011c) • AdV (Zhang et al., 2011c) • Newcastle disease virus (NDV) (Cheng et al., 2017) • *Listeria monocytogenes* (Parvatiyar et al., 2012)
DAI	STING	• HSV-1 (Furr et al., 2011)
IFI16	1. STING 2. ASC	• (1) Epstein-Barr virus (EBV) (Pisano et al., 2017) • (1) HSV-1 (Johnson et al., 2014; Merkl et al., 2018; Orzalli et al., 2012; Unterholzner et al., 2010) • (2) Kaposi's sarcoma-associated herpesvirus (KSHV) (Kerur et al., 2011)
Ku70	STING-dependent or independent	• Hepatitis B virus (HBV) (Li et al., 2016) • Human T lymphotropic virus type 1 (HTLV-1) (Wang et al., 2017) • HSV-2 (Zhang et al., 2011a)
MRE11	STING	• MRE11 is not required or induce a type I IFN response to HSV-1 or *Listeria monocytogenes* infection but plays a role in DNA damage recognition (Kondo et al., 2013)
LSm14A	STING	• HSV-1 (Li et al., 2012b)
RNA pol. III	RIG-I/ MAVS	• *Legionella pneumophila* (Chiu et al., 2009) • EBV (Ablasser et al., 2009) • Invertebrate iridescent virus 6 (IIV-6) (Ahlers et al., 2016) • Varicella zoster virus (VZV) (Ogunjimi et al., 2017)
LRRFIP1	β-Catenin	• Vesicular stomatitis virus (VSV) (Yang et al., 2010) • *Listeria monocytogenes* (Yang et al., 2010)
Sox2	TAB2, TAK1	• HPV (Martínez-Ramírez et al., 2017)
DHX9/36	MyD88	• HSV-1 (Kim et al., 2010a)
Rad50	CARD9	• Vaccinia virus (VV) (Alcamí and Smith, 1996; Roth et al., 2014) • AdV (Pancholi and Weitzman, 2018; Stracker et al., 2002)

Table 3 Summary of Host Proteins That Sense DNA Virus or Bacterial Infections—cont'd

DNA Sensor	DNA Sensor Associated Protein	Pathogens That Stimulate Sensor
AIM2	ASC	• *Francisella tularensis* (Fernandes-Alnemri et al., 2010; Jones et al., 2010) • *Listeria monocytogenes* (Kim et al., 2010a,b) • *Streptococcus pneumonia, Mycobacterium tuberculosis, Legionella pneumophila, Staphylococcus aureus* (Fang et al., 2011; Ge et al., 2012; Hanamsagar et al., 2014; Saiga et al., 2012) • *Francisella novicida* (Meunier et al., 2015) • *Chlamydia muridarum, Chlamydia trachomatis* (Finethy et al., 2015) • MCMV (Shi et al., 2015) • HPV (Milutin Gašperov et al., 2014) • Influenza virus (Schattgen et al., 2016; Zhang et al., 2017a)
TLR9	MyD88	• HSV-1/2 (Krug et al., 2004; Lundberg et al., 2003) • *Campylobacter jejuni, Klebsiella pneumonia, Staphylococcus aureus* (Bhan et al., 2007; Dalpke et al., 2006) • VZV (Yu et al., 2011) • HCMV (Varani et al., 2007) • MCMV (Krug et al., 2004; Puttur et al., 2016) • EBV (Fiola et al., 2010; Lim et al., 2006) • KSHV (West et al., 2011) • VV and Ectromelia virus (Samuelsson et al., 2008) • AdV (Appledorn et al., 2008; Basner-Tschakarjan et al., 2006)

3.2 STING-Dependent Sensors: DDX41, Ku70, MRE11, and LSm14A

3.2.1 DDX41

DDX41 is a member of the DEXDc helicase family that was recently identified as a DNA sensor in myeloid dendritic cells (mDCs). Knockdown of DDX41 by short-hairpin RNA prevents induction of type I IFN response via IRF3 and NF-κB activation after treatment with DNA or viral DNA infection but not RNA (Fullam and Schröder, 2013; Jiang et al., 2017; Zhang et al., 2011c). DDX41 can also bind directly to bacterial CDNs after which downstream signaling occurs via the recruitment of STING and

TBK1 to activate IRF3 (Omura et al., 2016; Parvatiyar et al., 2012). Two residues of DDX41 (Tyr364 and Tyr414) are necessary for its recognition of DNA, and activation of DDX41 depends on phosphorylation of its Tyr414 residue by BTK (Bruton's tyrosine kinase), which then facilitates binding to STING (Lee et al., 2015). It is also known that the E3 ligase TRIM21 interacts with the DEAD domain of DDX41 and that DDX41 is subsequently degraded via ubiquitination of its Lys9 and Lys115 residues. These results were corroborated by the observation that knockdown of TRIM21 results in over-induction of type I IFN and overexpression of TRIM21 results in reduced IFN-β production (Zhang et al., 2012). One study showed that TBK1, NF-κB, and IRF3 activation are mediated by direct interaction of DDX41 with STING in dendritic cells, bone marrow-derived DCs, and human monocytes after infection with HSV-1 or adenovirus (Zhang et al., 2011c). This study found that knockdown of DDX41 or STING in THP-1 cells resulted in lower IFN-β production after DNA treatment or infection with HSV-1. Another study identified chicken DDX41 as a DNA sensor which results in the production of IFN-β in a STING-dependent manner after DNA treatment or infection with Newcastle disease virus (Cheng et al., 2017). This study used a ssRNA virus to show the role of DDX41 in mediating an innate immune response; however, it lacked experiments investigating the role of DDX41 in the context of DNA viruses and bacterial infection (Jiang et al., 2017). Nonetheless, it is important to note that DDX41 can bind to CDNs such as those produced by *L. monocytogenes* (Parvatiyar et al., 2012). Both DNA- and CDN-sensing roles of DDX41 indicate that it may detect a variety of intracellular viruses and bacteria.

3.2.2 Ku70

Ku70 is a DNA repair subunit protein that binds to DNA double-strand break ends and helps repair DNA via the non-homologous end-joining (NHEJ) pathway (Mimori et al., 1986). Recently, Ku70 was identified as a cytosolic DNA sensor that induces the production of IFN-λ1, a type III IFN, in primary human cell lines after viral infection after translocating from nucleus to cytosol to interact with STING (Sui et al., 2017; Zhang et al., 2011a). IRF1, IRF3, and IRF7 are implicated in this Ku70/STING pathway. One study has found that the Ku70/80 complex can directly sense hepatitis B virus (HBV) DNA which results in recruitment of PARP1, activation and translocation of IRF1 to the nucleus, and upregulation of chemokine secretion (Li et al., 2016). The relevance of this study is that it shows the role of Ku70 in mediating an innate immune response of a

DNA virus. Similarly, another study has found that Ku70 can sense the DNA of human T lymphotropic virus type 1 (HTLV-1). Specifically, this study also found that knockdown of Ku70 led to decreased IRF3 phosphorylation and induction of IFN-β, TNF-α, and IFN-stimulated gene 56 (ISG56) (Wang et al., 2017). Importantly, Ku70 was shown to associate directly with STING to produce IFN-β. While HTLV-1 is a retrovirus, the involvement of Ku70 in detecting viral DNA indicates that Ku70 is an important DNA sensor that may act as reverse transcription intermediate and its role in inducing an innate immune response awaits further investigation. Other studies have shown that Ku70 induces a type III IFN response after HSV-2 infection independently of STING (Zhang et al., 2011a). Nonetheless, the binding of Ku70 to STING as well as mediating an IFN-λ1 response to cytosolic DNA and DNA viruses highlights ability of Ku70 to function both independently and dependently of STING.

3.2.3 MRE11

Meiotic recombination 11 homolog A (MRE11) is an exonuclease better known for its role in microhomology-mediated end-joining (MMEJ). Over-expression of MRE11 has been shown to cause mutations that lead to breast cancer (Sharma et al., 2015; Spehalski et al., 2017; Yuan et al., 2012). One study has shown that MRE11 can physically interact with cytosolic DNA, trigger STING translocation, and subsequently activate IRF3 via interaction with Rad50, another DNA repair protein. However, this response was not observed during infection with *Listeria* or HSV-1, suggesting that MRE11 may induce a type I IFN response specifically to DNA damage instead of pathogen defense (Kondo et al., 2013). The MRE11/Rad50/NBS1 (MRN) complex is known for its conserved role in DNA repair (Maser et al., 1997), yet in Kondo et al., NBS1 was shown to be dispensable for inducing STING trafficking upon treatment with exogenous DNA. An important function of the MRN complex is to induce a DNA damage response (DDR) from stimuli that results in DNA damage (Chapman and Jackson, 2008; He et al., 2012; Paull and Deshpande, 2014). How viruses persist despite active DDR remains an unanswered question. For example, how is HPV able to establish infection despite inducing a DDR, presumably when it causes double-stranded breaks during its replication cycle (Bristol et al., 2017; Kadaja et al., 2009). Therefore, a survey of the MRE11-STING pathway with multiple pathogens would help clarify the role of MRE11 in inducing an innate immune response via STING.

3.2.4 LSm14A

LSm14A is a processing body-associated sensor of viral RNA and DNA that was recently shown to induce a specific type I IFN response via induction of IFN-α, IFN-β, and IL-6 through regulation of STING activation in mouse dendritic cells but not mouse embryonic fibroblasts or macrophages. This indicates that LSm14A is a cell-specific sensor of DNA that functions through nuclear mRNA processing. Further investigation into this DNA sensor will clarify its role in DNA-sensing and how it may function. Currently, only HSV-1 has been shown to induce an innate immune response through LSm14A and STING (Li et al., 2012b; Liu et al., 2016).

3.3 STING-Independent DNA Sensors: TLR9, RNA Polymerase III, DHX9/DHX36, AIM2, IFI16, Sox2, LRRFIP1, and Rad50

3.3.1 TLR9

TLR9 was first identified by its ability to recognize unmethylated 2′-deoxyribo(cytidine-phosphate-guanosine) (CpG) DNA from bacteria, and viral DNA is also recognized by TLR9 (Bauer et al., 2001; Hemmi et al., 2000; Hochrein et al., 2004; Tabeta et al., 2004). Additionally, TLR9 detects ssDNA of at least 21 nucleotides, and methylated ssDNA or dsDNA only weakly activate TLR9 (Pohar et al., 2015a,b). However, the addition of oligonucleotides as short as two nucleotides augments TLR activation (Pohar et al., 2017). In support of this, it has been shown that the endonuclease DNaseII is also required for a robust TLR9-mediated response (Chan et al., 2015). Preference for microbial DNA rather than self-DNA is due to TLR9's location in the endolysosomes (Li et al., 2012a). TLR9 is localized in intracellular vesicles of the ER, lysosomes, and endosomes of resting cells. DNA recognition by TLR9 occurs after transport to endolysosomes, which is mediated by the multi-spanning protein UNC93B (Kawasaki et al., 2011; Latz et al., 2007; Tabeta et al., 2004, 2006). UNC93B physically interacts with TLR9 through its transmembrane domain 2 where the ectodomain of TLR9 is processed by cathepsins to become active (Ewald et al., 2011; Latz et al., 2007; Park et al., 2008; Sepulveda et al., 2009; Tabeta et al., 2006). TLR9 contains a large nonconserved Z-loop between LRR14 and LRR15 that is susceptible to cathepsin-mediated proteolysis. While processing of the Z-loop is required for subsequent TLR9 oligomerization, DNA binding to TLR9 is independent of the Z-loop (Ewald et al., 2011; Li et al., 2012a; Ohto et al., 2015). The chromatin protein HMGB1 (high mobility group box 1) enhances DNA recognition by TLR9 by bending it and bringing it closer to the RAGE (receptor for advanced glycation endproducts) receptor and delivering the HMGB1-DNA complex to early endosomes (Li et al., 2012a; Murugesapillai et al., 2017; Tian et al., 2007). Additionally, it has been recently shown that TLR9 has two DNA-binding sites, namely CpG and 5′-TCG binding sites, both of which contribute to the dimerization of TLR9 (Ohto et al., 2018). Additional activation of TLR9 is enhanced by endosome maturation and acidification (Latz et al., 2004; Wagner, 2004; Yasuda et al., 2005). After DNA binds to TLR9, there is subsequent induction of an innate immune response via two mechanisms. The first is through activation of NF-κB-dependent proinflammatory cytokines, and the second is through IRF7-dependent type I IFN induction. While both pathways are mediated by the adapter protein MyD88 (myeloid differentiation primary response 88) and the TNFR-associated factor TRAF6, IRF7-dependent IFN-α production requires TLR9 trafficking from endosomes to endolysosomes which is mediated by adaptor protein 3 (AP3) (Gohda et al., 2004; Medzhitov et al., 1998; Sasai et al., 2010). In contrast, TLR9 that is transported by UNC93B to the early endosome is cleaved and signals through NF-κB to induce proinflammatory cytokine genes that encode TNF-α, IL-6, and IL-12 (Sasai et al., 2010).

Pathogen-specific studies investigating the mechanism by which TLR9 induces an innate immune response were driven by the discovery that TLR9 recognizes unmethylated CpGs and that human CpGs are methylated and therefore immunologically inert. For example, the genome of HSV-1/2 is heavily unmethylated and CpG-rich, and thus induces a TLR9-dependent immune response in murine bone marrow-derived macrophages (BMDMs) (Lundberg et al., 2003). Bacteria are also capable of activating a TLR9-dependent innate immune response. For example, bacterial species such as *Campylobacter jejuni*, *K. pneumonia*, and *S. aureus* were shown to activate TLR9 based on the abundance of [CG] content. That is, higher CG base content resulted in increased activation of TLR9 and IL-8 production (Dalpke et al., 2006). One study has demonstrated that *TLR9*$^{-/-}$ mice infected with *K. pneumonia* had impaired activation and maturation of dendritic cells, reduced TNF-α induction, and lower production of TNF-α and IL-12. Interestingly, when dendritic cells from wild-type mice were intratracheally transferred into *TLR9*$^{-/-}$ mice, bacterial load was significantly reduced and TNF-α and IL-12 cytokine production increased (Bhan et al., 2007). Of note, human TLR9 is predominantly expressed in plasmacytoid dendritic cells and B cells, resulting in a strong type I IFN response via MyD88/IRF7 signaling. Conversely, murine TLR9 is expressed abundantly in myeloid immune cells and can also result in IFN-γ production and the recruitment of NK, αβ-, and γδ-T cells (Hartmann, 2017; Hornung et al., 2002; Krug et al., 2004; Walker et al., 2010). For example, it has been shown that CMV infection in mice activates a TLR9/MyD88 response that occurs selectively in $CD11^+$ dendritic cells (Krug et al., 2004; Puttur et al., 2016). Other DNA viruses that have been implicated in activating an innate immune response through TLR9 are Varicella zoster virus (VZV), Epstein-Barr virus (EBV), Kaposi sarcoma-associated herpesvirus (KSHV), vaccinia virus (VV), adenovirus (AdV), and human CMV (Appledorn et al., 2008; Basner-Tschakarjan et al., 2006; Fiola et al., 2010; Lim et al., 2006; Samuelsson et al., 2008; Varani et al., 2007; West et al., 2011; Yu et al., 2011). However, the majority of viruses will result in the activation of lymphocytes that can then initiate adaptive responses that differ between mouse and human models of infection. These species-specific differences highlight the importance of understanding how evolution of host defense mechanisms relates to TLR9-mediated immune responses.

3.3.2 RNA Polymerase III

RNA polymerase III transcribes AT-rich dsDNA into an RNA-containing 5′-triosphate moiety which can then be recognized by RIG-I (retinoic acid-inducible gene I) and induce a type I IFN response through IRF3, IRF7, and NF-κB (Ablasser et al., 2009; Chiu et al., 2009). The observation that RNA polymerase III can sense cytosolic DNA has helped researchers investigate the complex ways that pathogens avoid recognition by the innate immune system (Bauernfeind et al., 2010). While Chen's group showed that inhibition of RNA polymerase III resulted in abrogated IFN-β production from *Legionella pneumophila* infection, Hornung's group found that inhibition of RNA polymerase III resulted in abrogated IFN-α production from EBV infection (Ablasser et al., 2009, Chiu et al., 2009). Another study demonstrated that during infection with invertebrate iridescent virus 6 (IIV-6), an insect DNA virus, the mammalian host requires RNA polymerase III to produce IFN-β (Ahlers et al., 2016). Through whole exome sequencing of 21 human patients, a recent study found that different mutations in the RNA polymerase III gene can explain why some patients suffer from severe acute VZV infection (Ogunjimi et al., 2017). In contrast, sequencing of 222 patients suffering from herpesviral encephalitis did not result in identification of RNA polymerase III mutations, which may be explained by the presence of high AT base content in several genome islands of VZV and overall lower cGAS expression in blood cells (Gram et al., 2017; Ogunjimi et al., 2017).

Together, these studies indicate the important role of RNA polymerase III in mediating an innate immune response to DNA viruses.

3.3.3 DHX9/DHX36

DEAH box proteins 9 and 36 (DHX9 and DHX36) are cytosolic helicases that were found to bind to CpG viral DNAs but not RNAs in human primary dendritic cells (Kim et al., 2010b). Specifically, DHX9 and DHX36 sense CpG-oligodeoxynucleotides, CpG-B and CpG-A, respectively, through the adaptor protein MyD88. DHX9 and DHX36 both bind directly to the TIR domain of MyD88 independent of TLR9 signaling (Hochrein et al., 2004; Hokeness-Antonelli et al., 2007; Kim et al., 2010b; Ohnishi et al., 2009). Cytosolic binding of DHX9/DHX36 to CpG-B/A is corroborated by fractionation experiments that showed DHX9/36 are not present in any endosomal structures (Kim et al., 2010b). After binding to MyD88, DHX9 activates NF-κB, which then induces TNF-α and IL-6 production while DHX36 activates IRF7 and induces high IFN-α production. Interestingly, DHX9 and DHX36 were first identified as RNA helicases as they are part of the DExD/H box family, including proteins that have critical roles in RNA metabolism. However, DHX9 and DHX36 have also been shown to act as DNA helicases (Linder, 2006; Zhou et al., 2003). The role of DHX9 as an RNA or DNA helicase has been studied in the context of cancer such as colorectal and lung cancer, where DHX9 regulation is cancer-specific (He et al., 2017; Mi et al., 2016; Rahman et al., 2017; Sun et al., 2014). It is also known that DHX9 binds to viral dsRNA in myeloid dendritic cells, leading to the activation of NF-κB and IRF3, along with the production of IFN-α/β (Zhang et al., 2011d). Similarly, DHX36 has been shown to form a complex with DDX1 and DDX21 to function as a dsRNA sensor that uses the TRIF adaptor molecule to activate NF-κB and a type I IFN response in dendritic cells (Zhang et al., 2011b). As DNA sensors, DHX9/36 proteins have shown to mediate an innate immune response to HSV but not influenza virus (Kim et al., 2010b). These and other studies evidence suggests that the function of RNA and DNA sensing in DHX proteins depends both on cell type and viral genome that highlights the need for more in-depth investigation in this family of proteins.

3.3.4 AIM2 and IFI16

Cytosolic DNA can also bind directly to the AIM2 (absent in melanoma 2) inflammasome protein and activate ASC (apoptosis-associated speck-like protein containing a CARD), a critical component of the inflammasome complex (Fernandes-Alnemri et al., 2007, 2009; Muruve et al., 2008). AIM2 is member of an IFN-inducible HIN-200 family of proteins that contains an N-terminal pyrin domain and a C-terminal oligonucleotide/oligosaccharide-binding domain (Ludlow et al., 2005). Recruitment and activation of caspase-1 by AIM2 depends on the pyrin domain and direct interaction with the adaptor protein ASC. Direct binding of cytoplasmic DNA to AIM2 results in oligomerization which results in the formation of the oligomeric ASC pyroptosome that is required for caspase-1-dependent inflammatory cell death known as pyroptosis (Fernandes-Alnemri et al., 2007, 2009). While other inflammasome proteins such as NALP3 belong to the NLR protein family, AIM2 is not an NLR due to its unique pyrin domain needed for binding of host and pathogen-derived DNA in the cytosol which was first found to directly activate the inflammasome complex independent of NALP3 (Bürckstümmer et al., 2009; Fernandes-Alnemri et al., 2009; Halle et al., 2008; Pétrilli et al., 2007). Nevertheless, AIM2-dependent DNA binding also results in the maturation of proinflammatory cytokines like pro-IL-1β and pro-IL-18 into their active and secreted forms, IL-1β and IL-18 (Muruve et al., 2008; Pétrilli et al., 2007). A recent study has shown that AIM2-like receptor knockout mice respond differently to endogenous retroviral DNA which suggests that AIM2-like receptors might have a more important role in sensing endogenous DNA (Nakaya et al., 2017).

Several bacterial species activate the AIM2 receptor to induce an inflammasome-mediated immune response, such as *Francisella tularensis* and *L. monocytogenes* (Fernandes-Alnemri et al., 2010; Jones et al., 2010; Kim et al., 2010a). *Listeria* is known to activate several innate immune pathways, but in the context of a inflammasome-dependent caspase-1 activation response, AIM2 compensates for an inflammasome response in NLRC4- and NLRP3-deficient mouse macrophages (Kim et al., 2010a). Other bacterial species that have been shown to activate AIM2 include *Streptococcus pneumonia*, some *Mycobacterium* species, *L. pneumophila*, and *S. aureus* (Fang et al., 2011; Ge et al., 2012; Hanamsagar et al., 2014; Saiga et al., 2012). The disruption of bacterial vacuoles allows bacterial DNA to be recognized by AIM2. For example, *Francisella novicida* infection increases the expression of IRF1 which then induces the expression and activation of IFN-inducible GTPases called guanylate-binding proteins (GBPs) to disrupt the bacterial vacuole (Meunier et al., 2015). GBPs colocalize with the IFN-inducible protein IRGB10 to cause bacteriolysis and ultimate exposure of DNA that can be recognized by AIM2 (Man et al., 2015). Recently, *Chlamydia muridarum* and *trachomatis* have also been shown to employ GBPs to induce caspase-1 and caspase-11-mediated inflammasome responses via NLRP3 and AIM2 (Finethy et al., 2015). In contrast, the role of AIM2 in detecting viral DNA is much less explored. For example, mouse CMV and HPV have been shown to activate an AIM2-mediated inflammasome assembly but the mechanism whereby viral DNA is exposed in the cytosol remains unclear (Milutin Gašperov et al., 2014; Shi et al., 2015). Interestingly, recent studies have showed that influenza virus can induce lung damage that results in the release of host DNA and subsequent AIM2-mediated activation (Schattgen et al., 2016; Zhang et al., 2017a). However, Zhang et al. showed that AIM2-deficient human and mouse cells are still able to induce caspase-1 activation, which may be due to difference in viral dose, virus propagation source, or difference between cell culture and *in vivo* models. It will be interesting to see how further research into different host cell types, *in vivo* models and accessory proteins that mediate AIM2-dependent activation of caspase-1 will clarify pathogen and host-specific differences.

Another DNA sensor closely related to AIM2 is IFI16 (gamma-IFN-inducible protein Ifi-16 or IFN-inducible myeloid differentiation transcriptional activator), which is part of the pyrin and HIN-200 domain-containing protein family. In contrast to AIM2, which induces the assembly of an inflammasome complex, IFI16 is known to induce a type I IFN response upon binding to intracellular dsDNA (Trapani et al., 1994; Unterholzner et al., 2010). Knockdown of IFI16 has shown that it is required to maintain EBV latency in Akata cells, an EBV-producing cell line (Pisano et al., 2017). A more recent study has shown that IFI16-depleted human foreskin fibroblasts display increased replication of HSV-1 as well as expression of HSV-1 immediate-early, early, and late proteins, which confirms previous studies using siRNA to knockdown IFI16 (Johnson et al., 2014; Merkl et al., 2018). Although IFI16 is thought to induce a type I IFN response through STING, one study also links IFI16 to caspase-1 and ASC induction in response to KSHV, showing that the inflammasome also functions in the nucleus (Kerur et al., 2011). However, it has been shown that in primary fibroblasts, IFI16 is not required for an immune response to human CMV and that the AIM-like receptors are unnecessary for the response to immunostimulatory DNA (Gray et al., 2016). Further research into the specific pathway in which IFI16 mediates an innate immune response to different pathogens might indeed show more similarities between AIM2 and IFI16 as inflammasome assemblers and their specific roles in a cytosolic DNA-mediated immune response.

3.3.5 Sox2, LRRFIP1, and Rad50

Sex determining region Y-box 2 (Sox2) is a transcription factor that is expressed in the cytosol of neutrophils and that was recently discovered to bind bacterial DNA in a sequence-specific manner (Xia et al., 2015). Specifically, upon bacterial DNA stimulation, Sox2 interacts with TAB2, which causes the TAB2/TAK1 kinase complex to dimerize and lead to the activation of NF-κB and AP-1 signaling. A recent study analyzed the long control regions, which are important sites of viral replication regulation for HPV16, and found that overexpression that Sox2 can repress the *E6* and *E7* oncogene expression of HPV16 (Martínez-Ramírez et al., 2017). Although this study did not investigate innate immune induction by Sox2, this study highlights the role of Sox2 in mediating a DNA virus infection and will need further investigation to clarify its mechanism. While further research is needed to understand the DNA sensing pathway by Sox2 in humans, the discovery of Sox2 as a DNA sensor also highlights the conserved sequence-specific recognition system of foreign DNA used by eubacteria and archaea (Jinek et al., 2012; Xia et al., 2016).

Leucine-rich repeat flightless-interacting protein 1 (LRRFIP1) is a transcriptional repressor that binds to a GC-rich consensus sequence (5-AGCCCCCGGCG-3) and is thought to control smooth cell proliferation via platelet-derived growth factor repression, and positively regulate TLR signaling (Choe et al., 2013; Dai et al., 2009; Labbé et al., 2017). One study showed that LRRFIP1 can mediate a type I IFN response to VSV and *Listeria* infection via a β-catenin-dependent pathway (Yang et al., 2010). Specifically, this study showed that LRRFIP1 interacts and promotes β-catenin activation, resulting in the binding of β-catenin to IRF3. The LRRFIP1-β-catenin interaction results in the recruitment of the acetyltransferase p300 to the IFN-β-enhanceosome through IRF3. Importantly, Yang et al. showed that knockdown of LRRFIP1 decreased the expression of IFN-β mRNA induced by bacterial DNA in the cytosol but not by extracellular LPS. Similarly, the presence of LRRFIP1 increased production of IFN-β following *Listeria* and VSV infection. While further studies have examined the role of LRRFIP1 in inducing a type I IFN response during RNA virus infection, additional studies investigating the role of LRRFIP1 in sensing cytosolic DNA would be beneficial (Bagashev et al., 2010; Liu et al., 2015b). For example, while the crystal structure of LRRFIP1 shows that it remains highly extended while bound to DNA but aggregates at high concentrations of DNA (Nguyen and Modis, 2013), further studies examining the conformational changes of LRRFI1P1 from different DNA pathogens will help elucidate how DNA-binding mediates signaling through LRRFIP1.

As described previously, Rad50 is an important DNA repair protein in the MRN complex. Recently, a study has shown that Rad50 directly interacts with DNA to form a complex with the adaptor protein CARD9 and induce an immune response resulting in the production of IL-1β (Roth et al., 2014; Zhong et al., 2018). These groups observed that delivery of dsDNA or infection with vaccinia virus resulted in recruitment of Rad50 binds which recruited CARD9 and then the B-cell lymphoma/leukemia 10 protein (Bcl10) is recruited through its CARD-binding domain, the entire complex being necessary to activate NF-κB, increased transcription of pro-IL-1β, and IL-1β production. Importantly, this study also showed that activation of NF-κB and subsequent proinflammatory response was independent of an inflammasome-mediated immune response or type I IFN response via STING as seen by unaffected levels of caspase-1 and IRF3. The implications of this study highlight the importance of the MRN complex in sensing DNA and inducing an innate immune response through STING-dependent and independent mechanisms. Specifically, many viruses have developed strategies to avoid Rad50-mediated NF-κB activation and subsequent IL-1β production. For example, adenovirus is known to inhibit Rad50 signaling while several vaccinia virus strains produce a soluble IL-1 receptor that can bind the host-produce IL-1β and prevent fever (Alcamí and Smith, 1996; Stracker et al., 2002). A recent study has shown that even different serotypes of adenovirus have different effects in their mechanism of inhibiting MRN/ATM activation (Pancholi and Weitzman, 2018). Further investigation into the MRN proteins like MRE11 and Rad50 will help elucidate the pathogen-specific mechanisms by which host innate immune response is blocked by DNA viruses as well as create the possibility of viral targets.

In summary, a number of DNA-sensing proteins have been identified over the past decade: TLR9, DAI, AIM2, RNA polymerase III, LRRFIP1, DHX9/DHX36, IFI16, Ku70, DDX41, Sox2, and cGAS. Notably, DAI has recently been shown to recognize viral RNA in addition to its role in DNA sensing to induce necroptosis (Maelfait et al., 2017; Thapa et al., 2016). The ability of STING to potentiate signals from many of these DNA sensors and bind CDNs places it central to the DNA sensing pathway for viral and bacterial infections. Indeed, the wide diversity of proteins that can recognize DNA and induce an innate immune response shows that host defense mechanisms have a critical role in promoting protection and host survival. When these DNA-sensing mechanisms are dysfunctional, they can result in autoimmune and autoinflammatory disorders, a significant area of research driven by the need to discover new therapeutic targets. The next section will discuss autoimmune and autoinflammatory disorders in the context of DNA-sensing.

3.4 Autoimmune and Autoinflammatory Disorders Derived From Immune System Dysfunction

A critical application of innate and adaptive immunity in the context of human disease has been to dissect how autoimmune and autoinflammatory disorders occur as a result of immune system dysfunction. Autoinflammatory disorders are generally driven by innate immune components and do not actively rely on adaptive immune components. In contrast, autoimmunity generally refers to disorders that originate from defects in the innate immune system but require adaptive immune components such as lymphocyte influx or T and B-cell responses. A classic group of autoinflammatory disorders are the cryopyrin-associated periodic syndromes (CAPS), also known as familial cold autoinflammatory syndrome (FCAS), which result from gain-of-function mutations in the *NLRP3/CIAS1* gene (Cordero et al., 2018; Hoffman et al., 2001; Kanneganti et al., 2006; Li et al., 2017a). On the other hand, classic autoimmune disorders include Aicardi–Goutières syndrome (AGS), and although several mutations in AGS patients are rooted in innate immune dysfunction, AGS has both autoimmune and autoinflammatory components. Differences in disease progression have been observed between human and mouse models, illustrating species-specific differences in these disease phenotypes. In characterizing autoimmune and autoinflammatory diseases, it is important to dissect the origins of disease. Our understanding of autoimmune disorders relies on our knowledge of innate immune recognition proteins such as the DNA sensors described above. While not all autoimmune/inflammatory disorders have a common DNA-sensing origin, this section will focus on disorders that originate in failures of DNA sensing pathways.

3.5 DNA-Sensing Molecular Mechanisms of Diseases and Nucleic Acid Accumulation

A group of disorders that result from errors in the innate immune system that result in constitutive activation of type I IFN signaling are collectively termed type I interferonopathies. While there are different causes of type I interferonopathies, two classes can summarize the majority of those derived from errors in DNA-sensing proteins with an end result of unregulated production of IFNs. The first derives from mutations in genes encoding enzymes that regulate accumulation of nucleic acids that ultimately feed into the cGAS/STING pathway. The second results from constitutive activation or hypersensitivity of the cGAS/STING and RLR pathways. In the next subsections we describe how nucleic acid accumulation through mutations or dysfunction of genes encoding the proteins TREX1, RNase H2, SAMHD1, and STING results in autoimmune/autoinflammatory diseases such as AGS, SLE, Sjögren's syndrome, STING-associated vasculopathy with onset in infancy (SAVI), and familial chilblain lupus (Table 4).

Table 4 Summary of Polymorphisms in DNA Sensors and Their Associated Syndromes

Protein	Genetic Polymorphisms That Lead to Autoimmunity/Autoinflammation	Associated Syndrome/Pathology
TLR9	1. Loss of TLR9 leads to increased lymphocyte activation, type I IFN, and autoantibodies (Christensen et al., 2006) 2. Impaired TLR9 signaling in B cells with decreased CD19/21 (Gies et al., 2018)	1-2. Systemic lupus erythematosus
TREX1	1. Loss-of-function mutations in *TREX1* of AGS patients (Abe et al., 2014; Bailey et al., 2012; Crow, 2011; Crow et al., 2006a; Grieves et al., 2015; Lindahl et al., 2009; Namjou et al., 2011; Olivieri et al., 2013; Orebaugh et al., 2011; Uyur Yalçın et al., 2015) 2. TREX1-deficient mice have cGAMP accumulation and produce high amounts of proinflammatory cytokines (Ahn et al., 2012, 2014; Gao et al., 2015; Gray et al., 2015) 3. Missense *TREX1* mutations present in systemic lupus erythematosus patients (Lee-Kirsch et al., 2007)	1-3. Aicardi–Goutières syndrome, familial chilblain lupus, systemic lupus erythematosus

Continued

Table 4 Summary of Polymorphisms in DNA Sensors and Their Associated Syndromes—cont'd

Protein	Genetic Polymorphisms That Lead to Autoimmunity/Autoinflammation	Associated Syndrome/Pathology
RNase H2	1. Mutations in *RNASEH2A*, *RNASEH2B*, and *RNASH2C* lead to nucleic acid accumulation (Chon et al., 2013; Coffin et al., 2011; Günther et al., 2015; Kind et al., 2014; Pizzi et al., 2015; Ramantani et al., 2010; Reijns et al., 2011) 2. RNase H2-deficient mice are embryonic lethal; all subunits are necessary (Hiller et al., 2012; Reijns et al., 2012) 3. Mice that are RNASEH2A- and RNASEH2B-null have increased nucleic acid accumulation and increased ISG expression (Mackenzie et al., 2016; Pokatayev et al., 2016)	1-3. Aicardi–Goutières syndrome, lupus-like symptoms
SAMHD1	1. Over 16 missense, nonsense, and 12-nucleotide deletion mutations (e.g., AGS82, AGS92, and AGS91) have been identified in AGS patients and their families (Rice et al., 2009) 2. Mice deficient in *Samhd1* lack AGS symptoms but have increased ISG expression (Maelfait et al., 2016)	1. Aicardi–Goutières syndrome 2. Increased cGAS/STING signaling
STING	1. De novo gain-of-function mutations in *TMEM173* (Dobbs et al., 2015; Jeremiah et al., 2014; König et al., 2017; Konno et al., 2018; Liu et al., 2014, 2015a; Melki et al., 2017; Munoz et al., 2015; Picard et al., 2016; Warner et al., 2017) 2. Activation of STING via agonist leads to development of anti-STING antibodies, results in production of IFN-β in salivary gland cells (Papinska et al., 2018)	1. SAVI, familial chilblain lupus 2. Sjögren's syndrome

3.5.1 TLR9, TREX1, RNase H2, and SAMHD1

As introduced in the previous section, AGS is characterized as a heritable inflammatory disease that can lead to severe neurological disorders. In addition to the roles that *ADAR1* and MDA5 have in AGS, AGS has also been documented to be caused by genetic mutations encoding nucleic-acid-metabolizing proteins, such as TREX1, RNASEH2A-C (Ribonuclease H2), and SAMHD1 (Crow et al., 2006a,b; Rice et al., 2009, 2012, 2014). Mutations in three of these proteins, namely TREX1, RNase H2, and SAMHD1, result in the direct accumulation of nucleic acids and subsequent activation of the cGAS/STING signaling pathway. Dysfunctions in nucleic acid sensing are also found in SLE patients and include hyperactivity of TREX1 or TLR9 in B cells (Barrat et al., 2005; Christensen et al., 2005; Crispín et al., 2013; Lenert, 2010; Wu et al., 2009; Yu, 2006).

As described in the previous section focused on cytosolic RNA sensing, mice lacking TLR7 exhibited decreased SLE pathology, due to lower levels of autoantibodies and lymphocyte activation. In contrast, the loss of TLR9 results in increased SLE pathology, lymphocyte activation, type I IFN, and autoantibodies (Christensen et al., 2006). B cells from SLE patients display impaired TLR9 signaling and decreased levels of CD19 and CD21, while pDCs from SLE patients displayed normal TLR9 signaling. However, TLR7 signaling was normal in SLE B cells (Gies et al., 2018). Moreover, when lupus-prone *Sle1* mice were crossed with $TLR9^{-/-}$ mice, SLE pathology was exacerbated, and TLR7 expression was increased, leading to increased antigen presentation on DCs (Celhar et al., 2018). Similar results were also observed in $TLR9^{-/-}$ mice treated epicutaneously with imiquimod, which induces local inflammation, as compared to wild-type mice (Liu et al., 2018). Recently, a mechanism behind TLR9-mediated tolerance of SLE was shown to function through the AhR transcription factor, which is activated in apoptotic cells via TLR9 and drives anti-inflammatory IL-10 production. The loss of AhR also resulted in a more severe SLE phenotype (Shinde et al., 2018). Together, these studies highlight the protective role TLR9 in TLR7-mediated SLE and the role of apoptotic-cell DNA in promoting B-cell tolerance.

An important detector of cytosolic DNA is TREX1 (three prime pair exonuclease 1), whose major function is to digest ssDNA and dsDNA in the cytosol and prevent autoimmune activation (de Silva et al., 2007; Hoss, 1999; Lehtinen et al., 2008; Mazur and Perrino, 1999; Yang et al., 2007). Unlike other proofreading DNases, TREX1 is anchored to the ER by its C-terminal region (Chowdhury et al., 2006; Mazur and Perrino, 2001;

Richards et al., 2007; Stetson et al., 2008; Wolf et al., 2016). Indeed, the role of TREX1 in AGS has been documented extensively by several groups which have identified different loss-of-function mutations in the *TREX1* gene in AGS (Abe et al., 2014; Bailey et al., 2012; Crow, 2011; Crow et al., 2006a; Grieves et al., 2015; Lindahl et al., 2009; Namjou et al., 2011; Olivieri et al., 2013; Orebaugh et al., 2011; Uyur Yalçın et al., 2015). Mouse models of TREX1 deficiency have demonstrated that $TREX1^{-/-}$ mice have increased type I IFN signaling, inflammation in multiple tissues, and increased mortality (Gall et al., 2012; Peschke et al., 2016; Stetson et al., 2008; Xu et al., 2014a). Importantly, disease progression in $TREX1^{-/-}$ mice depends significantly on cGAS, STING, and subsequent IRF3 activation, due in part by the accumulation of 2′-3′-cGAMP in $TREX1^{-/-}$ mice. Importantly, double knockout cGAS/TREX1 mice lack the autoinflammatory phenotype due to the loss of the cGAS DNA sensor (Ahn et al., 2012; Gao et al., 2015; Gray et al., 2015). The origin of cytosolic DNA that leads to overstimulation of cGAS and STING has also been studied extensively (Ablasser et al., 2014). Some groups have shown that TREX1 may function specifically to be a detector of viral retroelements which is supported by its role in metabolizing the of HIV-derived DNA and overall HIV pathogenesis (Booiman et al., 2014; Pontillo et al., 2013; Wheeler et al., 2016; Yan et al., 2010). Indeed, Beck-Engeser et al. suggest that the treatment of AGS patients with retroelement inhibitors may is a potential therapy, but Achleitner et al. has shown that these types of drugs do not improve disease (Achleitner et al., 2017; Beck-Engeser et al., 2011). Alternatively, it could be that TREX1 functions primarily to prevent DNA damage and serves a dual purpose in feeding into the cGAS/STING pathway when it detects cytosolic DNA. This hypothesis is supported in part from evidence showing that TREX1 metabolizes cytosolic DNA and contributes to the prevention of genome instability (Ahn et al., 2014; Domínguez-Sánchez et al., 2011; Yang et al., 2007). Indeed, a recent study has shown that TREX1 has a nuclease-independent function in preventing L1-mediated retrotransposon-induced DNA damage, thus maintaining genome integrity (Li et al., 2017b).

AGS is also caused by mutations in alleles in the trimeric protein RNase H2, whose two main functions are to degrade RNA/DNA hybrids resulting from misincorporation of ribonucleotides and to cleave dsDNA at a phosphodiester linkage adjacent to a single ribonucleotide (Crow et al., 2006b; Reijns et al., 2012). It is now well-documented that mutations in *RNASEH2A*, *RNASEH2B*, and *RNASEH2C* lead to the abnormal accumulation of nucleic acids which result in AGS (Chon et al., 2013; Coffin et al., 2011; Günther et al., 2015; Hiller et al., 2012; Kind et al., 2014; Pizzi et al., 2015; Pokatayev et al., 2016; Ramantani et al., 2010; Reijns et al., 2011). RNase H2-deficient mouse models show that mutations in this complex are embryonic lethal, indicating a critical function of all subunits (Reijns et al., 2012). Nonetheless, mice expressing mutant *RNASEH2A* and *RNASEH2B* alleles show that these mutations result in the direct accumulation of nucleic acids that induce IFN through cGAS/STING signaling (Mackenzie et al., 2016; Pokatayev et al., 2016). Evidence suggests that the specific origin of nucleic acids that leads to RNase H2-mediated inflammatory disease may occur through DNA damage responses as well as recognition of retroelements (Bartsch et al., 2017, 2018; Günther et al., 2015; Zhao et al., 2018). Furthermore, two recent studies have also shown that DNA damage leading to the formation of micronuclei induces an IFN response through cGAS and that this occurs in frequently in the absence of *RNASEH2B* (Harding et al., 2017; Mackenzie et al., 2017).

A third and recently characterized protein that plays a role in AGS is SAMHD1 (SAM domain and HD domain-containing protein 1), which encodes an enzyme with phosphohydrolase activity as well as antiviral protection from HIV (Beloglazova et al., 2013; Goldstone et al., 2011; Powell et al., 2011; Rice et al., 2009). Specifically, SAMHD1 detects and cleaves deoxyribose adenine triphosphates (dNTPs) to prevent the reverse transcription of the HIV genome (Goldstone et al., 2011; Lahouassa et al., 2012). A recent study using SAMHD1-deficent mice showed these mice lack autoimmune phenotypes but are hyperactive in cGAS/STING signaling and induce the expression of type I IFN genes *Ifit1* and *Ifi44* (Behrendt et al., 2013; Maelfait et al., 2016). Additionally, SAMHD1 has also been implicated in DNA damage responses through its ability to maintain genome stability, digest ssDNA fragments at stalling replication forks, and inhibiting a cGAS/STING inflammatory response (Coquel et al., 2018; Kretschmer et al., 2015; Medeiros et al., 2018). A recent study has shown that SAMHD1 can suppress innate immune induction independently of its dNTPase activity. Specifically, SAMHD1 can mediate suppression of NF-κB through its interaction with NF-κB1/2 and preventing phosphorylation of IκBα, and also by interacting with IKKε to prevent IKKε-dependent phosphorylation of IRF7 (Chen et al., 2018). Importantly, this study suggests that because SAMHD1 negatively regulates an innate immune response, it may serve as a therapeutic target. Indeed, all of these recent findings highlight the need to continue studying the multiple mechanisms that TREX1, RNase H2, and SAMHD1 have in mediating an inflammatory response that can lead to AGS and other inflammatory diseases.

3.5.2 Sjögren's Syndrome

Sjögren's syndrome (SS) is an autoimmune disorder characterized by dry mouth and eyes due to reduced lacrimal and salivary gland secretion first described in 1933 by Henrik Sjögren (Mutlu and Scully, 1993). The molecular mechanisms behind SS are unclear as symptoms are similar to rheumatoid arthritis and many groups have grappled with defining indicators and causes (Baldini et al., 2018; Daniels and Fox, 1992; Wang et al., 2018; Yang et al., 2018). Studies have documented that the DNA sensor IFI16 is upregulated in SS patients and leads to increased antibodies against IFI16 (Alunno et al., 2015; Baer et al., 2016; Mondini et al., 2006). A recent study has implicated STING hyperactivation as a possible cause of SS, shown by the increase in circulating levels of the IFN-α, IFN-β, IL-6, and TNF-α proinflammatory cytokines in primary salivary gland cells of mice after DMXAA-mediated activation of STING. Notably, these mice developed antibodies to STING, and STING/TBK1/IRF3 signaling resulted in increased IFN-β production in the salivary glands. In addition, transfection with cGAMP in salivary glands also resulted in robust production on IFN-β which suggests that a possible mechanisms for SS might be due to the presence of cytosolic DNA that leads to the activation of cGAS and subsequent STING signaling (Papinska et al., 2018). Importantly, this is the first study to associate SS with accumulation of nucleic acids that activate the cGAS/STING signaling pathway, but further dissection of the pathway in this disorder will be needed to clarify a mechanism.

In summary, it is evident that accumulation of nucleic acids leads to autoinflammatory and autoimmune diseases such as AGS, SLE, and SS. The proteins TREX1, RNase H2, and SAMHD1 all have important roles in protecting the cell from hyperactive cytosolic nucleic acid signaling via the cGAS/STING signaling pathway. Further studies connecting cytosolic DNA presence to AGS, SLE, and SS will clarify how DNA damage responses and innate immune responses like cGAS/STING lead to autoimmune and autoinflammatory disorders.

3.6 Hypersensitivity of cGAS/STING Signaling

In addition to accumulation of nucleic acids, another important cause of autoimmune and autoinflammatory disorders is hypersensitivity of the cGAS/STING pathway. This type of dysfunction is due to mutations that result in the constitutive activation of STING and type I IFN in the presence or absence of ligands, namely CDNs. Two examples of such disorders are SAVI and familial chilblain lupus where mutations in the *TMEM173* gene encoding the STING protein result in constitutive production of IFN and proinflammatory cytokines. Lastly, we describe an example of a therapeutic option in a mouse model of MS that indirectly connects it to the cGAS/STING signaling pathway.

3.6.1 SAVI and Familial Chilblain Lupus

De novo, gain-of-function mutations in the *TMEM173* gene were documented to cause SAVI (Liu et al., 2014). The disease begins in the first few months of life and autoinflammatory symptoms include rash, flares, nodules, blistering of fingers, toes, nose, cheeks, fever, and joint pain (de Jesus et al., 2015; Jeremiah et al., 2014; Munoz et al., 2015; Picard et al., 2016). Other characteristics of SAVI include ulcerating skin lesions that resemble chilblain lupus. Indeed, sequencing of *TMEM173* in chilblain lupus patients has also revealed gain-of-function mutations in STING which lead to disease (König et al., 2017). Previously, the identified gain-of-function mutations in patient samples had involved one of four amino acids in STING at positions 147, 154, 155, or 166 which sequester STING in the ER, simulate ligand binding, resulting in increased production of IFNs (Dobbs et al., 2015; Liu et al., 2014; Zhang et al., 2013). However, substitutions in the amino acid residues 206, 281, and 284 of STING implicated a ligand-independent mechanism of STING activation (Melki et al., 2017). Additionally, a recent study in which whole exome sequencing was performed from a 9-month-old Ecuadorian boy displaying fever and a severe neck abscess revealed he had a missense heterozygous mutation resulting in the STING variant R284S which constitutively activates STING in the absence of CDNs (Konno et al., 2018). Interestingly, the chilblain lupus mutations that result in activation of STING and production of IFN-β are also induced in the absence of CDNs, indicating that other STING-trafficking mechanisms need to be investigated. For example, the constitutive production of IFN-β in SAVI has been thought to occur by activation of TBK1 and IRF3. However, a study generated STING N153S heterozygous knock-in mice and observed that these mutants develop disease symptoms independent of IRF3 which suggests that other type I IFN-independent mechanisms are involved causing diseased state or that species-specific differences exist molecularly (Warner et al., 2017).

As seen by the gain-of-function mutations in *TMEM173*, it is evident that overstimulation or activation of STING is enough to induce a diseased state. The use of JAK1/2 inhibitors is currently a growing therapeutic approach being used treat other inflammatory diseases such as rheumatoid arthritis and myelofibrosis that may also be used to treat SAVI. Indeed, a recent study showed that when the JAK inhibitor baricitinib was used to treat SAVI patients, resulting in amelioration of vasculitis (Sanchez et al., 2018). Nevertheless, many questions remain regarding JAK inhibitors such as their side effects which prevent FDA approval (Shreberk-Hassidim et al., 2017). Another therapeutic strategy lies in the development of STING-specific inhibitors to treat STING-mediated autoinflammatory diseases. In fact, covalent small-molecules have been developed that target a transmembrane domain of STING to block its palmitoylation and subsequent multimeric assembly at the Golgi (Haag et al., 2018). These molecules improve inflammatory disease in *TREX1*$^{-/-}$ mice and may soon be used for clinical trials in SAVI patients or other STING-mediated interferonopathies.

3.6.2 S6K1 and Multiple Sclerosis

One interesting therapeutic example for autoimmune disorders has been observed through the use of the mouse model of multiple sclerosis (MS), namely experimental autoimmune encephalitis (EAE). MS is an autoimmune disease that leaves scars, namely sclerosis, in the myelin sheath of multiple nerve fibers over time by the influx of T lymphocytes into the CNS (McFarland and Martin, 2007). Symptoms include muscle weakness, fatigue, difficulty balancing and walking, as well tremors, speech problems, and cognitive issues like cerebral and brainstem dysfunction (Landtblom et al., 2010; Minagar, 2014). There are varying degrees of MS ranging from subtle to severe that change over time as the disease progresses. Microarray analysis, large- and small-scale gene expression studies, and GWAS have been performed in patients and have revealed over 20 mutations that may lead to disease, some of which induce the production of cytokines by Th1 and Th17 cells (Munoz-Culla et al., 2013; Murugaiyan et al., 2011). There is no explicit evidence supporting DNA-sensing mechanisms involved in MS pathogenesis, however, there is one possible therapeutic option that indirectly connects to the cGAS/STING signaling pathway. Earlier, the ribosomal kinase S6K1 was described in the mechanism for cGAS/STING signaling where it forms a complex with STING and TBK1 to induce IRF3 activation in the context of cytosolic DNA sensing (Wang et al., 2016). Inhibition of S6K1 with a pan-ribosomal S6 kinase inhibitor BI-D1870 protected mice from EAE (Takada et al., 2016). Furthermore, when EAE mice were injected with DNA nanoparticles (DNPs) which led to selective activation of STING, subsequent IFN-α/β release, and overall suppression of EAE by reducing the recruitment of Th1 and Th17 cells into the CNS (Lemos et al., 2014). This therapeutic option mediated by DNPs suggests that other accessory proteins may indeed be able to serve as targets for other autoimmune or inflammatory diseases.

While type I interferonopathies are characterized by varying degrees of autoinflammation, autoimmunity, or immunodeficiency, only a few have been documented to be caused directly or indirectly by dysregulation of DNA-sensing mechanisms. Specifically, the molecular dissection of autoinflammatory and autoimmune disorders like AGS, SS, SAVI, and familial chilblain lupus has all revealed a commonality in DNA-sensing errors. Namely, mutations in the proteins TREX1, RNase H2, and SAMHD1 lead to nucleic acid accumulation while mutations in *TMEM173* lead to overstimulation of STING and subsequent overproduction of IFN-β. On the other hand, SLE and SMS are indirectly connected to DNA-sensing mechanisms but nonetheless represent the complexity of the mechanisms behind these disorders.

4. PROSPECTIVE

Detailed mechanistic understanding of innate immunity is crucial for the development of novel therapies to combat microbial infection and provide relief to those suffering from certain autoimmune pathologies. Already, it has been shown that the STING pathway can be exploited to reduce infectious load of a number of viral (Guo et al., 2017; Skouboe et al., 2018) and bacterial (Barker et al., 2013; Karaolis et al., 2007b; Rosen et al., 2017) pathogens. Additionally, STING-specific adjuvants can boost vaccine efficiency (Junkins et al., 2018; Škrnjug et al., 2014) and be used for cancer immunotherapy (Miyabe et al., 2014). Similarly, RIG-I agonists can be used to reduce viral burden (Bedard et al., 2012; Coch et al., 2017; Green et al., 2016; Nielsen et al., 2017; Pattabhi et al., 2015) and induce anti-tumor activity (Dassler-Plenker et al., 2016). Given these advances in developing small molecules and methods to activate innate immunity to reduce microbial burden, methods to inhibit hyperactive innate immune signaling may aid in the development of therapies for autoimmune disorders. Already, small molecules have been developed that target cGAS and reduce its activity (An et al., 2015; Hall et al., 2017; Vincent et al., 2017). Given the evolutionary arms race between pathogens and host, especially with respect to the co-evolution of DNA- and retro-viruses with humans (Elde and Malik, 2009), might it be possible to develop virally based strategies or molecules to inhibit hyperactive innate immunity? As the saying goes, "…keep your enemies closer."

ACKNOWLEDGMENTS

We thank the anonymous reviewers for their thoughtful comments and constructive criticism of our chapter contribution.

Research in the Goodman lab is funded by NIH Grant R21 AI128103 and funds from Washington State University. K.M.M. is supported by the Barry Goldwater Scholarship and Excellence in Education Foundation.

REFERENCES

Abe, T., Barber, G.N., 2014. Cytosolic-DNA-mediated, STING-dependent proinflammatory gene induction necessitates canonical NF-kappaB activation through TBK1. J. Virol. 88, 5328–5341.

Abe, J., Nakamura, K., Nishikomori, R., Kato, M., Mitsuiki, N., Izawa, K., Awaya, T., Kawai, T., Yasumi, T., Toyoshima, I., Hasegawa, K., Ohshima, Y., Hiragi, T., Sasahara, Y., Suzuki, Y., Kikuchi, M., Osaka, H., Ohya, T., Ninomiya, S., Fujikawa, S., Akasaka, M., Iwata, N., Kawakita, A., Funatsuka, M., Shintaku, H., Ohara, O., Ichinose, H., Heike, T., 2014. A nationwide survey of Aicardi-Goutieres syndrome patients identifies a strong association between dominant TREX1 mutations and chilblain lesions: Japanese cohort study. Rheumatology 53, 448–458.

Ablasser, A., Bauernfeind, F., Hartmann, G., Latz, E., Fitzgerald, K.A., Hornung, V., 2009. RIG-I-dependent sensing of poly(dA:dT) through the induction of an RNA polymerase III-transcribed RNA intermediate. Nat. Immunol. 10, 1065–1072.

Ablasser, A., Goldeck, M., Cavlar, T., Deimling, T., Witte, G., Röhl, I., Hopfner, K.-P., Ludwig, J., Hornung, V., 2013. cGAS produces a 2′-5′-linked cyclic dinucleotide second messenger that activates STING. Nature 498, 380–384.

Ablasser, A., Hemmerling, I., Schmid-Burgk, J.L., Behrendt, R., Roers, A., Hornung, V., 2014. TREX1 deficiency triggers cell-autonomous immunity in a cGAS-dependent manner. J. Immunol. 192, 5993–5997.

Achleitner, M., Kleefisch, M., Hennig, A., Peschke, K., Polikarpova, A., Oertel, R., Gabriel, B., Schulze, L., Lindeman, D., Gerbaulet, A., Fiebig, U., Lee-Kirsch, M.A., Roers, A., Behrendt, R., 2017. Lack of Trex1 causes systemic autoimmunity despite the presence of antiretroviral drugs. J. Immunol. 199, 2261–2269.

Aguirre, S., Fernandez-Sesma, A., 2017. Collateral damage during dengue virus infection: making sense of DNA by cGAS. J. Virol. 91, e01081-16.

Aguirre, S., Maestre, A.M., Pagni, S., Patel, J.R., Savage, T., Gutman, D., Maringer, K., Bernal-Rubio, D., Shabman, R.S., Simon, V., Rodriguez-Madoz, J.R., Mulder, L.C.F., Barber, G.N., Fernandez-Sesma, A., 2012. DENV inhibits type I IFN production in infected cells by cleaving human STING. PLoS Pathog. 8, e1002934.

Ahlers, L.R.H., Goodman, A.G., 2018. The immune responses of the animal hosts of west nile virus: a comparison of insects, birds, and mammals. Front. Cell. Infect. Microbiol. 8, 96.

Ahlers, L.R.H., Bastos, R.G., Hiroyasu, A., Goodman, A.G., 2016. Invertebrate iridescent virus 6, a DNA virus, stimulates a mammalian innate immune response through RIG-I-like receptors. PLoS One 11, e0166088.

Ahmad, S., Mu, X., Yang, F., Greenwald, E., Park, J.W., Jacob, E., Zhang, C.Z., Hur, S., 2018. Breaching self-tolerance to Alu duplex RNA underlies MDA5-mediated inflammation. Cell 172 (797–810), e13.

Ahn, J., Gutman, D., Saijo, S., Barber, G.N., 2012. STING manifests self DNA-dependent inflammatory disease. Proc. Natl. Acad. Sci. U.S.A. 109, 19386–19391.

Ahn, J., Ruiz, P., Barber, G.N., 2014. Intrinsic self-DNA triggers inflammatory disease dependent on STING. J. Immunol. 193, 4634–4642.

Alcamí, A., Smith, G.L., 1996. A mechanism for the inhibition of fever by a virus. Proc. Natl. Acad. Sci. U.S.A. 93, 11029–11034.

Alexopoulou, L., Holt, A.C., Medzhitov, R., Flavell, R.A., 2001. Recognition of double-stranded RNA and activation of NF-κB by Toll-like receptor 3. Nature 413, 732.

Alkanani, A.K., Rewers, M., Dong, F., Waugh, K., Gottlieb, P.A., Zipris, D., 2012. Dysregulated Toll-like receptor-induced interleukin-1beta and interleukin-6 responses in subjects at risk for the development of type 1 diabetes. Diabetes 61, 2525–2533.

Alunno, A., Caneparo, V., Carubbi, F., Bistoni, O., Caterbi, S., Bartoloni, E., Giacomelli, R., Gariglio, M., Landolfo, S., Gerli, R., 2015. Interferon gamma-inducible protein 16 in primary Sjögren's syndrome: a novel player in disease pathogenesis? Arthritis Res. Ther. 17, 208.

An, J., Woodward, J.J., Sasaki, T., Minie, M., Elkon, K.B., 2015. Cutting edge: antimalarial drugs inhibit IFN-beta production through blockade of cyclic GMP-AMP synthase-DNA interaction. J. Immunol. 194, 4089–4093.

Anghelina, D., Lam, E., Falck-Pedersen, E., 2016. Diminished innate antiviral response to adenovirus vectors in cGAS/STING-deficient mice minimally impacts adaptive immunity. J. Virol. 90, 5915–5927.

Angleró-Rodríguez, Y.I., Pantoja, P., Sariol, C.A., 2014. Dengue virus subverts the interferon induction pathway via NS2B/3 protease-IκB kinase epsilon interaction. Clin. Vaccine Immunol. 21, 29–38.

Appledorn, D.M., Patial, S., McBride, A., Godbehere, S., Van Rooijen, N., Parameswaran, N., Amalfitano, A., 2008. Adenovirus vector-induced innate inflammatory mediators, MAPK signaling, as well as adaptive immune responses are dependent upon both TLR2 and TLR9 in vivo. J. Immunol. 181, 2134–2144.

Arakelyan, A., Nersisyan, L., Poghosyan, D., Khondkaryan, L., Hakobyan, A., Löffler-Wirth, H., Melanitou, E., Binder, H., 2017. Autoimmunity and autoinflammation: a systems view on signaling pathway dysregulation profiles. PLoS One 12 (11), e0187572.

Arbuckle, M.R., McClain, M.T., Rubertone, M.V., Scofield, R.H., Dennis, G.J., James, J.A., Harley, J.B., 2003. Development of autoantibodies before the clinical onset of systemic lupus erythematosus. N. Engl. J. Med. 349, 1526–1533.

Arimoto, K.-I., Takahashi, H., Hishiki, T., Konishi, H., Fujita, T., Shimotohno, K., 2007. Negative regulation of the RIG-I signaling by the ubiquitin ligase RNF125. Proc. Natl. Acad. Sci. U.S.A. 104, 7500–7505.

Ashour, J., Laurent-Rolle, M., Shi, P.-Y., García-Sastre, A., 2009. NS5 of dengue virus mediates STAT2 binding and degradation. J. Virol. 83, 5408–5418.

Assmann, T.S., Brondani Lde, A., Bauer, A.C., Canani, L.H., Crispim, D., 2014. Polymorphisms in the TLR3 gene are associated with risk for type 1 diabetes mellitus. Eur. J. Endocrinol. 170, 519–527.

Awais, M., Wang, K., Lin, X., Qian, W., Zhang, N., Wang, C., Wang, K., Zhao, L., Fu, Z.F., Cui, M., 2017. TLR7 deficiency leads to TLR8 compensative regulation of immune response against JEV in mice. Front. Immunol. 8, 160.

Baer, A.N., Petri, M., Sohn, J., Rosen, A., Casciola-Rosen, L., 2016. Association of antibodies to interferon-inducible protein-16 with markers of more severe disease in primary Sjögren's syndrome: anti-IFI16 antibodies and severe disease in Sjögren's syndrome. Arthritis Care Res. 68, 254–260.

Baeuerle, P., Baltimore, D., 1988. I kappa B: a specific inhibitor of the NF-kappa B transcription factor. Science 242, 540–546.

Bagashev, A., Fitzgerald, M.C., LaRosa, D.F., Rose, P.P., Cherry, S., Johnson, A.C., Sullivan, K.E., 2010. Leucine-rich repeat (in flightless I) interacting protein-1 regulates a rapid type I interferon response. J. Interf. Cytokine Res. 30, 843–852.

Bailey, S.L., Harvey, S., Perrino, F.W., Hollis, T., 2012. Defects in DNA degradation revealed in crystal structures of TREX1 exonuclease mutations linked to autoimmune disease. DNA Repair 11, 65–73.

Balachandran, S., Roberts, P.C., Brown, L.E., Truong, H., Pattnaik, A.K., Archer, D.R., Barber, G.N., 2000. Essential role for the dsRNA-dependent protein kinase PKR in innate immunity to viral infection. Immunity 13, 129–141.

Baldini, C., Ferro, F., Mosca, M., Fallahi, P., Antonelli, A., 2018. The association of Sjögren syndrome and autoimmune thyroid disorders. Front. Endocrinol. 9, 121.

Barker, J.R., Koestler, B.J., Carpenter, V.K., Burdette, D.L., Waters, C.M., Vance, R.E., Valdivia, R.H., 2013. STING-dependent recognition of cyclic di-AMP mediates type I interferon responses during Chlamydia trachomatis infection. mBio 4, e00018-13.

Barrat, F.J., Meeker, T., Gregorio, J., Chan, J.H., Uematsu, S., Akira, S., Chang, B., Duramad, O., Coffman, R.L., 2005. Nucleic acids of mammalian origin can act as endogenous ligands for Toll-like receptors and may promote systemic lupus erythematosus. J. Exp. Med. 202, 1131–1139.

Barro, M., Patton, J.T., 2005. Rotavirus nonstructural protein 1 subverts innate immune response by inducing degradation of IFN regulatory factor 3. Proc. Natl. Acad. Sci. U.S.A 102, 4114–4119.

Bartsch, K., Knittler, K., Borowski, C., Rudnik, S., Damme, M., Aden, K., Spehlmann, M.E., Frey, N., Saftig, P., Chalaris, A., Rabe, B., 2017. Absence of RNase H2 triggers generation of immunogenic micronuclei removed by autophagy. Hum. Mol. Genet. 26, 3960–3972.

Bartsch, K., Damme, M., Regen, T., Becker, L., Garrett, L., Hölter, S.M., Knittler, K., Borowski, C., Waisman, A., Glatzel, M., Fuchs, H., Gailus-Durner, V., Hrabe de Angelis, M., Rabe, B., 2018. RNase H2 loss in murine astrocytes results in cellular defects reminiscent of nucleic acid-mediated autoinflammation. Front. Immunol. 9, 587.

Basner-Tschakarjan, E., Gaffal, E., O'Keeffe, M., Tormo, D., Limmer, A., Wagner, H., Hochrein, H., Tüting, T., 2006. Adenovirus efficiently transduces plasmacytoid dendritic cells resulting in TLR9-dependent maturation and IFN-α production. J. Gene Med. 8, 1300–1306.

Bauer, S., Kirschning, C.J., Häcker, H., Redecke, V., Hausmann, S., Akira, S., Wagner, H., Lipford, G.B., 2001. Human TLR9 confers responsiveness to bacterial DNA via species-specific CpG motif recognition. Proc. Natl. Acad. Sci. U.S.A 98, 9237–9242.

Bauernfeind, F., Ablasser, A., Kim, S., Bartok, E., Hornung, V., 2010. An unexpected role for RNA in the recognition of DNA by the innate immune system. RNA Biol. 7, 151–157.

B'Chir, W., Maurin, A.C., Carraro, V., Averous, J., Jousse, C., Muranishi, Y., Parry, L., Stepien, G., Fafournoux, P., Bruhat, A., 2013. The eIF2alpha/ATF4 pathway is essential for stress-induced autophagy gene expression. Nucleic Acids Res. 41, 7683–7699.

Beck-Engeser, G.B., Eilat, D., Wabl, M., 2011. An autoimmune disease prevented by anti-retroviral drugs. Retrovirology 8, 91.

Bedard, K.M., Wang, M.L., Proll, S.C., Loo, Y.M., Katze, M.G., Gale Jr., M., Iadonato, S.P., 2012. Isoflavone agonists of IRF-3 dependent signaling have antiviral activity against RNA viruses. J. Virol. 86, 7334–7344.

Beg, A.A., Baldwin, A.S., 1993. The I kappa B proteins: multifunctional regulators of Rel/NF-kappa B transcription factors. Genes Dev. 7, 2064–2070.

Behrendt, R., Schumann, T., Gerbaulet, A., Nguyen, L.A., Schubert, N., Alexopoulou, D., Berka, U., Lienenklaus, S., Peschke, K., Gibbert, K., Wittmann, S., Lindemann, D.,

Weiss, S., Dahl, A., Naumann, R., Dittmer, U., Kim, B., Mueller, W., Gramberg, T., Roers, A., 2013. Mouse SAMHD1 has antiretroviral activity and suppresses a spontaneous cell-intrinsic antiviral response. Cell Rep. 4, 689–696.

Beloglazova, N., Flick, R., Tchigvintsev, A., Brown, G., Popovic, A., Nocek, B., Yakunin, A.F., 2013. Nuclease activity of the human SAMHD1 protein implicated in the aicardi-goutières syndrome and HIV-1 restriction. J. Biol. Chem. 288, 8101–8110.

Bhan, U., Lukacs, N.W., Osterholzer, J.J., Newstead, M.W., Zeng, X., Moore, T.A., McMillan, T.R., Krieg, A.M., Akira, S., Standiford, T.J., 2007. TLR9 is required for protective innate immunity in gram-negative bacterial pneumonia: role of dendritic cells. J. Immunol. 179, 3937–3946.

Bharaj, P., Wang, Y.E., Dawes, B.E., Yun, T.E., Park, A., Yen, B., Basler, C.F., Freiberg, A.N., Lee, B., Rajsbaum, R., 2016. The matrix protein of nipah virus targets the E3-ubiquitin ligase TRIM6 to inhibit the IKKε kinase-mediated type-I IFN antiviral response. PLoS Pathog. 12, e1005880.

Bhat, N., Fitzgerald, K.A., 2014. Recognition of cytosolic DNA by cGAS and other STING-dependent sensors: HIGHLIGHTS. Eur. J. Immunol. 44, 634–640.

Booiman, T., Setiawan, L.C., Kootstra, N.A., 2014. Genetic variation in Trex1 affects HIV-1 disease progression. AIDS 28, 2517–2521.

Bowen, J.R., Quicke, K.M., Maddur, M.S., O'Neal, J.T., McDonald, C.E., Fedorova, N.B., Puri, V., Shabman, R.S., Pulendran, B., Suthar, M.S., 2017. Zika virus antagonizes type I interferon responses during infection of human dendritic cells. PLoS Pathog. 13, e1006164.

Bretin, A., Carrière, J., Dalmasso, G., Bergougnoux, A., B'Chir, W., Maurin, A.-C., Müller, S., Seibold, F., Barnich, N., Bruhat, A., Darfeuille-Michaud, A., Nguyen, H.T.T., 2016. Activation of the EIF2AK4-EIF2A/eIF2α-ATF4 pathway triggers autophagy response to Crohn disease-associated adherent-invasive Escherichia coli infection. Autophagy 12, 770–783.

Bristol, M., Das, D., Morgan, I., 2017. Why human papillomaviruses activate the DNA damage response (DDR) and how cellular and viral replication persists in the presence of DDR signaling. Viruses 9, 268.

Broquet, A.H., Hirata, Y., McAllister, C.S., Kagnoff, M.F., 2011. RIG-I/MDA5/MAVS are required to signal a protective IFN response in rotavirus-infected intestinal epithelium. J. Immunol. 186, 1618–1626.

Bürckstümmer, T., Baumann, C., Blüml, S., Dixit, E., Dürnberger, G., Jahn, H., Planyavsky, M., Bilban, M., Colinge, J., Bennett, K.L., Superti-Furga, G., 2009. An orthogonal proteomic-genomic screen identifies AIM2 as a cytoplasmic DNA sensor for the inflammasome. Nat. Immunol. 10, 266–272.

Burdette, D.L., Vance, R.E., 2013. STING and the innate immune response to nucleic acids in the cytosol. Nat. Immunol. 14, 19–26.

Burdette, D.L., Monroe, K.M., Sotelo-Troha, K., Iwig, J.S., Eckert, B., Hyodo, M., Hayakawa, Y., Vance, R.E., 2011. STING is a direct innate immune sensor of cyclic di-GMP. Nature 478, 515–518.

Butbul Aviel, Y., Mandel, H., Avitan Hersh, E., Bergman, R., Adiv, O.E., Luder, A., Brik, R., 2012. Prolidase deficiency associated with systemic lupus erythematosus (SLE): single site experience and literature review. Pediatr. Rheumatol. Online J. 10, 18.

Caballero, I.S., Honko, A.N., Gire, S.K., Winnicki, S.M., Melé, M., Gerhardinger, C., Lin, A.E., Rinn, J.L., Sabeti, P.C., Hensley, L.E., Connor, J.H., 2016. In vivo Ebola virus infection leads to a strong innate response in circulating immune cells. BMC Genomics 17, 707.

Cardenas, W.B., Loo, Y.M., Gale Jr., M., Hartman, A.L., Kimberlin, C.R., Martinez-Sobrido, L., Saphire, E.O., Basler, C.F., 2006. Ebola virus VP35 protein binds double-stranded RNA and inhibits alpha/beta interferon production induced by RIG-I signaling. J. Virol. 80, 5168–5178.

Celhar, T., Yasuga, H., Lee, H.Y., Zharkova, O., Tripathi, S., Thornhill, S.I., Lu, H.K., Au, B., Lim, L.H.K., Thamboo, T.P., Akira, S., Wakeland, E.K., Connolly, J.E., Fairhurst, A.M., 2018. TLR9 deficiency breaks tolerance to RNA-associated antigens and upregulates TLR7 protein in Sle1 mice. Arthritis Rheumatol. in press.

Chan, M.P., Onji, M., Fukui, R., Kawane, K., Shibata, T., Saitoh, S., Ohto, U., Shimizu, T., Barber, G.N., Miyake, K., 2015. DNase II-dependent DNA digestion is required for DNA sensing by TLR9. Nat. Commun. 6, 5853.

Chapman, J.R., Jackson, S.P., 2008. Phospho-dependent interactions between NBS1 and MDC1 mediate chromatin retention of the MRN complex at sites of DNA damage. EMBO Rep. 9, 795–801.

Chauhan, S., Mandell, M.A., Deretic, V., 2015. IRGM governs the core autophagy machinery to conduct antimicrobial defense. Mol. Cell 58, 507–521.

Chen, Q., Sun, L., Chen, Z.J., 2016. Regulation and function of the cGAS–STING pathway of cytosolic DNA sensing. Nat. Immunol. 17, 1142–1149.

Chen, S., Bonifati, S., Qin, Z., St. Gelais, C., Kodigepalli, K.M., Barrett, B.S., Kim, S.H., Antonucci, J.M., Ladner, K.J., Buzovetsky, O., Knecht, K.M., Xiong, Y., Yount, J.S., Guttridge, D.C., Santiago, M.L., Wu, L., 2018. SAMHD1 suppresses innate immune responses to viral infections and inflammatory stimuli by inhibiting the NF-κB and interferon pathways. Proc. Natl. Acad. Sci. U.S.A. 115, E3798–E3807.

Cheng, Y., Liu, Y., Wang, Y., Niu, Q., Gao, Q., Fu, Q., Ma, J., Wang, H., Yan, Y., Ding, C., Sun, J., 2017. Chicken DNA virus sensor DDX41 activates IFN-β signaling pathway dependent on STING. Dev. Comp. Immunol. 76, 334–342.

Chiu, Y.-H., MacMillan, J.B., Chen, Z.J., 2009. RNA polymerase III detects cytosolic DNA and induces type I interferons through the RIG-I pathway. Cell 138, 576–591.

Choe, N., Kwon, J.-S., Kim, J.-R., Eom, G.H., Kim, Y., Nam, K.-I., Ahn, Y., Kee, H.J., Kook, H., 2013. The microRNA miR-132 targets Lrrfip1 to block vascular smooth muscle cell proliferation and neointimal hyperplasia. Atherosclerosis 229, 348–355.

Choi, H.J., Park, A., Kang, S., Lee, E., Lee, T.A., Ra, E.A., Lee, J., Lee, S., Park, B., 2018. Human cytomegalovirus-encoded US9 targets MAVS and STING signaling to evade type I interferon immune responses. Nat. Commun. 9, 125.

Chon, H., Sparks, J.L., Rychlik, M., Nowotny, M., Burgers, P.M., Crouch, R.J., Cerritelli, S.M., 2013. RNase H2 roles in genome integrity revealed by unlinking its activities. Nucleic Acids Res. 41, 3130–3143.

Chowdhury, D., Beresford, P.J., Zhu, P., Zhang, D., Sung, J.-S., Demple, B., Perrino, F.W., Lieberman, J., 2006. The exonuclease TREX1 is in the SET complex and acts in concert with NM23-H1 to degrade DNA during granzyme A-mediated cell death. Mol. Cell 23, 133–142.

Christensen, S.R., Kashgarian, M., Alexopoulou, L., Flavell, R.A., Akira, S., Shlomchik, M.J., 2005. Toll-like receptor 9 controls anti-DNA autoantibody production in murine lupus. J. Exp. Med. 202, 321–331.

Christensen, S.R., Shupe, J., Nickerson, K., Kashgarian, M., Flavell, R.A., Shlomchik, M.J., 2006. Toll-like receptor 7 and TLR9 dictate autoantibody specificity and have opposing inflammatory and regulatory roles in a murine model of lupus. Immunity 25, 417–428.

Christensen, M.H., Jensen, S.B., Miettinen, J.J., Luecke, S., Prabakaran, T., Reinert, L.S., Mettenleiter, T., Chen, Z.J., Knipe, D.M., Sandri-Goldin, R.M., Enquist, L.W., Hartmann, R., Mogensen, T.H., Rice, S.A., Nyman, T.A., Matikainen, S., Paludan, S.R., 2016. HSV-1 ICP27 targets the TBK1-activated STING signalsome to inhibit virus-induced type I IFN expression. EMBO J. 35, 1385–1399.

Chung, H., Calis, J.J.A., Wu, X., Sun, T., Yu, Y., Sarbanes, S.L., Dao Thi, V.L., Shilvock, A.R., Hoffmann, H.H., Rosenberg, B.R., Rice, C.M., 2018. Human ADAR1 prevents endogenous RNA from triggering translational shutdown. Cell 172 (811–824), e14.

Ciancanelli, M.J., Volchkova, V.A., Shaw, M.L., Volchkov, V.E., Basler, C.F., 2009. Nipah virus sequesters inactive STAT1 in the nucleus via a P gene-encoded mechanism. J. Virol. 83, 7828–7841.
Civril, F., Deimling, T., de Oliveira Mann, C.C., Ablasser, A., Moldt, M., Witte, G., Hornung, V., Hopfner, K.-P., 2013. Structural mechanism of cytosolic DNA sensing by cGAS. Nature 498, 332–337.
Coch, C., Stumpel, J.P., Lilien-Waldau, V., Wohlleber, D., Kummerer, B.M., Bekeredjian-Ding, I., Kochs, G., Garbi, N., Herberhold, S., Schuberth-Wagner, C., Ludwig, J., Barchet, W., Schlee, M., Hoerauf, A., Bootz, F., Staeheli, P., Hartmann, G., Hartmann, E., 2017. RIG-I activation protects and rescues from lethal influenza virus infection and bacterial superinfection. Mol. Ther. 25, 2093–2103.
Coffin, S.R., Hollis, T., Perrino, F.W., 2011. Functional consequences of the RNase H2A subunit mutations that cause aicardi-goutières syndrome. J. Biol. Chem. 286, 16984–16991.
Coquel, F., Silva, M.-J., Técher, H., Zadorozhny, K., Sharma, S., Nieminuszczy, J., Mettling, C., Dardillac, E., Barthe, A., Schmitz, A.-L., Promonet, A., Cribier, A., Sarrazin, A., Niedzwiedz, W., Lopez, B., Costanzo, V., Krejci, L., Chabes, A., Benkirane, M., Lin, Y.-L., Pasero, P., 2018. SAMHD1 acts at stalled replication forks to prevent interferon induction. Nature 557, 57–61.
Cordero, M.D., Alcocer-Gómez, E., Ryffel, B., 2018. Gain of function mutation and inflammasome driven diseases in human and mouse models. J. Autoimmun. 91, 13–22.
Coutermarsh-Ott, S., Eden, K., Allen, I.C., 2016. Beyond the inflammasome: regulatory NOD-like receptor modulation of the host immune response following virus exposure. J. Gen. Virol. 97, 825–838.
Crispín, J.C., Hedrich, C.M., Tsokos, G.C., 2013. Gene-function studies in systemic lupus erythematosus. Nat. Rev. Rheumatol. 9, 476–484.
Crow, Y.J., 2011. Type I interferonopathies: a novel set of inborn errors of immunity: type I interferonopathies. Ann. N. Y. Acad. Sci. 1238, 91–98.
Crow, Y.J., Hayward, B.E., Parmar, R., Robins, P., Leitch, A., Ali, M., Black, D.N., van Bokhoven, H., Brunner, H.G., Hamel, B.C., Corry, P.C., Cowan, F.M., Frints, S.G., Klepper, J., Livingston, J.H., Lynch, S.A., Massey, R.F., Meritet, J.F., Michaud, J.L., Ponsot, G., Voit, T., Lebon, P., Bonthron, D.T., Jackson, A.P., Barnes, D.E., Lindahl, T., 2006a. Mutations in the gene encoding the 3′-5′ DNA exonuclease TREX1 cause Aicardi-Goutières syndrome at the AGS1 locus. Nat. Genet. 38, 917–920.
Crow, Y.J., Leitch, A., Hayward, B.E., Garner, A., Parmar, R., Griffith, E., Ali, M., Semple, C., Aicardi, J., Babul-Hirji, R., Baumann, C., Baxter, P., Bertini, E., Chandler, K.E., Chitayat, D., Cau, D., Déry, C., Fazzi, E., Goizet, C., King, M.D., Klepper, J., Lacombe, D., Lanzi, G., Lyall, H., Martínez-Frías, M.L., Mathieu, M., McKeown, C., Monier, A., Oade, Y., Quarrell, O.W., Rittey, C.D., Rogers, R.C., Sanchis, A., Stephenson, J.B.P., Tacke, U., Till, M., Tolmie, J.L., Tomlin, P., Voit, T., Weschke, B., Woods, C.G., Lebon, P., Bonthron, D.T., Ponting, C.P., Jackson, A.P., 2006b. Mutations in genes encoding ribonuclease H2 subunits cause Aicardi-Goutières syndrome and mimic congenital viral brain infection. Nat. Genet. 38, 910–916.
Crow, Y.J., Chase, D.S., Lowenstein Schmidt, J., Szynkiewicz, M., Forte, G.M.A., Gornall, H.L., Oojageer, A., Anderson, B., Pizzino, A., Helman, G., Abdel-Hamid, M.S., Abdel-Salam, G.M., Ackroyd, S., Aeby, A., Agosta, G., Albin, C., Allon-Shalev, S., Arellano, M., Ariaudo, G., Aswani, V., Babul-Hirji, R., Baildam, E.M., Bahi-Buisson, N., Bailey, K.M., Barnerias, C., Barth, M., Battini, R., Beresford, M.W., Bernard, G., Bianchi, M., Billette de Villemeur, T., Blair, E.M., Bloom, M., Burlina, A.B., Carpanelli, M.L., Carvalho, D.R., Castro-Gago, M., Cavallini, A., Cereda, C., Chandler, K.E., Chitayat, D.A., Collins, A.E., Sierra Corcoles, C.,

Cordeiro, N.J.V., Crichiutti, G., Dabydeen, L., Dale, R.C., D'Arrigo, S., De Goede, C.G.E.L., De Laet, C., De Waele, L.M.H., Denzler, I., Desguerre, I., Devriendt, K., Di Rocco, M., Fahey, M.C., Fazzi, E., Ferrie, C.D., Figueiredo, A., Gener, B., Goizet, C., Gowrinathan, N.R., Gowrishankar, K., Hanrahan, D., Isidor, B., Kara, B., Khan, N., King, M.D., Kirk, E.P., Kumar, R., Lagae, L., Landrieu, P., Lauffer, H., Laugel, V., La Piana, R., Lim, M.J., Lin, J.-P.S.M., Linnankivi, T., Mackay, M.T., Marom, D.R., Marques Lourenço, C., McKee, S.A., Moroni, I., Morton, J.E.V., Moutard, M.-L., Murray, K., Nabbout, R., Nampoothiri, S., Nunez-Enamorado, N., Oades, P.J., Olivieri, I., Ostergaard, J.R., Pérez-Dueñas, B., Prendiville, J.S., Ramesh, V., Rasmussen, M., Régal, L., Ricci, F., Rio, M., Rodriguez, D., et al., 2015. Characterization of human disease phenotypes associated with mutations in TREX1, RNASEH2A, RNASEH2B, RNASEH2C, SAMHD1, ADAR, and IFIH1. Am. J. Med. Genet. A 167A, 296–312.

Cui, Y., Li, M., Walton, K.D., Sun, K., Hanover, J.A., Furth, P.A., Hennighausen, L., 2001. The Stat3/5 locus encodes novel endoplasmic reticulum and helicase-like proteins that are preferentially expressed in normal and neoplastic mammary tissue. Genomics 78, 129–134.

Dai, P., Jeong, S.Y., Yu, Y., Leng, T., Wu, W., Xie, L., Chen, X., 2009. Modulation of TLR signaling by multiple MyD88-interacting partners including leucine-rich repeat Fli-I-interacting proteins. J. Immunol. 182, 3450–3460.

Dalpke, A., Frank, J., Peter, M., Heeg, K., 2006. Activation of toll-like receptor 9 by DNA from different bacterial species. Infect. Immun. 74, 940–946.

Dalrymple, N.A., Cimica, V., Mackow, E.R., 2015. Dengue virus NS proteins inhibit RIG-I/MAVS signaling by blocking TBK1/IRF3 phosphorylation: dengue virus serotype 1 NS4A is a unique interferon-regulating virulence determinant. mBio 6, e00553-15.

Daniels, T.E., Fox, P.C., 1992. Salivary and oral components of Sjögren's syndrome. Rheum. Dis. Clin. North Am. 18, 571–589.

Darnell, J.E., Kerr, I.M., Stark, G.R., 1994. Jak-STAT pathways and transcriptional activation in response to IFNs and other extracellular signaling proteins. Science 264, 1415–1421.

Dassler-Plenker, J., Reiners, K.S., van den Boorn, J.G., Hansen, H.P., Putschli, B., Barnert, S., Schuberth-Wagner, C., Schubert, R., Tuting, T., Hallek, M., Schlee, M., Hartmann, G., Pogge von Strandmann, E., Coch, C., 2016. RIG-I activation induces the release of extracellular vesicles with antitumor activity. Oncoimmunology 5, e1219827.

Davidson, S., Kaiko, G., Loh, Z., Lalwani, A., Zhang, V., Spann, K., Foo, S.Y., Hansbro, N., Uematsu, S., Akira, S., Matthaei, K.I., Rosenberg, H.F., Foster, P.S., Phipps, S., 2011. Plasmacytoid dendritic cells promote host defense against acute pneumovirus infection via the TLR7-MyD88-dependent signaling pathway. J. Immunol. 186, 5938–5948.

de Bruyn, M., Vermeire, S., 2017. NOD2 and bacterial recognition as therapeutic targets for Crohn's disease. Expert Opin. Ther. Targets 21, 1123–1139.

de Jesus, A.A., Canna, S.W., Liu, Y., Goldbach-Mansky, R., 2015. Molecular mechanisms in genetically defined autoinflammatory diseases: disorders of amplified danger signaling. Annu. Rev. Immunol. 33, 823–874.

de Silva, U., Choudhury, S., Bailey, S.L., Harvey, S., Perrino, F.W., Hollis, T., 2007. The crystal structure of TREX1 explains the 3′ nucleotide specificity and reveals a polyproline II helix for protein partnering. J. Biol. Chem. 282, 10537–10543.

Deane, J.A., Pisitkun, P., Barrett, R.S., Feigenbaum, L., Town, T., Ward, J.M., Flavell, R.A., Bolland, S., 2007. Control of toll-like receptor 7 expression is essential to restrict autoimmunity and dendritic cell proliferation. Immunity 27, 801–810.

Deretic, V., 2016. Autophagy in leukocytes and other cells: mechanisms, subsystem organization, selectivity, and links to innate immunity. J. Leukoc. Biol. 100, 969–978.

Desmet, C.J., Ishii, K.J., 2012. Nucleic acid sensing at the interface between innate and adaptive immunity in vaccination. Nat. Rev. Immunol. 12, 479–491.

Dever, T.E., 2002. Gene-specific regulation by general translation factors. Cell 108, 545–556.

Diebold, S.S., Kaisho, T., Hemmi, H., Akira, S., Reise Sousa, C., 2004. Innate antiviral responses by means of TLR7-mediated recognition of single-stranded RNA. Science 303, 1529–1531.

Diebold, S.S., Massacrier, C., Akira, S., Paturel, C., Morel, Y., Reis e Sousa, C., 2006. Nucleic acid agonists for toll-like receptor 7 are defined by the presence of uridine ribonucleotides. Eur. J. Immunol. 36, 3256–3267.

Diner, E.J., Burdette, D.L., Wilson, S.C., Monroe, K.M., Kellenberger, C.A., Hyodo, M., Hayakawa, Y., Hammond, M.C., Vance, R.E., 2013. The innate immune DNA sensor cGAS produces a noncanonical cyclic dinucleotide that activates human STING. Cell Rep. 3, 1355–1361.

Ding, Q., Huang, B., Lu, J., Liu, Y.-J., Zhong, J., 2012. Hepatitis C virus NS3/4A protease blocks IL-28 production. Eur. J. Immunol. 42, 2374–2382.

Ding, Q., Cao, X., Lu, J., Huang, B., Liu, Y.-J., Kato, N., Shu, H.-B., Zhong, J., 2013. Hepatitis C virus NS4B blocks the interaction of STING and TBK1 to evade host innate immunity. J. Hepatol. 59, 52–58.

Dobbs, N., Burnaevskiy, N., Chen, D., Gonugunta, V.K., Alto, N.M., Yan, N., 2015. STING activation by translocation from the ER is associated with infection and autoinflammatory disease. Cell Host Microbe 18, 157–168.

Dogusan, Z., Garcia, M., Flamez, D., Alexopoulou, L., Goldman, M., Gysemans, C., Mathieu, C., Libert, C., Eizirik, D.L., Rasschaert, J., 2008. Double-stranded RNA induces pancreatic beta-cell apoptosis by activation of the toll-like receptor 3 and interferon regulatory factor 3 pathways. Diabetes 57, 1236–1245.

Domínguez-Sánchez, M.S., Barroso, S., Gómez-González, B., Luna, R., Aguilera, A., 2011. Genome instability and transcription elongation impairment in human cells depleted of THO/TREX. PLoS Genet. 7, e1002386.

Dubensky, T.W., Kanne, D.B., Leong, M.L., 2013. Rationale, progress and development of vaccines utilizing STING-activating cyclic dinucleotide adjuvants. Ther. Adv. Vaccines 1, 131–143.

Dubois, E.L., Tuffanelli, D.L., 1964. Clinical manifestations of systemic lupus erythematosus: computer analysis of 520 cases. JAMA 190, 104–111.

Ebihara, H., Rockx, B., Marzi, A., Feldmann, F., Haddock, E., Brining, D., LaCasse, R.A., Gardner, D., Feldmann, H., 2011. Host response dynamics following lethal infection of rhesus macaques with Zaire ebolavirus. J. Infect. Dis. 204, S991–S999.

El-Bendary, M., Neamatallah, M., Elalfy, H., Besheer, T., Elkholy, A., El-Diasty, M., Elsareef, M., Zahran, M., El-Aarag, B., Gomaa, A., Elhammady, D., El-Setouhy, M., Hegazy, A., Esmat, G., 2018. The association of single nucleotide polymorphisms of Toll-like receptor 3, Toll-like receptor 7 and Toll-like receptor 8 genes with the susceptibility to HCV infection. Br. J. Biomed. Sci. 1–7.

Elde, N.C., Malik, H.S., 2009. The evolutionary conundrum of pathogen mimicry. Nat. Rev. Microbiol. 7, 787–797.

Errett, J.S., Suthar, M.S., McMillan, A., Diamond, M.S., Gale, M., 2013. The essential, nonredundant roles of RIG-I and MDA5 in detecting and controlling West Nile virus infection. J. Virol. 87, 11416–11425.

Ewald, S.E., Engel, A., Lee, J., Wang, M., Bogyo, M., Barton, G.M., 2011. Nucleic acid recognition by Toll-like receptors is coupled to stepwise processing by cathepsins and asparagine endopeptidase. J. Exp. Med. 208, 643–651.

Falik-Zaccai, T.C., Khayat, M., Luder, A., Frenkel, P., Magen, D., Brik, R., Gershoni-Baruch, R., Mandel, H., 2010. A broad spectrum of developmental delay in a large

cohort of prolidase deficiency patients demonstrates marked interfamilial and intrafamilial phenotypic variability. Am. J. Med. Genet. B Neuropsychiatr. Genet. 153B, 46–56.
Fan, H., Ren, D., Hou, Y., 2018. TLR7, a third signal for the robust generation of spontaneous germinal center B cells in systemic lupus erythematosus. Cell. Mol. Immunol. 15, 286–288.
Fang, R., Tsuchiya, K., Kawamura, I., Shen, Y., Hara, H., Sakai, S., Yamamoto, T., Fernandes-Alnemri, T., Yang, R., Hernandez-Cuellar, E., Dewamitta, S.R., Xu, Y., Qu, H., Alnemri, E.S., Mitsuyama, M., 2011. Critical roles of ASC inflammasomes in caspase-1 activation and host innate resistance to Streptococcus pneumoniae infection. J. Immunol. 187, 4890–4899.
Feng, Q., Hato, S.V., Langereis, M.A., Zoll, J., Virgen-Slane, R., Peisley, A., Hur, S., Semler, B.L., van Rij, R.P., van Kuppeveld, F.J.M., 2012. MDA5 detects the double-stranded RNA replicative form in picornavirus-infected cells. Cell Rep. 2, 1187–1196.
Fernandes-Alnemri, T., Wu, J., Yu, J.W., Datta, P., Miller, B., Jankowski, W., Rosenberg, S., Zhang, J., Alnemri, E.S., 2007. The pyroptosome: a supramolecular assembly of ASC dimers mediating inflammatory cell death via caspase-1 activation. Cell Death Differ. 14, 1590–1604.
Fernandes-Alnemri, T., Yu, J.-W., Datta, P., Wu, J., Alnemri, E.S., 2009. AIM2 activates the inflammasome and cell death in response to cytoplasmic DNA. Nature 458, 509–513.
Fernandes-Alnemri, T., Yu, J.-W., Juliana, C., Solorzano, L., Kang, S., Wu, J., Datta, P., McCormick, M., Huang, L., McDermott, E., Eisenlohr, L., Landel, C.P., Alnemri, E.S., 2010. The AIM2 inflammasome is critical for innate immunity to Francisella tularensis. Nat. Immunol. 11, 385–393.
Finethy, R., Jorgensen, I., Haldar, A.K., de Zoete, M.R., Strowig, T., Flavell, R.A., Yamamoto, M., Nagarajan, U.M., Miao, E.A., Coers, J., 2015. Guanylate binding proteins enable rapid activation of canonical and noncanonical inflammasomes in chlamydia-infected macrophages. Infect. Immun. 83, 4740–4749.
Fiola, S., Gosselin, D., Takada, K., Gosselin, J., 2010. TLR9 contributes to the recognition of EBV by primary monocytes and plasmacytoid dendritic cells. J. Immunol. 185, 3620–3631.
Fredericksen, B.L., Keller, B.C., Fornek, J., Katze, M.G., Gale, M., 2008. Establishment and maintenance of the innate antiviral response to west nile virus involves both RIG-I and MDA5 signaling through IPS-1. J. Virol. 82, 609–616.
Fu, Y.-Z., Su, S., Gao, Y.-Q., Wang, P.-P., Huang, Z.-F., Hu, M.-M., Luo, W.-W., Li, S., Luo, M.-H., Wang, Y.-Y., Shu, H.-B., 2017. Human cytomegalovirus tegument protein UL82 inhibits STING-mediated signaling to evade antiviral immunity. Cell Host Microbe 21, 231–243.
Fullam, A., Schröder, M., 2013. DExD/H-box RNA helicases as mediators of anti-viral innate immunity and essential host factors for viral replication. Biochim. Biophys. Acta 1829, 854–865.
Funabiki, M., Kato, H., Miyachi, Y., Toki, H., Motegi, H., Inoue, M., Minowa, O., Yoshida, A., Deguchi, K., Sato, H., Ito, S., Shiroishi, T., Takeyasu, K., Noda, T., Fujita, T., 2014. Autoimmune disorders associated with gain of function of the intracellular sensor MDA5. Immunity 40, 199–212.
Funke, B., Lasitschka, F., Roth, W., Penzel, R., Meuer, S., Saile, M., Gretz, N., Sido, B., Schirmacher, P., Autschbach, F., 2011. Selective downregulation of retinoic acid-inducible gene I within the intestinal epithelial compartment in Crohn's disease. Inflamm. Bowel Dis. 17, 1943–1954.
Furr, S.R., Chauhan, V.S., Moerdyk-Schauwecker, M.J., Marriott, I., 2011. A role for DNA-dependent activator of interferon regulatory factor in the recognition of herpes simplex virus type 1 by glial cells. J. Neuroinflammation 8, 99.

Gack, M.U., 2014. Mechanisms of RIG-I-like receptor activation and manipulation by viral pathogens. J. Virol. 88, 5213–5216.
Gale Jr., M.J., Korth, M.J., Tang, N.M., Tan, S.L., Hopkins, D.A., Dever, T.E., Polyak, S.J., Gretch, D.R., Katze, M.G., 1997. Evidence that hepatitis C virus resistance to interferon is mediated through repression of the PKR protein kinase by the nonstructural 5A protein. Virology 230, 217–227.
Gale, M., Tan, S.L., Wambach, M., Katze, M.G., 1996. Interaction of the interferon-induced PKR protein kinase with inhibitory proteins P58IPK and vaccinia virus K3L is mediated by unique domains: implications for kinase regulation. Mol. Cell. Biol. 16, 4172–4181.
Gall, A., Treuting, P., Elkon, K.B., Loo, Y.-M., Gale, M., Barber, G.N., Stetson, D.B., 2012. Autoimmunity initiates in nonhematopoietic cells and progresses via lymphocytes in an interferon-dependent autoimmune disease. Immunity 36, 120–131.
Gao, D., Wu, J., Wu, Y.T., Du, F., Aroh, C., Yan, N., Sun, L., Chen, Z.J., 2013a. Cyclic GMP-AMP synthase is an innate immune sensor of HIV and other retroviruses. Science 341, 903–906.
Gao, P., Ascano, M., Wu, Y., Barchet, W., Gaffney, B.L., Zillinger, T., Serganov, A.A., Liu, Y., Jones, R.A., Hartmann, G., Tuschl, T., Patel, D.J., 2013b. Cyclic [G(2′,5′)pA(3′,5′)p] is the metazoan second messenger produced by DNA-activated cyclic GMP-AMP synthase. Cell 153, 1094–1107.
Gao, P., Ascano, M., Zillinger, T., Wang, W., Dai, P., Serganov, A.A., Gaffney, B.L., Shuman, S., Jones, R.A., Deng, L., Hartmann, G., Barchet, W., Tuschl, T., Patel, D.J., 2013c. Structure-function analysis of STING activation by c[G(2′,5′)pA(3′,5′)p] and targeting by antiviral DMXAA. Cell 154, 748–762.
Gao, D., Li, T., Li, X.-D., Chen, X., Li, Q.-Z., Wight-Carter, M., Chen, Z.J., 2015. Activation of cyclic GMP-AMP synthase by self-DNA causes autoimmune diseases. Proc. Natl. Acad. Sci. U.S.A. 112, E5699–E5705.
Gao, H., Lin, L., Haq, I.U., Zeng, S.-M., 2016. Inhibition of NF-κB promotes autophagy via JNK signaling pathway in porcine granulosa cells. Biochem. Biophys. Res. Commun. 473, 311–316.
Garcia-Cattaneo, A., Gobert, F.-X., Müller, M., Toscano, F., Flores, M., Lescure, A., Del Nery, E., Benaroch, P., 2012. Cleavage of Toll-like receptor 3 by cathepsins B and H is essential for signaling. Proc. Natl. Acad. Sci. U.S.A. 109, 9053–9058.
Ge, J., Gong, Y.N., Xu, Y., Shao, F., 2012. Preventing bacterial DNA release and absent in melanoma 2 inflammasome activation by a Legionella effector functioning in membrane trafficking. Proc. Natl. Acad. Sci. U.S.A. 109, 6193–6198.
Gee, P., Chua, P.K., Gevorkyan, J., Klumpp, K., Najera, I., Swinney, D.C., Deval, J., 2008. Essential role of the N-terminal domain in the regulation of RIG-I ATPase activity. J. Biol. Chem. 283, 9488–9496.
Gies, V., Schickel, J.N., Jung, S., Joublin, A., Glauzy, S., Knapp, A.M., Soley, A., Poindron, V., Guffroy, A., Choi, J.Y., Gottenberg, J.E., Anolik, J.H., Martin, T., Soulas-Sprauel, P., Meffre, E., Korganow, A.S., 2018. Impaired TLR9 responses in B cells from patients with systemic lupus erythematosus. JCI Insight 3, 1–11.
Gohda, J., Matsumura, T., Inoue, J.-i., 2004. Cutting edge: TNFR-associated factor (TRAF) 6 is essential for MyD88-dependent pathway but Not Toll/IL-1 receptor domain-containing adaptor-inducing IFN-β (TRIF)-dependent pathway in TLR signaling. J. Immunol. 173, 2913–2917.
Goldstone, D.C., Ennis-Adeniran, V., Hedden, J.J., Groom, H.C.T., Rice, G.I., Christodoulou, E., Walker, P.A., Kelly, G., Haire, L.F., Yap, M.W., de Carvalho, L.P.S., Stoye, J.P., Crow, Y.J., Taylor, I.A., Webb, M., 2011. HIV-1 restriction factor SAMHD1 is a deoxynucleoside triphosphate triphosphohydrolase. Nature 480, 379–382.

Goodman, A.G., Smith, J.A., Balachandran, S., Perwitasari, O., Proll, S.C., Thomas, M.J., Korth, M.J., Barber, G.N., Schiff, L.A., Katze, M.G., 2007. The cellular protein P58IPK regulates influenza virus mrna translation and replication through a PKR-mediated mechanism. J. Virol. 81, 2221–2230.

Goodman, A.G., Fornek, J.L., Medigeshi, G.R., Perrone, L.A., Peng, X., Dyer, M.D., Proll, S.C., Knoblaugh, S.E., Carter, V.S., Korth, M.J., Nelson, J.A., Tumpey, T.M., Katze, M.G., 2009. P58IPK: A novel "CIHD" member of the host innate defense response against pathogenic virus infection. PLoS Pathog. 5, e100438.

Goodman, A.G., Tanner, B.C.W., Chang, S.T., Esteban, M., Katze, M.G., 2011. Virus infection rapidly activates the P58(IPK) pathway, delaying peak kinase activation to enhance viral replication. Virology 417, 27–36.

Gotoh, B., Takeuchi, K., Komatsu, T., Yokoo, J., 2003. The STAT2 activation process is a crucial target of Sendai virus C protein for the blockade of alpha interferon signaling. J. Virol. 77, 3360–3370.

Goubau, D., Schlee, M., Deddouche, S., Pruijssers, A.J., Zillinger, T., Goldeck, M., Schuberth, C., Van der Veen, A.G., Fujimura, T., Rehwinkel, J., Iskarpatyoti, J.A., Barchet, W., Ludwig, J., Dermody, T.S., Hartmann, G., Reis, E., Sousa, C., 2014. Antiviral immunity via RIG-I-mediated recognition of RNA bearing 5′-diphosphates. Nature 514, 372–375.

Graff, J.W., Ettayebi, K., Hardy, M.E., 2009. Rotavirus NSP1 inhibits NFkappaB activation by inducing proteasome-dependent degradation of beta-TrCP: a novel mechanism of IFN antagonism. PLoS Pathog. 5, e1000280.

Gram, A.M., Sun, C., Landman, S.L., Oosenbrug, T., Koppejan, H.J., Kwakkenbos, M.J., Hoeben, R.C., Paludan, S.R., Ressing, M.E., 2017. Human B cells fail to secrete type I interferons upon cytoplasmic DNA exposure. Mol. Immunol. 91, 225–237.

Grant, A., Ponia, S.S., Tripathi, S., Balasubramaniam, V., Miorin, L., Sourisseau, M., Schwarz, M.C., Sánchez-Seco, M.P., Evans, M.J., Best, S.M., García-Sastre, A., 2016. Zika virus targets human STAT2 to inhibit type I interferon signaling. Cell Host Microbe 19, 882–890.

Gray, E.E., Treuting, P.M., Woodward, J.J., Stetson, D.B., 2015. Cutting edge: cGAS is required for lethal autoimmune disease in the Trex1-deficient mouse model of Aicardi–Goutières syndrome. J. Immunol. 195, 1939–1943.

Gray, E.E., Winship, D., Snyder, J.M., Child, S.J., Geballe, A.P., Stetson, D.B., 2016. The AIM2-like receptors are dispensable for the interferon response to intracellular DNA. Immunity 45, 255–266.

Green, R.R., Wilkins, C., Pattabhi, S., Dong, R., Loo, Y., Gale Jr., M., 2016. Transcriptional analysis of antiviral small molecule therapeutics as agonists of the RLR pathway. Genom. Data 7, 290–292.

Grief, C., Galler, R., Cortes, L.M., Barth, O.M., 1997. Intracellular localisation of dengue-2 RNA in mosquito cell culture using electron microscopic in situ hybridisation. Arch. Virol. 142, 2347–2357.

Gries, C.M., Bruger, E.L., Moormeier, D.E., Scherr, T.D., Waters, C.M., Kielian, T., 2016. Cyclic di-AMP released from Staphylococcus aureus biofilm induces a macrophage type I interferon response. Infect. Immun. 84, 3564–3574.

Grieves, J.L., Fye, J.M., Harvey, S., Grayson, J.M., Hollis, T., Perrino, F.W., 2015. Exonuclease TREX1 degrades double-stranded DNA to prevent spontaneous lupus-like inflammatory disease. Proc. Natl. Acad. Sci. U.S.A. 112, 5117–5122.

Grimm, W.A., Messer, J.S., Murphy, S.F., Nero, T., Lodolce, J.P., Weber, C.R., Logsdon, M.F., Bartulis, S., Sylvester, B.E., Springer, A., Dougherty, U., Niewold, T.B., Kupfer, S.S., Ellis, N., Huo, D., Bissonnette, M., Boone, D.L., 2016. The Thr300Ala variant in ATG16L1 is associated with improved survival in human colorectal cancer and enhanced production of type I interferon. Gut 65, 456–464.

Günther, C., Kind, B., Reijns, M.A.M., Berndt, N., Martinez-Bueno, M., Wolf, C., Tüngler, V., Chara, O., Lee, Y.A., Hübner, N., Bicknell, L., Blum, S., Krug, C., Schmidt, F., Kretschmer, S., Koss, S., Astell, K.R., Ramantani, G., Bauerfeind, A., Morris, D.L., Cunninghame Graham, D.S., Bubeck, D., Leitch, A., Ralston, S.H., Blackburn, E.A., Gahr, M., Witte, T., Vyse, T.J., Melchers, I., Mangold, E., Nöthen, M.M., Aringer, M., Kuhn, A., Lüthke, K., Unger, L., Bley, A., Lorenzi, A., Isaacs, J.D., Alexopoulou, D., Conrad, K., Dahl, A., Roers, A., Alarcon-Riquelme, M.E., Jackson, A.P., Lee-Kirsch, M.A., 2015. Defective removal of ribonucleotides from DNA promotes systemic autoimmunity. J. Clin. Investig. 125, 413–424.

Guo, F., Tang, L., Shu, S., Sehgal, M., Sheraz, M., Liu, B., Zhao, Q., Cheng, J., Zhao, X., Zhou, T., Chang, J., Guo, J.T., 2017. Activation of stimulator of interferon genes in hepatocytes suppresses the replication of hepatitis B virus. Antimicrob. Agents Chemother. 61, 1–15.

Haag, S.M., Gulen, M.F., Reymond, L., Gibelin, A., Abrami, L., Decout, A., Heymann, M., Goot, F.G.V., Turcatti, G., Behrendt, R., Ablasser, A., 2018. Targeting STING with covalent small-molecule inhibitors. Nature 559, 269–273.

Habjan, M., Andersson, I., Klingström, J., Schümann, M., Martin, A., Zimmermann, P., Wagner, V., Pichlmair, A., Schneider, U., Mühlberger, E., Mirazimi, A., Weber, F., 2008. Processing of genome 5′ termini as a strategy of negative-strand RNA viruses to avoid RIG-I-dependent interferon induction. PLoS One 3, e2032.

Hall, J., Brault, A., Vincent, F., Weng, S., Wang, H., Dumlao, D., Aulabaugh, A., Aivazian, D., Castro, D., Chen, M., Culp, J., Dower, K., Gardner, J., Hawrylik, S., Golenbock, D., Hepworth, D., Horn, M., Jones, L., Jones, P., Latz, E., Li, J., Lin, L.L., Lin, W., Lin, D., Lovering, F., Niljanskul, N., Nistler, R., Pierce, B., Plotnikova, O., Schmitt, D., Shanker, S., Smith, J., Snyder, W., Subashi, T., Trujillo, J., Tyminski, E., Wang, G., Wong, J., Lefker, B., Dakin, L., Leach, K., 2017. Discovery of PF-06928215 as a high affinity inhibitor of cGAS enabled by a novel fluorescence polarization assay. PLoS One 12, e0184843.

Halle, A., Hornung, V., Petzold, G.C., Stewart, C.R., Monks, B.G., Reinheckel, T., Fitzgerald, K.A., Latz, E., Moore, K.J., Golenbock, D.T., 2008. The NALP3 inflammasome is involved in the innate immune response to amyloid-β. Nat. Immunol. 9, 857–865.

Hamel, R., Dejarnac, O., Wichit, S., Ekchariyawat, P., Neyret, A., Luplertlop, N., Perera-Lecoin, M., Surasombatpattana, P., Talignani, L., Thomas, F., Cao-Lormeau, V.-M., Choumet, V., Briant, L., Desprès, P., Amara, A., Yssel, H., Missé, D., 2015. Biology of Zika virus infection in human skin cells. J. Virol. 89, 8880–8896.

Hammerbeck, D.M., Burleson, G.R., Schuller, C.J., Vasilakos, J.P., Tomai, M., Egging, E., Cochran, F.R., Woulfe, S., Miller, R.L., 2007. Administration of a dual toll-like receptor 7 and toll-like receptor 8 agonist protects against influenza in rats. Antiviral Res. 73, 1–11.

Hampe, J., Cuthbert, A., Croucher, P.J., Mirza, M.M., Mascheretti, S., Fisher, S., Frenzel, H., King, K., Hasselmeyer, A., MacPherson, A.J., Bridger, S., van Deventer, S., Forbes, A., Nikolaus, S., Lennard-Jones, J.E., Foelsch, U.R., Krawczak, M., Lewis, C., Schreiber, S., Mathew, C.G., 2001a. Association between insertion mutation in NOD2 gene and Crohn's disease in German and British populations. Lancet 357, 1925–1928.

Hampe, J., Lynch, N.J., Daniels, S., Bridger, S., Macpherson, A.J., Stokkers, P., Forbes, A., Lennard-Jones, J.E., Mathew, C.G., Curran, M.E., Schreiber, S., 2001b. Fine mapping of the chromosome 3p susceptibility locus in inflammatory bowel disease. Gut 48, 191–197.

Hanamsagar, R., Aldrich, A., Kielian, T., 2014. Critical role for the AIM2 inflammasome during acute CNS bacterial infection. J. Neurochem. 129, 704–711.

Hansdottir, S., Monick, M.M., Lovan, N., Powers, L., Gerke, A., Hunninghake, G.W., 2010. Vitamin D decreases respiratory syncytial virus induction of NF-kappaB-linked chemokines and cytokines in airway epithelium while maintaining the antiviral state. J. Immunol. 184, 965–974.

Harding, H.P., Zeng, H., Zhang, Y., Jungries, R., Chung, P., Plesken, H., Sabatini, D.D., Ron, D., 2001. Diabetes mellitus and exocrine pancreatic dysfunction in perk −/− mice reveals a role for translational control in secretory cell survival. Mol. Cell 7, 1153–1163.

Harding, S.M., Benci, J.L., Irianto, J., Discher, D.E., Minn, A.J., Greenberg, R.A., 2017. Mitotic progression following DNA damage enables pattern recognition within micronuclei. Nature 548, 466–470.

Hartman, A.L., Ling, L., Nichol, S.T., Hibberd, M.L., 2008. Whole-genome expression profiling reveals that inhibition of host innate immune response pathways by Ebola virus can be reversed by a single amino acid change in the VP35 protein. J. Virol. 82, 5348–5358.

Hartmann, G., 2017. Nucleic acid immunity. In: Advances in Immunology. Elsevier.

He, J., Shi, L.Z., Truong, L.N., Lu, C.-S., Razavian, N., Li, Y., Negrete, A., Shiloach, J., Berns, M.W., Wu, X., 2012. Rad50 zinc hook is important for the Mre11 complex to bind chromosomal DNA double-stranded breaks and initiate various DNA damage responses. J. Biol. Chem. 287, 31747–31756.

He, L., Chen, Y., Wu, Y., Xu, Y., Zhang, Z., Liu, Z., 2017. Nucleic acid sensing pattern recognition receptors in the development of colorectal cancer and colitis. Cell. Mol. Life Sci. 74, 2395–2411.

Heil, F., Hemmi, H., Hochrein, H., Ampenberger, F., Kirschning, C., Akira, S., Lipford, G., Wagner, H., Bauer, S., 2004. Species-specific recognition of single-stranded RNA via toll-like receptor 7 and 8. Science 303, 1526–1529.

Hemmi, H., Takeuchi, O., Kawai, T., Kaisho, T., Sato, S., Sanjo, H., Matsumoto, M., Hoshino, K., Wagner, H., Takeda, K., Akira, S., 2000. A toll-like receptor recognizes bacterial DNA. Nature 408, 740–745.

Hiller, B., Achleitner, M., Glage, S., Naumann, R., Behrendt, R., Roers, A., 2012. Mammalian RNase H2 removes ribonucleotides from DNA to maintain genome integrity. J. Exp. Med. 209, 1419–1426.

Hochrein, H., Schlatter, B., O'Keeffe, M., Wagner, C., Schmitz, F., Schiemann, M., Bauer, S., Suter, M., Wagner, H., 2004. Herpes simplex virus type-1 induces IFN- production via Toll-like receptor 9-dependent and -independent pathways. Proc. Natl. Acad. Sci. U.S.A. 101, 11416–11421.

Hoffman, H.M., Mueller, J.L., Broide, D.H., Wanderer, A.A., Kolodner, R.D., 2001. Mutation of a new gene encoding a putative pyrin-like protein causes familial cold autoinflammatory syndrome and muckle-wells syndrome. Nat. Genet. 29, 301–305.

Hokeness-Antonelli, K.L., Crane, M.J., Dragoi, A.M., Chu, W.-M., Salazar-Mather, T.P., 2007. IFN-alphabeta-mediated inflammatory responses and antiviral defense in liver is TLR9-independent but MyD88-dependent during murine cytomegalovirus infection. J. Immunol. 179, 6176–6183.

Honda, K., Takaoka, A., Taniguchi, T., 2006. Type I interferon gene induction by the interferon regulatory factor family of transcription factors. Immunity 25, 349–360.

Horan, K.A., Hansen, K., Jakobsen, M.R., Holm, C.K., Soby, S., Unterholzner, L., Thompson, M., West, J.A., Iversen, M.B., Rasmussen, S.B., Ellermann-Eriksen, S., Kurt-Jones, E., Landolfo, S., Damania, B., Melchjorsen, J., Bowie, A.G., Fitzgerald, K.A., Paludan, S.R., 2013. Proteasomal degradation of herpes simplex virus capsids in macrophages releases DNA to the cytosol for recognition by DNA sensors. J. Immunol. 190, 2311–2319.

Hornung, V., Rothenfusser, S., Britsch, S., Krug, A., Jahrsdorfer, B., Giese, T., Endres, S., Hartmann, G., 2002. Quantitative expression of toll-like receptor 1-10 mRNA in cellular subsets of human peripheral blood mononuclear cells and sensitivity to CpG oligodeoxynucleotides. J. Immunol. 168, 4531–4537.

Hornung, V., Ellegast, J., Kim, S., Brzózka, K., Jung, A., Kato, H., Poeck, H., Akira, S., Conzelmann, K.-K., Schlee, M., Endres, S., Hartmann, G., 2006. 5′-Triphosphate RNA is the ligand for RIG-I. Science 314, 994–997.

Hoss, M., 1999. A human DNA editing enzyme homologous to the Escherichia coli DnaQ/MutD protein. EMBO J. 18, 3868–3875.

Huang, X., Yue, Y., Li, D., Zhao, Y., Qiu, L., Chen, J., Pan, Y., Xi, J., Wang, X., Sun, Q., Li, Q., 2016. Antibody-dependent enhancement of dengue virus infection inhibits RLR-mediated Type-I IFN-independent signalling through upregulation of cellular autophagy. Sci. Rep. 6, 22303.

Hugot, J.P., Chamaillard, M., Zouali, H., Lesage, S., Cezard, J.P., Belaiche, J., Almer, S., Tysk, C., O'Morain, C.A., Gassull, M., Binder, V., Finkel, Y., Cortot, A., Modigliani, R., Laurent-Puig, P., Gower-Rousseau, C., Macry, J., Colombel, J.F., Sahbatou, M., Thomas, G., 2001. Association of NOD2 leucine-rich repeat variants with susceptibility to Crohn's disease. Nature 411, 599–603.

Inohara, N., Ogura, Y., Chen, F.F., Muto, A., Nunez, G., 2001. Human Nod1 confers responsiveness to bacterial lipopolysaccharides. J. Biol. Chem. 276, 2551–2554.

Ishikawa, H., Barber, G.N., 2008. STING is an endoplasmic reticulum adaptor that facilitates innate immune signalling. Nature 455, 674–678.

Ishikawa, H., Ma, Z., Barber, G.N., 2009. STING regulates intracellular DNA-mediated, type I interferon-dependent innate immunity. Nature 461, 788–792.

Israel, A., 2010. The IKK complex, a central regulator of NF-κB activation. Cold Spring Harb. Perspect. Biol. 2, a000158.

Jang, M.-A., Kim, E.K., Now, H., Nguyen, N.T.H., Kim, W.-J., Yoo, J.-Y., Lee, J., Jeong, Y.-M., Kim, C.-H., Kim, O.-H., Sohn, S., Nam, S.-H., Hong, Y., Lee, Y.S., Chang, S.-A., Jang, S.Y., Kim, J.-W., Lee, M.-S., Lim, S.Y., Sung, K.-S., Park, K.-T., Kim, B.J., Lee, J.-H., Kim, D.-K., Kee, C., Ki, C.-S., 2015. Mutations in DDX58, which encodes RIG-I, cause atypical Singleton-Merten syndrome. Am. J. Hum. Genet. 96, 266–274.

Janssen, R., Bont, L., Siezen, C.L.E., Hodemaekers, H.M., Ermers, M.J., Doornbos, G., van 't Slot, R., Wijmenga, C., Goeman, J.J., Kimpen, J.L.L., van Houwelingen, H.C., Kimman, T.G., Hoebee, B., 2007. Genetic susceptibility to respiratory syncytial virus bronchiolitis is predominantly associated with innate immune genes. J. Infect. Dis. 196, 826–834.

Jeremiah, N., Neven, B., Gentili, M., Callebaut, I., Maschalidi, S., Stolzenberg, M.-C., Goudin, N., Frémond, M.-L., Nitschke, P., Molina, T.J., Blanche, S., Picard, C., Rice, G.I., Crow, Y.J., Manel, N., Fischer, A., Bader-Meunier, B., Rieux-Laucat, F., 2014. Inherited STING-activating mutation underlies a familial inflammatory syndrome with lupus-like manifestations. J. Clin. Investig. 124, 5516–5520.

Jiang, Y., Zhu, Y., Liu, Z.-J., Ouyang, S., 2017. The emerging roles of the DDX41 protein in immunity and diseases. Protein Cell 8, 83–89.

Jin, L., Waterman, P.M., Jonscher, K.R., Short, C.M., Reisdorph, N.A., Cambier, J.C., 2008. MPYS, a novel membrane tetraspanner, is associated with major histocompatibility complex class II and mediates transduction of apoptotic signals. Mol. Cell. Biol. 28, 5014–5026.

Jinek, M., Chylinski, K., Fonfara, I., Hauer, M., Doudna, J.A., Charpentier, E., 2012. A programmable dual-RNA-guided DNA endonuclease in adaptive bacterial immunity. Science 337, 816–821.

Johnson, K.E., Bottero, V., Flaherty, S., Dutta, S., Singh, V.V., Chandran, B., 2014. IFI16 restricts HSV-1 replication by accumulating on the HSV-1 genome, repressing HSV-1 gene expression, and directly or indirectly modulating histone modifications. PLoS Pathog. 10, e1004503.

Jones, J.W., Kayagaki, N., Broz, P., Henry, T., Newton, K., O'Rourke, K., Chan, S., Dong, J., Qu, Y., Roose-Girma, M., Dixit, V.M., Monack, D.M., 2010. Absent in melanoma 2 is required for innate immune recognition of Francisella tularensis. Proc. Natl. Acad. Sci. U.S.A. 107, 9771–9776.
Jounai, N., Takeshita, F., Kobiyama, K., Sawano, A., Miyawaki, A., Xin, K.-Q., Ishii, K.J., Kawai, T., Akira, S., Suzuki, K., Okuda, K., 2007. The Atg5 Atg12 conjugate associates with innate antiviral immune responses. Proc. Natl. Acad. Sci. U.S.A. 104, 14050–14055.
Junjhon, J., Pennington, J.G., Edwards, T.J., Perera, R., Lanman, J., Kuhn, R.J., 2014. Ultrastructural characterization and three-dimensional architecture of replication sites in dengue virus-infected mosquito cells. J. Virol. 88, 4687–4697.
Junkins, R.D., Gallovic, M.D., Johnson, B.M., Collier, M.A., Watkins-Schulz, R., Cheng, N., David, C.N., McGee, C.E., Sempowski, G.D., Shterev, I., McKinnon, K., Bachelder, E.M., Ainslie, K.M., Ting, J.P.Y., 2018. A robust microparticle platform for a STING-targeted adjuvant that enhances both humoral and cellular immunity during vaccination. J. Control. Release 270, 1–13.
Kadaja, M., Isok-Paas, H., Laos, T., Ustav, E., Ustav, M., 2009. Mechanism of genomic instability in cells infected with the high-risk human papillomaviruses. PLoS Pathog. 5, e1000397.
Kadowaki, N., Antonenko, S., Lau, J.Y.-N., Liu, Y.-J., 2000. Natural interferon α/β-producing cells link innate and adaptive immunity. J. Exp. Med. 192, 219–226.
Kalamvoki, M., Roizman, B., 2014. HSV-1 degrades, stabilizes, requires, or is stung by STING depending on ICP0, the US3 protein kinase, and cell derivation. Proc. Natl. Acad. Sci. U.S.A. 111, E611–E617.
Kang, D.-C., Gopalkrishnan, R.V., Lin, L., Randolph, A., Valerie, K., Pestka, S., Fisher, P.B., 2004. Expression analysis and genomic characterization of human melanoma differentiation associated gene-5, mda-5: a novel type I interferon-responsive apoptosis-inducing gene. Oncogene 23, 1789–1800.
Kanneganti, T.-D., Özören, N., Body-Malapel, M., Amer, A., Park, J.-H., Franchi, L., Whitfield, J., Barchet, W., Colonna, M., Vandenabeele, P., Bertin, J., Coyle, A., Grant, E.P., Akira, S., Núñez, G., 2006. Bacterial RNA and small antiviral compounds activate caspase-1 through cryopyrin/Nalp3. Nature 440, 233–236.
Karaolis, D.K.R., Means, T.K., Yang, D., Takahashi, M., Yoshimura, T., Muraille, E., Philpott, D., Schroeder, J.T., Hyodo, M., Hayakawa, Y., Talbot, B.G., Brouillette, E., Malouin, F., 2007a. Bacterial c-di-GMP is an immunostimulatory molecule. J. Immunol. 178, 2171–2181.
Karaolis, D.K.R., Newstead, M.W., Zeng, X., Hyodo, M., Hayakawa, Y., Bhan, U., Liang, H., Standiford, T.J., 2007b. Cyclic di-GMP stimulates protective innate immunity in bacterial pneumonia. Infect. Immun. 75, 4942–4950.
Kasama, Y., Saito, M., Takano, T., Nishimura, T., Satoh, M., Wang, Z., Ali, S.N.E.S., Harada, S., Kohara, M., Tsukiyama-Kohara, K., 2012. Translocase of outer mitochondrial membrane 70 induces interferon response and is impaired by hepatitis C virus NS3. Virus Res. 163, 405–409.
Kash, J.C., Mühlberger, E., Carter, V., Grosch, M., Perwitasari, O., Proll, S.C., Thomas, M.J., Weber, F., Klenk, H.-D., Katze, M.G., 2006. Global suppression of the host antiviral response by Ebola- and Marburgviruses: increased antagonism of the type I interferon response is associated with enhanced virulence. J. Virol. 80, 3009–3020.
Kato, H., Takeuchi, O., Sato, S., Yoneyama, M., Yamamoto, M., Matsui, K., Uematsu, S., Jung, A., Kawai, T., Ishii, K.J., Yamaguchi, O., Otsu, K., Tsujimura, T., Koh, C.-S., Reis e Sousa, C., Matsuura, Y., Fujita, T., Akira, S., 2006. Differential roles of MDA5 and RIG-I helicases in the recognition of RNA viruses. Nature 441, 101–105.

Kato, H., Takeuchi, O., Mikamo-Satoh, E., Hirai, R., Kawai, T., Matsushita, K., Hiiragi, A., Dermody, T.S., Fujita, T., Akira, S., 2008. Length-dependent recognition of double-stranded ribonucleic acids by retinoic acid-inducible gene-I and melanoma differentiation-associated gene 5. J. Exp. Med. 205, 1601–1610.

Katze, M.G., DeCorato, D., Safer, B., Galabru, J., Hovanessian, A.G., 1987. Adenovirus VAI RNA complexes with the 68 000 Mr protein kinase to regulate its autophosphorylation and activity. EMBO J. 6, 689–697.

Kawai, T., Akira, S., 2009. The roles of TLRs, RLRs and NLRs in pathogen recognition. Int. Immunol. 21, 317–337.

Kawai, T., Sato, S., Ishii, K.J., Coban, C., Hemmi, H., Yamamoto, M., Terai, K., Matsuda, M., Inoue, J., Uematsu, S., Takeuchi, O., Akira, S., 2004. Interferon-alpha induction through Toll-like receptors involves a direct interaction of IRF7 with MyD88 and TRAF6. Nat. Immunol. 5, 1061–1068.

Kawai, T., Takahashi, K., Sato, S., Coban, C., Kumar, H., Kato, H., Ishii, K.J., Takeuchi, O., Akira, S., 2005. IPS-1, an adaptor triggering RIG-I- and Mda5-mediated type I interferon induction. Nat. Immunol. 6, 981–988.

Kawasaki, T., Kawai, T., Akira, S., 2011. Recognition of nucleic acids by pattern-recognition receptors and its relevance in autoimmunity: nucleic acid recognition and autoimmunity. Immunol. Rev. 243, 61–73.

Kerur, N., Veettil, M.V., Sharma-Walia, N., Bottero, V., Sadagopan, S., Otageri, P., Chandran, B., 2011. IFI16 acts as a nuclear pathogen sensor to induce the inflammasome in response to kaposi sarcoma-associated herpesvirus infection. Cell Host Microbe 9, 363–375.

Kim, S., Bauernfeind, F., Ablasser, A., Hartmann, G., Fitzgerald, K.A., Latz, E., Hornung, V., 2010a. Listeria monocytogenes is sensed by the NLRP3 and AIM2 inflammasome. Eur. J. Immunol. 40, 1545–1551.

Kim, T., Pazhoor, S., Bao, M., Zhang, Z., Hanabuchi, S., Facchinetti, V., Bover, L., Plumas, J., Chaperot, L., Qin, J., Liu, Y.J., 2010b. Aspartate-glutamate-alanine-histidine box motif (DEAH)/RNA helicase A helicases sense microbial DNA in human plasmacytoid dendritic cells. Proc. Natl. Acad. Sci. U.S.A. 107, 15181–15186.

Kind, B., Muster, B., Staroske, W., Herce, H.D., Sachse, R., Rapp, A., Schmidt, F., Koss, S., Cardoso, M.C., Lee-Kirsch, M.A., 2014. Altered spatio-temporal dynamics of RNase H2 complex assembly at replication and repair sites in Aicardi–Goutières syndrome. Hum. Mol. Genet. 23, 5950–5960.

Kitajewski, J., Schneider, R.J., Safer, B., Munemitsu, S.M., Samuel, C.E., Thimmappaya, B., Shenk, T., 1986. Adenovirus VAI RNA antagonizes the antiviral action of interferon by preventing activation of the interferon-induced eIF-2 alpha kinase. Cell 45, 195–200.

Klein Klouwenberg, P., Tan, L., Werkman, W., van Bleek, G.M., Coenjaerts, F., 2009. The role of Toll-like receptors in regulating the immune response against respiratory syncytial virus. Crit. Rev. Immunol. 29, 531–550.

Knip, M., Veijola, R., Virtanen, S.M., Hyöty, H., Vaarala, O., Akerblom, H.K., 2005. Environmental triggers and determinants of type 1 diabetes. Diabetes 54 (Suppl. 2), S125–S136.

Komuro, A., Horvath, C.M., 2006. RNA- and virus-independent inhibition of antiviral signaling by RNA helicase LGP2. J. Virol. 80, 12332–12342.

Kondo, T., Kobayashi, J., Saitoh, T., Maruyama, K., Ishii, K.J., Barber, G.N., Komatsu, K., Akira, S., Kawai, T., 2013. DNA damage sensor MRE11 recognizes cytosolic double-stranded DNA and induces type I interferon by regulating STING trafficking. Proc. Natl. Acad. Sci. U.S.A. 110, 2969–2974.

König, N., Fiehn, C., Wolf, C., Schuster, M., Cura Costa, E., Tüngler, V., Alvarez, H.A., Chara, O., Engel, K., Goldbach-Mansky, R., Günther, C., Lee-Kirsch, M.A., 2017. Familial chilblain lupus due to a gain-of-function mutation in STING. Ann. Rheum. Dis. 76, 468–472.

Konno, H., Chinn, I.K., Hong, D., Orange, J.S., Lupski, J.R., Mendoza, A., Pedroza, L.A., Barber, G.N., 2018. Pro-inflammation associated with a gain-of-function mutation (R284S) in the innate immune sensor STING. Cell Rep. 23, 1112–1123.

Kranzusch, P.J., Lee, A.S.-Y., Berger, J.M., Doudna, J.A., 2013. Structure of human cGAS reveals a conserved family of second-messenger enzymes in innate immunity. Cell Rep. 3, 1362–1368.

Kretschmer, S., Wolf, C., König, N., Staroske, W., Guck, J., Häusler, M., Luksch, H., Nguyen, L.A., Kim, B., Alexopoulou, D., Dahl, A., Rapp, A., Cardoso, M.C., Shevchenko, A., Lee-Kirsch, M.A., 2015. SAMHD1 prevents autoimmunity by maintaining genome stability. Ann. Rheum. Dis. 74, e17.

Krug, A., French, A.R., Barchet, W., Fischer, J.A.A., Dzionek, A., Pingel, J.T., Orihuela, M.M., Akira, S., Yokoyama, W.M., Colonna, M., 2004. TLR9-dependent recognition of MCMV by IPC and DC generates coordinated cytokine responses that activate antiviral NK cell function. Immunity 21, 107–119.

Kumar, A., Haque, J., Lacoste, J., Hiscott, J., Williams, B.R., 1994. Double-stranded RNA-dependent protein kinase activates transcription factor NF-kappa B by phosphorylating I kappa B. Proc. Natl. Acad. Sci. U.S.A. 91, 6288–6292.

Kumar, H., Kawai, T., Akira, S., 2009. Toll-like receptors and innate immunity. Biochem. Biophys. Res. Commun. 388, 621–625.

Kumar, M., Roe, K., Nerurkar, P.V., Namekar, M., Orillo, B., Verma, S., Nerurkar, V.R., 2012. Impaired virus clearance, compromised immune response and increased mortality in type 2 diabetic mice infected with west nile virus. PLoS One 7, e44682.

Kumar, A., Hou, S., Airo, A.M., Limonta, D., Mancinelli, V., Branton, W., Power, C., Hobman, T.C., 2016. Zika virus inhibits type-I interferon production and downstream signaling. EMBO Rep. 17, 1766–1775.

Kurien, B.T., D'Sousa, A., Bruner, B.F., Gross, T., James, J.A., Targoff, I.N., Maier-Moore, J.S., Harley, I.T., Wang, H., Scofield, R.H., 2013. Prolidase deficiency breaks tolerance to lupus-associated antigens. Int. J. Rheum. Dis. 16, 674–680.

Labbé, P., Faure, E., Lecointe, S., Le Scouarnec, S., Kyndt, F., Marrec, M., Le Tourneau, T., Offmann, B., Duplaà, C., Zaffran, S., Schott, J.J., Merot, J., 2017. The alternatively spliced LRRFIP1 isoform-1 is a key regulator of the Wnt/β-catenin transcription pathway. Biochim. Biophys Acta 1864, 1142–1152.

Ladiges, W.C., Knoblaugh, S.E., Morton, J.F., Korth, M.J., Sopher, B.L., Baskin, C.R., MacAuley, A., Goodman, A.G., LeBoeuf, R.C., Katze, M.G., 2005. Pancreatic beta-cell failure and diabetes in mice with a deletion mutation of the endoplasmic reticulum molecular chaperone gene P58IPK. Diabetes 54, 1074–1081.

Lahaye, X., Satoh, T., Gentili, M., Cerboni, S., Conrad, C., Hurbain, I., El Marjou, A., Lacabaratz, C., Lelièvre, J.-D., Manel, N., 2013. The capsids of HIV-1 and HIV-2 determine immune detection of the viral cDNA by the innate sensor cGAS in dendritic cells. Immunity 39, 1132–1142.

Lahouassa, H., Daddacha, W., Hofmann, H., Ayinde, D., Logue, E.C., Dragin, L., Bloch, N., Maudet, C., Bertrand, M., Gramberg, T., Pancino, G., Priet, S., Canard, B., Laguette, N., Benkirane, M., Transy, C., Landau, N.R., Kim, B., Margottin-Goguet, F., 2012. SAMHD1 restricts the replication of human immunodeficiency virus type 1 by depleting the intracellular pool of deoxynucleoside triphosphates. Nat. Immunol. 13, 223–228.

Lam, E., Falck-Pedersen, E., 2014. Unabated adenovirus replication following activation of the cGAS/STING-dependent antiviral response in human cells. J. Virol. 88, 14426–14439.

Lam, E., Stein, S., Falck-Pedersen, E., 2014. Adenovirus detection by the cGAS/STING/TBK1 DNA sensing cascade. J. Virol. 88, 974–981.

Landtblom, A.-M., Fazio, P., Fredrikson, S., Granieri, E., 2010. The first case history of multiple sclerosis: Augustus d'Esté (1794–1848). Neurol. Sci. 31, 29–33.

Latz, E., Schoenemeyer, A., Visintin, A., Fitzgerald, K.A., Monks, B.G., Knetter, C.F., Lien, E., Nilsen, N.J., Espevik, T., Golenbock, D.T., 2004. TLR9 signals after translocating from the ER to CpG DNA in the lysosome. Nat. Immunol. 5, 190–198.

Latz, E., Verma, A., Visintin, A., Gong, M., Sirois, C.M., Klein, D.C.G., Monks, B.G., McKnight, C.J., Lamphier, M.S., Duprex, W.P., Espevik, T., Golenbock, D.T., 2007. Ligand-induced conformational changes allosterically activate Toll-like receptor 9. Nat. Immunol. 8, 772–779.

Lazear, H.M., Govero, J., Smith, A.M., Platt, D.J., Fernandez, E., Miner, J.J., Diamond, M.S., 2016. A mouse model of zika virus pathogenesis. Cell Host Microbe 19, 720–730.

Lee, T.G., Tomita, J., Hovanessian, A.G., Katze, M.G., 1990. Purification and partial characterization of a cellular inhibitor of the interferon-induced protein kinase of Mr 68,000 from influenza virus-infected cells. Proc. Natl. Acad. Sci. U.S.A. 87, 6208–6212.

Lee, T.G., Tomita, J., Hovanessian, A.G., Katze, M.G., 1992. Characterization and regulation of the 58,000-dalton cellular inhibitor of the interferon-induced, dsRNA-activated protein kinase. J. Biol. Chem. 267, 14238–14243.

Lee, A.S., Ghoreishi, M., Cheng, W.K., Chang, T.Y., Zhang, Y.Q., Dutz, J.P., 2011. Toll-like receptor 7 stimulation promotes autoimmune diabetes in the NOD mouse. Diabetologia 54, 1407–1416.

Lee, I.-K., Hsieh, C.-J., Chen, R.-F., Yang, Z.-S., Wang, L., Chen, C.-M., Liu, C.-F., Huang, C.-H., Lin, C.-Y., Chen, Y.-H., Yang, K.D., Liu, J.-W., 2013. Increased production of interleukin-4, interleukin-10, and granulocyte-macrophage colony-stimulating factor by type 2 diabetes' mononuclear cells infected with dengue virus, but not increased intracellular viral multiplication. Biomed. Res. Int. 2013, 965853.

Lee, K.-G., Kim, S.S.-Y., Kui, L., Voon, D.C.-C., Mauduit, M., Bist, P., Bi, X., Pereira, N.A., Liu, C., Sukumaran, B., Rénia, L., Ito, Y., Lam, K.-P., 2015. Bruton's tyrosine kinase phosphorylates DDX41 and activates its binding of dsDNA and STING to initiate type 1 interferon response. Cell Rep. 10, 1055–1065.

Lee-Kirsch, M.A., Gong, M., Chowdhury, D., Senenko, L., Engel, K., Lee, Y.A., de Silva, U., Bailey, S.L., Witte, T., Vyse, T.J., Kere, J., Pfeiffer, C., Harvey, S., Wong, A., Koskenmies, S., Hummel, O., Rohde, K., Schmidt, R.E., Dominiczak, A.F., Gahr, M., Hollis, T., Perrino, F.W., Lieberman, J., Hübner, N., 2007. Mutations in the gene encoding the 3'-5' DNA exonuclease TREX1 are associated with systemic lupus erythematosus. Nat. Genet. 39 (9), 1065–1077.

Lehtinen, D.A., Harvey, S., Mulcahy, M.J., Hollis, T., Perrino, F.W., 2008. The TREX1 double-stranded DNA degradation activity is defective in dominant mutations associated with autoimmune disease. J. Biol. Chem. 283, 31649–31656.

Lemos, H., Huang, L., Chandler, P.R., Mohamed, E., Souza, G.R., Li, L., Pacholczyk, G., Barber, G.N., Hayakawa, Y., Munn, D.H., Mellor, A.L., 2014. Activation of the STING adaptor attenuates experimental autoimmune encephalitis. J. Immunol. 192, 5571–5578.

Lenert, P., 2010. Nucleic acid sensing receptors in systemic lupus erythematosus: development of novel DNA- and/or RNA-like analogues for treating lupus. Clin. Exp. Immunol. 161 (2), 208–222.

Leung, L.W., Park, M.S., Martinez, O., Valmas, C., Lopez, C.B., Basler, C.F., 2011. Ebolavirus VP35 suppresses IFN production from conventional but not plasmacytoid dendritic cells. Immunol. Cell Biol. 89, 792–802.

Li, X.-D., Sun, L., Seth, R.B., Pineda, G., Chen, Z.J., 2005. Hepatitis C virus protease NS3/4A cleaves mitochondrial antiviral signaling protein off the mitochondria to evade innate immunity. Proc. Natl. Acad. Sci. U.S.A. 102, 17717–17722.

Li, X.-L., Ezelle, H.J., Hsi, T.Y., Hassel, B.A., 2011. A central role for RNA in the induction and biological activities of type 1 interferons. Wiley Interdiscip. Rev. RNA 2, 58–78.

Li, Y., Berke, I.C., Modis, Y., 2012a. DNA binding to proteolytically activated TLR9 is sequence-independent and enhanced by DNA curvature: ligand-binding study of cleaved TLR9. EMBO J. 31, 919–931.

Li, Y., Chen, R., Zhou, Q., Xu, Z., Li, C., Wang, S., Mao, A., Zhang, X., He, W., Shu, H.B., 2012b. LSm14A is a processing body-associated sensor of viral nucleic acids that initiates cellular antiviral response in the early phase of viral infection. Proc. Natl. Acad. Sci. U.S.A. 109, 11770–11775.

Li, X.D., Wu, J., Gao, D., Wang, H., Sun, L., Chen, Z.J., 2013. Pivotal roles of cGAS-cGAMP signaling in antiviral defense and immune adjuvant effects. Science 341, 1390–1394.

Li, Y., Wu, Y., Zheng, X., Cong, J., Liu, Y., Li, J., Sun, R., Tian, Z.G., Wei, H.M., 2016. Cytoplasm-translocated Ku70/80 complex sensing of HBV DNA induces hepatitis-associated chemokine secretion. Front. Immunol. 7, 569.

Li, C., Tan, X., Zhang, J., Li, S., Mo, W., Han, T., Kuang, W., Zhou, Y., Deng, J., 2017a. Gene mutations and clinical phenotypes in 15 Chinese children with cryopyrin-associated periodic syndrome (CAPS). Sci. China Life Sci. 60, 1436–1444.

Li, P., Du, J., Goodier, J.L., Hou, J., Kang, J., Kazazian Jr., H.H., Zhao, K., Yu, X.F., 2017b. Aicardi-Goutieres syndrome protein TREX1 suppresses L1 and maintains genome integrity through exonuclease-independent ORF1p depletion. Nucleic Acids Res. 45, 4619–4631.

Li, Y., Wilson, H.L., Kiss-Toth, E., 2017c. Regulating STING in health and disease. J. Inflamm. 14, 11.

Liang, D., Xiao-Feng, H., Guan-Jun, D., Er-Ling, H., Sheng, C., Ting-Ting, W., Qin-Gang, H., Yan-Hong, N., Ya-Yi, H., 2015. Activated STING enhances Tregs infiltration in the HPV-related carcinogenesis of tongue squamous cells via the c-jun/CCL22 signal. Biochim. Biophys. Acta 1852, 2494–2503.

Liao, Y.R., Li, Z.J., Zeng, P., Lan, Y.Q., 2017. TLR7 deficiency contributes to attenuated diabetic retinopathy via inhibition of inflammatory response. Biochem. Biophys. Res. Commun. 493, 1136–1142.

Liddicoat, B.J., Piskol, R., Chalk, A.M., Ramaswami, G., Higuchi, M., Hartner, J.C., Li, J.B., Seeburg, P.H., Walkley, C.R., 2015. RNA editing by ADAR1 prevents MDA5 sensing of endogenous dsRNA as nonself. Science 349, 1115–1120.

Lieu, K.G., Marsh, G.A., Wang, L.-F., Netter, H.J., 2015. The non-pathogenic Henipavirus Cedar paramyxovirus phosphoprotein has a compromised ability to target STAT1 and STAT2. Antiviral Res. 124, 69–76.

Lim, W.H., Kireta, S., Russ, G.R., Coates, P.T.H., 2006. Human plasmacytoid dendritic cells regulate immune responses to Epstein-Barr virus (EBV) infection and delay EBV-related mortality in humanized NOD-SCID mice. Blood 109, 1043–1050.

Lin, J., Kumari, S., Kim, C., Van, T.M., Wachsmuth, L., Polykratis, A., Pasparakis, M., 2016. RIPK1 counteracts ZBP1-mediated necroptosis to inhibit inflammation. Nature 540, 124–128.

Lincez, P.J., Shanina, I., Horwitz, M.S., 2015. Reduced expression of the MDA5 Gene IFIH1 prevents autoimmune diabetes. Diabetes 64, 2184–2193.

Lindahl, T., Barnes, D.E., Yang, Y.-G., Robins, P., 2009. Biochemical properties of mammalian TREX1 and its association with DNA replication and inherited inflammatory disease. Biochem. Soc. Trans. 37, 535–538.

Linder, P., 2006. Dead-box proteins: a family affair—active and passive players in RNP-remodeling. Nucleic Acids Res. 34, 4168–4180.

Lio, C.-W.J., McDonald, B., Takahashi, M., Dhanwani, R., Sharma, N., Huang, J., Pham, E., Benedict, C.A., Sharma, S., 2016. cGAS-STING signaling regulates initial innate control of cytomegalovirus infection. J. Virol. 90, 7789–7797.

Lisnevskaia, L., Murphy, G., Isenberg, D., 2014. Systemic lupus erythematosus. Lancet 384, 1878–1888.

Liu, S., Wang, H., Jin, Y., Podolsky, R., Reddy, M.V.P.L., Pedersen, J., Bode, B., Reed, J., Steed, D., Anderson, S., Yang, P., Muir, A., Steed, L., Hopkins, D., Huang, Y., Purohit, S., Wang, C.-Y., Steck, A.K., Montemari, A., Eisenbarth, G., Rewers, M., She, J.-X., 2009. IFIH1 polymorphisms are significantly associated with type 1 diabetes and IFIH1 gene expression in peripheral blood mononuclear cells. Hum. Mol. Genet. 18, 358–365.

Liu, Y., Jesus, A.A., Marrero, B., Yang, D., Ramsey, S.E., Montealegre Sanchez, G.A., Tenbrock, K., Wittkowski, H., Jones, O.Y., Kuehn, H.S., Lee, C.-C.R., DiMattia, M.A., Cowen, E.W., Gonzalez, B., Palmer, I., DiGiovanna, J.J., Biancotto, A., Kim, H., Tsai, W.L., Trier, A.M., Huang, Y., Stone, D.L., Hill, S., Kim, H.J., St. Hilaire, C., Gurprasad, S., Plass, N., Chapelle, D., Horkayne-Szakaly, I., Foell, D., Barysenka, A., Candotti, F., Holland, S.M., Hughes, J.D., Mehmet, H., Issekutz, A.C., Raffeld, M., McElwee, J., Fontana, J.R., Minniti, C.P., Moir, S., Kastner, D.L., Gadina, M., Steven, A.C., Wingfield, P.T., Brooks, S.R., Rosenzweig, S.D., Fleisher, T.A., Deng, Z., Boehm, M., Paller, A.S., Goldbach-Mansky, R., 2014. Activated STING in a vascular and pulmonary syndrome. N. Engl. J. Med. 371, 507–518.

Liu, S., Cai, X., Wu, J., Cong, Q., Chen, X., Li, T., Du, F., Ren, J., Wu, Y.T., Grishin, N.V., Chen, Z.J., 2015a. Phosphorylation of innate immune adaptor proteins MAVS, STING, and TRIF induces IRF3 activation. Science 347, aaa2630.

Liu, Y., Zou, Z., Zhu, B., Hu, Z., Zeng, P., Wu, L., 2015b. LRRFIP1 inhibits hepatitis C virus replication by inducing type I interferon in hepatocytes. Hepat. Mon. 15, e28473.

Liu, T.-T., Yang, Q., Li, M., Zhong, B., Ran, Y., Liu, L.-L., Yang, Y., Wang, Y.-Y., Shu, H.-B., 2016. LSm14A plays a critical role in antiviral immune responses by regulating MITA level in a cell-specific manner. J. Immunol. 196, 5101–5111.

Liu, Y., Seto, N.L., Carmona-Rivera, C., Kaplan, M.J., 2018. Accelerated model of lupus autoimmunity and vasculopathy driven by toll-like receptor 7/9 imbalance. Lupus Sci. Med. 5, e000259.

Lloyd, R.M., Shatkin, A.J., 1992. Translational stimulation by reovirus polypeptide sigma 3: substitution for VAI RNA and inhibition of phosphorylation of the alpha subunit of eukaryotic initiation factor 2. J. Virol. 66, 6878–6884.

Loeb, M., Eskandarian, S., Rupp, M., Fishman, N., Gasink, L., Patterson, J., Bramson, J., Hudson, T.J., Lemire, M., 2011. Genetic variants and susceptibility to neurological complications following West Nile virus infection. J. Infect. Dis. 204, 1031–1037.

Lönnrot, M., Korpela, K., Knip, M., Ilonen, J., Simell, O., Korhonen, S., Savola, K., Muona, P., Simell, T., Koskela, P., Hyöty, H., 2000. Enterovirus infection as a risk factor for beta-cell autoimmunity in a prospectively observed birth cohort: the Finnish diabetes prediction and prevention study. Diabetes 49, 1314–1318.

Loo, Y.-M., Gale, M., 2011. Immune signaling by RIG-I-like receptors. Immunity 34, 680–692.

Loo, Y.-M., Fornek, J., Crochet, N., Bajwa, G., Perwitasari, O., Martinez-Sobrido, L., Akira, S., Gill, M.A., García-Sastre, A., Katze, M.G., Gale, M., 2008. Distinct RIG-I and MDA5 signaling by RNA viruses in innate immunity. J. Virol. 82, 335–345.

Lu, Y., Wambach, M., Katze, M.G., Krug, R.M., 1995. Binding of the influenza virus NS1 protein to double-stranded RNA inhibits the activation of the protein kinase that phosphorylates the elF-2 translation initiation factor. Virology 214, 222–228.

Lubick, K.J., Robertson, S.J., McNally, K.L., Freedman, B.A., Rasmussen, A.L., Taylor, R.T., Walts, A.D., Tsuruda, S., Sakai, M., Ishizuka, M., Boer, E.F., Foster, E.C., Chiramel, A.I., Addison, C.B., Green, R., Kastner, D.L., Katze, M.G., Holland, S.M., Forlino, A., Freeman, A.F., Boehm, M., Yoshii, K., Best, S.M., 2015. Flavivirus antagonism of type I interferon signaling reveals prolidase as a regulator of IFNAR1 surface expression. Cell Host Microbe 18, 61–74.

Ludlow, L.E.A., Johnstone, R.W., Clarke, C.J.P., 2005. The HIN-200 family: more than interferon-inducible genes? Exp. Cell Res. 308, 1–17.

Lundberg, P., Welander, P., Han, X., Cantin, E., 2003. Herpes simplex virus type 1 DNA is immunostimulatory in vitro and in vivo. J. Virol. 77, 11158–11169.

Lundberg, M., Krogvold, L., Kuric, E., Dahl-Jorgensen, K., Skog, O., 2016. Expression of interferon-stimulated genes in insulitic pancreatic islets of patients recently diagnosed with type 1 diabetes. Diabetes 65, 3104–3110.

Luo, H., Winkelmann, E.R., Fernandez-Salas, I., Li, L., Mayer, S.V., Danis-Lozano, R., Sanchez-Casas, R.M., Vasilakis, N., Tesh, R., Barrett, A.D., Weaver, S.C., Wang, T., 2018. Zika, dengue and yellow fever viruses induce differential anti-viral immune responses in human monocytic and first trimester trophoblast cells. Antiviral Res. 151, 55–62.

Luthra, P., Ramanan, P., Mire, C.E., Weisend, C., Tsuda, Y., Yen, B., Liu, G., Leung, D.W., Geisbert, T.W., Ebihara, H., Amarasinghe, G.K., Basler, C.F., 2013. Mutual antagonism between the Ebola virus VP35 protein and the RIG-I activator PACT determines infection outcome. Cell Host Microbe 14, 74–84.

Mackenzie, J.M., Jones, M.K., Young, P.R., 1996. Immunolocalization of the dengue virus nonstructural glycoprotein NS1 suggests a role in viral RNA replication. Virology 220, 232–240.

Mackenzie, K.J., Carroll, P., Lettice, L., Tarnauskaitė, Ž., Reddy, K., Dix, F., Revuelta, A., Abbondati, E., Rigby, R.E., Rabe, B., Kilanowski, F., Grimes, G., Fluteau, A., Devenney, P.S., Hill, R.E., Reijns, M.A.M., Jackson, A.P., 2016. Ribonuclease H2 mutations induce a cGAS/STING-dependent innate immune response. EMBO J. 35, 831–844.

Mackenzie, K.J., Carroll, P., Martin, C.-A., Murina, O., Fluteau, A., Simpson, D.J., Olova, N., Sutcliffe, H., Rainger, J.K., Leitch, A., Osborn, R.T., Wheeler, A.P., Nowotny, M., Gilbert, N., Chandra, T., Reijns, M.A.M., Jackson, A.P., 2017. cGAS surveillance of micronuclei links genome instability to innate immunity. Nature 548, 461–465.

Maelfait, J., Bridgeman, A., Benlahrech, A., Cursi, C., Rehwinkel, J., 2016. Restriction by SAMHD1 limits cGAS/STING-dependent innate and adaptive immune responses to HIV-1. Cell Rep. 16, 1492–1501.

Maelfait, J., Liverpool, L., Bridgeman, A., Ragan, K.B., Upton, J.W., Rehwinkel, J., 2017. Sensing of viral and endogenous RNA by ZBP1/DAI induces necroptosis. EMBO J. 36, 2529–2543.

Man, S.M., Karki, R., Malireddi, R.K.S., Neale, G., Vogel, P., Yamamoto, M., Lamkanfi, M., Kanneganti, T.-D., 2015. The transcription factor IRF1 and guanylate-binding proteins target activation of the AIM2 inflammasome by Francisella infection. Nat. Immunol. 16, 467–475.

Mannion, N.M., Greenwood, S.M., Young, R., Cox, S., Brindle, J., Read, D., Nellaker, C., Vesely, C., Ponting, C.P., McLaughlin, P.J., Jantsch, M.F., Dorin, J., Adams, I.R., Scadden, A.D., Ohman, M., Keegan, L.P., O'Connell, M.A., 2014. The RNA-editing enzyme ADAR1 controls innate immune responses to RNA. Cell Rep. 9, 1482–1494.

Maringer, K., Fernandez-Sesma, A., 2014. Message in a bottle: lessons learned from antagonism of STING signalling during RNA virus infection. Cytokine Growth Factor Rev. 25, 669–679.

Marr, N., Wang, T.-I., Kam, S.H.Y., Hu, Y.S., Sharma, A.A., Lam, A., Markowski, J., Solimano, A., Lavoie, P.M., Turvey, S.E., 2014. Attenuation of respiratory syncytial virus-induced and RIG-I-dependent type I IFN responses in human neonates and very young children. J. Immunol. 192, 948–957.

Martínez-Ramírez, I., del-Castillo-Falconi, V., Mitre-Aguilar, I., Amador-Molina, A., Carrillo-García, A., Langley, E., Zentella-Dehesa, A., Soto-Reyes, E., García-Carrancá,

A., Herrera, L., Lizano, M., 2017. SOX2 as a new regulator of HPV16 transcription. Viruses 9, 175.
Maser, R.S., Monsen, K.J., Nelms, B.E., Petrini, J.H., 1997. hMre11 and hRad50 nuclear foci are induced during the normal cellular response to DNA double-strand breaks. Mol. Cell. Biol. 17, 6087–6096.
Mathes, E., O'Dea, E.L., Hoffmann, A., Ghosh, G., 2008. NF-κB dictates the degradation pathway of IκBα. EMBO J. 27, 1357–1367.
Mathieu, C., Guillaume, V., Volchkova, V.A., Pohl, C., Jacquot, F., Looi, R.Y., Wong, K.T., Legras-Lachuer, C., Volchkov, V.E., Lachuer, J., Horvat, B., 2012. Non-structural Nipah virus C protein regulates both the early host proinflammatory response and viral virulence. J. Virol. 86, 10766–10775.
Matsushima, N., Tanaka, T., Enkhbayar, P., Mikami, T., Taga, M., Yamada, K., Kuroki, Y., 2007. Comparative sequence analysis of leucine-rich repeats (LRRs) within vertebrate toll-like receptors. BMC Genomics 8, 124.
Mazur, D.J., Perrino, F.W., 1999. Identification and expression of the TREX1 and TREX2 cDNA sequences encoding mammalian 3′→5′ exonucleases. J. Biol. Chem. 274, 19655–19660.
Mazur, D.J., Perrino, F.W., 2001. Excision of 3′ termini by the Trex1 and TREX2 3′→5′ exonucleases: characterization of the recombinant proteins. J. Biol. Chem. 276, 17022–17029.
McCartney, S.A., Vermi, W., Lonardi, S., Rossini, C., Otero, K., Calderon, B., Gilfillan, S., Diamond, M.S., Unanue, E.R., Colonna, M., 2011. RNA sensor-induced type I IFN prevents diabetes caused by a β cell–tropic virus in mice. J. Clin. Invest. 121, 1497–1507.
McElroy, A.K., Erickson, B.R., Flietstra, T.D., Rollin, P.E., Nichol, S.T., Towner, J.S., Spiropoulou, C.F., 2014. Ebola hemorrhagic fever: novel biomarker correlates of clinical outcome. J. Infect. Dis. 210, 558–566.
McFarland, H.F., Martin, R., 2007. Multiple sclerosis: a complicated picture of autoimmunity. Nat. Immunol. 8, 913–919.
McWhirter, S.M., Barbalat, R., Monroe, K.M., Fontana, M.F., Hyodo, M., Joncker, N.T., Ishii, K.J., Akira, S., Colonna, M., Chen, Z.J., Fitzgerald, K.A., Hayakawa, Y., Vance, R.E., 2009. A host type I interferon response is induced by cytosolic sensing of the bacterial second messenger cyclic-di-GMP. J. Exp. Med. 206, 1899–1911.
Medeiros, A.C., Soares, C.S., Coelho, P.O., Vieira, N.A., Baqui, M.M.A., Teixeira, F.R., Gomes, M.D., 2018. DNA damage response signaling does not trigger redistribution of SAMHD1 to nuclear foci. Biochem. Biophys. Res. Commun. 499, 790–796.
Medzhitov, R., Preston-Hurlburt, P., Kopp, E., Stadlen, A., Chen, C., Ghosh, S., Janeway, C.A., 1998. MyD88 is an adaptor protein in the hToll/IL-1 receptor family signaling pathways. Mol. Cell 2, 253–258.
Melchjorsen, J., Jensen, S.B., Malmgaard, L., Rasmussen, S.B., Weber, F., Bowie, A.G., Matikainen, S., Paludan, S.R., 2005. Activation of innate defense against a paramyxovirus is mediated by RIG-I and TLR7 and TLR8 in a cell-type-specific manner. J. Virol. 79, 12944–12951.
Melki, I., Rose, Y., Uggenti, C., Van Eyck, L., Frémond, M.-L., Kitabayashi, N., Rice, G.I., Jenkinson, E.M., Boulai, A., Jeremiah, N., Gattorno, M., Volpi, S., Sacco, O., Terheggen-Lagro, S.W.J., Tiddens, H.A.W.M., Meyts, I., Morren, M.-A., De Haes, P., Wouters, C., Legius, E., Corveleyn, A., Rieux-Laucat, F., Bodemer, C., Callebaut, I., Rodero, M.P., Crow, Y.J., 2017. Disease-associated mutations identify a novel region in human STING necessary for the control of type I interferon signaling. J. Allergy Clin. Immunol. 140. 543–552.e5.
Melville, M.W., Tan, S.L., Wambach, M., Song, J., Morimoto, R.I., Katze, M.G., 1999. The cellular inhibitor of the PKR protein kinase, P58(IPK), is an influenza virus-activated co-chaperone that modulates heat shock protein 70 activity. J. Biol. Chem. 274, 3797–3803.

Merkl, P.E., Orzalli, M., Knipe, D.M., 2018. Mechanisms of host IFI16, PML and Daxx protein restriction of herpes simplex virus 1 replication. J. Virol. 92, e00057-18.
Meunier, E., Wallet, P., Dreier, R.F., Costanzo, S., Anton, L., Rühl, S., Dussurgey, S., Dick, M.S., Kistner, A., Rigard, M., Degrandi, D., Pfeffer, K., Yamamoto, M., Henry, T., Broz, P., 2015. Guanylate-binding proteins promote activation of the AIM2 inflammasome during infection with Francisella novicida. Nat. Immunol. 16, 476–484.
Mi, J., Ray, P., Liu, J., Kuan, C.-T., Xu, J., Hsu, D., Sullenger, B.A., White, R.R., Clary, B.M., 2016. In vivo selection against human colorectal cancer xenografts identifies an aptamer that targets RNA helicase protein DHX9. Mol. Ther. Nucleic Acids 5, e315.
Miller, S., Kastner, S., Krijnse-Locker, J., Buhler, S., Bartenschlager, R., 2007. The non-structural protein 4A of dengue virus is an integral membrane protein inducing membrane alterations in a 2K-regulated manner. J. Biol. Chem. 282, 8873–8882.
Milutin Gašperov, N., Farkas, S.A., Nilsson, T.K., Grce, M., 2014. Epigenetic activation of immune genes in cervical cancer. Immunol. Lett. 162, 256–257.
Mimori, T., Hardin, J.A., Steitz, J.A., 1986. Characterization of the DNA-binding protein antigen Ku recognized by autoantibodies from patients with rheumatic disorders. J. Biol. Chem. 261, 2274–2278.
Minagar, A., 2014. Multiple sclerosis: an overview of clinical features, pathophysiology, neuroimaging, and treatment options. Colloq. Ser. Integr. Syst. Physiol. Mol. Funct. 6, 1–117.
Miyabe, H., Hyodo, M., Nakamura, T., Sato, Y., Hayakawa, Y., Harashima, H., 2014. A new adjuvant delivery system 'cyclic di-GMP/YSK05 liposome' for cancer immunotherapy. J. Control. Release 184, 20–27.
Mondini, M., Vidali, M., Andrea, M.D., Azzimonti, B., Airò, P., D'Ambrosio, R., Riboldi, P., Meroni, P.L., Albano, E., Shoenfeld, Y., Gariglio, M., Landolfo, S., 2006. A novel autoantigen to differentiate limited cutaneous systemic sclerosis from diffuse cutaneous systemic sclerosis: the interferon-inducible gene IFI16. Arthritis Rheum. 54, 3939–3944.
Moretti, J., Roy, S., Bozec, D., Martinez, J., Chapman, J.R., Ueberheide, B., Lamming, D.W., Chen, Z.J., Horng, T., Yeretssian, G., Green, D.R., Blander, J.M., 2017. STING senses microbial viability to orchestrate stress-mediated autophagy of the endoplasmic reticulum. Cell 171. 809–823.e13.
Moriyama, M., Kato, N., Otsuka, M., Shao, R.-X., Taniguchi, H., Kawabe, T., Omata, M., 2007. Interferon-beta is activated by hepatitis C virus NS5B and inhibited by NS4A, NS4B, and NS5A. Hepatol. Int. 1, 302–310.
Morosky, S.A., Zhu, J., Mukherjee, A., Sarkar, S.N., Coyne, C.B., 2011. Retinoic acid-induced gene-I (RIG-I) associates with nucleotide-binding oligomerization domain-2 (NOD2) to negatively regulate inflammatory signaling. J. Biol. Chem. 286, 28574–28583.
Morrison, J., Laurent-Rolle, M., Maestre, A.M., Rajsbaum, R., Pisanelli, G., Simon, V., Mulder, L.C.F., Fernandez-Sesma, A., García-Sastre, A., 2013. Dengue virus co-opts UBR4 to degrade STAT2 and antagonize type I interferon signaling. PLoS Pathog. 9, e1003265.
Mukherjee, S., Lukacs, N.W., 2013. Innate immune responses to respiratory syncytial virus infection. Curr. Top. Microbiol. Immunol. 372, 139–154.
Munoz, J., Rodière, M., Jeremiah, N., Rieux-Laucat, F., Oojageer, A., Rice, G.I., Rozenberg, F., Crow, Y.J., Bessis, D., 2015. Stimulator of interferon genes-associated vasculopathy with onset in infancy: a mimic of childhood granulomatosis with polyangiitis. JAMA Dermatol. 151, 872.
Munoz-Culla, M., Irizar, H., Otaegui, D., 2013. The genetics of multiple sclerosis: review of current and emerging candidates. Appl. Clin. Genet. 6, 63–73.

Muñoz-Jordán, J.L., Sánchez-Burgos, G.G., Laurent-Rolle, M., García-Sastre, A., 2003. Inhibition of interferon signaling by dengue virus. Proc. Natl. Acad. Sci. U.S.A. 100, 14333–14338.

Murakami, Y., Fukui, R., Motoi, Y., Kanno, A., Shibata, T., Tanimura, N., Saitoh, S., Miyake, K., 2014. Roles of the cleaved N-terminal TLR3 fragment and cell surface TLR3 in double-stranded RNA sensing. J. Immunol. 193, 5208–5217.

Murali, A., Li, X., Ranjith-Kumar, C.T., Bhardwaj, K., Holzenburg, A., Li, P., Kao, C.C., 2008. Structure and function of LGP2, a DEX(D/H) helicase that regulates the innate immunity response. J. Biol. Chem. 283, 15825–15833.

Murayama, G., Furusawa, N., Chiba, A., Yamaji, K., Tamura, N., Miyake, S., 2017. Enhanced IFN-alpha production is associated with increased TLR7 retention in the lysosomes of palasmacytoid dendritic cells in systemic lupus erythematosus. Arthritis Res. Ther. 19, 234.

Murugaiyan, G., Beynon, V., Mittal, A., Joller, N., Weiner, H.L., 2011. Silencing MicroRNA-155 ameliorates experimental autoimmune encephalomyelitis. J. Immunol. 187, 2213–2221.

Murugesapillai, D., McCauley, M.J., Maher, L.J., Williams, M.C., 2017. Single-molecule studies of high-mobility group B architectural DNA bending proteins. Biophys. Rev. 9, 17–40.

Muruve, D.A., Pétrilli, V., Zaiss, A.K., White, L.R., Clark, S.A., Ross, P.J., Parks, R.J., Tschopp, J., 2008. The inflammasome recognizes cytosolic microbial and host DNA and triggers an innate immune response. Nature 452, 103–107.

Mutlu, S., Scully, C., 1993. The person behind the eponym: Henrik Sjogren (1899–1986). J. Oral Pathol. Med. 22, 439.

Nakaya, Y., Lilue, J., Stavrou, S., Moran, E.A., Ross, S.R., 2017. AIM2-like receptors positively and negatively regulate the interferon response induced by cytosolic DNA. mBio 8, e00944-17.

Namjou, B., Kothari, P.H., Kelly, J.A., Glenn, S.B., Ojwang, J.O., Adler, A., Alarcón-Riquelme, M.E., Gallant, C.J., Boackle, S.A., Criswell, L.A., Kimberly, R.P., Brown, E., Edberg, J., Stevens, A.M., Jacob, C.O., Tsao, B.P., Gilkeson, G.S., Kamen, D.L., Merrill, J.T., Petri, M., Goldman, R.R., Vila, L.M., Anaya, J.M., Niewold, T.B., Martin, J., Pons-Estel, B.A., Sabio, J.M., Callejas, J.L., Vyse, T.J., Bae, S.C., Perrino, F.W., Freedman, B.I., Scofield, R.H., Moser, K.L., Gaffney, P.M., James, J.A., Langefeld, C.D., Kaufman, K.M., Harley, J.B., Atkinson, J.P., 2011. Evaluation of the TREX1 gene in a large multi-ancestral lupus cohort. Genes Immun. 12, 270–279.

Nasirudeen, A.M.A., Wong, H.H., Thien, P., Xu, S., Lam, K.-P., Liu, D.X., 2011. RIG-I, MDA5 and TLR3 synergistically play an important role in restriction of dengue virus infection. PLoS Negl. Trop. Dis. 5, e926.

Nejentsev, S., Walker, N., Riches, D., Egholm, M., Todd, J.A., 2009. Rare variants of IFIH1, a gene implicated in antiviral responses, protect against type 1 diabetes. Science 324, 387.

Newton, K., Wickliffe, K.E., Maltzman, A., Dugger, D.L., Strasser, A., Pham, V.C., Lill, J.R., Roose-Girma, M., Warming, S., Solon, M., Ngu, H., Webster, J.D., Dixit, V.M., 2016. RIPK1 inhibits ZBP1-driven necroptosis during development. Nature 540, 129–133.

Nguyen, J.B., Modis, Y., 2013. Crystal structure of the dimeric coiled-coil domain of the cytosolic nucleic acid sensor LRRFIP1. J. Struct. Biol. 181, 82–88.

Nicholas, S.A., Bubnov, V.V., Yasinska, I.M., Sumbayev, V.V., 2011. Involvement of xanthine oxidase and hypoxia-inducible factor 1 in Toll-like receptor 7/8-mediated activation of caspase 1 and interleukin-1β. Cell. Mol. Life Sci. 68, 151–158.

Nielsen, A.E., Hantho, J.D., Mancini, R.J., 2017. Synthetic agonists of NOD-like, RIG-I-like, and C-type lectin receptors for probing the inflammatory immune response. Future Med. Chem. 9, 1345–1360.

Nitta, S., Sakamoto, N., Nakagawa, M., Kakinuma, S., Mishima, K., Kusano-Kitazume, A., Kiyohashi, K., Murakawa, M., Nishimura-Sakurai, Y., Azuma, S., Tasaka-Fujita, M., Asahina, Y., Yoneyama, M., Fujita, T., Watanabe, M., 2013. Hepatitis C virus NS4B protein targets STING and abrogates RIG-I-mediated type I interferon-dependent innate immunity. Hepatology 57, 46–58.

Novak, N., Koch, S., Allam, J.P., Bieber, T., 2010. Dendritic cells: bridging innate and adaptive immunity in atopic dermatitis. J. Allergy Clin. Immunol. 125, 50–59.

Ogunjimi, B., Zhang, S.-Y., Sørensen, K.B., Skipper, K.A., Carter-Timofte, M., Kerner, G., Luecke, S., Prabakaran, T., Cai, Y., Meester, J., Bartholomeus, E., Bolar, N.A., Vandeweyer, G., Claes, C., Sillis, Y., Lorenzo, L., Fiorenza, R.A., Boucherit, S., Dielman, C., Heynderickx, S., Elias, G., Kurotova, A., Auwera, A.V., Verstraete, L., Lagae, L., Verhelst, H., Jansen, A., Ramet, J., Suls, A., Smits, E., Ceulemans, B., Van Laer, L., Plat Wilson, G., Kreth, J., Picard, C., Von Bernuth, H., Fluss, J., Chabrier, S., Abel, L., Mortier, G., Fribourg, S., Mikkelsen, J.G., Casanova, J.-L., Paludan, S.R., Mogensen, T.H., 2017. Inborn errors in RNA polymerase III underlie severe varicella zoster virus infections. J. Clin. Investig. 127, 3543–3556.

Ogura, Y., Bonen, D.K., Inohara, N., Nicolae, D.L., Chen, F.F., Ramos, R., Britton, H., Moran, T., Karaliuskas, R., Duerr, R.H., Achkar, J.P., Brant, S.R., Bayless, T.M., Kirschner, B.S., Hanauer, S.B., Nunez, G., Cho, J.H., 2001a. A frameshift mutation in NOD2 associated with susceptibility to Crohn's disease. Nature 411, 603–606.

Ogura, Y., Inohara, N., Benito, A., Chen, F.F., Yamaoka, S., Nunez, G., 2001b. Nod2, a Nod1/Apaf-1 family member that is restricted to monocytes and activates NF-kappaB. J. Biol. Chem. 276, 4812–4818.

Ohman, L., Franzén, L., Rudolph, U., Birnbaumer, L., Hörnquist, E.H., 2002. Regression of Peyer's patches in G alpha i2 deficient mice prior to colitis is associated with reduced expression of Bcl-2 and increased apoptosis. Gut 51, 392–397.

Ohnishi, H., Tochio, H., Kato, Z., Orii, K.E., Li, A., Kimura, T., Hiroaki, H., Kondo, N., Shirakawa, M., 2009. Structural basis for the multiple interactions of the MyD88 TIR domain in TLR4 signaling. Proc. Natl. Acad. Sci. U.S.A. 106, 10260–10265.

Ohto, U., Shibata, T., Tanji, H., Ishida, H., Krayukhina, E., Uchiyama, S., Miyake, K., Shimizu, T., 2015. Structural basis of CpG and inhibitory DNA recognition by Toll-like receptor 9. Nature 520, 702–705.

Ohto, U., Ishida, H., Shibata, T., Sato, R., Miyake, K., Shimizu, T., 2018. Toll-like receptor 9 contains two DNA binding sites that function cooperatively to promote receptor dimerization and activation. Immunity 48. 649-658.e4.

Oikarinen, S., Tauriainen, S., Hober, D., Lucas, B., Vazeou, A., Sioofy-Khojine, A., Bozas, E., Muir, P., Honkanen, H., Ilonen, J., Knip, M., Keskinen, P., Saha, M.-T., Huhtala, H., Stanway, G., Bartsocas, C., Ludvigsson, J., Taylor, K., Hyöty, H., VirDiab Study, G., 2014. Virus antibody survey in different European populations indicates risk association between coxsackievirus B1 and type 1 diabetes. Diabetes 63, 655–662.

Okahira, S., Nishikawa, F., Nishikawa, S., Akazawa, T., Seya, T., Matsumoto, M., 2005. Interferon-β induction through toll-like receptor 3 depends on double-stranded RNA structure. DNA Cell Biol. 24, 614–623.

Olivieri, I., Cattalini, M., Tonduti, D., Piana, R.L., Uggetti, C., Galli, J., Meini, A., Tincani, A., Moratto, D., Fazzi, E., Balottin, U., Orcesi, S., 2013. Dysregulation of the immune system in Aicardi-Goutières syndrome: another example in a TREX1-mutated patient. Lupus 22, 1064–1069.

Omura, H., Oikawa, D., Nakane, T., Kato, M., Ishii, R., Ishitani, R., Tokunaga, F., Nureki, O., 2016. Structural and functional analysis of DDX41: a bispecific immune receptor for DNA and cyclic dinucleotide. Sci. Rep. 6.

Opoku-Temeng, C., Zhou, J., Zheng, Y., Su, J., Sintim, H.O., 2016. Cyclic dinucleotide (c-di-GMP, c-di-AMP, and cGAMP) signalings have come of age to be inhibited by small molecules. Chem. Commun. 52, 9327–9342.

Orebaugh, C.D., Fye, J.M., Harvey, S., Hollis, T., Perrino, F.W., 2011. The TREX1 exonuclease R114H mutation in Aicardi-Goutières syndrome and lupus reveals dimeric structure requirements for DNA degradation activity. J. Biol. Chem. 286, 40246–40254.
Orzalli, M.H., DeLuca, N.A., Knipe, D.M., 2012. Nuclear IFI16 induction of IRF-3 signaling during herpesviral infection and degradation of IFI16 by the viral ICP0 protein. Proc. Natl. Acad. Sci. U.S.A. 109 (44), E3008–E3017.
Oshiumi, H., Matsumoto, M., Funami, K., Akazawa, T., Seya, T., 2003. TICAM-1, an adaptor molecule that participates in toll-like receptor 3-mediated interferon-β induction. Nat. Immunol. 4, 161–167.
Palomino-Morales, R.J., Oliver, J., Gómez-García, M., López-Nevot, M.A., Rodrigo, L., Nieto, A., Alizadeh, B.Z., Martín, J., 2009. Association of ATG16L1 and IRGM genes polymorphisms with inflammatory bowel disease: a meta-analysis approach. Genes Immun. 10, 356–364.
Paludan, S.R., Bowie, A.G., 2013. Immune sensing of DNA. Immunity 38, 870–880.
Pancholi, N.J., Weitzman, M.D., 2018. Serotype-specific restriction of wild-type adenoviruses by the cellular Mre11-Rad50-Nbs1 complex. Virology 518, 221–231.
Panda, S.K., Facchinetti, V., Voynova, E., Hanabuchi, S., Karnell, J.L., Hanna, R.N., Kolbeck, R., Sanjuan, M.A., Ettinger, R., Liu, Y.J., 2018. Galectin-9 inhibits TLR7-mediated autoimmunity in murine lupus models. J. Clin. Invest. 128, 1873–1887.
Pandit, R.A., Chen, C.-J., Butt, T.A., Islam, N., 2013. Identification and analysis of a novel mutation in PEPD gene in two Kashmiri siblings with prolidase enzyme deficiency. Gene 516, 316–319.
Papinska, J., Bagavant, H., Gmyrek, G.B., Sroka, M., Tummala, S., Fitzgerald, K.A., Deshmukh, U.S., 2018. Activation of stimulator of interferon genes (STING) and Sjögren syndrome. J. Dent. Res. 97, 893–900.
Park, B., Brinkmann, M.M., Spooner, E., Lee, C.C., Kim, Y.-M., Ploegh, H.L., 2008. Proteolytic cleavage in an endolysosomal compartment is required for activation of Toll-like receptor 9. Nat. Immunol. 9, 1407–1414.
Parvatiyar, K., Zhang, Z., Teles, R.M., Ouyang, S., Jiang, Y., Iyer, S.S., Zaver, S.A., Schenk, M., Zeng, S., Zhong, W., Liu, Z.-J., Modlin, R.L., Liu, Y.-J., Cheng, G., 2012. The helicase DDX41 recognizes the bacterial secondary messengers cyclic di-GMP and cyclic di-AMP to activate a type I interferon immune response. Nat. Immunol. 13, 1155–1161.
Pattabhi, S., Wilkins, C.R., Dong, R., Knoll, M.L., Posakony, J., Kaiser, S., Mire, C.E., Wang, M.L., Ireton, R.C., Geisbert, T.W., Bedard, K.M., Iadonato, S.P., Loo, Y.M., Gale Jr., M., 2015. Targeting innate immunity for antiviral therapy through small molecule agonists of the RLR pathway. J. Virol. 90, 2372–2387.
Paull, T.T., Deshpande, R.A., 2014. The Mre11/Rad50/Nbs1 complex: recent insights into catalytic activities and ATP-driven conformational changes. Exp. Cell Res. 329, 139–147.
Peschke, K., Achleitner, M., Frenzel, K., Gerbaulet, A., Ada, S.R., Zeller, N., Lienenklaus, S., Lesche, M., Poulet, C., Naumann, R., Dahl, A., Ravens, U., Günther, C., Müller, W., Knobeloch, K.-P., Prinz, M., Roers, A., Behrendt, R., 2016. Loss of Trex1 in dendritic cells is sufficient to trigger systemic autoimmunity. J. Immunol. 197, 2157–2166.
Pestal, K., Funk, C.C., Snyder, J.M., Price, N.D., Treuting, P.M., Stetson, D.B., 2015. Isoforms of RNA-editing enzyme ADAR1 independently control nucleic acid sensor MDA5-driven autoimmunity and multi-organ development. Immunity 43, 933–944.
Pétrilli, V., Dostert, C., Muruve, D.A., Tschopp, J., 2007. The inflammasome: a danger sensing complex triggering innate immunity. Curr. Opin. Immunol. 19, 615–622.
Picard, C., Thouvenin, G., Kannengiesser, C., Dubus, J.-C., Jeremiah, N., Rieux-Laucat, F., Crestani, B., Belot, A., Thivolet-Béjui, F., Secq, V., Ménard, C., Reynaud-Gaubert, M., Reix, P., 2016. Severe pulmonary fibrosis as the first manifestation of interferonopathy (TMEM173 mutation). Chest 150, e65–e71.

Pichlmair, A., Schulz, O., Tan, C.P., Näslund, T.I., Liljeström, P., Weber, F., Reis e Sousa, C., 2006. RIG-I-mediated antiviral responses to single-stranded RNA bearing 5′-phosphates. Science 314, 997–1001.

Pine, R., Decker, T., Kessler, D.S., Levy, D.E., Darnell, J.E., 1990. Purification and cloning of interferon-stimulated gene factor 2 (ISGF2): ISGF2 (IRF-1) can bind to the promoters of both beta interferon- and interferon-stimulated genes but is not a primary transcriptional activator of either. Mol. Cell. Biol. 10, 2448–2457.

Pinto, A.K., Ramos, H.J., Wu, X., Aggarwal, S., Shrestha, B., Gorman, M., Kim, K.Y., Suthar, M.S., Atkinson, J.P., Gale Jr., M., Diamond, M.S., 2014. Deficient IFN signaling by myeloid cells leads to MAVS-dependent virus-induced sepsis. PLoS Pathog. 10, e1004086.

Piret, J., Carbonneau, J., Rhéaume, C., Baz, M., Boivin, G., 2018. Predominant role of IPS-1 over TRIF adaptor proteins in early innate immune response against Zika virus in mice. J. Gen. Virol. 99, 209–218.

Pisano, G., Roy, A., Ahmed Ansari, M., Kumar, B., Chikoti, L., Chandran, B., 2017. Interferon-γ-inducible protein 16 (IFI16) is required for the maintenance of Epstein-Barr virus latency. Virol. J. 14, 221.

Pizzi, S., Sertic, S., Orcesi, S., Cereda, C., Bianchi, M., Jackson, A.P., Lazzaro, F., Plevani, P., Muzi-Falconi, M., 2015. Reduction of hRNase H2 activity in Aicardi–Goutières syndrome cells leads to replication stress and genome instability. Hum. Mol. Genet. 24, 649–658.

Plantinga, T.S., Joosten, L.A., van der Meer, J.W., Netea, M.G., 2012. Modulation of inflammation by autophagy: consequences for Crohn's disease. Curr. Opin. Pharmacol. 12, 497–502.

Pohar, J., Kuznik Krajnik, A., Jerala, R., Bencina, M., 2015a. Minimal sequence requirements for oligodeoxyribonucleotides activating human TLR9. J. Immunol. 194, 3901–3908.

Pohar, J., Lainscek, D., Fukui, R., Yamamoto, C., Miyake, K., Jerala, R., Bencina, M., 2015b. Species-specific minimal sequence motif for oligodeoxyribonucleotides activating mouse TLR9. J. Immunol. 195, 4396–4405.

Pohar, J., Lainscek, D., Ivicak-Kocjan, K., Cajnko, M.M., Jerala, R., Bencina, M., 2017. Short single-stranded DNA degradation products augment the activation of Toll-like receptor 9. Nat. Commun. 8, 15363.

Pokatayev, V., Hasin, N., Chon, H., Cerritelli, S.M., Sakhuja, K., Ward, J.M., Morris, H.D., Yan, N., Crouch, R.J., 2016. RNase H2 catalytic core Aicardi-Goutières syndrome-related mutant invokes cGAS–STING innate immune-sensing pathway in mice. J. Exp. Med. 213, 329–336.

Pontillo, A., Girardelli, M., Catamo, E., Duarte, A.J., Crovella, S., 2013. Polymorphisms in TREX1 and susceptibility to HIV-1 infection. Int. J. Immunogenet. 40, 492–494.

Powell, R.D., Holland, P.J., Hollis, T., Perrino, F.W., 2011. Aicardi-Goutières syndrome gene and HIV-1 restriction factor SAMHD1 is a dGTP-regulated deoxynucleotide triphosphohydrolase. J. Biol. Chem. 286, 43596–43600.

Prins, K.C., Binning, J.M., Shabman, R.S., Leung, D.W., Amarasinghe, G.K., Basler, C.F., 2010. Basic residues within the ebolavirus VP35 protein are required for its viral polymerase cofactor function. J. Virol. 84, 10581–10591.

Puttur, F., Francozo, M., Solmaz, G., Bueno, C., Lindenberg, M., Gohmert, M., Swallow, M., Tufa, D., Jacobs, R., Lienenklaus, S., Kühl, A.A., Borkner, L., Cicin-Sain, L., Holzmann, B., Wagner, H., Berod, L., Sparwasser, T., 2016. Conventional dendritic cells confer protection against mouse cytomegalovirus infection via TLR9 and MyD88 signaling. Cell Rep. 17, 1113–1127.

Qi, F., Wang, D., Liu, J., Zeng, S., Xu, L., Hu, H., Liu, B., 2015. Respiratory macrophages and dendritic cells mediate respiratory syncytial virus-induced IL-33 production in TLR3- or TLR7-dependent manner. Int. Immunopharmacol. 29, 408–415.

Qian, F., Thakar, J., Yuan, X., Nolan, M., Murray, K.O., Lee, W.T., Wong, S.J., Meng, H., Fikrig, E., Kleinstein, S.H., Montgomery, R.R., 2014. Immune markers associated with host susceptibility to infection with West Nile virus. Viral Immunol. 27, 39–47.

Qin, L., Ren, L., Zhou, Z., Lei, X., Chen, L., Xue, Q., Liu, X., Wang, J., Hung, T., 2011. Rotavirus nonstructural protein 1 antagonizes innate immune response by interacting with retinoic acid inducible gene I. Virol. J. 8, 526.

Rahman, M.M., Bagdassarian, E., Ali, M.A.M., McFadden, G., 2017. Identification of host DEAD-box RNA helicases that regulate cellular tropism of oncolytic Myxoma virus in human cancer cells. Sci. Rep. 7, 15710.

Ramantani, G., Kohlhase, J., Hertzberg, C., Innes, A.M., Engel, K., Hunger, S., Borozdin, W., Mah, J.K., Ungerath, K., Walkenhorst, H., Richardt, H.-H., Buckard, J., Bevot, A., Siegel, C., von Stülpnagel, C., Ikonomidou, C., Thomas, K., Proud, V., Niemann, F., Wieczorek, D., Häusler, M., Niggemann, P., Baltaci, V., Conrad, K., Lebon, P., Lee-Kirsch, M.A., 2010. Expanding the phenotypic spectrum of lupus erythematosus in Aicardi-Goutières syndrome. Arthritis Rheum. 62, 1469–1477.

Ramaswamy, M., Shi, L., Monick, M.M., Hunninghake, G.W., Look, D.C., 2004. Specific inhibition of type I interferon signal transduction by respiratory syncytial virus. Am. J. Respir. Cell Mol. Biol. 30, 893–900.

Ramaswamy, M., Shi, L., Varga, S.M., Barik, S., Behlke, M.A., Look, D.C., 2006. Respiratory syncytial virus nonstructural protein 2 specifically inhibits type I interferon signal transduction. Virology 344, 328–339.

Ramos, H.J., Lanteri, M.C., Blahnik, G., Negash, A., Suthar, M.S., Brassil, M.M., Sodhi, K., Treuting, P.M., Busch, M.P., Norris, P.J., Gale Jr., M.G., 2012. IL-1β signaling promotes CNS-intrinsic immune control of West Nile virus infection. PLoS Pathog. 8, e1003039.

Rasaiyaah, J., Tan, C.P., Fletcher, A.J., Price, A.J., Blondeau, C., Hilditch, L., Jacques, D.A., Selwood, D.L., James, L.C., Noursadeghi, M., Towers, G.J., 2013. HIV-1 evades innate immune recognition through specific cofactor recruitment. Nature 503, 402–405.

Rasschaert, J., Ladriere, L., Urbain, M., Dogusan, Z., Katabua, B., Sato, S., Akira, S., Gysemans, C., Mathieu, C., Eizirik, D.L., 2005. Toll-like receptor 3 and STAT-1 contribute to double-stranded RNA+ interferon-gamma-induced apoptosis in primary pancreatic beta-cells. J. Biol. Chem. 280, 33984–33991.

Reijns, M.A.M., Bubeck, D., Gibson, L.C.D., Graham, S.C., Baillie, G.S., Jones, E.Y., Jackson, A.P., 2011. The structure of the human RNase H2 complex defines key interaction interfaces relevant to enzyme function and human disease. J. Biol. Chem. 286, 10530–10539.

Reijns, M.A.M., Rabe, B., Rigby, R.E., Mill, P., Astell, K.R., Lettice, L.A., Boyle, S., Leitch, A., Keighren, M., Kilanowski, F., Devenney, P.S., Sexton, D., Grimes, G., Holt, I.J., Hill, R.E., Taylor, M.S., Lawson, K.A., Dorin, J.R., Jackson, A.P., 2012. Enzymatic removal of ribonucleotides from DNA is essential for mammalian genome integrity and development. Cell 149, 1008–1022.

Rice, G., Patrick, T., Parmar, R., Taylor, C.F., Aeby, A., Aicardi, J., Artuch, R., Montalto, S.A., Bacino, C.A., Barroso, B., Baxter, P., Benko, W.S., Bergmann, C., Bertini, E., Biancheri, R., Blair, E.M., Blau, N., Bonthron, D.T., Briggs, T., Brueton, L.A., Brunner, H.G., Burke, C.J., Carr, I.M., Carvalho, D.R., Chandler, K.E., Christen, H.-J., Corry, P.C., Cowan, F.M., Cox, H., D'Arrigo, S., Dean, J., De Laet, C., De Praeter, C., Déry, C., Ferrie, C.D., Flintoff, K., Frints, S.G.M., Garcia-Cazorla, A., Gener, B., Goizet, C., Goutières, F., Green, A.J., Guët, A., Hamel, B.C.J., Hayward, B.E., Heiberg, A., Hennekam, R.C., Husson, M., Jackson, A.P., Jayatunga, R., Jiang, Y.-H., Kant, S.G., Kao, A., King, M.D., Kingston, H.M., Klepper, J., van der Knaap, M.S., Kornberg, A.J., Kotzot, D., Kratzer, W., Lacombe, D., Lagae, L., Landrieu, P.G., Lanzi, G.,

Leitch, A., Lim, M.J., Livingston, J.H., Lourenco, C.M., Lyall, E.G.H., Lynch, S.A., Lyons, M.J., Marom, D., McClure, J.P., McWilliam, R., Melancon, S.B., Mewasingh, L.D., Moutard, M.-L., Nischal, K.K., Østergaard, J.R., Prendiville, J., Rasmussen, M., Rogers, R.C., Roland, D., Rosser, E.M., Rostasy, K., Roubertie, A., Sanchis, A., Schiffmann, R., Scholl-Bürgi, S., Seal, S., Shalev, S.A., Corcoles, C.S., Sinha, G.P., Soler, D., Spiegel, R., Stephenson, J.B.P., Tacke, U., Tan, T.Y., Till, M., Tolmie, J.L., et al., 2007. Clinical and molecular phenotype of Aicardi-Goutières syndrome. Am. J. Hum. Genet. 81, 713–725.

Rice, G.I., Bond, J., Asipu, A., Brunette, R.L., Manfield, I.W., Carr, I.M., Fuller, J.C., Jackson, R.M., Lamb, T., Briggs, T.A., Ali, M., Gornall, H., Couthard, L.R., Aeby, A., Attard-Montalto, S.P., Bertini, E., Bodemer, C., Brockmann, K., Brueton, L.A., Corry, P.C., Desguerre, I., Fazzi, E., Cazorla, A.G., Gener, B., Hamel, B.C.J., Heiberg, A., Hunter, M., van der Knaap, M.S., Kumar, R., Lagae, L., Landrieu, P.G., Lourenco, C.M., Marom, D., McDermott, M.F., van der Merwe, W., Orcesi, S., Prendiville, J.S., Rasmussen, M., Shalev, S.A., Soler, D.M., Shinawi, M., Spiegel, R., Tan, T.Y., Vanderver, A., Wakeling, E.L., Wassmer, E., Whittaker, E., Lebon, P., Stetson, D.B., Bonthron, D.T., Crow, Y.J., 2009. Mutations involved in Aicardi-Goutières syndrome implicate SAMHD1 as regulator of the innate immune response. Nat. Genet. 41, 829–832.

Rice, G.I., Kasher, P.R., Forte, G.M.A., Mannion, N.M., Greenwood, S.M., Szynkiewicz, M., Dickerson, J.E., Bhaskar, S.S., Zampini, M., Briggs, T.A., Jenkinson, E.M., Bacino, C.A., Battini, R., Bertini, E., Brogan, P.A., Brueton, L.A., Carpanelli, M., De Laet, C., de Lonlay, P., del Toro, M., Desguerre, I., Fazzi, E., Garcia-Cazorla, À., Heiberg, A., Kawaguchi, M., Kumar, R., Lin, J.-P.S.M., Lourenco, C.M., Male, A.M., Marques, W., Mignot, C., Olivieri, I., Orcesi, S., Prabhakar, P., Rasmussen, M., Robinson, R.A., Rozenberg, F., Schmidt, J.L., Steindl, K., Tan, T.Y., van der Merwe, W.G., Vanderver, A., Vassallo, G., Wakeling, E.L., Wassmer, E., Whittaker, E., Livingston, J.H., Lebon, P., Suzuki, T., McLaughlin, P.J., Keegan, L.P., O'Connell, M.A., Lovell, S.C., Crow, Y.J., 2012. Mutations in ADAR1 cause Aicardi-Goutières syndrome associated with a type I interferon signature. Nat. Genet. 44, 1243–1248.

Rice, G.I., del Toro Duany, Y., Jenkinson, E.M., Forte, G.M.A., Anderson, B.H., Ariaudo, G., Bader-Meunier, B., Baildam, E.M., Battini, R., Beresford, M.W., Casarano, M., Chouchane, M., Cimaz, R., Collins, A.E., Cordeiro, N.J.V., Dale, R.C., Davidson, J.E., De Waele, L., Desguerre, I., Faivre, L., Fazzi, E., Isidor, B., Lagae, L., Latchman, A.R., Lebon, P., Li, C., Livingston, J.H., Lourenço, C.M., Mancardi, M.M., Masurel-Paulet, A., McInnes, I.B., Menezes, M.P., Mignot, C., O'Sullivan, J., Orcesi, S., Picco, P.P., Riva, E., Robinson, R.A., Rodriguez, D., Salvatici, E., Scott, C., Szybowska, M., Tolmie, J.L., Vanderver, A., Vanhulle, C., Vieira, J.P., Webb, K., Whitney, R.N., Williams, S.G., Wolfe, L.A., Zuberi, S.M., Hur, S., Crow, Y.J., 2014. Gain-of-function mutations in IFIH1 cause a spectrum of human disease phenotypes associated with upregulated type I interferon signaling. Nat. Genet. 46, 503–509.

Richards, A., van den Maagdenberg, A.M.J.M., Jen, J.C., Kavanagh, D., Bertram, P., Spitzer, D., Liszewski, M.K., Barilla-LaBarca, M.-L., Terwindt, G.M., Kasai, Y., McLellan, M., Grand, M.G., Vanmolkot, K.R.J., de Vries, B., Wan, J., Kane, M.J., Mamsa, H., Schäfer, R., Stam, A.H., Haan, J., de Jong, P.T.V.M., Storimans, C.W., van Schooneveld, M.J., Oosterhuis, J.A., Gschwendter, A., Dichgans, M., Kotschet, K.E., Hodgkinson, S., Hardy, T.A., Delatycki, M.B., Hajj-Ali, R.A., Kothari, P.H., Nelson, S.F., Frants, R.R., Baloh, R.W., Ferrari, M.D., Atkinson, J.P., 2007. C-terminal truncations in human 3′-5′ DNA exonuclease TREX1 cause autosomal dominant retinal vasculopathy with cerebral leukodystrophy. Nat. Genet. 39, 1068–1070.

Rodriguez, J.J., Cruz, C.D., Horvath, C.M., 2004. Identification of the nuclear export signal and STAT-binding domains of the Nipah virus V protein reveals mechanisms underlying interferon evasion. J. Virol. 78, 5358–5367.
Rodriguez-Madoz, J.R., Belicha-Villanueva, A., Bernal-Rubio, D., Ashour, J., Ayllon, J., Fernandez-Sesma, A., 2010. Inhibition of the type I interferon response in human dendritic cells by dengue virus infection requires a catalytically active NS2B3 complex. J. Virol. 84, 9760–9774.
Roosendaal, J., Westaway, E.G., Khromykh, A., Mackenzie, J.M., 2006. Regulated cleavages at the West Nile virus NS4A-2K-NS4B junctions play a major role in rearranging cytoplasmic membranes and Golgi trafficking of the NS4A protein. J. Virol. 80, 4623–4632.
Rosen, D.A., Twentyman, J., Hunstad, D.A., 2017. High levels of cyclic Di-GMP in Klebsiella pneumoniae attenuate virulence in the lung. Infect. Immun. 86, e00647-17.
Roth, S., Rottach, A., Lotz-Havla, A.S., Laux, V., Muschaweckh, A., Gersting, S.W., Muntau, A.C., Hopfner, K.-P., Jin, L., Vanness, K., Petrini, J.H.J., Drexler, I., Leonhardt, H., Ruland, J., 2014. Rad50-CARD9 interactions link cytosolic DNA sensing to IL-1β production. Nat. Immunol. 15, 538–545.
Rothenfusser, S., Goutagny, N., DiPerna, G., Gong, M., Monks, B.G., Schoenemeyer, A., Yamamoto, M., Akira, S., Fitzgerald, K.A., 2005. The RNA helicase Lgp2 inhibits TLR-independent sensing of viral replication by retinoic acid-inducible gene-I. J. Immunol. 175, 5260–5268.
Rutsch, F., MacDougall, M., Lu, C., Buers, I., Mamaeva, O., Nitschke, Y., Rice, G.I., Erlandsen, H., Kehl, H.G., Thiele, H., Nürnberg, P., Höhne, W., Crow, Y.J., Feigenbaum, A., Hennekam, R.C., 2015. A specific IFIH1 gain-of-function mutation causes Singleton-Merten syndrome. Am. J. Hum. Genet. 96, 275–282.
Saiga, H., Kitada, S., Shimada, Y., Kamiyama, N., Okuyama, M., Makino, M., Yamamoto, M., Takeda, K., 2012. Critical role of AIM2 in mycobacterium tuberculosis infection. Int. Immunol. 24, 637–644.
Saito, T., Hirai, R., Loo, Y.M., Owen, D., Johnson, C.L., Sinha, S.C., Akira, S., Fujita, T., Gale Jr., M., 2007. Regulation of innate antiviral defenses through a shared repressor domain in RIG-I and LGP2. Proc. Natl. Acad. Sci. U.S.A. 104, 582–587.
Samuelsson, C., Hausmann, J., Lauterbach, H., Schmidt, M., Akira, S., Wagner, H., Chaplin, P., Suter, M., O'Keeffe, M., Hochrein, H., 2008. Survival of lethal poxvirus infection in mice depends on TLR9, and therapeutic vaccination provides protection. J. Clin. Investig. 118, 1776–1784.
Sanchez, G.A.M., Reinhardt, A., Ramsey, S., Wittkowski, H., Hashkes, P.J., Berkun, Y., Schalm, S., Murias, S., Dare, J.A., Brown, D., Stone, D.L., Gao, L., Klausmeier, T., Foell, D., de Jesus, A.A., Chapelle, D.C., Kim, H., Dill, S., Colbert, R.A., Failla, L., Kost, B., O'Brien, M., Reynolds, J.C., Folio, L.R., Calvo, K.R., Paul, S.M., Weir, N., Brofferio, A., Soldatos, A., Biancotto, A., Cowen, E.W., Digiovanna, J.J., Gadina, M., Lipton, A.J., Hadigan, C., Holland, S.M., Fontana, J., Alawad, A.S., Brown, R.J., Rother, K.I., Heller, T., Brooks, K.M., Kumar, P., Brooks, S.R., Waldman, M., Singh, H.K., Nickeleit, V., Silk, M., Prakash, A., Janes, J.M., Ozen, S., Wakim, P.G., Brogan, P.A., Macias, W.L., Goldbach-Mansky, R., 2018. JAK1/2 inhibition with baricitinib in the treatment of autoinflammatory interferonopathies. J. Clin. Invest. 128, 3041–3052.
Santin, I., Moore, F., Grieco, F.A., Marchetti, P., Brancolini, C., Eizirik, D.L., 2012. USP18 is a key regulator of the interferon-driven gene network modulating pancreatic beta cell inflammation and apoptosis. Cell Death Dis. 3, e419.
Sarmiento, L., Cubas-Dueñas, I., Cabrera-Rode, E., 2013. Evidence of association between type 1 diabetes and exposure to enterovirus in cuban children and adolescents. MEDICC Rev. 15, 29–32.

Sarvestani, S.T., Tate, M.D., Moffat, J.M., Jacobi, A.M., Behlke, M.A., Miller, A.R., Beckham, S.A., McCoy, C.E., Chen, W., Mintern, J.D., O'Keeffe, M., John, M., Williams, B.R.G., Gantier, M.P., 2014. Inosine-mediated modulation of RNA sensing by toll-like receptor 7 (TLR7) and TLR8. J. Virol. 88, 799–810.

Sasai, M., Linehan, M.M., Iwasaki, A., 2010. Bifurcation of Toll-like receptor 9 signaling by adaptor protein 3. Science 329, 1530–1534.

Satoh, T., Kato, H., Kumagai, Y., Yoneyama, M., Sato, S., Matsushita, K., Tsujimura, T., Fujita, T., Akira, S., Takeuchi, O., 2010. LGP2 is a positive regulator of RIG-I- and MDA5-mediated antiviral responses. Proc. Natl. Acad. Sci. U.S.A. 107, 1512–1517.

Sauer, J.-D., Sotelo-Troha, K., von Moltke, J., Monroe, K.M., Rae, C.S., Brubaker, S.W., Hyodo, M., Hayakawa, Y., Woodward, J.J., Portnoy, D.A., Vance, R.E., 2011. The N-ethyl-N-nitrosourea-induced Goldenticket mouse mutant reveals an essential function of STING in the in vivo interferon response to Listeria monocytogenes and cyclic dinucleotides. Infect. Immun. 79, 688–694.

Schattgen, S.A., Gao, G., Kurt-Jones, E.A., Fitzgerald, K.A., 2016. Cutting edge: DNA in the lung microenvironment during influenza virus infection tempers inflammation by engaging the DNA sensor AIM2. J. Immunol. 196, 29–33.

Schijf, M.A., Lukens, M.V., Kruijsen, D., van Uden, N.O., Garssen, J., Coenjaerts, F.E., Van't Land, B., van Bleek, G.M., 2013. Respiratory syncytial virus induced type I IFN production by pDC is regulated by RSV-infected airway epithelial cells, RSV-exposed monocytes and virus specific antibodies. PLoS One 8, e81695.

Schlaepfer, E., Speck, R.F., 2008. Anti-HIV activity mediated by natural killer and CD8+ cells after toll-like receptor 7/8 triggering. PLoS One 3, e1999.

Schmelzer, L., Smitka, M., Wolf, C., Lucas, N., Tüngler, V., Hahn, G., Tzschach, A., Di Donato, N., Lee-Kirsch, M.A., von der Hagen, M., 2018. Variable clinical phenotype in two siblings with Aicardi-Goutières syndrome type 6 and a novel mutation in the ADAR gene. Eur. J. Paediatr. Neurol. 22, 186–189.

Schuberth-Wagner, C., Ludwig, J., Bruder, A.K., Herzner, A.M., Zillinger, T., Goldeck, M., Schmidt, T., Schmid-Burgk, J.L., Kerber, R., Wolter, S., Stumpel, J.P., Roth, A., Bartok, E., Drosten, C., Coch, C., Hornung, V., Barchet, W., Kummerer, B.M., Hartmann, G., Schlee, M., 2015. A conserved histidine in the RNA sensor RIG-I controls immune tolerance to N1-2′O-methylated self RNA. Immunity 43, 41–51.

Seleme, M.C., Lei, W., Burg, A.R., Goh, K.Y., Metz, A., Steele, C., Tse, H.M., 2012. Dysregulated TLR3-dependent signaling and innate immune activation in superoxide-deficient macrophages from nonobese diabetic mice. Free Radic. Biol. Med. 52, 2047–2056.

Sepulveda, F.E., Maschalidi, S., Colisson, R., Heslop, L., Ghirelli, C., Sakka, E., Lennon-Duménil, A.-M., Amigorena, S., Cabanie, L., Manoury, B., 2009. Critical role for asparagine endopeptidase in endocytic toll-like receptor signaling in dendritic cells. Immunity 31, 737–748.

Seth, R.B., Sun, L., Ea, C.-K., Chen, Z.J., 2005. Identification and characterization of MAVS, a mitochondrial antiviral signaling protein that activates NF-kappaB and IRF 3. Cell 122, 669–682.

Shang, G., Zhu, D., Li, N., Zhang, J., Zhu, C., Lu, D., Liu, C., Yu, Q., Zhao, Y., Xu, S., Gu, L., 2012. Crystal structures of STING protein reveal basis for recognition of cyclic di-GMP. Nat. Struct. Mol. Biol. 19, 725–727.

Sharma, S., tenOever, B.R., Grandvaux, N., Zhou, G.-P., Lin, R., Hiscott, J., 2003. Triggering the interferon antiviral response through an IKK-related pathway. Science 300, 1148–1151.

Sharma, S., Javadekar, S.M., Pandey, M., Srivastava, M., Kumari, R., Raghavan, S.C., 2015. Homology and enzymatic requirements of microhomology-dependent alternative end joining. Cell Death Dis. 6, e1697.

Shaw, M.L., García-Sastre, A., Palese, P., Basler, C.F., 2004. Nipah virus V and W proteins have a common STAT1-binding domain yet inhibit STAT1 activation from the cytoplasmic and nuclear compartments, respectively. J. Virol. 78, 5633–5641.

Shi, Y., An, J., Liang, J., Hayes, S.E., Sandusky, G.E., Stramm, L.E., Yang, N.N., 1999. Characterization of a mutant pancreatic eIF-2α kinase, PEK, and co-localization with somatostatin in islet delta cells. J. Biol. Chem. 274, 5723–5730.

Shi, X., Dong, Y., Li, Y., Zhao, Z., Li, H., Qiu, S., Li, Y., Guo, W., Qiao, Y., 2015. Inflammasome activation in mouse inner ear in response to MCMV induced hearing loss. J. Otol. 10, 143–149.

Shibasaki, S., Imagawa, A., Tauriainen, S., Iino, M., Oikarinen, M., Abiru, H., Tamaki, K., Seino, H., Nishi, K., Takase, I., Okada, Y., Uno, S., Murase-Mishiba, Y., Terasaki, J., Makino, H., Shimomura, I., Hyoty, H., Hanafusa, T., 2010. Expression of toll-like receptors in the pancreas of recent-onset fulminant type 1 diabetes. Endocr. J. 57, 211–219.

Shigemoto, T., Kageyama, M., Hirai, R., Zheng, J., Yoneyama, M., Fujita, T., 2009. Identification of loss of function mutations in human genes encoding RIG-I and MDA5. J. Biol. Chem. 284, 13348–13354.

Shinde, R., Hezaveh, K., Halaby, M.J., Kloetgen, A., Chakravarthy, A., da Silva Medina, T., Deol, R., Manion, K.P., Baglaenko, Y., Eldh, M., Lamorte, S., Wallace, D., Chodisetti, S.B., Ravishankar, B., Liu, H., Chaudhary, K., Munn, D.H., Tsirigos, A., Madaio, M., Gabrielsson, S., Touma, Z., Wither, J., De Carvalho, D.D., McGaha, T.L., 2018. Apoptotic cell-induced AhR activity is required for immunological tolerance and suppression of systemic lupus erythematosus in mice and humans. Nat. Immunol. 19, 571–582.

Shreberk-Hassidim, R., Ramot, Y., Zlotogorski, A., 2017. Janus kinase inhibitors in dermatology: a systematic review. J. Am. Acad. Dermatol. 76. 745-753.e19.

Siezen, C.L.E., Bont, L., Hodemaekers, H.M., Ermers, M.J., Doornbos, G., van't Slot, R., Wijmenga, C., Houwelingen, H.C.V., Kimpen, J.L.L., Kimman, T.G., Hoebee, B., Janssen, R., 2009. Genetic susceptibility to respiratory syncytial virus bronchiolitis in preterm children is associated with airway remodeling genes and innate immune genes. Pediatr. Infect. Dis. J. 28, 333.

Skonieczna, K., Wozniacka, A., Czajkowski, R., Styczynski, J., Krenska, A., Robak, E., Gawrych, M., Kaszewski, S., Wysocki, M., Grzybowski, T., 2018. X-linked TLR7 gene polymorphisms are associated with diverse immunological conditions but not with discoid lupus erythematosus in polish patients. Postepy Dermatol. Alergol. 35, 26–32.

Skouboe, M.K., Knudsen, A., Reinert, L.S., Boularan, C., Lioux, T., Perouzel, E., Thomsen, M.K., Paludan, S.R., 2018. STING agonists enable antiviral cross-talk between human cells and confer protection against genital herpes in mice. PLoS Pathog. 14, e1006976.

Škrnjug, I., Guzmán, C.A., Ruecker, C., 2014. Cyclic GMP-AMP displays mucosal adjuvant activity in mice. PLoS One 9, e110150.

Smith, E.J., Marie, I., Prakash, A., Garcia-Sastre, A., Levy, D.E., 2001. IRF3 and IRF7 phosphorylation in virus-infected cells does not require double-stranded RNA-dependent protein kinase R or Ikappa B kinase but is blocked by Vaccinia virus E3L protein. J. Biol. Chem. 276, 8951–8957.

Spehalski, E., Capper, K.M., Smith, C.J., Morgan, M.J., Dinkelmann, M., Buis, J., Sekiguchi, J.M., Ferguson, D.O., 2017. MRE11 promotes tumorigenesis by facilitating resistance to oncogene-induced replication stress. Cancer Res. 77, 5327–5338.

Stabell, A.C., Meyerson, N.R., Gullberg, R.C., Gilchrist, A.R., Webb, K.J., Old, W.M., Perera, R., Sawyer, S.L., 2018. Dengue viruses cleave STING in humans but not in nonhuman primates, their presumed natural reservoir. eLife 7, e31919.

Stetson, D.B., Ko, J.S., Heidmann, T., Medzhitov, R., 2008. Trex1 prevents cell-intrinsic initiation of autoimmunity. Cell 134, 587–598.
Stone, A.E., Mitchell, A., Brownell, J., Miklin, D.J., Golden-Mason, L., Polyak, S.J., Gale Jr., M.J., Rosen, H.R., 2014. Hepatitis C virus core protein inhibits interferon production by a human plasmacytoid dendritic cell line and dysregulates interferon regulatory factor-7 and signal transducer and activator of transcription (STAT) 1 protein expression. PLoS One 9, e95627.
Stoppelenburg, A.J., von Hegedus, J.H., Huis in't Veld, R., Bont, L., Boes, M., 2014. Defective control of vitamin D receptor-mediated epithelial STAT1 signalling predisposes to severe respiratory syncytial virus bronchiolitis. J. Pathol. 232, 57–64.
Stracker, T.H., Carson, C.T., Weitzman, M.D., 2002. Adenovirus oncoproteins inactivate the Mre11-Rad50-NBS1 DNA repair complex. Nature 418, 348–352.
Su, C., Zheng, C., 2017. Herpes simplex virus 1 abrogates the cGAS/STING-mediated cytosolic DNA-sensing pathway via its virion host shutoff protein, UL41. J. Virol. 91, e02414–e02416.
Sugai, A., Sato, H., Takayama, I., Yoneda, M., Kai, C., 2017. Nipah and Hendra virus nucleoproteins inhibit nuclear accumulation of signal transducer and activator of transcription 1 (STAT1) and STAT2 by interfering with their complex formation. J. Virol. 91, e01136-17.
Sui, H., Zhou, M., Imamichi, H., Jiao, X., Sherman, B.T., Lane, H.C., Imamichi, T., 2017. STING is an essential mediator of the Ku70-mediated production of IFN-λ1 in response to exogenous DNA. Sci. Signal. 10, eaah5054.
Sumpter, R., Loo, Y.-M., Foy, E., Li, K., Yoneyama, M., Fujita, T., Lemon, S.M., Gale, M., 2005. Regulating intracellular antiviral defense and permissiveness to Hepatitis C virus RNA replication through a cellular RNA helicase, RIG-I. J. Virol. 79, 2689–2699.
Sun, W., Li, Y., Chen, L., Chen, H., You, F., Zhou, X., Zhou, Y., Zhai, Z., Chen, D., Jiang, Z., 2009. ERIS, an endoplasmic reticulum IFN stimulator, activates innate immune signaling through dimerization. Proc. Natl. Acad. Sci. U.S.A. 106, 8653–8658.
Sun, L., Wu, J., Du, F., Chen, X., Chen, Z.J., 2013. Cyclic GMP-AMP synthase is a cytosolic DNA sensor that activates the type I interferon pathway. Science 339, 786–791.
Sun, Z., Wang, L., Eckloff, B.W., Deng, B., Wang, Y., Wampfler, J.A., Jang, J., Wieben, E.D., Jen, J., You, M., Yang, P., 2014. Conserved recurrent gene mutations correlate with pathway deregulation and clinical outcomes of lung adenocarcinoma in never-smokers. BMC Med. Genomics 7, 32.
Sunthamala, N., Thierry, F., Teissier, S., Pientong, C., Kongyingyoes, B., Tangsiriwatthana, T., Sangkomkamhang, U., Ekalaksananan, T., 2014. E2 Proteins of high risk human papillomaviruses down-modulate STING and IFN-κ transcription in keratinocytes. PLoS One 9, e91473.
Suthar, M.S., Ma, D.Y., Thomas, S., Lund, J.M., Zhang, N., Daffis, S., Rudensky, A.Y., Bevan, M.J., Clark, E.A., Kaja, M.-K., Diamond, M.S., Gale, M., 2010. IPS-1 is essential for the control of West Nile virus infection and immunity. PLoS Pathog. 6, e1000757.
Synofzik, M., Haack, T.B., Kopajtich, R., Gorza, M., Rapaport, D., Greiner, M., Schönfeld, C., Freiberg, C., Schorr, S., Holl, R.W., Gonzalez, M.A., Fritsche, A., Fallier-Becker, P., Zimmermann, R., Strom, T.M., Meitinger, T., Züchner, S., Schüle, R., Schöls, L., Prokisch, H., 2014. Absence of BiP co-chaperone DNAJC3 causes diabetes mellitus and multisystemic neurodegeneration. Am. J. Hum. Genet. 95, 689–697.
Szretter, K.J., Daffis, S., Patel, J., Suthar, M.S., Klein, R.S., Gale Jr., M., Diamond, M.S., 2010. The innate immune adaptor molecule MyD88 restricts West Nile virus replication and spread in neurons of the central nervous system. J. Virol. 84, 12125–12138.
Tabeta, K., Georgel, P., Janssen, E., Du, X., Hoebe, K., Crozat, K., Mudd, S., Shamel, L., Sovath, S., Goode, J., Alexopoulou, L., Flavell, R.A., Beutler, B., 2004. Toll-like

receptors 9 and 3 as essential components of innate immune defense against mouse cytomegalovirus infection. Proc. Natl. Acad. Sci. U.S.A. 101, 3516–3521.

Tabeta, K., Hoebe, K., Janssen, E.M., Du, X., Georgel, P., Crozat, K., Mudd, S., Mann, N., Sovath, S., Goode, J., Shamel, L., Herskovits, A.A., Portnoy, D.A., Cooke, M., Tarantino, L.M., Wiltshire, T., Steinberg, B.E., Grinstein, S., Beutler, B., 2006. The Unc93b1 mutation 3d disrupts exogenous antigen presentation and signaling via Toll-like receptors 3, 7 and 9. Nat. Immunol. 7, 156–164.

Takada, I., Yogiashi, Y., Makishima, M., 2016. The ribosomal S6 kinase inhibitor BI-D1870 ameliorated experimental autoimmune encephalomyelitis in mice. Immunobiology 221, 188–192.

Takahama, M., Akira, S., Saitoh, T., 2018. Autophagy limits activation of the inflammasomes. Immunol. Rev. 281, 62–73.

Takaoka, A., Wang, Z., Choi, M.K., Yanai, H., Negishi, H., Ban, T., Lu, Y., Miyagishi, M., Kodama, T., Honda, K., Ohba, Y., Taniguchi, T., 2007. DAI (DLM-1/ZBP1) is a cytosolic DNA sensor and an activator of innate immune response. Nature 448, 501–505.

Tal, M.C., Sasai, M., Lee, H.K., Yordy, B., Shadel, G.S., Iwasaki, A., 2009. Absence of autophagy results in reactive oxygen species-dependent amplification of RLR signaling. Proc. Natl. Acad. Sci. U.S.A. 106, 2770–2775.

Tan, E.M., Schur, P.H., Carr, R.I., Kunkel, H.G., 1966. Deoxybonucleic acid (DNA) and antibodies to DNA in the serum of patients with systemic lupus erythematosus. J. Clin. Investig. 45, 1732–1740.

Tanaka, Y., Chen, Z.J., 2012. STING specifies IRF3 phosphorylation by TBK1 in the cytosolic DNA signaling pathway. Sci. Signal. 5, ra20.

Tao, J., Zhang, X.-W., Jin, J., Du, X.-X., Lian, T., Yang, J., Zhou, X., Jiang, Z., Su, X.-D., 2017. Nonspecific DNA binding of cGAS N terminus promotes cGAS activation. J. Immunol. 198, 3627–3636.

Tatematsu, M., Nishikawa, F., Seya, T., Matsumoto, M., 2013. Toll-like receptor 3 recognizes incomplete stem structures in single-stranded viral RNA. Nat. Commun. 4, 1833.

Thapa, R.J., Ingram, J.P., Ragan, K.B., Nogusa, S., Boyd, D.F., Benitez, A.A., Sridharan, H., Kosoff, R., Shubina, M., Landsteiner, V.J., Andrake, M., Vogel, P., Sigal, L.J., tenOever, B.R., Thomas, P.G., Upton, J.W., Balachandran, S., 2016. DAI senses influenza a virus genomic RNA and activates RIPK3-dependent cell death. Cell Host Microbe 20, 674–681.

Tian, J., Avalos, A.M., Mao, S.-Y., Chen, B., Senthil, K., Wu, H., Parroche, P., Drabic, S., Golenbock, D., Sirois, C., Hua, J., An, L.L., Audoly, L., La Rosa, G., Bierhaus, A., Naworth, P., Marshak-Rothstein, A., Crow, M.K., Fitzgerald, K.A., Latz, E., Kiener, P.A., Coyle, A.J., 2007. Toll-like receptor 9-dependent activation by DNA-containing immune complexes is mediated by HMGB1 and RAGE. Nat. Immunol. 8, 487–496.

Toscano, F., Estornes, Y., Virard, F., Garcia-Cattaneo, A., Pierrot, A., Vanbervliet, B., Bonnin, M., Ciancanelli, M.J., Zhang, S.-Y., Funami, K., Seya, T., Matsumoto, M., Pin, J.-J., Casanova, J.-L., Renno, T., Lebecque, S., 2013. Cleaved/associated TLR3 represents the primary form of the signaling receptor. J. Immunol. 190, 764–773.

Trapani, J., Dawson, M., Apostolidis, V., Browne, K., 1994. Genomic organization of IFI16, an interferon-inducible gene whose expression is associated with human myeloid cell differentiation: correlation of predicted protein domains with exon organization. Immunogenetics 40, 415–424.

Uchida, L., Espada-Murao, L.A., Takamatsu, Y., Okamoto, K., Hayasaka, D., Yu, F., Nabeshima, T., Buerano, C.C., Morita, K., 2014. The dengue virus conceals double-stranded RNA in the intracellular membrane to escape from an interferon response. Sci. Rep. 4, 7395.

Uchikawa, E., Lethier, M., Malet, H., Brunel, J., Gerlier, D., Cusack, S., 2016. Structural analysis of dsRNA binding to anti-viral pattern recognition receptors LGP2 and MDA5. Mol. Cell 62, 586–602.

Unterholzner, L., Keating, S.E., Baran, M., Horan, K.A., Jensen, S.B., Sharma, S., Sirois, C.M., Jin, T., Latz, E., Xiao, T.S., Fitzgerald, K.A., Paludan, S.R., Bowie, A.G., 2010. IFI16 is an innate immune sensor for intracellular DNA. Nat. Immunol. 11, 997–1004.

Uyur Yalçın, E., Maraş Genç, H., Kara, B., 2015. Clinical and neuroradiologic variability of Aicardi-Goutiéres syndrome: two siblings with RNASEH2C mutation and a boy with TREX1 mutation. Turk. J. Pediatr. 57, 504–508.

Uzri, D., Greenberg, H.B., 2013. Characterization of rotavirus RNAs that activate innate immune signaling through the RIG-I-like receptors. PLoS One 8, e69825.

van Drunen Littel-van den Hurk, S., Watkiss, E.R., 2012. Pathogenesis of respiratory syncytial virus. Curr. Opin. Virol. 2, 300–305.

Varani, S., Cederarv, M., Feld, S., Tammik, C., Frascaroli, G., Landini, M.P., Soderberg-Naucler, C., 2007. Human cytomegalovirus differentially controls B Cell and T Cell responses through effects on plasmacytoid dendritic cells. J. Immunol. 179, 7767–7776.

Venkataraman, T., Valdes, M., Elsby, R., Kakuta, S., Caceres, G., Saijo, S., Iwakura, Y., Barber, G.N., 2007. Loss of DExD/H box RNA helicase LGP2 manifests disparate antiviral responses. J. Immunol. 178, 6444–6455.

Vincent, J., Adura, C., Gao, P., Luz, A., Lama, L., Asano, Y., Okamoto, R., Imaeda, T., Aida, J., Rothamel, K., Gogakos, T., Steinberg, J., Reasoner, S., Aso, K., Tuschl, T., Patel, D.J., Glickman, J.F., Ascano, M., 2017. Small molecule inhibition of cGAS reduces interferon expression in primary macrophages from autoimmune mice. Nat. Commun. 8, 750.

Wagner, H., 2004. The immunobiology of the TLR9 subfamily. Trends Immunol. 25, 381–386.

Walker, W.E., Booth, C.J., Goldstein, D.R., 2010. TLR9 and IRF3 cooperate to induce a systemic inflammatory response in mice injected with liposome:DNA. Mol. Ther. 18, 775–784.

Wang, T., Town, T., Alexopoulou, L., Anderson, J.F., Fikrig, E., Flavell, R.A., 2004. Toll-like receptor 3 mediates West Nile virus entry into the brain causing lethal encephalitis. Nat. Med. 10, 1366–1373.

Wang, Y., Zhang, H.-X., Sun, Y.-P., Liu, Z.-X., Liu, X.-S., Wang, L., Lu, S.-Y., Kong, H., Liu, Q.-L., Li, X.-H., Lu, Z.-Y., Chen, S.-J., Chen, Z., Bao, S.-S., Dai, W., Wang, Z.-G., 2007. Rig-I-/- mice develop colitis associated with downregulation of G alpha i2. Cell Res. 17, 858–868.

Wang, F., Alain, T., Szretter, K.J., Stephenson, K., Pol, J.G., Atherton, M.J., Hoang, H.-D., Fonseca, B.D., Zakaria, C., Chen, L., Rangwala, Z., Hesch, A., Chan, E.S.Y., Tuinman, C., Suthar, M.S., Jiang, Z., Ashkar, A.A., Thomas, G., Kozma, S.C., Gale, M., Fitzgerald, K.A., Diamond, M.S., Mossman, K., Sonenberg, N., Wan, Y., Lichty, B.D., 2016. S6K-STING interaction regulates cytosolic DNA-mediated activation of the transcription factor IRF3. Nat. Immunol. 17, 514–522.

Wang, J., Kang, L., Song, D., Liu, L., Yang, S., Ma, L., Guo, Z., Ding, H., Wang, H., Yang, B., 2017. Ku70 senses HTLV-1 DNA and modulates HTLV-1 replication. J. Immunol. 199, 2475–2482.

Wang, J.J., Reed, J.H., Colella, A.D., Russell, A.J., Murray-Brown, W., Chataway, T.K., Jackson, K.J.L., Goodnow, C.C., Gordon, T.P., 2018. Molecular profiling and clonal tracking of secreted rheumatoid factors in primary Sjögren's syndrome. Arthritis Rheumatol. 1–9. in press.

Warner, J.D., Irizarry-Caro, R.A., Bennion, B.G., Ai, T.L., Smith, A.M., Miner, C.A., Sakai, T., Gonugunta, V.K., Wu, J., Platt, D.J., Yan, N., Miner, J.J., 2017.

STING-associated vasculopathy develops independently of IRF3 in mice. J. Exp. Med. 214, 3279–3292.
Watson, R.O., Bell, S.L., MacDuff, D.A., Kimmey, J.M., Diner, E.J., Olivas, J., Vance, R.E., Stallings, C.L., Virgin, H.W., Cox, J.S., 2015. The cytosolic sensor cGAS detects mycobacterium tuberculosis DNA to induce type I interferons and activate autophagy. Cell Host Microbe 17, 811–819.
West, J.A., Gregory, S.M., Sivaraman, V., Su, L., Damania, B., 2011. Activation of plasmacytoid dendritic cells by Kaposi's sarcoma-associated herpesvirus. J. Virol. 85, 895–904.
Wheeler, L.A., Trifonova, R.T., Vrbanac, V., Barteneva, N.S., Liu, X., Bollman, B., Onofrey, L., Mulik, S., Ranjbar, S., Luster, A.D., Tager, A.M., Lieberman, J., 2016. TREX1 knockdown induces an interferon response to HIV that delays viral infection in humanized mice. Cell Rep. 15, 1715–1727.
Wies, E., Wang, M.K., Maharaj, N.P., Chen, K., Zhou, S., Finberg, R.W., Gack, M.U., 2013. Dephosphorylation of the RNA sensors RIG-I and MDA5 by the phosphatase PP1 is essential for innate immune signaling. Immunity 38, 437–449.
Wolf, C., Rapp, A., Berndt, N., Staroske, W., Schuster, M., Dobrick-Mattheuer, M., Kretschmer, S., König, N., Kurth, T., Wieczorek, D., Kast, K., Cardoso, M.C., Günther, C., Lee-Kirsch, M.A., 2016. RPA and Rad51 constitute a cell intrinsic mechanism to protect the cytosol from self DNA. Nat. Commun. 7, 11752.
Wolf, S.J., Theros, J., Reed, T.J., Liu, J., Grigorova, I.L., Martinez-Colon, G., Jacob, C.O., Hodgin, J.B., Kahlenberg, J.M., 2018. TLR7-mediated lupus nephritis is independent of type I IFN signaling. J. Immunol. 201, 393–405.
Woodward, J.J., Iavarone, A.T., Portnoy, D.A., 2010. c-di-AMP secreted by intracellular Listeria monocytogenes activates a host type I interferon response. Science 328, 1703–1705.
Wu, O., Chen, G.P., Chen, H., Li, X.P., Xu, J.H., Zhao, S.S., Sheng, J., Feng, J.B., Cai, J., Fang, X.H., Zhang, W.H., Li, L.H., Zhang, N., Li, J., Li, J.J., Pan, F.M., Wang, C.Z., Ye, D.Q., 2009. The expressions of toll-like receptor 9 and T-bet in circulating B and T cells in newly diagnosed, untreated systemic lupus erythematosus and correlations with disease activity and laboratory data in a Chinese population. Immunobiology 214, 392–402.
Wu, J., Sun, L., Chen, X., Du, F., Shi, H., Chen, C., Chen, Z.J., 2013. Cyclic GMP-AMP is an endogenous second messenger in innate immune signaling by cytosolic DNA. Science 339, 826–830.
Wu, C., Thalhamer, T., Franca, R.F., Xiao, S., Wang, C., Hotta, C., Zhu, C., Hirashima, M., Anderson, A.C., Kuchroo, V.K., 2014. Galectin-9-CD44 interaction enhances stability and function of adaptive regulatory T cells. Immunity 41, 270–282.
Xia, P., Wang, S., Ye, B., Du, Y., Huang, G., Zhu, P., Fan, Z., 2015. Sox2 functions as a sequence-specific DNA sensor in neutrophils to initiate innate immunity against microbial infection. Nat. Immunol. 16, 366–375.
Xia, P., Wang, S., Gao, P., Gao, G., Fan, Z., 2016. DNA sensor cGAS-mediated immune recognition. Protein Cell 7, 777–791.
Xia, H., Luo, H., Shan, C., Muruato, A.E., Nunes, B.T.D., Medeiros, D.B.A., Zou, J., Xie, X., Giraldo, M.I., Vasconcelos, P.F.C., Weaver, S.C., Wang, T., Rajsbaum, R., Shi, P.-Y., 2018. An evolutionary NS1 mutation enhances Zika virus evasion of host interferon induction. Nat. Commun. 9, 414.
Xie, G., Welte, T., Wang, J., Whiteman, M.C., Wicker, J.A., Saxena, V., Cong, Y., Barrett, A.D., Wang, T., 2013. A West Nile virus NS4B-P38G mutant strain induces adaptive immunity via TLR7-MyD88-dependent and independent signaling pathways. Vaccine 31, 4143–4151.

Xu, J., Zoltick, P.W., Gamero, A.M., Gallucci, S., 2014a. TLR ligands up-regulate Trex1 expression in murine conventional dendritic cells through type I interferon and NF-κB-dependent signaling pathways. J. Leukoc. Biol. 96, 93–103.

Xu, X., Zheng, J., Zheng, K., Hou, Y., Zhao, F., Zhao, D., 2014b. Respiratory syncytial virus NS1 protein degrades STAT2 by inducing SOCS1 expression. Intervirology 57, 65–73.

Xu, H., Su, C., Pearson, A., Mody, C.H., Zheng, C., 2017. Herpes simplex virus 1 UL24 abrogates the DNA sensing signal pathway by inhibiting NF-κB activation. J. Virol. 91, e00025-17.

Yamaguchi, M., Kitagawa, Y., Zhou, M., Itoh, M., Gotoh, B., 2014. An anti-interferon activity shared by paramyxovirus C proteins: inhibition of toll-like receptor 7/9-dependent alpha interferon induction. FEBS Lett. 588, 28–34.

Yan, W., Frank, C.L., Korth, M.J., Sopher, B.L., Novoa, I., Ron, D., Katze, M.G., 2002. Control of PERK eIF2α kinase activity by the endoplasmic reticulum stress-induced molecular chaperone P58IPK. Proc. Nat. Acad. Sci. U.S.A. 99, 15920–15925.

Yan, N., Regalado-Magdos, A.D., Stiggelbout, B., Lee-Kirsch, M.A., Lieberman, J., 2010. The cytosolic exonuclease TREX1 inhibits the innate immune response to human immunodeficiency virus type 1. Nat. Immunol. 11, 1005–1013.

Yang, Y.-G., Lindahl, T., Barnes, D.E., 2007. Trex1 exonuclease degrades ssDNA to prevent chronic checkpoint activation and autoimmune disease. Cell 131, 873–886.

Yang, P., An, H., Liu, X., Wen, M., Zheng, Y., Rui, Y., Cao, X., 2010. The cytosolic nucleic acid sensor LRRFIP1 mediates the production of type I interferon via a β-catenin-dependent pathway. Nat. Immunol. 11, 487–494.

Yang, J.Y., Kim, M.S., Kim, E., Cheon, J.H., Lee, Y.S., Kim, Y., Lee, S.H., Seo, S.U., Shin, S.H., Choi, S.S., Kim, B., Chang, S.Y., Ko, H.J., Bae, J.W., Kweon, M.N., 2016. Enteric viruses ameliorate gut inflammation via toll-like receptor 3 and toll-like receptor 7-mediated interferon-beta production. Immunity 44, 889–900.

Yang, H., Bian, S., Chen, H., Wang, L., Zhao, L., Zhang, X., Zhao, Y., Zeng, X., Zhang, F., 2018. Clinical characteristics and risk factors for overlapping rheumatoid arthritis and Sjögren's syndrome. Sci. Rep. 8, 6180.

Yasuda, K., Ogawa, Y., Yamane, I., Nishikawa, M., Takakura, Y., 2005. Macrophage activation by a DNA/cationic liposome complex requires endosomal acidification and TLR9-dependent and -independent pathways. J. Leukoc. Biol. 77, 71–79.

Ye, R., Su, C., Xu, H., Zheng, C., 2017. Herpes simplex virus 1 ubiquitin-specific protease UL36 abrogates NF-κB activation in DNA sensing signal pathway. J. Virol. 91, e02417-16.

Yen, B., Mulder, L.C., Martinez, O., Basler, C.F., 2014. Molecular basis for ebolavirus VP35 suppression of human dendritic cell maturation. J. Virol. 88, 12500–12510.

Yildiz, S., Alpdundar, E., Gungor, B., Kahraman, T., Bayyurt, B., Gursel, I., Gursel, M., 2015. Enhanced immunostimulatory activity of cyclic dinucleotides on mouse cells when complexed with a cell-penetrating peptide or combined with CpG: innate immunity. Eur. J. Immunol. 45, 1170–1179.

Ylipaasto, P., Klingel, K., Lindberg, A.M., Otonkoski, T., Kandolf, R., Hovi, T., Roivainen, M., 2004. Enterovirus infection in human pancreatic islet cells, islet tropism in vivo and receptor involvement in cultured islet beta cells. Diabetologia 47, 225–239.

Yoneyama, M., Suhara, W., Fukuhara, Y., Sato, M., Ozato, K., Fujita, T., 1996. Autocrine amplification of type I interferon gene expression mediated by interferon stimulated gene factor 3 (ISGF3). J. Biochem. 120, 160–169.

Yoneyama, M., Kikuchi, M., Natsukawa, T., Shinobu, N., Imaizumi, T., Miyagishi, M., Taira, K., Akira, S., Fujita, T., 2004. The RNA helicase RIG-I has an essential function in double-stranded RNA-induced innate antiviral responses. Nat. Immunol. 5, 730–737.

Yoneyama, M., Kikuchi, M., Matsumoto, K., Imaizumi, T., Miyagishi, M., Taira, K., Foy, E., Loo, Y.-M., Gale, M., Akira, S., Yonehara, S., Kato, A., Fujita, T., 2005. Shared and unique functions of the DExD/H-Box helicases RIG-I, MDA5, and LGP2 in antiviral innate immunity. J. Immunol. 175, 2851–2858.

Yu, P., 2006. Toll-like receptor 9-independent aggravation of glomerulonephritis in a novel model of SLE. Int. Immunol. 18, 1211–1219.

Yu, H.-R., Huang, H.-C., Kuo, H.-C., Sheen, J.-M., Ou, C.-Y., Hsu, T.-Y., Yang, K.D., 2011. IFN-α production by human mononuclear cells infected with varicella-zoster virus through TLR9-dependent and -independent pathways. Cell. Mol. Immunol. 8, 181–188.

Yu, C.-Y., Chang, T.-H., Liang, J.-J., Chiang, R.-L., Lee, Y.-L., Liao, C.-L., Lin, Y.-L., 2012. Dengue virus targets the adaptor protein MITA to subvert host innate immunity. PLoS Pathog. 8, e1002780.

Yuan, S.-S.F., Hou, M.-F., Hsieh, Y.-C., Huang, C.-Y., Lee, Y.-C., Chen, Y.-J., Lo, S., 2012. Role of MRE11 in cell proliferation, tumor invasion, and DNA repair in breast cancer. J. Natl. Cancer Inst. 104, 1485–1502.

Zeng, W., Sun, L., Jiang, X., Chen, X., Hou, F., Adhikari, A., Xu, M., Chen, Z.J., 2010. Reconstitution of the RIG-I pathway reveals a signaling role of unanchored polyubiquitin chains in innate immunity. Cell 141, 315–330.

Zhang Huifang, M., Qiu, Y., Ye, X., Hemida Maged, G., Hanson, P., Yang, D., 2013. P58IPK inhibits coxsackievirus-induced apoptosis via the PI3K/Akt pathway requiring activation of ATF6a and subsequent upregulation of mitofusin 2. Cell. Microbiol. 16, 411–424.

Zhang, X., Brann, T.W., Zhou, M., Yang, J., Oguariri, R.M., Lidie, K.B., Imamichi, H., Huang, D.W., Lempicki, R.A., Baseler, M.W., Veenstra, T.D., Young, H.A., Lane, H.C., Imamichi, T., 2011a. Cutting edge: Ku70 is a novel cytosolic DNA sensor that induces type III rather than type I IFN. J. Immunol. 186, 4541–4545.

Zhang, Z., Kim, T., Bao, M., Facchinetti, V., Jung, S.Y., Ghaffari, A.A., Qin, J., Cheng, G., Liu, Y.-J., 2011b. DDX1, DDX21, and DHX36 helicases form a complex with the adaptor molecule TRIF to sense dsRNA in dendritic cells. Immunity 34, 866–878.

Zhang, Z., Yuan, B., Bao, M., Lu, N., Kim, T., Liu, Y.-J., 2011c. The helicase DDX41 senses intracellular DNA mediated by the adaptor STING in dendritic cells. Nat. Immunol. 12, 959–965.

Zhang, Z., Yuan, B., Lu, N., Facchinetti, V., Liu, Y.J., 2011d. DHX9 pairs with IPS-1 to sense double-stranded RNA in myeloid dendritic cells. J. Immunol. 187, 4501–4508.

Zhang, Z., Bao, M., Lu, N., Weng, L., Yuan, B., Liu, Y.-J., 2012. The E3 ubiquitin ligase TRIM21 negatively regulates the innate immune response to intracellular double-stranded DNA. Nat. Immunol. 14, 172–178.

Zhang, X., Shi, H., Wu, J., Zhang, X., Sun, L., Chen, C., Chen, Z.J., 2013. Cyclic GMP-AMP containing mixed phosphodiester linkages is an endogenous high-affinity ligand for STING. Mol. Cell 51, 226–235.

Zhang, Y., Yeruva, L., Marinov, A., Prantner, D., Wyrick, P.B., Lupashin, V., Nagarajan, U.M., 2014. The DNA sensor, cyclic GMP–AMP synthase, is essential for induction of IFN-β during Chlamydia trachomatis infection. J. Immunol. 193, 2394–2404.

Zhang, D., Su, C., Zheng, C., 2016. Herpes simplex virus 1 serine protease VP24 blocks the DNA-sensing signal pathway by abrogating activation of interferon regulatory factor 3. J. Virol. 90, 5824–5829.

Zhang, H., Luo, J., Alcorn, J.F., Chen, K., Fan, S., Pilewski, J., Liu, A., Chen, W., Kolls, J.K., Wang, J., 2017a. AIM2 inflammasome is critical for influenza-induced lung injury and mortality. J. Immunol. 198, 4383–4393.

Zhang, H.-L., Ye, H.-Q., Liu, S.-Q., Deng, C.-L., Li, X.-D., Shi, P.-Y., Zhang, B., 2017b. West Nile virus NS1 antagonizes interferon beta production by targeting RIG-I and MDA5. J. Virol. 91, e02396-16.
Zhao, H., Zhu, M., Limbo, O., Russell, P., 2018. RNase H eliminates R-loops that disrupt DNA replication but is nonessential for efficient DSB repair. EMBO Rep. 19, e45335.
Zheng, C., 2018. Evasion of cytosolic DNA-stimulated innate immune responses by herpes simplex virus 1. J. Virol. 92, e00099-17.
Zhong, B., Yang, Y., Li, S., Wang, Y.Y., Li, Y., Diao, F., Lei, C., He, X., Zhang, L., Tien, P., Shu, H.B., 2008. The adaptor protein MITA links virus-sensing receptors to IRF3 transcription factor activation. Immunity 29, 538–550.
Zhong, X., Chen, B., Yang, L., Yang, Z., 2018. Molecular and physiological roles of the adaptor protein CARD9 in immunity. Cell Death Dis. 9.
Zhou, K., Choe, K.-T., Zaidi, Z., Wang, Q., Mathews, M.B., Lee, C.-G., 2003. RNA helicase A interacts with dsDNA and topoisomerase IIalpha. Nucleic Acids Res. 31, 2253–2260.

CHAPTER THREE

Nucleic Acid Induced Interferon and Inflammasome Responses in Regulating Host Defense to Gastrointestinal Viruses

Hanne Dubois[*,†,‡], Geert van Loo[†,§], Andy Wullaert[*,†,‡,1]
[*]Department of Internal Medicine, Ghent University, Ghent, Belgium
[†]VIB-UGent Center for Inflammation Research, VIB, Ghent, Belgium
[‡]Ghent Gut Inflammation Group (GGIG), Ghent University, Ghent, Belgium
[§]Department of Biomedical Molecular Biology, Ghent University, Ghent, Belgium
[1]Corresponding author: e-mail address: andy.wullaert@ugent.be

Contents

Abstract

The gut bacterial and fungal communities residing in the gastrointestinal tract have undisputed far-reaching effects in regulating host health. In the meantime, however, metagenomic sequencing efforts are revealing enteric viruses as the most abundant

International Review of Cell and Molecular Biology, Volume 345
ISSN 1937-6448
https://doi.org/10.1016/bs.ircmb.2018.08.003

dimension of the intestinal gut ecosystem, and the first gut virome-wide association studies showed that inflammatory bowel disease as well as type 1 diabetes could be linked to the presence or absence of particular viral inhabitants in the intestine. In line with the genetic component of these human diseases, mouse model studies demonstrated how beneficial functions of a resident virus can switch to detrimental inflammatory effects in a genetically predisposed host. Such viral-induced intestinal immune disturbances are also recapitulated by several gastrointestinal infectious viruses such as rotavirus and human norovirus. This wide range of viral effects on intestinal immunity emphasizes the need for understanding the innate immune responses to gastrointestinal viruses. Numerous nucleic acid sensors such as DexD/H helicases and AIM2 serve as cytosolic viral guardians to induce antiviral interferon and/or pro-inflammatory inflammasome responses. In both cases, pioneering examples are emerging in which RNA helicases cooperate with particular Nod-like receptors to trigger these cellular responses to enteric viruses. Here we summarize the reported beneficial versus detrimental effects of enteric viruses in the intestinal immune system, and we zoom in on the mechanisms through which sensing of nucleic acids from these enteric viruses trigger interferon and inflammasome responses.

1. THE VIRAL DIMENSION OF THE GUT MICROBIOME

A vast number of studies over the last decade demonstrated that the phylogenetic composition of the gut microbiome as well as its collective metabolic functional output crucially impact on host health. Next-generation sequencing revolutions enabling the 16S DNA based identification of enteric bacteria sparked numerous microbiome-wide studies that correlated particular bacterial composition shifts termed dysbiosis with human disease. The presence of specific bacteria with presumed pathobiont characteristics and/or the absence of specific bacterial species with suspected beneficial effects such as the production of short-chain fatty acids have both been linked to a wide range of human immune, metabolic and neurodegenerative diseases as well as to cancer and its resilience to immunotherapy (Gilbert et al., 2016; Gopalakrishnan et al., 2018)

As this now undisputed role of enteric bacteria in modulating host health was being established, deep sequencing of the internal-transcribed-spacer (ITS) regions of fungal ribosomal genes exposed the enteric mycobiome as a second dimension of the gut microbiome (Nash et al., 2017). Recent studies indicated that also many fungi residing in our gastrointestinal tract affect host health. For instance, following depletion of the resident gut bacteria, commensal fungi could functionally substitute for these microbiota and as such protected the host from increased susceptibility to colitis and influenza infection (Jiang et al., 2017). In addition, changes in the gastrointestinal fungi composition were recently shown to result in severe colitis in mice (Leonardi et al., 2018), demonstrating that—as shown on multiple occasions for commensal gut bacteria—also fungal dysbiosis can be linked to disease development in the host.

Although the above and also numerous other studies firmly established how a sophisticated interplay between the human host and the bacterial and fungal ecosystems inhabiting its intestines is continuously balancing host health and disease, the gut virome represents a third dimension of the gut microbiome that may very well outnumber the gastrointestinal bacteria and fungi and for which the effects on host health are only beginning to emerge. Indeed, apart from bacteria and fungi, human feces contains $>10^9$ viral particles per gram. This hidden part of our microbiota includes viruses infecting each domain of life (Bacteria, Archaea and Eukarya). The latter include the human host itself, in which enteric viruses can produce acute-, persistent- or latent-like infections and in which certain viruses can integrate into the host genome (Cadwell, 2015; Carding et al., 2017; Virgin, 2014).

In recent years, metagenomic fecal DNA sequencing studies started to systematically uncover the enormous complexity of the gut virome. While many viral sequences retrieved from intestinal contents cannot be taxonomically assigned and thus represent as yet uncharacterized viruses, it is clear from the mapped viral sequences that bacteriophages are by far the predominant component of the viral microbiome dimension (Carding et al., 2017; Manrique et al., 2016; Minot et al., 2013). Illustrative of the mass abundance of bacteriophages in the intestinal ecosystem, a single *Bacteroidetes*-infecting phage family was recently characterized for which the founding phage on its own represented up to 22% of all reads in the human gut metagenome project (Dutilh et al., 2014; Yutin et al., 2018). However, despite their massive abundance, thus far no host immune responses have been identified that are specifically directed toward gastrointestinal bacteriophages. Therefore, this review will focus on pattern recognition receptor (PRR) initiated responses directed at the eukaryotic viruses residing in the intestines.

Although the number of eukaryotic viruses present in the gut are fewer than bacteriophages (Minot et al., 2013; Reyes et al., 2010), they could be detected in metagenomics studies as well as by PCR-based fecal shedding analyses in healthy individuals. For instance, viruses from the *Picobirnaviridae, Adenoviridae, Anelloviridae* and *Astroviridae* families and species such as *bocavirus, rotavirus, enterovirus* and *sapovirus* were present in the microbiome of healthy children (Kapusinszky et al., 2012; Zhao et al., 2017). These observations demonstrate that intestinal colonization with eukaryotic viruses, some of which known to have pathogenic potential, can be tolerated without apparent symptomatic disease and thus may be considered "commensal" gastrointestinal viruses. Similarly, specific murine norovirus (MNV) strains were found to persistently infect the intestines of wild-type mice without causing pathology (Baldridge et al., 2016). However, host genetic deficiencies in anti-viral interferon (IFN) responses or in the autophagy machinery turned this virus-host relationship sour, as in these cases MNV could cause lethality or increased susceptibility to intestinal inflammation, respectively (Cadwell et al., 2010; Karst, 2003). These genetic studies in mice elegantly demonstrated that although lifelong asymptomatic viral presence within the gastrointestinal immune system is possible, it requires tight control by the mucosal immune system. Therefore, continuous monitoring of the gut virome by means of PRR signaling induced cytokine and IFN responses is essential for maintaining intestinal immune homeostasis, while also being responsible for protecting the host against occasional intestinal invasions by true pathogenic viruses. PRR-mediating sensing of gastrointestinal viruses leading to cytokine and IFN responses will therefore form the focus of this review.

2. THE GASTROINTESTINAL VIROME IN REGULATING HOST HEALTH

As observed for the other gut microbiota dimensions, the gut virome comprises core components shared among multiple individuals (Carding et al., 2017; Manrique et al., 2016; Reyes et al., 2010), but significant inter-individual differences exist and gut virome diversity is influenced by age as well as by environmental factors such as diet composition (Lim et al., 2015; Minot et al., 2011; Reyes et al., 2015). Owing to their abundance, it is clear that the potential influence of such alterations in the gut virome composition on host health cannot be underestimated. Indeed, large metagenomics efforts have shown that the gut virome composition of human inflammatory bowel disease (IBD) and type 1 diabetes (T1D) patients differs from healthy individuals (Norman et al., 2015; Zhao et al., 2017). As opposed to a decreasing richness of the bacterial microbiota typically observed in IBD patients, both Crohn's disease (CD) and ulcerative colitis (UC) patients displayed an increased richness in their fecal virome when compared to healthy individuals (Norman et al., 2015). While this increased viral richness was associated mainly with elevated numbers of the *Caudovirales* bacteriophages, Manrique et al. subsequently found that the fecal virome communities of these CD and UC patients harbored reduced numbers of the core bacteriophages commonly found in healthy subjects (Manrique et al., 2016). Although the exact viruses involved remain to be identified, these observations suggest that IBD development may be associated with the loss of particular core bacteriophages with beneficial functions for intestinal health, allowing a bloom in other bacteriophages that result in the observed increased virome richness in IBD patients. In contrast to the richer gut virome observed in IBD patients, the intestinal viromes of individuals developing T1D were found to be less diverse than the ones from healthy controls, with the latter harboring significantly more viruses of the *Circoviridae* family (Zhao et al., 2017). Together, these large virome-wide association studies thus highlighted the possibility that particular viruses in the gastrointestinal tract, such as *Circoviridae* in T1D, may exert beneficial functions for preserving host health. However, as many of the host PRR responses to bacteriophages and also several other enteric viruses remain to be identified, it is currently not clear how the observed correlations between skewed gut virome compositions and human disease development could be linked with altered innate immune responses to these intestinal viruses.

Nevertheless, several animal model studies have illustrated how particular enteric viruses can offer benefits for host health. For instance, a landmark study in mice showed that MNV mono-colonization of germfree (GF) or antibiotics (Abx)-treated mice reverted the gastrointestinal architectural and immune defects normally associated with depletion of commensal bacteria (Kernbauer et al., 2014). Indeed, MNV colonization restored small intestinal crypt-villus morphologies, increased lymphocyte numbers and functioning with regards to IFNγ and IgA production, reverted the type 2 versus type 3 innate lymphoid cell (ILC) balances, and restored Paneth cell functioning in these mice (Kernbauer et al., 2014). MNV colonization could thus functionally replace the gut homeostatic effects of the resident commensal bacteria, and was able to do so without causing disease or overt inflammation in the host. Moreover, beyond basal gastrointestinal immune homeostasis, MNV inoculation also protected Abx-treated mice against immune disturbances such as caused by *Citrobacter rodentium* infection or caused by dextran sodium sulfate (DSS)-induced tissue damage. Although the precise mechanisms of action require further investigation, experiments in mice lacking the IFNα receptor 1 (IFNAR1) showed that type I IFN signaling was needed for the beneficial effects of MNV and thus indicate the importance of PRR-based responses in mediating its effects on the host (Kernbauer et al., 2014). The observation that MNV colonization could protect the host from subsequent gastrointestinal *C. rodentium* infection was reminiscent of previous studies showing protective effects of viral infections against future challenges. For instance, mice latently infected with the gammaherpesvirus 68 or murine cytomegalovirus (mCMV) were protected against bacterial infection with either *Listeria monocytogenes* or *Yersinia pestis* (Barton et al., 2007). These mouse model studies provided the proof of concept that resident individual viruses can exert beneficial effects for the host immune system both in homeostasis and upon infectious challenges.

Support for a beneficial role of the overall community of gut resident viruses in intestinal homeostasis was provided by Yang et al., who showed that depleting the gut virome through administrating an antiviral drug cocktail increased the sensitivity of mice to DSS-induced colitis (Yang et al., 2016). Administration of an inactivated rotavirus or combined triggering of Toll-like receptor (TLR)3 and TLR7 attenuated DSS-induced colitis, suggesting that viral RNA sensing through TLR3/7 may represent the molecular mode of action by which enteric viruses prevent excessive intestinal inflammation. In line with this notion as well as with the reported requirement for type I IFN signaling in the beneficial functions of MNV (Kernbauer et al., 2014), plasmacytoid dendritic cells (DCs) isolated from DSS-inflamed colons produced IFNβ upon combined TLR3/7 stimulation (Yang et al., 2016). Supported by several other studies showing protective effects of type I IFNs in the induction of DSS-induced acute colitis (Katakura et al., 2005; Rauch et al., 2014; Sainathan et al., 2012; Vijay-Kumar et al., 2007), these findings suggested that the resident gut virome prevents colitis through TLR3/7-induced type I IFN production. Interestingly, the combination of antiviral drugs used in the Yang et al. study decreased the overall viral load in the intestine but increased the level of bacteriophage *Caudovirales*, similar to the observations in human IBD patients (Norman et al., 2015). Moreover, IBD patients bearing combined mutations in the *Tlr3* and *Tlr7* genes had higher hospitalization rates, suggesting that also in humans viral RNA triggering of these PRRs may be protective (Yang et al., 2016).

Although the above human observational and mouse experimental studies clearly indicate that particular members of the gut virome may help in maintaining intestinal immune homeostasis, it should be noted that there is only a fine line between the beneficial functions of a given virus and the potential detrimental outcomes of its presence. For instance, in mice with decreased expression levels of the autophagy executer *Atg16l1*, for which mutations in humans predispose to CD development, MNV colonization increased host sensitivity to DSS colitis (Cadwell et al., 2008, 2010). In addition, while remaining asymptomatic in wild-type mice, MNV colonization worsened intestinal inflammation in IL-10-deficient mice (Basic et al., 2014). As opposed to its beneficial role in maturing the intestinal immune system of wild-type mice (Kernbauer et al., 2014), these MNV studies illustrate how the host genetic status profoundly impacts on the response of the intestinal immune system to a gastrointestinal virus. Moreover, while gut virome depletion revealed an overall colitis preventing effect of the resident enteric viruses (Yang et al., 2016), mice with a latent mCMV infection developed more severe DSS colitis than control mice (Onyeagocha et al., 2009), illustrating the delicate balance between the either beneficial or detrimental effects of the intestinal viral ecosystem on the host.

Taken together, all of the above observations show that intestinal health depends on complex interactions between enteric viruses, host genetics, the mucosal immune system and other yet unidentified pathways. This complex host-viral cross talk emphasizes the need for better understanding of the innate immune mechanisms recognizing and responding to enteric viruses, which will be discussed further in this review.

3. GASTROINTESTINAL VIRAL INFECTIONS

In contrast to the above described potential beneficial actions of gastrointestinal viruses for host health, several enteric viruses can elicit acute gastroenteritis. Enteric viruses are mostly transmitted via the fecal-oral transmission route, and their shedding in the stool can continue for a long time after the gastroenteritis symptoms resolved. Depending on the conditions of infection, viral gastroenteritis patients can suffer from abdominal pain, vomiting, nausea and diarrhea potentially leading to serious dehydration. However, in some cases viral gastroenteritis remains asymptomatic or evolves into a chronic infection. The latter implies continuous viral shedding in the stool, which creates a reservoir for future outbreaks of newly developed pathogenic strains (Doerflinger et al., 2017; Graves, 2013; Shortland et al., 2014). Viral gastroenteritis pathologies normally are self-limiting and generally resolve within 14 days, but they can turn life-threatening for immunocompromised patients and for infants in developing countries

(Mcclarren et al., 2011). In fact, the World Health Organization (WHO) reported diarrheal diseases as the second leading cause of death worldwide in children under 5 years of age, estimated to kill more than half a million of these children each year (World Health Organization, 2017). Due to difficulties in their detection, the role for enteric viruses in eliciting diarrheal diseases was underestimated for a long time. However, after Kapikian identified in 1972 for the first time viral particles in the stool of a patient with diarrhea (Kapikian et al., 1972), many enteric viruses capable of causing diarrheal disease were described (Bartsch et al., 2016; Bok and Green, 2012; Tsolenyanu et al., 2014), of which the most common ones are introduced below.

3.1 Rotavirus

Rotavirus is a non-enveloped, double-stranded (ds)RNA virus of the *Reoviridae* family causing intestinal disease mainly in children. Rotavirus infection symptoms generally last up to 4 days and are represented in vomiting, diarrhea and low grade fever. Rotavirus infects and replicates in small intestinal epithelial cells, after which subsequent transcytosis into the lamina propria via the M cells can potentially lead to systemic spread and in some cases may evoke extra-intestinal disease manifestations (Blutt et al., 2003; Rivero-Calle et al., 2016). However, induction of an intestinal anti-rotavirus IgA response eventually clears the infection and induces protective immunity (Blutt and Conner, 2013; Blutt et al., 2012). Rotavirus is the only enteric virus for which (since 2006) two vaccines are available that in the meantime have been licensed by >100 countries worldwide. Given that in 2013 rotavirus infections were estimated to have caused 215,000 deaths in children below 5 years, routine use of the rotavirus vaccines is highly recommended by the WHO (Tate et al., 2009, 2012, 2016).

3.2 Norovirus

Human noroviruses (NoV) are non-enveloped single-stranded (ss)RNA viruses of the *Caliciviridae* family classified in six genogroups, of which genogroups I, II and IV are the ones infecting humans (Kroneman et al., 2013; Lopman et al., 2004). In comparison to other enteric viruses, NoV causes a shorter period of diarrhea, but it is often accompanied by abdominal pain, headache and vomiting. Human NoV infection is the most common non-bacterial cause of gastroenteritis as well as the most common cause of food-borne gastroenteritis worldwide, accounting for an estimated 684 million cases per year. Norovirus infections also cause >200,000 deaths annually, mostly affecting children in developing countries below 5 years of age (Lopman et al., 2016). Despite this huge global health and socio-economic impact, no vaccines or specific treatments for NoV infection are currently available.

3.3 Reovirus

Next to rotavirus, reovirus is another *Reoviridae* member of which particular strains can infect the gastrointestinal tract (Tai et al., 2005). Enteric infection with the dsRNA reovirus normally does not provoke pathology in humans or adult mice, but infecting mice with this seemingly innocuous virus was found to trigger inflammatory responses to dietary antigens causing loss of oral tolerance in a manner that at least partially relied on type I IFN responses (Bouziat et al., 2017). Therefore, despite not classifying as a typical gastro-enteritis causative agent, it is interesting to include the mechanisms underlying host IFN responses to reovirus in this review.

3.4 Astrovirus

Astroviruses are positive-sense ssRNA viruses belonging to the *Astroviridae* family (Monroe et al., 1993). After rotavirus, astrovirus is thought to be the second most common instigator of viral gastroenteritis in children (De Benedictis et al., 2011). Although the mechanisms of astrovirus-induced intestinal pathogenesis have not been fully elucidated, a capsid protein was suggested to act as an enterotoxin that damages the intestinal epithelium and as such to provoke diarrhea (Cortez et al., 2017). However, while astroviruses were shown to induce type I IFN production (Marvin et al., 2016), the viral ligands nor the host receptors initiating this antiviral response have been identified.

3.5 Other Gastrointestinal Infectious Viruses

Besides the above well-known instigators of viral gastroenteritis, other opportunistic viruses in some cases are capable of infecting the gastrointestinal tract and inducing pathology. For instance, *Coronaviridae* have long been considered solely as causative agents of common colds, but an outbreak of severe acute respiratory syndrome coronavirus (SARS-CoV) in China revealed diarrhea in up to 38% of the infected patients, presumably caused by toxins secreted during viral replication in enterocytes (Hui et al., 2004; Rota et al., 2003). However, SARS-CoV gastroenteritis incidence is rather low and no mortality has been reported (Ding et al., 2004; Guan et al., 2004; Hon et al., 2003; Wang and Chang, 2004). In addition, immunosuppressed patients can suffer from atypical viral gastroenteritis caused by viruses that normally do not infect a healthy gastrointestinal tract such as human Cytomegalovirus, Epstein-Barr virus, Enterovirus and Coxsackievirus (Li et al., 2009; Ryan et al., 2012; Tapparel et al., 2013; Weber et al., 1999; Yuen et al., 1998).

4. GENERAL PRINCIPLES IN ANTIVIRAL IFN AND INFLAMMASOME RESPONSES

The above described collection of gastrointestinal viruses with diverse effects on the host mucosal immune system, ranging from beneficial actions in immune maturation and homeostasis to deleterious effects potentially causing lethal diarrhea, imply the need for sophisticated host viral recognition systems that in some cases sustain the symbiotic relationship but in other cases initiate immediate antiviral responses to clear the infection. Antiviral immune responses are initiated by germline encoded PRRs that scan the cytosolic and endosomal environments for viral pathogen-associated molecular patterns (PAMPs). The obligate intracellular viral infectious life cycle including genome replication, protein production and virion assembly implies the cytosolic presence of viral nucleic acids as well as proteins as potential PRR triggers. Although endosomal presence of viral nucleic acids is detected by TLR3, TLR7/8 and TLR9 (Lester and Li, 2014), in this review we will mainly focus on PRR-induced signaling pathways induced by viral ssRNA, dsRNA and dsDNA molecules residing in the cytosol. Several cytosolic PRRs have evolved to discriminate non-self viral nucleic acids from host nucleic acids in order to maintain tolerance to the latter (Diebold et al., 2006; Haas et al., 2008; Rutz et al., 2004) While discriminating viral DNA from host DNA is based mainly on their cytosolic versus nuclear presence, cytosolic non-self RNA sensors can detect a wide range of specific molecular patterns that are absent in host RNA and that have been reviewed elsewhere (Chow et al., 2018). Together, coordinated intracellular signaling pathways collectively initiated by these viral nucleic acid receptors will result in the activation of the NF-κB transcription factor leading to production of multiple pro-inflammatory mediators, as well as in an antiviral response characterized by the production of type I as well as type III IFNs (Schlee and Hartmann, 2016; Teijaro, 2016; Zanoni et al., 2017).

The type I IFNs are defined as IFNs that signal through the IFNAR complex, and comprise several IFNα subtypes, IFNβ, IFN-δ, IFN-ε, IFN-κ, IFN-ω, IFN-τ and IFN-ζ. Among type I IFNs, mainly IFNα/β are known to play key roles in limiting viral replication and spread. Autocrine and paracrine IFNα/β signaling will trigger a JAK/STAT signaling cascade leading to STAT1/2 phosphorylation and hetero-dimerization. These STAT heterodimers translocate to the nucleus to induce the expression of numerous interferon stimulated genes (ISG) that aid in restricting viral replication, inhibit protein synthesis and prime neighboring cells for a viral attack (Kisseleva et al., 2002; Wang et al., 2017b). In addition, antiviral IFNα/β signaling can induce apoptosis, thereby destroying the viral replicative niche (Balachandran et al., 2000; Mattei et al., 2010).

Type III IFNs (IFNλ) exert antiviral effects analogous to the ones exerted by IFNα/β owing to the similar JAK/STAT signal transduction pathway and IFN response gene expression they can induce upon binding their cellular receptor. However, type III IFNs (in humans comprising IFNλ1, IFNλ2, IFNλ3 and IFNλ4; while in mice only IFNλ2 and IFNλ3 are active) do not bind the IFNAR but instead signal through a receptor complex composed of the IL-10Rβ subunit and the specific IFNλ receptor 1 (IFNλR1) subunit. Binding to the latter defines type III IFNs. As the expression pattern of the IFNλR1 subunit is mainly restricted to epithelial cells, IFNλ is particularly important in responding to gastrointestinal antiviral responses, as it was shown to control the persistent character of several enteric viruses (Ingle et al., 2018; Lee and Baldridge, 2017).

Besides inducing these antiviral IFN responses, in some cases viral nucleic acid receptors induce signaling toward activation of host inflammasomes, mostly in cooperation with cytosolic Nod-like receptor (NLR) family members (Lupfer and Kanneganti, 2013; Man et al., 2017). The inflammasome is a multi-protein complex in which activation of its catalytic subunit caspase-1 mediates maturation and secretion of IL-1β and IL-18 cytokines (Dubois et al., 2016). While IL-1β exerts a broad diversity of functions within the innate immune response including lymphocyte activation and leukocyte transmigration, IL-18 is important for orchestrating the ensuing adaptive immune response by promoting IFNγ production (Dinarello, 2009). In addition, inflammasome activation can provoke a lytic form of cell death termed pyroptosis through cleavage of its Gasdermin D (GSDMD) substrate that forms pores in the cellular membrane (Aglietti et al., 2016; Ding et al., 2016; Zhang et al., 2017). Apart from eliminating the replicative niche of intracellularly-replicating pathogens such as viruses, the lytic nature of pyroptotic cells also invokes extracellular release of danger-associated molecular patterns (DAMPs) such as IL-1α and HMGB1 that assist in coordinating inflammatory responses (Aachoui et al., 2013; Casson et al., 2013; Lamkanfi et al., 2010).

Thus, alongside inducing antiviral IFN responses aimed at limiting viral dissemination, activating the inflammasome is an important cellular response to cytosolic viral nucleic acids that promotes inflammatory reactions. Moreover, IFN and inflammasome pathways are intimately linked. For instance, transcription of caspase-11 is controlled by type I IFN signaling (Rathinam et al., 2012), suggesting that anti-viral IFN responses could prime neighboring cells for executing non-canonical inflammasome activation in response to Gram-negative bacteria. Conversely, several studies have shown that type I IFNs can also downregulate inflammasome responses. For instance, type I IFNs were shown to reduce Nlrp3 inflammasome mediated production of mature IL-1β. This could in part be explained through type I IFNs activating STAT1, which in turn lead to production of IL-10 that decreased pro-IL-1β levels in an autocrine manner (Guarda et al., 2011). In addition to these IL-10 dependent effects, type I IFNs were also shown to induce the production of 25-hydroxycholesterol (25-HC) that inhibits inflammasome-mediated inflammatory responses both in the priming and in the activation stages. Indeed, on the one hand 25-HC was reported to restrain pro-IL-1β expression on the transcriptional level (Reboldi et al., 2014), while on the other hand it was also reported to maintain mitochondrial integrity and as such to prevent the cytosolic release of mitochondrial DNA that could otherwise activate the absent in melanoma 2 (AIM2) inflammasome (Dang et al., 2017). Therefore, it is clear that the concerted actions of type I IFNs and inflammasome responses induced by gastrointestinal viruses could influence mucosal immunity in multiple autocrine and paracrine ways. Below, we will outline the several classes of viral nucleic acid recognizing PRRs and highlight in more detail the ones that have been implicated in IFN and inflammasome responses to gastrointestinal viruses.

5. CYTOSOLIC VIRAL NUCLEIC ACID RECEPTORS INDUCING IFN AND INFLAMMASOME RESPONSES

5.1 Viral RNA Recognition by DExD/H Helicases

Cytosolic viral ssRNA and dsRNA molecules are recognized by particular members of a large family of DexD/H-box containing helicases, among which RIG-I-like receptors (RLRs) have been best characterized

(Hornung et al., 2006; Kato et al., 2006; Yoneyama et al., 2015). Members of the DexD/H helicase protein family all share a conserved DexD/H motif. This motif varies between DEAD and DEAH amino acid sequences, which defines the DDX and DHX helicases, respectively. Regardless, although the DEAD-box containing helicases have a preference for ATP, all DexD/H helicases can bind RNA structures to induce conformational changes upon ATP/NTP hydrolysis. This confers many of these proteins host RNA household functions such as in aiding RNA transcription, editing, splicing, export, translation, and turnover, as well as in ribosome biogenesis. However, some DexD/H helicases have evolved to recognize non-self RNA. The best known example is retinoic acid-inducible gene (RIG)-I, encoded by the *Ddx58* gene, which is the prototypical member of the RLRs (Chow et al., 2018; Fullam and Schroder, 2013; Linder, 2006).

5.1.1 RIG-I-Like Receptors (RLRs)

The RLRs are DexD/H helicases that stably bind cytosolic viral ssRNA or dsRNA to induce production of antiviral IFNs and pro-inflammatory cytokines. Members of this family include RIG-I, melanoma differentiation-associated gene 5 (MDA-5) and laboratory of genetics and physiology 2 (LGP2) (Cui et al., 2001; Kang et al., 2002; Yoneyama et al., 2004). Both RIG-I and MDA-5 contain an RNA-binding DExD/H helicase domain and two N-terminal caspase-recruitment domains (CARDs), which serve for interacting with the signaling adaptor protein mitochondrial anti-viral signaling (MAVS) (Andrejeva et al., 2004; Seth et al., 2005; Yoneyama et al., 2004). As it lacks CARDs, LGP2 does not have signaling capacities like RIG-I and MDA-5, but rather functions as an accessory protein that can regulate RLR signaling either positively or negatively (Komuro and Horvath, 2006; Venkataraman et al., 2007; Yoneyama et al., 2005). RIG-I or MDA-5 mediated binding of viral RNA initiates CARD-CARD interactions between the respective RLR and MAVS. The latter then propagates a signaling cascade culminating in phosphorylation of IFN responsive factor (IRF)3 and IRF7, which translocate to the nucleus for inducing transcription of IRF responsive genes. In addition, RLR-induced MAVS signaling will also lead to NF-κB activation to induce transcription of several pro-inflammatory genes (Reikine et al., 2014). RIG-I and MDA-5 together enable broad-range antiviral responses because they recognize different RNA structures. Although the exact ranges and structural underpinnings of the differential viral RNA recognition modes by these RLRs are not entirely clear yet and have been reviewed elsewhere (Chow et al., 2018), in general RIG-I was reported to bind short dsRNA stretches as well as ssRNA bearing 5′ di- or tri-phosphate ends, while MDA-5 prefers binding to long dsRNA sequences (Goubau et al., 2014; Hornung et al., 2006; Kato et al., 2006, 2008).

5.1.2 Other DExD/H Helicases Involved in Cytosolic Viral RNA Recognition

Apart from the above RLRs, several other DexD/H helicases were suggested to directly recognize viral RNA and to initiate cellular anti-viral responses (Fullam and Schroder, 2013). For instance, the DDX1-DDX21-DHX36 protein complex was reported to bind reovirus and influenza dsRNA via the helicase domain of DDX1, after which DDX21 and DHX36 enabled interaction with the TRIF adaptor protein to initiate downstream signaling (Zhang et al., 2011a). Other reports proposed also DDX3 to be capable of directly sensing viral RNA in the cytosol, although some observations indicated that this RNA helicase could be involved as a more general scaffold in RLR-mediated antiviral signaling cascades. For instance, DDX3 was shown to interact with RNA molecules from diverse viral origins suggesting its role as an RNA sensor (Gringhuis et al., 2017), but it was also reported to mediate RIG-I-induced signaling by promoting the assembly of MAVS signaling complexes (Gu et al., 2017), rendering it hard to discriminate whether DDX3 functions as a RIG-I-independent viral RNA sensor or as a signal transducer. Finally, although a function for DDX60 has been proposed as an upstream RNA sensor facilitating RIG-I-induced signaling (Oshiumi et al., 2015), these findings were later questioned by DDX60-deficient cells and mice that did not show impaired IFN responses to a wide variety of viral RNA triggers (Goubau et al., 2015). While it is thus clear that the individual roles of multiple non-RLR DexD/H helicases in viral RNA induced responses require additional investigational efforts, we will further highlight the functions of three particular DexD/H helicases—DHX9, DHX15 and DHX33—that were suggested to collaborate with distinct cytosolic NLRs in order to mount IFN or inflammasome responses to gastrointestinal viruses (summarized in Fig. 1). However, it should be mentioned that for each of these three cases also TLRs or "classical" RLRs have been suggested to participate in sensing viral nucleic acids, as summarized in Table 1. Therefore, a certain level of redundancy and/or cell type specificity can be envisaged in recognizing and responding to gastrointestinal viruses.

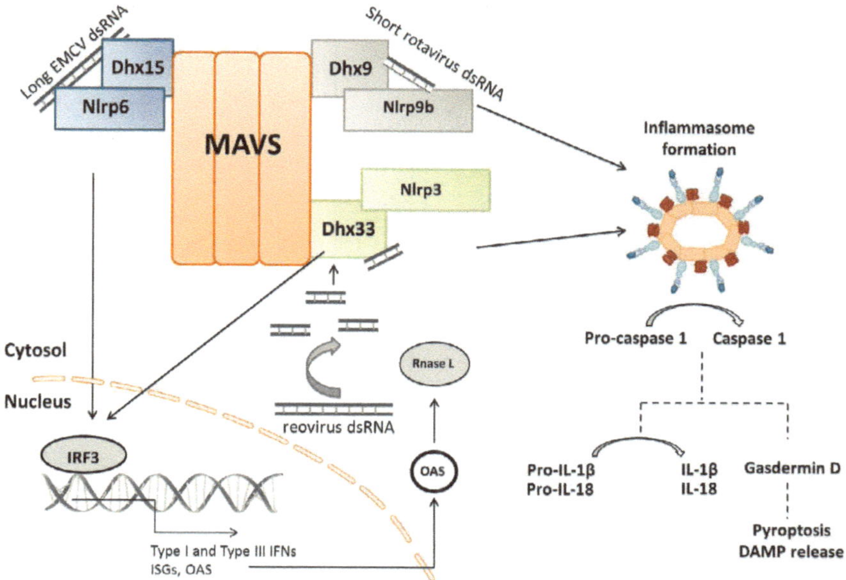

Fig. 1 Scheme of DExD/H helicase—NLR collaborations in cytosolic viral dsRNA induced IFN and inflammasome responses to gastrointestinal viruses. Dhx15-Nlrp6 complex formation enables recognition of EMCV long dsRNA stretches, upon which interaction with MAVS initiates downstream signaling toward IRF3-mediated IFN responses. Conversely, the Dhx9-Nlrp9b and Dhx33-Nlrp3 complexes facilitate dsRNA-induced inflammasome activation. Dhx9-Nlrp9b complex formation facilitates recognition of short dsRNA molecules from rotavirus infection, leading to inflammasome activation. Dhx33 recognizes short dsRNA cleavage products generated by OAS-activated RNase L. Dhx33 is then able to initiate signaling toward IFN responses in an NLR-independent manner by interacting with MAVS, while Dhx33 interaction with Nlrp3 is required to activate the inflammasome. See text for further details.

5.2 DexD/H Helicase: NLR Collaborations in Responding to Gastrointestinal Viruses

5.2.1 Dhx9-Nlrp9b Mediated Inflammasome Responses to Rotavirus Infection

The acute but transient nature of rotavirus-induced gastroenteritis suggests that specific innate immune responses are capable of rapidly controlling the infection. Initial studies on host responses to rotavirus infection focused on the role of TLRs. Administration of rotavirus dsRNA to mice resulted in small intestinal inflammation in a manner dependent on the endosomal dsRNA sensor TLR3 (Zhou et al., 2007). TLR3-deficient mice were subsequently found to be hyper-susceptible to rotavirus infection at adult age but not in the neonatal stage of life (Pott et al., 2012). Although this

Table 1 Overview of Potential Redundancies and/or Cell Type Specificities of Host Receptors Initiating Viral RNA Signaling in Response to Gastrointestinal Rotavirus, Norovirus and Reovirus Infections

Enteric Virus	Nucleic Acid PAMP	Host PRR	Host Adaptors and Downstream Effect	Human/Mouse Cell Type or Model System in Which the Response Was Demonstrated	References
Rotavirus	Endosomal dsRNA	TLR3	TRIF, type I IFNs	Adult mouse IECs	Zhou et al. (2007) and Pott et al. (2012)
	Cytosolic dsRNA	RIG-I, MDA-5	MAVS, type I IFNs	Human transformed IECs, MAVS$^{-/-}$ mice	Broquet et al. (2011)
	Cytosolic short dsRNA stretches	Dhx9	Nlrp9b, ASC, inflammasome	Adult mouse IECs	Zhu et al. (2017)
Murine Norovirus	Cytosolic dsRNA	MDA-5	MAVS, type I IFN and cytokine responses	Mouse bone marrow derived DCs	McCartney et al. (2008)
	Cytosolic long dsRNA stretches	Dhx15, suggested based on ECMV studies	Suggested Nlrp6—MAVS—type I IFN response based on ECMV studies	Higher MNV loads in Nlrp6$^{-/-}$ mice upon oral infection	Wang et al. (2015b)

Reovirus	Cytosolic dsRNA (suggested to be RNAse L-processed based on influenza dsRNA studies)	DHX33	NLRP3, ASC, inflammasome	Human THP-1 macrophages, human primary monocyte-derived macrophages	Mitoma et al. (2013) and Chakrabarti et al. (2015)
	Cytosolic dsRNA	Dhx33	MAVS, type I IFNs	Murine splenic DC cell line D2SC, mouse bone marrow derived DCs, and mouse embryonic fibroblasts	Liu et al. (2014)
	Cytosolic dsRNA	RIG-I and MDA-5	MAVS, type I IFNs	Mouse bone marrow derived cDCs and mouse embryonic fibroblasts	Kato et al. (2008)
	Cytosolic dsRNA	RIG-I	MAVS, type I IFNs	Human 293 T cells	Holm et al. (2007)
	Cytosolic dsRNA	DDX1/ DDX21/ DHX36	TRIF, type I IFNs and TNF	Murine splenic DC cell line D2SC	Zhang et al. (2011a)

observation showing a crucial role for TLR3 specifically in adulthood suggested that instead RLR sensing of rotavirus RNA could be essential at neonatal stages, another study also using adult mice showed that abrogating RLR signaling by deleting MAVS rendered mice unable to induce antiviral IFN responses leading to increased rotavirus burdens in these MAVS-deficient mice (Broquet et al., 2011). Together, these studies suggested non-redundant protective roles for TLR3- and RLR-mediated IFN responses in murine rotavirus infection.

In addition to IFNs, a later study pointed at a potentially important role for inflammasomes in preventing rotavirus infection, as intraperitoneal injection of mice with IL-18 prior to gastrointestinal rotavirus infection diminished subsequent fecal shedding of the virus (Zhang et al., 2014). Since IL-18 requires maturation by the inflammasome to obtain its biological activity, this study suggested that perhaps endogenous inflammasome activation by the virus could evoke host protective responses. This hypothesis was confirmed by Zhu et al., who showed that rotavirus infection in mice lacking the inflammasome components ASC or caspase-1 resulted in higher small intestinal viral titers and a higher diarrhea incidence (Zhu et al., 2017). The authors identified Nlrp9b as the NLR responsible for initiating rotavirus-induced inflammasome responses, as Nlrp9b$^{-/-}$ mice phenocopied the response of ASC- or caspase-1-deficient mice upon rotavirus infection. Moreover, consistent with the fact that rotavirus replicates in intestinal epithelial cells (IECs), also IEC-specific deletion of Nlrp9b or caspase-1 aggravated rotavirus gastroenteritis. Rotavirus dsRNA-induced Nlrp9b inflammasome activation resulted both in IL-18 release and in GSDMD-dependent pyroptosis. Interestingly, GSDMD-deficient mice but not IL-18-deficient mice displayed increased susceptibility to rotavirus infection when compared with wild-type mice (Zhu et al., 2017). This suggests that beyond inducing IL-18 release, Nlrp9b inflammasome activation limits intestinal rotavirus burdens by inducing pyroptosis, perhaps to destroy the viral intracellular replicative niche or to release additional DAMPs needed for augmenting host antiviral responses. Similar to the reported age-related increase in TLR3 expression (Pott et al., 2012), also intestinal expression levels of ASC and Nlrp9b increased as mice grew older (Zhu et al., 2017), suggesting that potentiating these two non-redundant modes of viral dsRNA recognition may explain how rotavirus susceptibility declines with age.

In order to identify the molecular mechanism by which Nlrp9b senses rotavirus presence in IECs, immunoprecipitation studies were performed that revealed an interaction between the human NLRP9 and DHX9 proteins (Zhu et al., 2017). The latter was previously shown to act as an RNA-sensing DexD/H helicase during reovirus and influenza infection, after which it interacted with MAVS and as such mediated production of IFNs and cytokines (Zhang et al., 2011c). Aligning with this already reported DHX9 function, DHX9-NLRP9 complexes were found to bind short dsRNA stretches more efficiently than NLPR9 itself. In addition, primary IEC organoid cultures derived from Dhx9$^{-/-}$ mice were defective in rotavirus induced inflammasome responses and allowed higher rotavirus replication rates than wild-type organoids. Thus, this study provided an example of a DexD/H helicase—NLR collaboration, showing that Dhx9 assists in rotavirus dsRNA sensing allowing Nlrp9b to induce the assembly and activation of an inflammasome that restricts viral replication by triggering pyroptosis (see Fig. 1) (Zhu et al., 2017).

5.2.2 Dhx15-Nlrp6 Mediated IFN Responses to EMCV and MNV Infection

A second example of DexD/H helicase-assisted functioning of NLR-mediated antiviral immunity in the gut was provided by a study using an oral encephalomyocarditis virus (EMCV) infection model. Mice lacking the NLR family member Nlrp6 showed higher viral burdens in their intestines and increased mortality upon oral EMCV infection as compared to wild-type mice, indicating a role for Nlrp6 in limiting EMCV replication in the intestinal mucosa (Wang et al., 2015b). Oral EMCV infection of Nlrp6$^{-/-}$ mice was associated with lower intestinal IFN and ISG levels compared to wild-type mice, showing that Nlrp6 was crucially contributing to antiviral IFN responses in this model. Consistent with this suggested role of Nlrp6 in mediating IFN responses, co-immunoprecipitation experiments showed that Nlrp6 was able to interact with MAVS. Similar experiments then revealed the ATP-dependent RNA helicase Dhx15 as a direct binding partner of Nlrp6 (Wang et al., 2015b). As human DHX15 had previously been shown already to bind EMCV dsRNA (Mosallanejad et al., 2014), this suggested that Nlrp6 binding to this DexD/H helicase could increase their cumulative affinity for EMCV dsRNA. Indeed, similar to the above observations for the DHX9/NLRP9 complex, the interaction of Dhx15 and Nlrp6 resulted in improved dsRNA binding and facilitated complex formation with MAVS to induce IFN responses (see Fig. 1). However, in contrast to the preferential binding of DHX9/NLRP9 to short dsRNA, the Dhx15/Nlrp6 interaction resulted in a preference for binding high-molecular weight dsRNA (Wang et al., 2015b; Zhu et al., 2017).

Accordingly, unlike Nlrp6$^{-/-}$ mice, Nlrp9b$^{-/-}$ mice were not more sensitive to oral EMCV infection than wild-type mice (Zhu et al., 2017).

Interestingly, apart from EMCV, also MNV was described as subject to Nlrp6-dependent antiviral immune signaling. In the absence of Nlrp6, oral MNV infection resulted in higher fecal and intestinal epithelial MNV loads, suggesting a role for Nlrp6 in clearing this virus from the gut (Wang et al., 2015b). However, unlike for EMCV, no information was provided by Wang et al. on the role of Dhx15 in Nlrp6 responses to MNV, nor on the effects of Nlrp6 on MNV-induced IFN responses (Wang et al., 2015b). In fact, IFN responses to MNV were previously reported to depend entirely on MDA-5-mediated recognition of its dsRNA. Indeed, in contrast to wild-type as well as TLR3$^{-/-}$ DCs, MDA-5-deficient DCs did not mount cytokine or IFN responses to MNV infection (McCartney et al., 2008). This observation showing that MDA-5 is crucial for initiating MNV dsRNA-induced IFN responses seems contradictory with the suggested similar role for Nlrp6, although it should be noted that MDA-5 and Nlrp6 have differential expression patterns. Indeed, given its major expression in IECs, Nlrp6 might represent the intestinal epithelial sensor for MNV dsRNA, while MDA-5 could initiate MNV antiviral defense in immune cells. However, although human NoV can readily infect human IECs (Ettayebi et al., 2016), the preferred murine target cells of MNV are immune cells such as macrophages, DCs and B cells (Jones et al., 2014; Wobus et al., 2004). Only recently, a potential effect of MNV in IECs was discovered. After several studies consolidated a major role for IFNλ signaling in controlling the persistency of intestinal MNV colonization (Baldridge et al., 2015b, 2017), a particular persistent MNV-CR6 strain was found to dodge the intestinal antiviral effects of IFNλ in a tiny subpopulation of IECs, which it used as a reservoir to persist *in vivo* (Lee et al., 2017). These MNV-CR6-infected cells accounted for only 0.0003% of the IECs, and were recently identified by Wilen et al. as specialized intestinal epithelial Tuft cells (Wilen et al., 2018). Considering the mainly IEC-restricted Nlrp6 expression pattern, it is puzzling how Nlrp6 could exert its antiviral effects on MNV loads from these very rare Tuft cells. Since MNV infection of Tuft cells was linked to secretion of IL-4 and IL-25 (Wilen et al., 2018), it will be interesting to see whether the Nlrp6 effects acting in IECs could potentially restrict intestinal MNV replication through modulating the expression of such type 2 cytokines thereby indirectly affecting intestinal immune cells.

5.2.3 DHX33-Nlrp3 Mediated Inflammasome Responses to Reovirus

Similar to the above Dhx9-Nlrp9b and Dhx15-Nlrp6 cooperative antiviral responses, also the DexD/H RNA helicase DHX33 was reported to relay viral RNA signaling through a cytosolic NLR. Indeed, Mitoma et al. provided evidence that human DHX33 recognizes reovirus dsRNA leading to NLRP3 inflammasome activation (Mitoma et al., 2013). The NLRP3 inflammasome is well known as an innate immune effector triggered by a broad range of stimuli including enteric bacterial pathogens such as *Vibrio cholerae* and *C. rodentium* (Song-Zhao et al., 2014; Toma et al., 2010). However, reovirus was the first viral intestinal trigger suggested to activate the NLRP3 inflammasome through it dsRNA. Upon transfection of reoviral dsRNA in human macrophages, DHX33 was proposed to bind the reovirus dsRNA via its helicase domain, after which the DEAD domain of DHX33 directly interacts with the NACHT domain of NLRP3 leading to activation of the inflammasome. Knock-down of DHX33 expression abrogated this reovirus dsRNA induced NLRP3 inflammasome activation (Mitoma et al., 2013). Similar Dhx33 knock-down experiments in murine DCs and fibroblasts later suggested that Dhx33 also initiates MAVS-mediated but RIG-I- and MDA-5-independent IFN responses upon reovirus infection (Liu et al., 2014), arguing that also in mice Dhx33 could have a role in responding to reovirus infections. In contrast, also classical RLR pathways were described in reovirus infected murine cells, as deleting both RIG-I and MDA-5 prevented the induction of type I IFNs after reovirus infection of conventional DCs and upon transfection of reovirus genomic RNA in mouse embryonic fibroblasts (Kato et al., 2008). Furthermore, adding to the complexity of comparing reovirus responses in different murine or human cell types, knockdown experiments in human 293T cells showed that only RIG-I but not MDA-5 was responsible for inducing IFN responses to this virus (Holm et al., 2007). Moreover, as mentioned before, also a DDX1/DDX21/DHX36 complex was suggested already as a receptor recognizing reoviral dsRNA in murine DCs via both TRIF and MAVS (Zhang et al., 2011a). Although these observations illustrate the many potential redundancies and different cell type specific responses to this virus (Table 1), it is clear that DHX33 can contribute at least in some cases to host responses to reoviral dsRNA.

Although the exact mechanisms how DHX33 links viral dsRNA to both IFN and inflammasome signaling, an independent study on the role of the 2′,5′-oligoadenylate synthetase (OAS)-RNase L signaling cascade during influenza infection suggested an additional layer of complexity in this response (see Fig. 1) (Chakrabarti et al., 2015). Upon viral infection, OAS is one of the antiviral ISGs expressed. This enzyme mediates synthesis of 2′,5′-linked oligonucleotides, which activate the host RNase L to cleave viral dsRNA into smaller fragments. DHX33 knock-down experiments in human macrophages suggested that DHX33 recognizes such RNase L generated influenza dsRNA cleavage products, thereby promoting its interaction with both MAVS and Nlrp3 (Banerjee, 2016; Chakrabarti et al., 2015). Thus, although genetic *in vivo* confirmation is still needed, the above *in vitro* DHX33 knock-down studies suggest that this DexD/H helicase performs a dual role in viral dsRNA-induced IFN responses and Nlrp3 inflammasome activation.

5.3 Cytosolic Viral DNA Receptors Leading to IFN and Inflammasome Responses

Intracellular viral DNA can be detected by several host DNA sensors. A central theme upon cytoplasmic DNA sensing is cyclic GMP-AMP synthase (cGAS)-mediated production of cGAMP, which can bind and activate stimulator of interferon gene (STING) as an essential signaling intermediate in antiviral IFN responses (Tao et al., 2016). Triggering of STING-mediated IFN responses have been suggested for several upstream cytosolic DNA sensors, among which the DexD/H helicase DDX41 (Zhang et al., 2011b), the IFNγ-inducible protein (IFI)16 (Unterholzner et al., 2010), and DNA-dependent activator of IFN regulatory factors (DAI, encoded by *Zbp1*) (Takaoka et al., 2007). In addition, multiple other STING-dependent as well as -independent DNA sensors have been proposed and were reviewed elsewhere (Xia et al., 2016). However, since to our knowledge these IFN response inducing intracellular DNA sensors have not been implicated yet in the response to gastrointestinal viruses, these PRRs are beyond the scope of this review.

In contrast, AIM2—a cytosolic dsDNA sensor that does not induce an IFN response but instead leads to inflammasome activation (Fernandes-Alnemri et al., 2009; Hornung et al., 2009)—was implicated in the host response to mCMV (Rathinam et al., 2010), of which the human CMV counterpart causes gastroenteritis in immunocompromised patients (Weber et al., 1999). Although mCMV was administered intraperitoneally in this study and intestinal immune responses were not studied, AIM2- as well as ASC-deficient mice showed impaired IL-18 responses consistent with defective inflammasome activation upon mCMV infection (Rathinam et al., 2010). Insufficient systemic IL-18 levels in these AIM2- and ASC-deficient mice correlated with a lack of NK cell dependent IFNγ production and higher viral loads (Rathinam et al., 2010). This observations thus illustrates the *in vivo* requirement for AIM2-mediated mCMV dsDNA sensing to mount strong innate and adaptive immune responses for viral control, suggesting that perhaps impaired AIM2 responses contribute to the sensitivity of immunocompromised individuals to hCMV-induced gastroenteritis.

6. CONCLUDING REMARKS

The above outlined viral nucleic acid recognition systems and signaling pathways that can be triggered by enteric viruses illustrate the complexity of the coordinated cellular response to the presence of a viral guest in its cytosol. While many of the cytosolic DexD/H helicases need further investigation on their individual role in responding to non-self nucleic acids, it is already clear that many of them will couple to other PRR systems in order to fine-tune and diversify their actions. The outlined examples of the Dhx9/Nlrp9b, Dhx15/Nlrp6 and the even more complex RNase L/Dhx33/Nlrp3 viral RNA signaling complexes (see Fig. 1) nicely illustrate this notion. In the described DexD/H helicase—NLR collaborations it is interesting to note that Dhx9 and Dhx33 engage NLRs to activate inflammasomes, while Dhx15 instructs an NLR to induce IFN responses. Although it is tempting to speculate that interaction with particular DexD/H helicases could modulate NLR functions, these differences could also simply reflect inherently different functions of the respective NLR family members. For instance, while Nlrp6 was relaying Dhx15 EMCV dsRNA sensing to IFN responses, inflammasome responses upon EMCV infection were in fact reported to depend on MDA-5 mediated sensing of its dsRNA leading to Nlrp3 activation (Poeck et al., 2010). This illustrates how different NLRs can induce either IFN or inflammasome responses to a single virus, although one needs to bear in mind that also cell type specific functions of the respective NLRs remain possible. Finally, it deserves mentioning that although we focused on the already described viral RNA/DNA initiated responses, several enteric viruses including SARS-CoV, Enterovirus 71 and Coxsackievirus B3 were reported to activate the Nlrp3 inflammasome upon host recognition of a viral-derived protein (Chen et al., 2017; Lei et al., 2017; Li et al., 2017; Nieto-Torres et al., 2015; Wang et al., 2015a, 2017a). Upon integrating also these viral protein induced signaling cascades the picture will appear even more complex.

Due to the multicellular organization of the gastrointestinal tract and the diversity of its inhabitants, an intense cross talk between the bacterial, fungal and viral dimensions of the microbiome as well as with the host ultimately determines intestinal homeostasis. For instance, prior antibiotics treatment leading to depletion of the gut microbiota was shown to decrease rotavirus, reovirus as well as MNV infectivity in mice (Jones et al., 2014; Kuss et al., 2011; Uchiyama et al., 2014). In the latter case, histo-blood group antigens derived from enteric bacteria were shown to act as stimulatory agents allowing MNV as well as human NoV to infect B cells (Jones et al., 2014). In addition, the resident gut bacterial community prevents persistent MNV colonization in mice by suppressing type III IFN responses (Baldridge et al., 2015b). These interactions demonstrate how the bacterial gut microbiota can affect host immune responses to a viral trigger both directly and indirectly. Moreover, given the vast amount of intestinal bacteriophages that depend on the presence of their specific bacterial hosts, there is no doubt that future research in this area will reveal additional ways by which alterations in the gut microbiota composition determine the composition of the gut virome and hence also host anti-viral responses.

With regards to viral-host interactions, the described beneficial effects of MNV colonization in wild-type GF intestines versus its detrimental effects in autophagy-defective or IL-10$^{-/-}$ mice illustrate how host genetics influence intestinal immune responses to resident viruses (Basic et al., 2014; Cadwell et al., 2010; Kernbauer et al., 2014). The altered gut virome compositions observed in CD, UC and T1D patients (Norman et al., 2015; Zhao et al., 2017) predict that many more of such disbalanced virus-host interaction may contribute to disease development. Indeed, similar to how failures in discriminating self versus non-self nucleic acids can provoke autoimmune diseases (Kato and Fujita, 2015), perhaps diseases such IBD and T1D derive from sustained responses to nucleic acids from pathogenic viruses, or from inappropriate responses to viral nucleic acids derived from the endogenous gut virome. In this respect, it is interesting to note that T1D is associated with gastrointestinal enterovirus infections and that a polymorphism in the *IFIH1* gene encoding MDA-5 was shown to be associated with the severity of enterovirus 70 gastrointestinal infection (Cinek et al., 2012;

Oikarinen et al., 2012). Moreover, the exact same *IFIH1* polymorphism also associates to T1D development, and the gut virome composition is associated with T1D development (Smyth et al., 2006; Zhao et al., 2017). Therefore, although cause and consequence are always difficult to discriminate, it is tempting to speculate that there could indeed be link between altered viral dsRNA induced MDA-5 responses, enteric viral presence and T1D development. Adding to this suspicion, the success of fecal transfers in IBD patients has recently been associated with the gut virome composition (Conceição-Neto et al., 2017; Zuo et al., 2018). While at the moment they are mere correlations, there is no doubt that these observations will spark many future efforts investigating the response of the intestinal immune system to "commensal" and infectious viral nucleic acids.

ACKNOWLEDGMENTS

Research in the A.W. lab is supported by the Odysseus Grant G.0C49.13 N, and the research Grants 3G.0447.18 and 3G.0448.18 from the Fund for Scientific Research-Flanders. We apologize for the studies that could not be cited due to space restrictions.

REFERENCES

Aachoui, Y., Leaf, I.A., Hagar, J.A., Fontana, M.F., Campos, C.G., Zak, D.E., Tan, M.H., Cotter, P.A., Vance, R.E., Aderem, A., Miao, E.A., 2013. Caspase-11 protects against bacteria that escape the vacuole. Science 339, 975–978.

Aglietti, R.A., Estevez, A., Gupta, A., Ramirez, M.G., Liu, P.S., Kayagaki, N., Ciferri, C., Dixit, V.M., Dueber, E.C., 2016. GsdmD p30 elicited by caspase-11 during pyroptosis forms pores in membranes. Proc. Natl. Acad. Sci. U.S.A. 113, 7858–7863.

Andrejeva, J., Childs, K.S., Young, D.F., Carlos, T.S., Stock, N., Goodbourn, S., Randall, R.E., 2004. The V proteins of paramyxoviruses bind the IFN-inducible RNA helicase, mda-5, and inhibit its activation of the IFN-beta promoter. Proc. Natl. Acad. Sci. U.S.A. 101, 17264–17269.

Balachandran, S., Roberts, P.C., Kipperman, T., Bhalla, K.N., Compans, R.W., Archer, D.R., Barber, G.N., 2000. Alpha/beta interferons potentiate virus-induced apoptosis through activation of the FADD/caspase-8 death signaling pathway. J. Virol. 74, 1513–1523.

Baldridge, M.T., Nice, T.J., McCune, B.T., Yokoyama, C.C., Kambal, A., Wheadon, M., Diamond, M.S., Ivanova, Y., Artyomov, M., Virgin, H.W., 2015b. Commensal microbes and interferon-lambda determine persistence of enteric murine norovirus infection. Science 347, 266–269.

Baldridge, M.T., Turula, H., Wobus, C.E., 2016. Norovirus regulation by host and microbe. Trends Mol. Med. 22, 1047–1059.

Baldridge, M.T., Lee, S., Brown, J.J., McAllister, N., Urbanek, K., Dermody, T.S., Nice, T.J., Virgin, H.W., 2017. Expression of Ifnlr1 on intestinal epithelial cells is critical to the antiviral effects of IFN-lambda against norovirus and reovirus. J. Virol. 91, 02079–16.

Banerjee, S., 2016. RNase L and the NLRP3-inflammasome: an old merchant in a new trade. Cytokine Growth Factor Rev. 29, 63–70.

Barton, E.S., White, D.W., Cathelyn, J.S., Brett-McClellan, K.A., Engle, M., Diamond, M.S., Miller, V.L., Virgin, H.W.T., 2007. Herpesvirus latency confers symbiotic protection from bacterial infection. Nature 447, 326–329.

Bartsch, S.M., Lopman, B.A., Ozawa, S., Hall, A.J., Lee, B.Y., 2016. Global economic burden of norovirus gastroenteritis. PLoS One 11, e0151219.

Basic, M., Keubler, L.M., Buettner, M., Achard, M., Breves, G., Schroder, B., Smoczek, A., Jorns, A., Wedekind, D., Zschemisch, N.H., Gunther, C., Neumann, D., Lienenklaus, S., Weiss, S., Hornef, M.W., Mahler, M., Bleich, A., 2014. Norovirus triggered microbiota-driven mucosal inflammation in interleukin 10-deficient mice. Inflamm. Bowel Dis. 20, 431–443.

Blutt, S.E., Conner, M.E., 2013. The gastrointestinal frontier: IgA and viruses. Front. Immunol. 4, 402.

Blutt, S.E., Kirkwood, C.D., Parreño, V., Warfield, K.L., Ciarlet, M., Estes, M.K., Bok, K., Bishop, R.F., Conner, M.E., 2003. Rotavirus antigenaemia and viraemia: a common event? Lancet 362, 1445–1449.

Blutt, S.E., Miller, A.D., Salmon, S.L., Metzger, D.W., Conner, M.E., 2012. IgA is important for clearance and critical for protection from rotavirus infection. Mucosal Immunol. 5, 712–719.

Bok, K., Green, K.Y., 2012. Norovirus gastroenteritis in immunocompromised patients. N. Engl. J. Med. 367, 2126–2132.

Bouziat, R., Hinterleitner, R., Brown, J.J., Stencel-Baerenwald, J.E., Ikizler, M., Mayassi, T., Meisel, M., Kim, S.M., Discepolo, V., Pruijssers, A.J., Ernest, J.D., Iskarpatyoti, J.A., Costes, L.M., Lawrence, I., Palanski, B.A., Varma, M., Zurenski, M.A., Khomandiak, S., McAllister, N., Aravamudhan, P., Boehme, K.W., Hu, F., Samsom, J.N., Reinecker, H.C., Kupfer, S.S., Guandalini, S., Semrad, C.E., Abadie, V., Khosla, C., Barreiro, L.B., Xavier, R.J., Ng, A., Dermody, T.S., Jabri, B., 2017. Reovirus infection triggers inflammatory responses to dietary antigens and development of celiac disease. Science 356, 44–50.

Broquet, A.H., Hirata, Y., McAllister, C.S., Kagnoff, M.F., 2011. RIG-I/MDA5/MAVS are required to signal a protective IFN response in rotavirus-infected intestinal epithelium. J. Immunol. 186, 1618–1626.

Cadwell, K., 2015. The virome in host health and disease. Immunity 42, 805–813.

Cadwell, K., Liu, J.Y., Brown, S.L., Miyoshi, H., Loh, J., Lennerz, J.K., Kishi, C., Kc, W., Carrero, J.A., Hunt, S., Stone, C.D., Brunt, E.M., Xavier, R.J., Sleckman, B.P., Li, E., Mizushima, N., Stappenbeck, T.S., Virgin, H.W.T., 2008. A key role for autophagy and the autophagy gene Atg16l1 in mouse and human intestinal Paneth cells. Nature 456, 259–263.

Cadwell, K., Patel, K.K., Maloney, N.S., Liu, T.C., Ng, A.C., Storer, C.E., Head, R.D., Xavier, R., Stappenbeck, T.S., Virgin, H.W., 2010. Virus-plus-susceptibility gene interaction determines Crohn's disease gene Atg16L1 phenotypes in intestine. Cell 141, 1135–1145.

Carding, S.R., Davis, N., Hoyles, L., 2017. Review article: the human intestinal virome in health and disease. Aliment. Pharmacol. Ther. 46, 800–815.

Casson, C.N., Copenhaver, A.M., Zwack, E.E., Nguyen, H.T., Strowig, T., Javdan, B., Bradley, W.P., Fung, T.C., Flavell, R.A., Brodsky, I.E., Shin, S., 2013. Caspase-11 activation in response to bacterial secretion systems that access the host cytosol. PLoS Pathog. 9, e1003400.

Chakrabarti, A., Banerjee, S., Franchi, L., Loo, Y.M., Gale Jr., M., Nunez, G., Silverman, R.H., 2015. RNase L activates the NLRP3 inflammasome during viral infections. Cell Host Microbe 17, 466–477.

Chen, W., Foo, S.S., Zaid, A., Teng, T.S., Herrero, L.J., Wolf, S., Tharmarajah, K., Vu, L.D., Van Vreden, C., Taylor, A., Freitas, J.R., Li, R.W., Woodruff, T.M.,

Gordon, R., Ojcius, D.M., Nakaya, H.I., Kanneganti, T.D., O'neill, L.A.J., Robertson, A.A.B., King, N.J., Suhrbier, A., Cooper, M.A., Ng, L.F.P., Mahalingam, S., 2017. Specific inhibition of NLRP3 in chikungunya disease reveals a role for inflammasomes in alphavirus-induced inflammation. Nat. Microbiol. 2, 1435–1445.

Chow, K.T., Gale Jr., M., Loo, Y.-M., 2018. RIG-I and other RNA sensors in antiviral immunity. Annu. Rev. Immunol. 36, 667–694.

Cinek, O., Tapia, G., Witso, E., Kramna, L., Holkova, K., Rasmussen, T., Stene, L.C., Ronningen, K.S., 2012. Enterovirus RNA in peripheral blood may be associated with the variants of rs1990760, a common type 1 diabetes associated polymorphism in IFIH1. PLoS One 7, e48409.

Conceição-Neto, N., Deboutte, W., Dierckx, T., Machiels, K., Wang, J., Yinda, K.C., Maes, P., Van Ranst, M., Joossens, M., Raes, J., Vermeire, S., Matthijnssens, J., 2017. Low eukaryotic viral richness is associated with faecal microbiota transplantation success in patients with UC. Gut 67, 1558–1559.

Cortez, V., Meliopoulos, V.A., Karlsson, E.A., Hargest, V., Johnson, C., Schultz-Cherry, S., 2017. Astrovirus biology and pathogenesis. Annu. Rev. Virol. 4, 327–348.

Cui, Y., Li, M., Walton, K.D., Sun, K., Hanover, J.A., Furth, P.A., Hennighausen, L., 2001. The Stat3/5 locus encodes novel endoplasmic reticulum and helicase-like proteins that are preferentially expressed in normal and neoplastic mammary tissue. Genomics 78, 129–134.

Dang, E.V., McDonald, J.G., Russell, D.W., Cyster, J.G., 2017. Oxysterol restraint of cholesterol synthesis prevents AIM2 inflammasome activation. Cell 171, 1057–1071.e11.

De Benedictis, P., Schultz-Cherry, S., Burnham, A., Cattoli, G., 2011. Astrovirus infections in humans and animals—molecular biology, genetic diversity, and interspecies transmissions. Infect. Genet. Evol. 11, 1529–1544.

Diebold, S.S., Massacrier, C., Akira, S., Paturel, C., Morel, Y., Reis E Sousa, C., 2006. Nucleic acid agonists for Toll-like receptor 7 are defined by the presence of uridine ribonucleotides. Eur. J. Immunol. 36, 3256–3267.

Dinarello, C.A., 2009. Immunological and inflammatory functions of the interleukin-1 family. Annu. Rev. Immunol. 27, 519–550.

Ding, Y., He, L., Zhang, Q., Huang, Z., Che, X., Hou, J., Wang, H., Shen, H., Qiu, L., Li, Z., Geng, J., Cai, J., Han, H., Li, X., Kang, W., Weng, D., Liang, P., Jiang, S., 2004. Organ distribution of severe acute respiratory syndrome (SARS) associated coronavirus (SARS-CoV) in SARS patients: implications for pathogenesis and virus transmission pathways. J. Pathol. 203, 622–630.

Ding, J., Wang, K., Liu, W., She, Y., Sun, Q., Shi, J., Sun, H., Wang, D.C., Shao, F., 2016. Pore-forming activity and structural autoinhibition of the gasdermin family. Nature 535, 111–116.

Doerflinger, S.Y., Weichert, S., Koromyslova, A., Chan, M., Schwerk, C., Adam, R., Jennewein, S., Hansman, G.S., Schroten, H., 2017. Human norovirus evolution in a chronically infected host. mSphere 2, 00352–16.

Dubois, H., Wullaert, A., Lamkanfi, M., 2016. General strategies in inflammasome biology. Curr. Top. Microbiol. Immunol. 397, 1–22.

Dutilh, B.E., Cassman, N., McNair, K., Sanchez, S.E., Silva, G.G., Boling, L., Barr, J.J., Speth, D.R., Seguritan, V., Aziz, R.K., Felts, B., Dinsdale, E.A., Mokili, J.L., Edwards, R.A., 2014. A highly abundant bacteriophage discovered in the unknown sequences of human faecal metagenomes. Nat. Commun. 5, 4498.

Ettayebi, K., Crawford, S.E., Murakami, K., Broughman, J.R., Karandikar, U., Tenge, V.R., Neill, F.H., Blutt, S.E., Zeng, X.-L., Qu, L., Kou, B., Opekun, A.R., Burrin, D., Graham, D.Y., Ramani, S., Atmar, R.L., Estes, M.K., 2016. Replication of human noroviruses in stem cell-derived human enteroids. Science 353, 1387–1393.

Fernandes-Alnemri, T., Yu, J.W., Datta, P., Wu, J., Alnemri, E.S., 2009. AIM2 activates the inflammasome and cell death in response to cytoplasmic DNA. Nature 458, 509–513.

Fullam, A., Schroder, M., 2013. DExD/H-box RNA helicases as mediators of anti-viral innate immunity and essential host factors for viral replication. Biochim. Biophys. Acta 1829, 854–865.

Gilbert, J.A., Quinn, R.A., Debelius, J., Xu, Z.Z., Morton, J., Garg, N., Jansson, J.K., Dorrestein, P.C., Knight, R., 2016. Microbiome-wide association studies link dynamic microbial consortia to disease. Nature 535, 94–103.

Gopalakrishnan, V., Helmink, B.A., Spencer, C.N., Reuben, A., Wargo, J.A., 2018. The influence of the gut microbiome on cancer, immunity, and cancer immunotherapy. Cancer Cell 33, 570–580.

Goubau, D., Schlee, M., Deddouche, S., Pruijssers, A.J., Zillinger, T., Goldeck, M., Schuberth, C., Van Der Veen, A.G., Fujimura, T., Rehwinkel, J., Iskarpatyoti, J.A., Barchet, W., Ludwig, J., Dermody, T.S., Hartmann, G., Reis, E.S.C., 2014. Antiviral immunity via Rig-I-mediated recognition of RNA bearing 5′-diphosphates. Nature 514, 372–375.

Goubau, D., Van Der Veen, A.G., Chakravarty, P., Lin, R., Rogers, N., Rehwinkel, J., Deddouche, S., Rosewell, I., Hiscott, J., Reis, E.S.C., 2015. Mouse superkiller-2-like helicase DDX60 is dispensable for type I IFN induction and immunity to multiple viruses. Eur. J. Immunol. 45, 3386–3403.

Graves, N.S., 2013. Acute gastroenteritis. Prim. Care 40, 727–741.

Gringhuis, S.I., Hertoghs, N., Kaptein, T.M., Zijlstra-Willems, E.M., Sarrami-Forooshani, R., Sprokholt, J.K., Van Teijlingen, N.H., Kootstra, N.A., Booiman, T., Van Dort, K.A., Ribeiro, C.M., Drewniak, A., Geijtenbeek, T.B., 2017. HIV-1 blocks the signaling adaptor MAVS to evade antiviral host defense after sensing of abortive HIV-1 RNA by the host helicase DDX3. Nat. Immunol. 18, 225–235.

Gu, L., Fullam, A., McCormack, N., Hohn, Y., Schroder, M., 2017. DDX3 directly regulates TRAF3 ubiquitination and acts as a scaffold to co-ordinate assembly of signalling complexes downstream from MAVS. Biochem. J. 474, 571–587.

Guan, Y., Peiris, J.S.M., Zheng, B., Poon, L.L.M., Chan, K.H., Zeng, F.Y., Chan, C.W.M., Chan, M.N., Chen, J.D., Chow, K.Y.C., Hon, C.C., Hui, K.H., Li, J., Li, V.Y.Y., Wang, Y., Leung, S.W., Yuen, K.Y., Leung, F.C., 2004. Molecular epidemiology of the novel coronavirus that causes severe acute respiratory syndrome. Lancet 363, 99–104.

Guarda, G., Braun, M., Staehli, F., Tardivel, A., Mattmann, C., Forster, I., Farlik, M., Decker, T., Du Pasquier, R.A., Romero, P., Tschopp, J., 2011. Type I interferon inhibits interleukin-1 production and inflammasome activation. Immunity 34, 213–223.

Haas, T., Metzger, J., Schmitz, F., Heit, A., Muller, T., Latz, E., Wagner, H., 2008. The DNA sugar backbone 2′ deoxyribose determines toll-like receptor 9 activation. Immunity 28, 315–323.

Holm, G.H., Zurney, J., Tumilasci, V., Leveille, S., Danthi, P., Hiscott, J., Sherry, B., Dermody, T.S., 2007. Retinoic acid-inducible gene-I and interferon-beta promoter stimulator-1 augment proapoptotic responses following mammalian reovirus infection via interferon regulatory factor-3. J. Biol. Chem. 282, 21953–21961.

Hon, K.L.E., Leung, C.W., Cheng, W.T.F., Chan, P.K.S., Chu, W.C.W., Kwan, Y.W., Li, A.M., Fong, N.C., Ng, P.C., Chiu, M.C., Li, C.K., Tam, J.S., Fok, T.F., 2003. Clinical presentations and outcome of severe acute respiratory syndrome in children. Lancet 361, 1701–1703.

Hornung, V., Ellegast, J., Kim, S., Brzozka, K., Jung, A., Kato, H., Poeck, H., Akira, S., Conzelmann, K.K., Schlee, M., Endres, S., Hartmann, G., 2006. 5′-Triphosphate RNA is the ligand for RIG-I. Science 314, 994–997.

Hornung, V., Ablasser, A., Charrel-Dennis, M., Bauernfeind, F., Horvath, G., Caffrey, D.R., Latz, E., Fitzgerald, K.A., 2009. AIM2 recognizes cytosolic dsDNA and forms a caspase-1-activating inflammasome with ASC. Nature 458, 514–518.

Hui, D.S., Chan, M.C., Wu, A.K., Ng, P.C., 2004. Severe acute respiratory syndrome (SARS): epidemiology and clinical features. Postgrad. Med. J. 80, 373–381.

Ingle, H., Peterson, S.T., Baldridge, M.T., 2018. Distinct effects of type I and III interferons on enteric viruses. Viruses 10, 46.

Jiang, T.T., Shao, T.Y., Ang, W.X.G., Kinder, J.M., Turner, L.H., Pham, G., Whitt, J., Alenghat, T., Way, S.S., 2017. Commensal fungi recapitulate the protective benefits of intestinal bacteria. Cell Host Microbe 22, 809–816.e4.

Jones, M.K., Watanabe, M., Zhu, S., Graves, C.L., Keyes, L.R., Grau, K.R., Gonzalez-Hernandez, M.B., Iovine, N.M., Wobus, C.E., Vinje, J., Tibbetts, S.A., Wallet, S.M., Karst, S.M., 2014. Enteric bacteria promote human and mouse norovirus infection of B cells. Science 346, 755–759.

Kang, D.C., Gopalkrishnan, R.V., Wu, Q., Jankowsky, E., Pyle, A.M., Fisher, P.B., 2002. mda-5: an interferon-inducible putative RNA helicase with double-stranded RNA-dependent ATPase activity and melanoma growth-suppressive properties. Proc. Natl. Acad. Sci. U.S.A. 99, 637–642.

Kapikian, A.Z., Wyatt, R.G., Dolin, R., Thornhill, T.S., Kalica, A.R., Chanock, R.M., 1972. Visualization by immune electron microscopy of a 27-nm particle associated with acute infectious nonbacterial gastroenteritis. J. Virol. 10, 1075–1081.

Kapusinszky, B., Minor, P., Delwart, E., 2012. Nearly constant shedding of diverse enteric viruses by two healthy infants. J. Clin. Microbiol. 50, 3427–3434.

Karst, S.M.E.A., 2003. Stat1-dependent innate immunity to a Norwalk-like virus. Science 299, 1575–1578.

Katakura, K., Lee, J., Rachmilewitz, D., Li, G., Eckmann, L., Raz, E., 2005. Toll-like receptor 9-induced type I IFN protects mice from experimental colitis. J. Clin. Invest. 115, 695–702.

Kato, H., Fujita, T., 2015. RIG-I-like receptors and autoimmune diseases. Curr. Opin. Immunol. 37, 40–45.

Kato, H., Takeuchi, O., Sato, S., Yoneyama, M., Yamamoto, M., Matsui, K., Uematsu, S., Jung, A., Kawai, T., Ishii, K.J., Yamaguchi, O., Otsu, K., Tsujimura, T., Koh, C.S., Reis E Sousa, C., Matsuura, Y., Fujita, T., Akira, S., 2006. Differential roles of MDA5 and RIG-I helicases in the recognition of RNA viruses. Nature 441, 101–105.

Kato, H., Takeuchi, O., Mikamo-Satoh, E., Hirai, R., Kawai, T., Matsushita, K., Hiiragi, A., Dermody, T.S., Fujita, T., Akira, S., 2008. Length-dependent recognition of double-stranded ribonucleic acids by retinoic acid-inducible gene-I and melanoma differentiation-associated gene 5. J. Exp. Med. 205, 1601–1610.

Kernbauer, E., Ding, Y., Cadwell, K., 2014. An enteric virus can replace the beneficial function of commensal bacteria. Nature 516, 94–98.

Kisseleva, T., Bhattacharya, S., Braunstein, J., Schindler, C.W., 2002. Signaling through the JAK/STAT pathway, recent advances and future challenges. Gene 285, 1–24.

Komuro, A., Horvath, C.M., 2006. RNA- and virus-independent inhibition of antiviral signaling by RNA helicase LGP2. J. Virol. 80, 12332–12342.

Kroneman, A., Vega, E., Vennema, H., Vinje, J., White, P.A., Hansman, G., Green, K., Martella, V., Katayama, K., Koopmans, M., 2013. Proposal for a unified norovirus nomenclature and genotyping. Arch. Virol. 158, 2059–2068.

Kuss, S.K., Best, G.T., Etheredge, C.A., Pruijssers, A.J., Frierson, J.M., Hooper, L.V., Dermody, T.S., Pfeiffer, J.K., 2011. Intestinal microbiota promote enteric virus replication and systemic pathogenesis. Science 334, 249–252.

Lamkanfi, M., Sarkar, A., Vande Walle, L., Vitari, A.C., Amer, A.O., Wewers, M.D., Tracey, K.J., Kanneganti, T.D., Dixit, V.M., 2010. Inflammasome-dependent release of the alarmin HMGB1 in endotoxemia. J. Immunol. 185, 4385–4392.

Lee, S., Baldridge, M.T., 2017. Interferon-lambda: a potent regulator of intestinal viral infections. Front. Immunol. 8, 749.

Lee, S., Wilen, C.B., Orvedahl, A., McCune, B.T., Kim, K.W., Orchard, R.C., Peterson, S.T., Nice, T.J., Baldridge, M.T., Virgin, H.W., 2017. Norovirus cell tropism is determined by combinatorial action of a viral non-structural protein and host cytokine. Cell Host Microbe 22, 449–459.

Lei, X., Zhang, Z., Xiao, X., Qi, J., He, B., Wang, J., 2017. Enterovirus 71 inhibits pyroptosis through cleavage of gasdermin D. J. Virol. 91, e01069–17.

Leonardi, I., Li, X., Semon, A., Li, D., Doron, I., Putzel, G., Bar, A., Prieto, D., Rescigno, M., McGovern, D.P.B., Pla, J., Iliev, I.D., 2018. CX3CR1$^+$ mononuclear phagocytes control immunity to intestinal fungi. Science 359, 232–236.

Lester, S.N., Li, K., 2014. Toll-like receptors in antiviral innate immunity. J. Mol. Biol. 426, 1246–1264.

Li, L., Victoria, J., Kapoor, A., Blinkova, O., Wang, C., Babrzadeh, F., Mason, C.J., Pandey, P., Triki, H., Bahri, O., Oderinde, B.S., Baba, M.M., Bukbuk, D.N., Besser, J.M., Bartkus, J.M., Delwart, E.L., 2009. A novel picornavirus associated with gastroenteritis. J. Virol. 83, 12002–12006.

Li, Z., Wang, H., Chen, Y., Niu, J., Guo, Q., Leng, Q., Huang, Z., Deng, Z., Meng, G., 2017. Interleukin-18 protects mice from enterovirus 71 infection. Cytokine 96, 132–137.

Lim, E.S., Zhou, Y., Zhao, G., Bauer, I.K., Droit, L., Ndao, I.M., Warner, B.B., Tarr, P.I., Wang, D., Holtz, L.R., 2015. Early life dynamics of the human gut virome and bacterial microbiome in infants. Nat. Med. 21, 1228–1234.

Linder, P., 2006. Dead-box proteins: a family affair—active and passive players in RNP-remodeling. Nucleic Acids Res. 34, 4168–4180.

Liu, Y., Lu, N., Yuan, B., Weng, L., Wang, F., Liu, Y.J., Zhang, Z., 2014. The interaction between the helicase DHX33 and IPS-1 as a novel pathway to sense double-stranded RNA and RNA viruses in myeloid dendritic cells. Cell. Mol. Immunol. 11, 49–57.

Lopman, B.A., Reacher, M.H., Vipond, I.B., Sarangi, J., Brown, D.W.G., 2004. Clinical manifestation of norovirus gastroenteritis in health care settings. Clin. Infect. Dis. 39, 318–324.

Lopman, B.A., Steele, D., Kirkwood, C.D., Parashar, U.D., 2016. The vast and varied global burden of norovirus: prospects for prevention and control. PLoS Med. 13, e1001999.

Lupfer, C., Kanneganti, T.-D., 2013. The expanding role of NLRs in antiviral immunity. Immunol. Rev. 255, 13–24.

Man, S.M., Karki, R., Kanneganti, T.D., 2017. Molecular mechanisms and functions of pyroptosis, inflammatory caspases and inflammasomes in infectious diseases. Immunol. Rev. 277, 61–75.

Manrique, P., Bolduc, B., Walk, S.T., Van Der Oost, J., De Vos, W.M., Young, M.J., 2016. Healthy human gut phageome. Proc. Natl. Acad. Sci. U.S.A. 113, 10400–10405.

Marvin, S.A., Huerta, C.T., Sharp, B., Freiden, P., Cline, T.D., Schultz-Cherry, S., 2016. Type I interferon response limits astrovirus replication and protects against increased barrier permeability in vitro and in vivo. J. Virol. 90, 1988–1996.

Mattei, F., Schiavoni, G., Tough, D.F., 2010. Regulation of immune cell homeostasis by type I interferons. Cytokine Growth Factor Rev. 21, 227–236.

McCartney, S.A., Thackray, L.B., Gitlin, L., Gilfillan, S., Virgin, H.W., Colonna, M., 2008. MDA-5 recognition of a murine norovirus. PLoS Pathog. 4, e1000108.

Mcclarren, R.L., Lynch, B., Nyayapati, N., 2011. Acute infectious diarrhea. Prim. Care 38, 539–564. ix.

Minot, S., Sinha, R., Chen, J., Li, H., Keilbaugh, S.A., Wu, G.D., Lewis, J.D., Bushman, F.D., 2011. The human gut virome: inter-individual variation and dynamic response to diet. Genome Res. 21, 1616–1625.

Minot, S., Bryson, A., Chehoud, C., Wu, G.D., Lewis, J.D., Bushman, F.D., 2013. Rapid evolution of the human gut virome. Proc. Natl. Acad. Sci. U.S.A. 110, 12450–12455.

Mitoma, H., Hanabuchi, S., Kim, T., Bao, M., Zhang, Z., Sugimoto, N., Liu, Y.J., 2013. The DHX33 RNA helicase senses cytosolic RNA and activates the NLRP3 inflammasome. Immunity 39, 123–135.

Monroe, S.S., Jiang, B., Stine, S.E., Koopmans, M., Glass, R.I., 1993. Subgenomic RNA sequence of human astrovirus supports classification of Astroviridae as a new family of RNA viruses. J. Virol. 67, 3611–3614.

Mosallanejad, K., Sekine, Y., Ishikura-Kinoshita, S., Kumagai, K., Nagano, T., Matsuzawa, A., Takeda, K., Naguro, I., Ichijo, H., 2014. The DEAH-box RNA helicase DHX15 activates NF-kappaB and MAPK signaling downstream of MAVS during antiviral responses. Sci. Signal. 7, ra40.

Nash, A.K., Auchtung, T.A., Wong, M.C., Smith, D.P., Gesell, J.R., Ross, M.C., Stewart, C.J., Metcalf, G.A., Muzny, D.M., Gibbs, R.A., Ajami, N.J., Petrosino, J.F., 2017. The gut mycobiome of the Human Microbiome Project healthy cohort. Microbiome 5, 153.

Nieto-Torres, J.L., Verdia-Baguena, C., Jimenez-Guardeno, J.M., Regla-Nava, J.A., Castano-Rodriguez, C., Fernandez-Delgado, R., Torres, J., Aguilella, V.M., Enjuanes, L., 2015. Severe acute respiratory syndrome coronavirus E protein transports calcium ions and activates the NLRP3 inflammasome. Virology 485, 330–339.

Norman, J.M., Handley, S.A., Baldridge, M.T., Droit, L., Liu, C.Y., Keller, B.C., Kambal, A., Monaco, C.L., Zhao, G., Fleshner, P., Stappenbeck, T.S., McGovern, D.P., Keshavarzian, A., Mutlu, E.A., Sauk, J., Gevers, D., Xavier, R.J., Wang, D., Parkes, M., Virgin, H.W., 2015. Disease-specific alterations in the enteric virome in inflammatory bowel disease. Cell 160, 447–460.

Oikarinen, M., Tauriainen, S., Oikarinen, S., Honkanen, T., Collin, P., Rantala, I., MÄki, M., Kaukinen, K., HyÖty, H., 2012. Type 1 diabetes is associated with enterovirus infection in gut mucosa. Diabetes 61, 687–691.

Onyeagocha, C., Hossain, M.S., Kumar, A., Jones, R.M., Roback, J., Gewirtz, A.T., 2009. Latent cytomegalovirus infection exacerbates experimental colitis. Am. J. Pathol. 175, 2034–2042.

Oshiumi, H., Miyashita, M., Okamoto, M., Morioka, Y., Okabe, M., Matsumoto, M., Seya, T., 2015. DDX60 is involved in RIG-I-dependent and independent antiviral responses, and its function is attenuated by virus-induced EGFR activation. Cell Rep. 11, 1193–1207.

Poeck, H., Bscheider, M., Gross, O., Finger, K., Roth, S., Rebsamen, M., Hannesschlager, N., Schlee, M., Rothenfusser, S., Barchet, W., Kato, H., Akira, S., Inoue, S., Endres, S., Peschel, C., Hartmann, G., Hornung, V., Ruland, J., 2010. Recognition of RNA virus by RIG-I results in activation of CARD9 and inflammasome signaling for interleukin 1 beta production. Nat. Immunol. 11, 63–69.

Pott, J., Stockinger, S., Torow, N., Smoczek, A., Lindner, C., Mcinerney, G., Backhed, F., Baumann, U., Pabst, O., Bleich, A., Hornef, M.W., 2012. Age-dependent TLR3 expression of the intestinal epithelium contributes to rotavirus susceptibility. PLoS Pathog. 8, e1002670.

Rathinam, V.A., Jiang, Z., Waggoner, S.N., Sharma, S., Cole, L.E., Waggoner, L., Vanaja, S.K., Monks, B.G., Ganesan, S., Latz, E., Hornung, V., Vogel, S.N., Szomolanyi-Tsuda, E., Fitzgerald, K.A., 2010. The AIM2 inflammasome is essential for host defense against cytosolic bacteria and DNA viruses. Nat. Immunol. 11, 395–402.

Rathinam, V.A., Vanaja, S.K., Waggoner, L., Sokolovska, A., Becker, C., Stuart, L.M., Leong, J.M., Fitzgerald, K.A., 2012. TRIF licenses caspase-11-dependent NLRP3 inflammasome activation by gram-negative bacteria. Cell 150, 606–619.

Rauch, I., Hainzl, E., Rosebrock, F., Heider, S., Schwab, C., Berry, D., Stoiber, D., Wagner, M., Schleper, C., Loy, A., Urich, T., Muller, M., Strobl, B., Kenner, L., Decker, T., 2014. Type I interferons have opposing effects during the emergence and recovery phases of colitis. Eur. J. Immunol. 44, 2749–2760.

Reboldi, A., Dang, E.V., McDonald, J.G., Liang, G., Russell, D.W., Cyster, J.G., 2014. Inflammation. 25-hydroxycholesterol suppresses interleukin-1-driven inflammation downstream of type I interferon. Science 345, 679–684.

Reikine, S., Nguyen, J.B., Modis, Y., 2014. Pattern recognition and signaling mechanisms of RIG-I and MDA5. Front. Immunol. 5, 342.

Reyes, A., Haynes, M., Hanson, N., Angly, F.E., Heath, A.C., Rohwer, F., Gordon, J.I., 2010. Viruses in the faecal microbiota of monozygotic twins and their mothers. Nature 466, 334–338.

Reyes, A., Blanton, L.V., Cao, S., Zhao, G., Manary, M., Trehan, I., Smith, M.I., Wang, D., Virgin, H.W., Rohwer, F., Gordon, J.I., 2015. Gut DNA viromes of Malawian twins discordant for severe acute malnutrition. Proc. Natl. Acad. Sci. U.S.A. 112, 11941–11946.

Rivero-Calle, I., Gomez-Rial, J., Martinon-Torres, F., 2016. Systemic features of rotavirus infection. J. Infect. 72 (Suppl), S98–S105.

Rota, P.A., Oberste, M.S., Monroe, S.S., Nix, W.A., Campagnoli, R., Icenogle, J.P., PeÑaranda, S., Bankamp, B., Maher, K., Chen, M.-H., Tong, S., Tamin, A., Lowe, L., Frace, M., Derisi, J.L., Chen, Q., Wang, D., Erdman, D.D., Peret, T.C.T., Burns, C., Ksiazek, T.G., Rollin, P.E., Sanchez, A., Liffick, S., Holloway, B., Limor, J., Mccaustland, K., Olsen-Rasmussen, M., Fouchier, R., GÜnther, S., Osterhaus, A.D.M.E., Drosten, C., Pallansch, M.A., Anderson, L.J., Bellini, W.J., 2003. Characterization of a novel coronavirus associated with severe acute respiratory syndrome. Science 300, 1394–1399.

Rutz, M., Metzger, J., Gellert, T., Luppa, P., Lipford, G.B., Wagner, H., Bauer, S., 2004. Toll-like receptor 9 binds single-stranded CpG-DNA in a sequence- and pH-dependent manner. Eur. J. Immunol. 34, 2541–2550.

Ryan, J.L., Shen, Y.J., Morgan, D.R., Thorne, L.B., Kenney, S.C., Dominguez, R.L., Gulley, M.L., 2012. Epstein-Barr virus infection is common in inflamed gastrointestinal mucosa. Dig. Dis. Sci. 57, 1887–1898.

Sainathan, S.K., Bishnupuri, K.S., Aden, K., Luo, Q., Houchen, C.W., Anant, S., Dieckgraefe, B.K., 2012. Toll-like receptor-7 ligand Imiquimod induces type I interferon and antimicrobial peptides to ameliorate dextran sodium sulfate-induced acute colitis. Inflamm. Bowel Dis. 18, 955–967.

Schlee, M., Hartmann, G., 2016. Discriminating self from non-self in nucleic acid sensing. Nat. Rev. Immunol. 16, 566–580.

Seth, R.B., Sun, L., Ea, C.K., Chen, Z.J., 2005. Identification and characterization of MAVS, a mitochondrial antiviral signaling protein that activates NF-kappaB and IRF 3. Cell 122, 669–682.

Shortland, A., Chettle, J., Archer, J., Wood, K., Bailey, D., Goodfellow, I., Blacklaws, B.A., Heeney, J.L., 2014. Pathology caused by persistent murine norovirus infection. J. Gen. Virol. 95, 413–422.

Smyth, D.J., Cooper, J.D., Bailey, R., Field, S., Burren, O., Smink, L.J., Guja, C., Ionescu-Tirgoviste, C., Widmer, B., Dunger, D.B., Savage, D.A., Walker, N.M., Clayton, D.G., Todd, J.A., 2006. A genome-wide association study of nonsynonymous SNPs identifies a type 1 diabetes locus in the interferon-induced helicase (IFIH1) region. Nat. Genet. 38, 617.

Song-Zhao, G.X., Srinivasan, N., Pott, J., Baban, D., Frankel, G., Maloy, K.J., 2014. Nlrp3 activation in the intestinal epithelium protects against a mucosal pathogen. Mucosal Immunol. 7, 763–774.
Tai, J.H., Williams, J.V., Edwards, K.M., Wright, P.F., Crowe, J.J.E., Dermody, T.S., 2005. Prevalence of reovirus-specific antibodies in young children in Nashville, Tennessee. J. Infect. Dis. 191, 1221–1224.
Takaoka, A., Wang, Z., Choi, M.K., Yanai, H., Negishi, H., Ban, T., Lu, Y., Miyagishi, M., Kodama, T., Honda, K., Ohba, Y., Taniguchi, T., 2007. DAI (DLM-1/ZBP1) is a cytosolic DNA sensor and an activator of innate immune response. Nature 448, 501–505.
Tao, J., Zhou, X., Jiang, Z., 2016. cGAS-cGAMP-STING: the three musketeers of cytosolic DNA sensing and signaling. IUBMB Life 68, 858–870.
Tapparel, C., Siegrist, F., Petty, T.J., Kaiser, L., 2013. Picornavirus and enterovirus diversity with associated human diseases. Infect. Genet. Evol. 14, 282–293.
Tate, J.E., Rheingans, R.D., O'Reilly, C.E., Obonyo, B., Burton, D.C., Tornheim, J.A., Adazu, K., Jaron, P., Ochieng, B., Kerin, T., Calhoun, L., Hamel, M., Laserson, K., Breiman, R.F., Feikin, D.R., Mintz, E.D., Widdowson, M.A., 2009. Rotavirus disease burden and impact and cost-effectiveness of a rotavirus vaccination program in Kenya. J. Infect. Dis. 200 (Suppl. 1), S76–S84.
Tate, J.E., Burton, A.H., Boschi-Pinto, C., Steele, A.D., Duque, J., Parashar, U.D., 2012. 2008 Estimate of worldwide rotavirus-associated mortality in children younger than 5 years before the introduction of universal rotavirus vaccination programmes: a systematic review and meta-analysis. Lancet Infect. Dis. 12, 136–141.
Tate, J.E., Burton, A.H., Boschi-Pinto, C., Parashar, U.D., World Health Organization-Coordinated Global Rotavirus Surveillance, N, 2016. Global, regional, and national estimates of rotavirus mortality in children <5 years of age, 2000–2013. Clin. Infect. Dis. 62 (Suppl. 2), S96–S105.
Teijaro, J.R., 2016. Pleiotropic roles of type 1 interferons in antiviral immune responses. In: ALT, F.W. (Ed.), Advances in Immunology. Academic Press. Chapter 5.
Toma, C., Higa, N., Koizumi, Y., Nakasone, N., Ogura, Y., McCoy, A.J., Franchi, L., Uematsu, S., Sagara, J., Taniguchi, S., Tsutsui, H., Akira, S., Tschopp, J., Nunez, G., Suzuki, T., 2010. Pathogenic vibrio activate NLRP3 inflammasome via cytotoxins and Tlr/nucleotide-binding oligomerization domain-mediated NF-kappa B signaling. J. Immunol. 184, 5287–5297.
Tsolenyanu, E., Seheri, M., Dagnra, A., Djadou, E., Tigossou, S., Nyaga, M., Adjeoda, E., Armah, G., Mwenda, J.M., Atakouma, Y., 2014. Surveillance for rotavirus gastroenteritis in children less than 5 years of age in Togo. Pediatr. Infect. Dis. J. 33 (Suppl. 1), S14–S18.
Uchiyama, R., Chassaing, B., Zhang, B., Gewirtz, A.T., 2014. Antibiotic treatment suppresses rotavirus infection and enhances specific humoral immunity. J. Infect. Dis. 210, 171–182.
Unterholzner, L., Keating, S.E., Baran, M., Horan, K.A., Jensen, S.B., Sharma, S., Sirois, C.M., Jin, T., Latz, E., Xiao, T.S., Fitzgerald, K.A., Paludan, S.R., Bowie, A.G., 2010. IFI16 is an innate immune sensor for intracellular DNA. Nat. Immunol. 11, 997–1004.
Venkataraman, T., Valdes, M., Elsby, R., Kakuta, S., Caceres, G., Saijo, S., Iwakura, Y., Barber, G.N., 2007. Loss of DExD/H box RNA helicase LGP2 manifests disparate antiviral responses. J. Immunol. 178, 6444–6455.
Vijay-Kumar, M., Wu, H., Aitken, J., Kolachala, V.L., Neish, A.S., Sitaraman, S.V., Gewirtz, A.T., 2007. Activation of toll-like receptor 3 protects against DSS-induced acute colitis. Inflamm. Bowel Dis. 13, 856–864.
Virgin, H.W., 2014. The virome in mammalian physiology and disease. Cell 157, 142–150.

Wang, J.T., Chang, S.C., 2004. Severe acute respiratory syndrome. Curr. Opin. Infect. Dis. 17, 143–148.

Wang, H., Lei, X., Xiao, X., Yang, C., Lu, W., Huang, Z., Leng, Q., Jin, Q., HE, B., Meng, G., Wang, J., 2015a. Reciprocal regulation between enterovirus 71 and the NLRP3 inflammasome. Cell Rep. 12, 42–48.

Wang, P., Zhu, S., Yang, L., Cui, S., Pan, W., Jackson, R., Zheng, Y., Rongvaux, A., Sun, Q., Yang, G., Gao, S., Lin, R., You, F., Flavell, R., Fikrig, E., 2015b. Nlrp6 regulates intestinal antiviral innate immunity. Science 350, 826–830.

Wang, W., Xiao, F., Wan, P., Pan, P., Zhang, Y., Liu, F., Wu, K., Liu, Y., Wu, J., 2017a. EV71 3D protein binds with NLRP3 and enhances the assembly of inflammasome complex. PLoS Pathog. 13, e1006123.

Wang, W., Xu, L., SU, J., Peppelenbosch, M.P., Pan, Q., 2017b. Transcriptional regulation of antiviral interferon-stimulated genes. Trends Microbiol. 25, 573–584.

Weber, R., Ledergerber, B., Zbinden, R., et al., 1999. Enteric infections and diarrhea in human immunodeficiency virus-infected persons: prospective community-based cohort study. Arch. Intern. Med. 159, 1473–1480.

Wilen, C.B., Lee, S., Hsieh, L.L., Orchard, R.C., Desai, C., Hykes, B.L., Mcallaster, M.R., Balce, D.R., Feehley, T., Brestoff, J.R., Hickey, C.A., Yokoyama, C.C., Wang, Y.-T., Macduff, D.A., Kreamalmayer, D., Howitt, M.R., Neil, J.A., Cadwell, K., Allen, P.M., Handley, S.A., Van Lookeren Campagne, M., Baldridge, M.T., Virgin, H.W., 2018. Tropism for tuft cells determines immune promotion of norovirus pathogenesis. Science 360, 204–208.

Wobus, C.E., Karst, S.M., Thackray, L.B., Chang, K.O., Sosnovtsev, S.V., Belliot, G., Krug, A., Mackenzie, J.M., Green, K.Y., Virgin, H.W., 2004. Replication of Norovirus in cell culture reveals a tropism for dendritic cells and macrophages. PLoS Biol. 2, e432.

World Health Organization, 2017. Diarrhoeal Disease. [Online]. Accessed 2 May 2017.

Xia, P., Wang, S., Gao, P., Gao, G., Fan, Z., 2016. DNA sensor cGAS-mediated immune recognition. Protein Cell 7, 777–791.

Yang, J.Y., Kim, M.S., Kim, E., Cheon, J.H., Lee, Y.S., Kim, Y., Lee, S.H., Seo, S.U., Shin, S.H., Choi, S.S., Kim, B., Chang, S.Y., Ko, H.J., Bae, J.W., Kweon, M.N., 2016. Enteric viruses ameliorate gut inflammation via Toll-like receptor 3 and Toll-like receptor 7-mediated interferon-beta production. Immunity 44, 889–900.

Yoneyama, M., Kikuchi, M., Natsukawa, T., Shinobu, N., Imaizumi, T., Miyagishi, M., Taira, K., Akira, S., Fujita, T., 2004. The RNA helicase RIG-I has an essential function in double-stranded RNA-induced innate antiviral responses. Nat. Immunol. 5, 730–737.

Yoneyama, M., Kikuchi, M., Matsumoto, K., Imaizumi, T., Miyagishi, M., Taira, K., Foy, E., Loo, Y.M., Gale, M., Akira, S., Yonehara, S., Kato, A., Fujita, T., 2005. Shared and unique functions of the DexD/H-box helicases RIG-I, MDA5, and LGP2 in antiviral innate immunity. J. Immunol. 175, 2851–2858.

Yoneyama, M., Onomoto, K., Jogi, M., Akaboshi, T., Fujita, T., 2015. Viral RNA detection by RIG-I-like receptors. Curr. Opin. Immunol. 32, 48–53.

Yuen, K.Y., Woo, P.C., Liang, R.H., Chiu, E.K., Chen, F.F., Wong, S.S., Lau, Y.L., Ha, S.Y., Peiris, J.S., Siau, H., Chan, T.K., 1998. Clinical significance of alimentary tract microbes in bone marrow transplant recipients. Diagn. Microbiol. Infect. Dis. 30, 75–81.

Yutin, N., Makarova, K.S., Gussow, A.B., Krupovic, M., Segall, A., Edwards, R.A., Koonin, E.V., 2018. Discovery of an expansive bacteriophage family that includes the most abundant viruses from the human gut. Nat. Microbiol. 3, 38–46.

Zanoni, I., Granucci, F., Broggi, A., 2017. Interferon (IFN)-lambda takes the helm: immunomodulatory roles of type III IFNs. Front. Immunol. 8, 1661.

Zhang, Z., Kim, T., Bao, M., Facchinetti, V., Jung, S.Y., Ghaffari, A.A., Qin, J., Cheng, G., Liu, Y.J., 2011a. DDX1, DDX21, and DHX36 helicases form a complex with the adaptor molecule TRIF to sense dsRNA in dendritic cells. Immunity 34, 866–878.

Zhang, Z., Yuan, B., Bao, M., Lu, N., Kim, T., Liu, Y.J., 2011b. The helicase DDX41 senses intracellular DNA mediated by the adaptor STING in dendritic cells. Nat. Immunol. 12, 959–965.

Zhang, Z., Yuan, B., Lu, N., Facchinetti, V., Liu, Y.J., 2011c. DHX9 pairs with IPS-1 to sense double-stranded RNA in myeloid dendritic cells. J. Immunol. 187, 4501–4508.

Zhang, B., Chassaing, B., Shi, Z., Uchiyama, R., Zhang, Z., Denning, T.L., Crawford, S.E., Pruijssers, A.J., Iskarpatyoti, J.A., Estes, M.K., Dermody, T.S., Ouyang, W., Williams, I.R., Vijay-Kumar, M., Gewirtz, A.T., 2014. Viral infection. Prevention and cure of rotavirus infection via TLR5/Nlrc4-mediated production of Il-22 and IL-18. Science 346, 861–865.

Zhang, Y., Chen, X., Gueydan, C., Han, J., 2017. Plasma membrane changes during programmed cell deaths. Cell Res. 28, 9–21.

Zhao, G., Vatanen, T., Droit, L., Park, A., Kostic, A.D., Poon, T.W., Vlamakis, H., Siljander, H., Härkönen, T., Hämäläinen, A.-M., Peet, A., Tillmann, V., Ilonen, J., Wang, D., Knip, M., Xavier, R.J., Virgin, H.W., 2017. Intestinal virome changes precede autoimmunity in type I diabetes-susceptible children. Proc. Natl. Acad. Sci. U.S.A. 114, E6166–E6175.

Zhou, R., Wei, H., Sun, R., Tian, Z., 2007. Recognition of double-stranded RNA by TLR3 induces severe small intestinal injury in mice. J. Immunol. 178, 4548–4556.

Zhu, S., Ding, S., Wang, P., Wei, Z., Pan, W., Palm, N.W., Yang, Y., Yu, H., Li, H.B., Wang, G., Lei, X., de Zoete, M.R., Zhao, J., Zheng, Y., Chen, H., Zhao, Y., Jurado, K.A., Feng, N., Shan, L., Kluger, Y., Lu, J., Abraham, C., Fikrig, E., Greenberg, H.B., Flavell, R.A., 2017. Nlrp9b inflammasome restricts rotavirus infection in intestinal epithelial cells. Nature 546, 667–670.

Zuo, T., Wong, S.H., Lam, K., Lui, R., Cheung, K., Tang, W., Ching, J.Y.L., Chan, P.K.S., Chan, M.C.W., Wu, J.C.Y., Chan, F.K.L., Yu, J., Sung, J.J.Y., Ng, S.C., 2018. Bacteriophage transfer during faecal microbiota transplantation in *Clostridium difficile* infection is associated with treatment outcome. Gut 67, 634–643.

FURTHER READING

World Health Organization, 2009. Meeting of the immunization Strategic Advisory Group of Experts. April 2009—conclusions and recommendations. Wkly. Epidemiol. Rec. 84, 220–236.

Baldridge, M.T., Nice, T.J., McCune, B.T., Yokoyama, C.C., Kambal, A., Wheadon, M., Diamond, M.S., Ivanova, Y., Artyomov, M., Virgin, H.W., 2015a. Commensal microbes and interferon-lambda determine persistence of enteric murine norovirus infection. Science 347, 266–269.

CHAPTER FOUR

Genome Damage Sensing Leads to Tissue Homeostasis in *Drosophila*

Chaitali Khan[2,3], Sonia Muliyil[3,4], B.J. Rao[1,5]
Department of Biological Sciences, Tata Institute of Fundamental Research, Mumbai, India
[1]Corresponding author: e-mail addresses: bjrao@iisertirupati.ac.in; bjrao@tifr.res.in

Contents

[2] Department of Genetics, Albert Einstein College of Medicine, 1300 Morris Park Avenue, Bronx, NY 10461, United States.
[3] Equal contribution.
[4] Sir William Dunn School of Pathology, University of Oxford, South Parks Road, Oxford OX1 3RE, United Kingdom.
[5] Indian Institute of Science Education and Research (IISER) Tirupati, Transit Campus: Sree Rama Engineering College, Karkambadi Road, Mangalam, Tirupati 517507, Andhra Pradesh, India.

International Review of Cell and Molecular Biology, Volume 345
ISSN 1937-6448
https://doi.org/10.1016/bs.ircmb.2018.12.001

Abstract

DNA repair is a critical cellular process required for the maintenance of genomic integrity. It is now well appreciated that cells employ several DNA repair pathways to take care of distinct types of DNA damage. It is also well known that a cascade of signals namely DNA damage response or DDR is activated in response to DNA damage which comprise cellular responses, such as cell cycle arrest, DNA repair and cell death, if the damage is irreparable. There is also emerging literature suggesting a cross-talk between DNA damage signaling and several signaling networks within a cell. Moreover, cell death players themselves are also well known to engage in processes outside their canonical function of apoptosis. This chapter attempts to build a link between DNA damage, DDR and signaling from the studies mainly conducted in mammals and *Drosophila* model systems, with a special emphasis on their relevance in overall tissue homeostasis and development.

1. DNA DAMAGE, RESPONSE AND REPAIR: A CONSERVED THEME FROM MAMMALS TO FLIES

DNA (Deoxyribonucleic acid) serves as a repository of genetic information in most of the prokaryotic and eukaryotic cells. It is estimated that every cell in the human body on an average incurs $\sim 10^4$–10^5 DNA damages per day from sources arising within the cell or from the external environment. Damages within the DNA can result in impairment of cellular process, such as replication, transcription or even lead to specific genomic alterations, which in turn contribute to pathological conditions, aging and cancer (Bohgaki et al., 2010; Curtin, 2012). The type of DNA lesions can range from as simple as base modifications or base-pairing mismatches, to something more severe like breaks in one or both the strands; these lesions are resolved by a network of DNA repair mechanisms (Hakem, 2008). Irrespective of the kind of DNA lesion or the repair pathway employed, presence of DNA damages initiates a cascade of cellular response known as DNA damage Response (DDR), which senses the DNA damage, signals its presence, and mediates its repair. DDR serves to coordinate the cellular response with DNA repair, by transiently arresting the cell cycle to allow for efficient DNA damage repair or signal cells to activate apoptosis-under conditions of severe damage (reviewed in Jackson, 2002). This section first introduces the kind of DNA lesions, their sources and repair pathways in mammalian system. In this section, DNA damage signaling will be discussed in the context of Double Strand Breaks (DSBs). The last section will attempt to compare and contrast the mammalian and *Drosophila* DNA damage repair pathways and signaling.

1.1 DNA Damage Repair and Response in Mammals

Cells are endowed with distinct molecular pathways dedicated to repair various kinds of DNA damages. Table 1 summarizes the nature of DNA lesions, genotoxins and repair mechanisms along with respective molecular players.

1.1.1 Direct Repair

Direct repair pathway involves the reversal of damaged DNA bases by a single enzymatic reaction (reviewed in Sedgwick et al., 2007). Some of the simple modification such as O-6-methylguanine, 1-methyladenine and 3-methylcytosine caused by methylating agents are repaired by the direct repair pathway. For instance, removal of O-6-methyl-guanine is carried out by O-6-methylguanine-DNA methyltransferase (MGMT). MGMT, also known as AGT, removes the DNA adducts by transferring the alkyl group from the oxygen in the DNA to a cysteine residue in its active site (Pegg et al., 1995). Whereas, 1-methyladenine (1meA) and 3-methylcytosine (3meC) are removed by another enzyme called dioxygenase AlkB homologues 2 and 3 (ABH2 and ABH3). Both ABH2 and ABH3 can repair 1meA and 3meC; however, ABH2 prefers double-stranded DNA, whereas ABH3 favors single-stranded DNA and RNA (Duncan et al., 2002). Although mutations in direct repair pathway are not associated with severe consequences on survival, they lead to increased predisposition to several types of cancers (Glassner et al., 1999).

1.1.2 Mismatch Repair

The MMR pathway plays an important role in repairing mismatches, which are small insertions and deletions that take place during DNA replication and recombination arising from errors of DNA polymerases. The MMR process involves a complex interplay of MMR-specific proteins with the replication and/or recombination machinery. It is activated by the binding of the mismatch-recognition factors, MutSα and MutSβ, to substrates that contain base mismatches and insertion/deletion loops. The ATP-dependent recruitment of MutL homologues to the mismatch-bound complex is followed by the exonuclease-mediated degradation of the error-containing strand. Once the mismatch has been removed, re-synthesis of the degraded region by a DNA polymerase, followed by sealing of the remaining nick by DNA ligase, completes the repair process (Jiricny, 2006; Modrich, 1991).

Table 1 Type of DNA Lesion, Sources, Repair Pathway, and the Key Molecular Players

Source(s)	Nature of DNA Lesion	Repair Pathway	Key Molecular Players
Methylation of DNA by *S*-adenosyl-methionine (SAM)	– O-6-Methyl-guanine – 1-Methyl adenine – 3-Methyl cytosine	Direct Repair	MGMT, ABH2, ABH3
Replication and recombination errors	– Base mismatch – Insertion, deletion	Mismatch Repair (MMR)	hMSH2, hMSH3, hMSH6, hMLH1, hMLH3, hPMS1, and hPMS2, PCNA, RPA, LIG III, Pol-delta
Alkylating agents	– Abasic sites – 8-Oxoguanine – Single strand break	Base excision repair (BER)	Ogg1 and Mutyh, APE1, Polβ, XRCC1/ligase III, Ligase 1
– UV-light-oxygen radical	– 6-4 photoproduct – Pyrimidine dimers – Bulky adducts	Nucleotide excision repair (NER)	GG NER: XPC–RAD23B and DDB1–DDB2/XPE TC-NER: CSA/ERCC8 and CSB/ERCC6
– X-ray oxygen radical – UV-light oxygen radical	– Double strand breaks – Inter-strand crosslink	Recombinational Repair (HR-NHEJ)	HR: MRN, 53BP1, CtIP, ATM, RPA, RAD51, RAD51-paralogs, RAD54, Brca2, GEN1, MUS81 NHEJ: Ku70/80, DNAPKC, XRCC4, LigIV, XLF, APLF

Several homologues of the bacterial MMR genes *MutS* and *MutL* have been identified in yeast and mammals. Human cells express three MutS homologous (hMSH2, hMSH3, and hMSH6) and four MutL homologous (hMLH1, hMLH3, hPMS1, and hPMS2) (Alani et al., 1997; Genschel et al., 1998). Additionally, other proteins required for human MMR such as PCNA and RPA are also part of DNA metabolism (reviewed in Jiricny, 2006). By removing biosynthetic errors from newly synthesized DNA, mismatch repair (MMR) improves the fidelity of DNA replication by several orders of magnitude. Defects in MMR are associated with genome-wide instability, predisposition to certain types of cancer, including hereditary non-polyposis colorectal cancer, resistance to certain chemotherapeutic agents, and abnormalities in meiosis and sterility in mammalian systems (Li, 2007; Sijmons and Hofstra, 2016).

1.1.3 Base Excision Repair

Base excision repair (BER) corrects small base lesions that do not significantly distort the DNA helix structure. These damages typically arise from deamination, oxidation, or methylation. Moreover, significant numbers of the damage also result from spontaneous decay of DNA; exposure to environmental chemicals, radiation, or treatment with cytostatic drugs can also lead to base modifications. BER is initiated by a DNA glycosylase that recognizes and removes the damaged base, leaving an abasic site; at least 11 distinct DNA glycosylases are known in mammalian, with different base recognition specificity (reviewed in Statistik, 2014). Abasic site is further processed by short-patch repair or long-patch repair. The short-patch BER sub-pathway typically replaces a single nucleotide, whereas the long-patch sub-pathway results in the incorporation of 2–13 nucleotides (Kim and Wilson, 2012). Base removal is followed by strand incision of the apurinic or apyrimidinic (AP) site by the apurinic/apyrimidinic endonuclease-1 (APE1). The newly generated gap is by DNA polymerase-β (Polβ) in case of the short-patch BER sub-pathway, and by Polβ and Polε or δ in case of the long-patch BER sub-pathway. Strand ligation is carried out by the X-ray repair cross-complementing protein-1 (XRCC1)/ligase III (LigIII) complex in the case of the short-patch sub-pathway. For the long-patch BER sub-pathway, other proteins are involved in the repair process before the ligation takes place. Thus, 5′-flap endonuclease (FEN1), Poly (ADP-ribose) polymerase (PARP1), and LigaseI (LigI) participate in the DNA synthesis–ligation step, and displace the 2- to 13-base DNA flap (Kim and Wilson, 2012).

1.1.4 Nucleotide Excision Repair

The NER pathway is a multi-step process that serves to repair a variety of DNA damages, including DNA lesions caused by UV radiation, mutagenic chemicals, or chemotherapeutic drugs that form bulky DNA adducts. UV lesions especially the photo-products (pyrimidine dimers) and the bulky adducts are repaired by the Nucleotide Excision Repair (NER) pathway (reviewed in Leibeling et al., 2006). NER sub-pathways that have been identified are as follows: the global genome NER (GG-NER) that detects and removes lesions throughout the genome, and the transcription-coupled NER (TC-NER), which repairs actively transcribed genes. NER begins with the recognition of the DNA lesion, followed by incisions at sites flanking the DNA lesion, and culminates in the removal of the oligonucleotide containing the DNA lesion. Ligation of a newly synthesized oligonucleotide, complementary to the pre-existing strand, serves to fill the gap, thus ending the NER process. GG-NER and TC-NER involve several common proteins and proceed through the same repair steps, except during recognition of the DNA lesion. In GG-NER this recognition involves the XPC–RAD23B and DDB1–DDB2/XPE proteins, whereas recognition in TC-NER is mediated by cockayne syndrome group A (CSA) (ERCC8) and CSB (ERCC6) (Fousteri and Mullenders, 2008). NER is an important DNA-repair pathway and its impairment is associated with growth defects, excessive UV sensitivity, and in certain cases, increased skin cancer (Marteijn et al., 2014; Stein et al., 1989).

1.1.5 DSB Repair and DNA Damage Signaling

Of all the DNA lesions, the most deleterious is thought to be the DSBs, breaks in the backbone of both strands of the DNA at least 10 base pairs or less apart. DSBs are caused by exposure to Ionizing radiation (IR) such as X-rays or gamma-rays (Cadet et al., 2008). In mammalian cells, DSBs are repaired by two major pathways: The Homologous Recombination (HR) and the Non-Homologous End Joining (NHEJ) (Jackson and Bartek, 2009). Depending upon the cellular context, the DSB ends can either be recognized by the MRE11/Rad50/NBS1 (MRN) or the Ku70/Ku80 complexes. The initial recognition and binding of DSB sensors decides the choice of repair pathway; binding of MRN poise the breaks for repair via HR, whereas binding of Ku70/80 leads to NHEJ (reviewed in Ciccia and Elledge, 2010; Sancar et al., 2004).

The MRN complex plays an essential role in initiating the DSB signaling by recruiting the apical kinase Ataxia-Telangiectasia-Mutated (ATM) via interactions with its N-terminal domain (Lamarche et al., 2010). In undamaged cells, ATM stays as inactive dimers or multimers; however, its recruitment to DSB sites via the MRN complex causes its auto-phosphorylation and dissociation into active monomers. The ATM kinase plays an essential role in the generation of γH2ax foci by phosphorylating H2Ax at Ser-139, which serves as one of the early hallmarks of initiation of DSB signaling. γH2ax foci in turn serve as the docking site for recruiting repair factors and aid in facilitating repair (Paull et al., 2000; Podhorecka et al., 2010). ATM kinase also phosphorylates other downstream candidates such as Nijmegen breakage syndrome protein 1 (NBS1), (p53 Binding Protein 1) 53BP1, (C-terminal Interacting Protein) CtIP to promote their efficient docking and repair functions (Kobayashi et al., 2004). Resection of the 5′-terminated DNA strand at DSBs by MRN also leads to the formation of 3′-overhanged ssDNA, initially coated by RPA. In a subsequent step, RPA is replaced by the key recombination protein RAD51. RAD51 and ssDNA form a nucleoprotein filament; this catalyzes by homology search by pairing with the homologous template strand invasion (Baumann and West, 1998). The transient interaction of the RAD51 nucleoprotein filament with the template dsDNA is referred to as a synaptic complex. This ultimately leads to the displacement of one of the template DNA strands forming a displacement loop (D-loop, also called joint molecule) intermediate. The activity of RAD51 is also promoted by a group of proteins termed RAD51 paralogs: RAD51B, RAD51C, RAD51D, XRCC2, and XRCC3 (Suwaki et al., 2011). RAD51 is removed from the heteroduplex dsDNA of the D-loop structure by RAD54, a dsDNA translocase that uses its ATPase-powered motor activity to displace RAD51 from dsDNA. RAD54 thus promotes D-loop stability, which allows DNA synthesis and facilitates canonical HR (Heyer et al., 2006). Synthesis of DNA creates a Holliday junction which is resolved by the action of structure-selective nucleases: yeast Mus81-Mms4 (human MUS81-EME1) and Yen1 (human GEN1) (Matos and West, 2014).

The initial step in NHEJ is the recognition and binding of the Ku heterodimer to the DSB; the Ku heterodimer is composed of the Ku70 and Ku80 subunits. Once the Ku heterodimer is bound to the DSB ends, it then serves as a scaffold to recruit the other NHEJ factors to the damage site. The Ku heterodimer either directly or indirectly binds to the main NHEJ factors, including DNA-PKcs, X-ray cross complementing protein 4 (XRCC4), DNA Ligase IV, XRCC4-like factor (XLF), and Aprataxin-and-PNK-like factor (APLF). The recruitment of these factors is followed by bridging of the DNA ends, end stability, DNA end processing and finally the ligation of the broken ends and dissolution of the NHEJ complex (Davis and Chen, 2013). Schematic shows the comparison of HR and NHEJ pathway (Fig. 1).

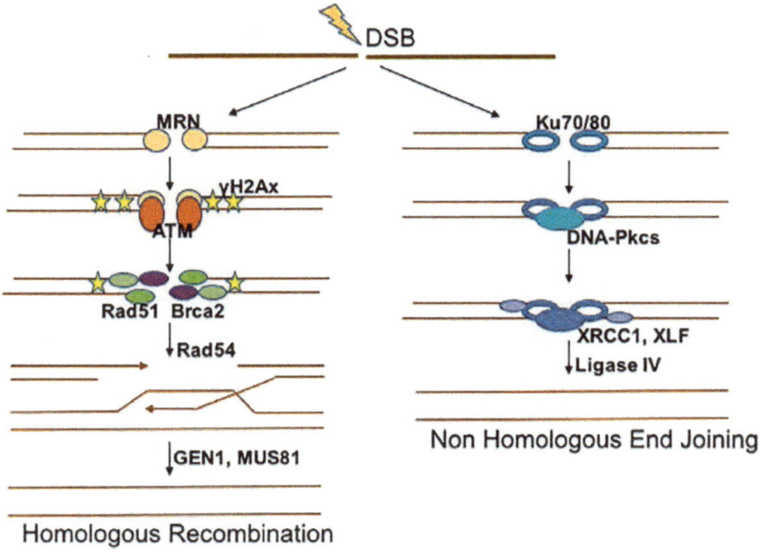

Fig. 1 The molecular players of HR and NHEJ pathway. Schematic showing the comparison of HR and NHEJ repair pathways. HR utilizes the information from the homologous chromosome to repair the lesion and involves multiple steps and players. In contrast, NHEJ involves the joining of broken strand together with the help of few players.

The HR pathway is relatively slow, but it serves as an error proof mechanism for repairing the DSBs. In contrast to HR, NHEJ is a relatively fast, but an error prone mechanism. NHEJ repair pathway involves simple steps of end processing and end-annealing, which are performed by a relatively small number of molecular players compared to HR (Mao et al., 2008, 2017; Orii et al., 2006). The importance of HR in DNA repair can be realized by the fact that mutation in its key players leads to lethality during early stages of development of mice (Suzuki et al., 1997; Tsuzuki et al., 1996). On the other hand, the absence of NHEJ players appears to be less critical in mice model carrying mutations in components of NHEJ survive past adulthood (Gu et al., 1997; Jhappan et al., 1997). As can be envisaged, the usage of the HR versus NHEJ pathways also varies between proliferative or differentiated tissues within an organism. NHEJ serves as the major repair pathway in post-differentiated tissues such as the nervous system, whereas actively dividing cells rely on HR (Lee and McKinnon, 2007; Orii et al., 2006).

1.1.6 Cellular Response to DNA damage

The members of Phosphatidylinositol-3 kinase-related kinases (PIKK) family are known to play an important role in relaying DNA damage signaling from break sites. This includes Ataxia-Telangiectasia-Mutated (ATM), Ataxia Telangiectasia and Rad3 Related (ATR) and DNA-dependent Protein Kinase catalytic subunit (DNA-PKcs) (Falck et al., 2005; Flynn and Zou, 2011; Shiloh, 2006; Shiloh and Ziv, 2013). These kinases phosphorylate a number of their targets at Ser/Thr residues, thereby regulating distinct steps of DSB signaling. The function of ATM and DNA-PKcs is important at the DSBs, whereas ATR plays a central role in signaling at SSBs and stalled replication forks (Flynn and Zou, 2011). DNA damage signaling is to coordinate cellular response along with DNA repair; the cascade of events initiated in the presence of DNA damage being defined as DNA Damage Response (DDR) (Jackson, 2002; Jackson and Bartek, 2009). Two of the best studied targets of ATM/ATR are the protein kinases CHK1 and CHK2 which, together with ATM and ATR, decrease the cyclin-dependent kinase (CDK) activity. Inhibition of CDKs slows down or arrests cell-cycle progression at the G1–S, intra-S and G2–M cell-cycle checkpoints, which is thought to increase the time available for DNA repair, before replication or mitosis ensues. Moreover, another arm of DDR serves to remove damaged and un-repairable cells from the system by triggering cell death (Kaina, 2003; Roos and Kaina, 2006). Fig. 2 summarizes the key steps of DDR taking DSB site as an example.

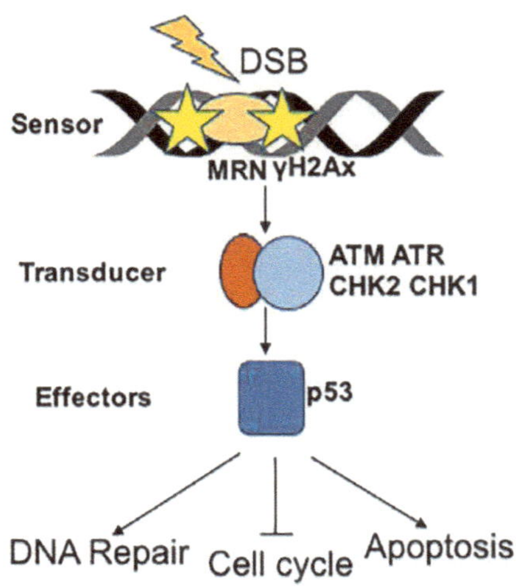

Fig. 2 DNA damage response. A schematic of DDR at DSB site, which consists of sensing of DNA damage by the sensor (MRN), leading to transduction of the signal by apical kinases (ATM, ATR, ChK2 and CHK1), eventually activating the effectors (p53) which regulate the cellular response.

To mediate the decision between cell cycle arrest, cell death or even senescence, cells rely upon another candidate of the DDR pathway, the tumor suppressor gene p53. p53 is activated by Chk2 kinase via phosphorylation, leading to its stabilization and nuclear translocation. Once stabilized and activated, p53 regulates the transcription of genes required for cell cycle arrest, DNA repair and cell death (Levine et al., 2006; Oren, 1999). The extent of DNA damage dictates the levels of p53 activation, which in turn decides the outcome of cellular fate. While low levels of p53 promote cell cycle arrest by activating p21, contrarily, higher levels of p53 cause expression of pro-apoptotic genes tipping the cell toward a suicidal program (Li et al., 2011). Interestingly, the p53 paralogs p63 and p73 can regulate cellular response independent of p53, and often have distinct roles in regulating cellular function and even differentiation (Pietsch et al., 2008). The role of p53 dependent and independent pathways in DNA damage induced cell death will be discussed in greater detail in the next few sections.

1.2 DNA Damage Sensing, Response and Repair Pathways in *Drosophila*

Analysis of the *Drosophila* genome reveals that most of the repair pathways—base-excision, nucleotide excision and DSB repair (NHEJ and HR) as well as their players—are conserved across mammals and *Drosophila* (Sekelsky, 2017; Sekelsky et al., 2000). Table 2 shows a comparison of *Drosophila* HR repair genes with those of humans. Surprisingly, some of the key

Table 2 The Conservation Between HR Pathway in Humans and *Drosophila*

Human Genes	*Drosophila* Homologues
Rad51	spn-A
RAD51C	spn-D
RAD51D	Rad51D
RAD54L	okr
MRE11, RAD50, NBS1	Mre11, rad50, nbs
BLM	Blm
Brac2	Brca2
Rad52	—
BRCA1	—

players such as DNAPKcs and Rad52 were found to be missing from the *Drosophila* genome (Sekelsky et al., 2000).

Moreover, in some *Drosophila* contexts, the requirement of DNA repair genes appears to be less critical, for example, a null mutation of *Drosophila brca2* gene leads to only subtle meiotic defects with almost no effect on survival, which is quite contrasting from its mammalian orthologue (Klovstad et al., 2008; Suzuki et al., 1997). Similarly, mutations in HR players namely Rad51, Rad54 and Rad51-paralogs display no apparently striking defect during somatic development (Ghabrial et al., 1998; Staeva-Vieira et al., 2003). This suggests a redundancy among different repair pathways and perhaps even the molecular players of these repair pathways. *Drosophila* ATM and ATR kinases also appear to regulate checkpoint response by modulating Chk1 and Chk2 kinases, similar to their mammalian counterparts (Bi et al., 2005a; Hari et al., 1995; Sibon et al., 1999). However, the functional requirement of the apical kinase ATM/tefu and ATR/Mei-4 seems to be slightly different from its mammalian orthologues, such that ATR plays a critical role during the later steps of DSB repair, independent of its function in the checkpoint response (Bi et al., 2005a; LaRocque et al., 2007). Another difference lies in the sensitivity of *Drosophila* ATM and ATR to the levels of DNA damage faced by the cell. Interestingly, ATM appears to play an important role during low levels of DNA damage, whereas ATR is thought to be critical during higher levels of DNA damage (Bi et al., 2005b). On a similar note, the role of the γH2ax signaling in *Drosophila* is also known to be less critical as opposed to γH2ax signaling in the mammalian context, as evident from the fact that exposure of H2av (*Drosophila* H2ax homologue) mutant to IR leads to only a twofold increase in cell death in wing discs with no apparent lethality (Madigan et al., 2002).

Moreover, the basis for this functional difference is also reflected in their sequences, for example, sequence comparison of dBRCA2 with that of hBRCA2 shows conservation with respect to the BRCT repeats and DNA binding domains, whereas the rest of the sequence shows no similarity (Chakraborty et al., 2015). A similar kind of sequence shuffling was also found in both the ATM and ATR genes. It is hypothesized that in comparison to ATM, ATR plays a central role in DDR response, and perhaps this difference might stem from the changes in their sequence during evolution (Bi et al., 2005b). These studies pointed out that the differences observed in *Drosophila* DNA repair pathway can be attributed to changes in genomic sequence, perhaps as a result of divergent evolution.

It is being increasingly appreciated known that genomic instability is the leading cause of cancer. The cause of cancer has been ascribed to abnormal cellular signaling leading to uncontrollable cell growth. Due to lack of viable mouse models of DNA repair mutants, it is hard to study the effects of the absence of some of these players during development, and even in cancer progression. Interestingly, most of the *Drosophila* DNA repair mutant(s) survive until adulthood, thus providing an opportunity to study them during development. It is also interesting to note that most of the *Drosophila* DSB repair genes give rise to defective patterning during embryogenesis causing ventralization of the egg (the phenotype was also called as the spindle-like (*spn*)). In fact, they were isolated in a screen conducted to uncover egg patterning defects in *Drosophila* (Ghabrial et al., 1998; Tearle and Nusslein-Volhard, 1987). The mechanism of the spindle (*spn*) phenotype in DNA repair mutants was linked to abnormal DNA damage signaling arising from defective meiotic crossovers, causing persistence of Chk2/mnk activation. The activated Chk2/mnk in turn leads to an abnormal EGFR signaling, as a result of mislocalization of EGFR ligand "gurken" (Abdu et al., 2002, 2003; Barbosa et al., 2007; Staeva-Vieira et al., 2003). These studies provide an elegant example of how crosstalk with the DDR can have an impact on tissue signaling, and overall development. Next section will focus on how DDR is linked to cell death, and the role of DNA damage-induced cell death during organismal development.

2. TISSUE SCULPTING BY CELL DEATH DURING GENOMIC STRESS

The two main strategies employed by cells in response to DNA damage happen to be repair versus death. The choice is often dictated by the extent or nature of the damage, as well as the cellular context. Some of these factors include p53 status, DNA repair capacity and death-receptor responsiveness. Importantly, these factors also crosstalk with cellular signaling pathways, and are known to be closely intertwined with cancer initiation and progression. This section will introduce the key cell death pathways and its respective players in mammals as well as in *Drosophila*.

2.1 Cell Death Pathways in Mammals

There are two main apoptotic pathways: the extrinsic or death receptor pathway and the intrinsic or mitochondrial pathway. Both the pathways are initially triggered in a different manner but eventually converge on the same terminal, or execution step of caspase activation. The intrinsic apoptotic pathway is mainly governed by the Bcl-2 family of proteins, which regulates mitochondrial permeability and the release of cytochrome *c* (Cory and Adams, 2002; Kuwana et al., 2002). Bcl-2 family members are classified on the basis of the Bcl-2 homology (BH) domains (BH1, BH2, BH3 and BH4). The BH3 domain is conserved among all the members, and is required for hetero-dimerization, and it is the minimum requirement for the pro-apoptotic function (Kelekar and Thompson, 1998). On the basis of their functions, the Bcl-2 family can be classified into three distinct classes: (1) pro-survival members, which are structurally most similar to Bcl-2, e.g., Bcl-XL; (2) pro-apoptotic proteins, BAX and BAK, structurally most similar to Bcl-2 and Bcl-XL functioning via antagonizing the pro-survival functions of other Bcl-2 members; and (3) the pro-apoptotic "BH3-only" proteins, BID, BIM, PUMA, NOXA, BAD, etc. (Lomonosova and Chinnadurai, 2009). In the classical mitochondrial apoptotic pathway, BH3-only proteins are thought to be the central regulators that connect cell death cues to other BCL-2 family members. It is known that "BH3-only" proteins localize to the mitochondria and interact with anti-apoptotic Bcl-2 family members, resulting in the activation of caspase-9 and cell death execution.

The extrinsic apoptotic pathway is regulated through the induction of genes encoding three transmembrane proteins: Fas, DR5 and PERP. The cell-surface receptor Fas (CD95), also a member of the TNF-R family of receptors, and its ligand Fas constitute the core components of the extrinsic death pathway (Muzio et al., 1998; Nagata and Golstein, 1995). p53 activates the extrinsic pathway by inducing the expression of *Fas* via regulation of elements in its promoter (Müller et al., 1998). Another candidate of this family is the receptor *DR5*, which encodes for the death-domain-containing receptor for TNF-related apoptosis-inducing ligand (TRAIL). In response to apoptotic assaults, *DR5* expression is upregulated, which in turn promotes cell death via caspase-8 (Dixit and Ashkenazi, 1998; Wu et al., 1997). Lastly, another candidate of the extrinsic pathway happens to be PERP (p53 apoptosis effector related to PMP-22), a tetraspan membrane protein. PERP was also identified as a p53 transcriptional target, and has been shown to activate the extrinsic pathway of cell death (Attardi et al., 2000).

2.2 Cell Death Pathways in *Drosophila*

The apoptotic cascade and its various components are relatively well characterized and found to be conserved between mammals and *Drosophila*. Fig. 3 shows a simple schematic of *Drosophila* cell death pathway.

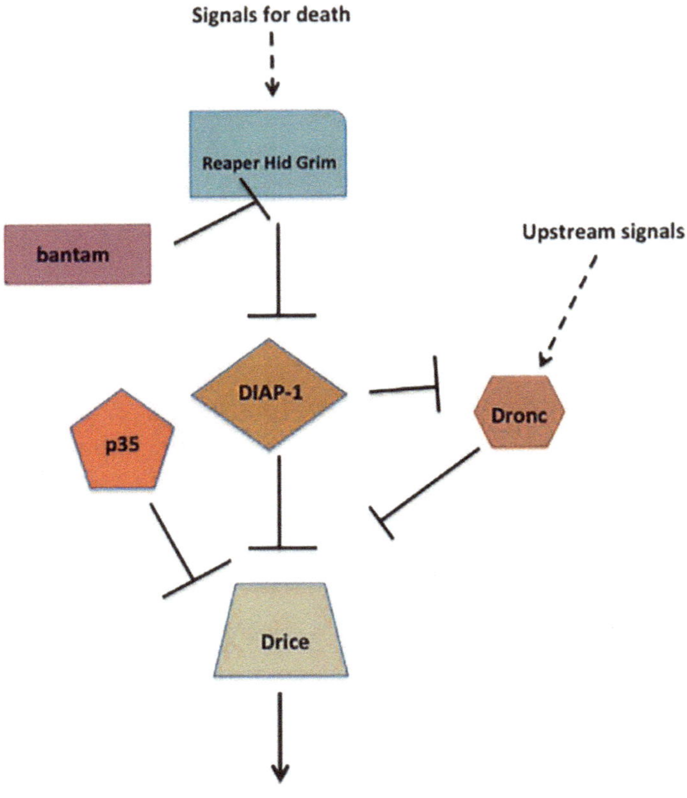

Fig. 3 Programmed cell death pathway in *Drosophila*. A schematic of intrinsic apoptosis pathway in fruit fly. Apoptotic signal can cause the activation of pro-apoptotic gene Rpr, hid and grim, which in turn blocks the DIAP-1 thus releasing imitator caspase Dronc free to activate downstream effector caspases Drice and Dcp1.

In *Drosophila*, there are a set of pro-apoptotic genes namely *reaper* (*rpr*), *head involution defective* (*hid*) and *grim* which constitute the RHG complex (Zhou and Steller, 2003). These in turn are known to inhibit DIAP-1 (Drosophila inhibitor of apoptotic proteins), one of the *Drosophila* homologues of a conserved class of proteins called Inhibitor(s) of Apoptosis (IAP) containing RING domains. DIAP-1 in turn is known to down-regulate the activity of both initiator and effector caspases through its RING domain which is responsible for their ubiquitination (Chai et al., 2001; Hawkins et al., 2000; Tenev et al., 2002; Wang et al., 1999). RHG proteins bind DIAP1 through BIR1 and/or BIR2 (Baculovirus IAP Repeat), and this interaction is mediated by a short sequence known as the IAP-binding motif (IBM).

Caspases bind to these same BIR grooves; however, binding of Rpr, hid or Grim displaces them from the bound grooves (Chai et al., 2001; Vucic et al., 1998; Yan et al., 2004). Inactivation of DIAP-1 by RHG complex sets the initiator caspase Dronc free, which in conjunction with the cytosolic adapter protein (Apoptosis Promoting Factor-1) Apaf-1 forms the apoptosome, leading to an activation of its downstream targets—the executioner caspases *Drosophila*-ICE (Drice) and *Drosophila* caspase-1 (Dcp1) leading to cell death (Chew et al., 2004). Other regulators like p35, a baculoviral inhibitor of apoptotic protein, can inhibit the effector caspase, Drice, when expressed in flies (Hay et al., 1994). Similar to mammals, *Drosophila* also utilizes extrinsic pathway of apoptosis, which involves stimulation via the receptors Fas and Tumor Necrosis Receptor Factor (TNFR) (Moreno et al., 2002).

2.3 DNA Damage-Induced Cell Death

Defects in the repair of DNA lesions often result in cellular death (Roos and Kaina, 2006). As discussed earlier, these lesions mostly arise as a result of exposure to genotoxins from external environment or from stresses originating from within the cell (endogenous). These involve base methylation, base N-alkylations, thymidine dimers, bulky DNA adducts, DNA cross-links, DNA single strand breaks (SSBs) and DNA double strand breaks (DSBs) (Curtin, 2012). Tumor suppressor gene p53 mediates cell death in multiple instances; however, the requirement of p53 can be bypassed by employing p53-independent pathways. The next section elaborates more on the roles and modes of p53-dependent and independent pathways in DNA damage induced cell death in both mammals and *Drosophila*.

2.3.1 Role of p53 in DNA Damage Induced Cell Death

Exposure to a variety of cellular stresses triggers the activation of tumor suppressor gene *p53*. These stresses include DNA damage, oncogene activation, ribosomal stress, cell–cell contacts and hypoxia (Horn and Vousden, 2007). p53 activation and cellular outcome are dictated by multiple factors, nature of the assault, cell type, co-regulators of p53 and its post-translational modifications (reviewed in Horn and Vousden, 2007; Kastenhuber and Lowe, 2017). In response to DNA damage, p53 serves a tumor surveillance role by promoting DNA repair and cell cycle arrest, or by activating the apoptosis program to remove the damaged cells (Liu and Kulesz-martin, 2001). Thus, it comes as no surprise that more than 50% of the tumors arise as a result of mutations in the *p53* gene (Harris and Hollstein, 1993). Mutant p53 not only loses its tumor suppressive functions, but often gains oncogenic functions that provide cells with growth and survival advantages (Muller and Vousden, 2014; Rivlin et al., 2011).

p53 functions as a transcription factor by binding to the promoters and introns of its target genes. The search for a p53 consensus binding site using an unbiased affinity-binding approach revealed that p53 binds to the tandem repeats (separated by 0–13 bases) of the consensus 5′-Pu-Pu-Pu-C-(A/T)-(T/A)-G-Pyr-Pyr-Pyr-3′, where Pu is purine and Pyr is pyrimidine (El-Deiry et al., 1992). Since then, hundreds of p53 target genes have been identified, containing one or more consensus sites in their 5′ promoter regions, introns and in certain cases exons (Riley et al., 2008). Upon DNA damage, p53 activates cell death by upregulating a number of genes (e.g., BAX, PUMA, NOXA) involved in the intrinsic cell death pathways. p53 can also activate the extrinsic apoptotic pathway through the induction of genes encoding three transmembrane proteins: Fas, DR5 and PERP (Attardi et al., 2000).

In early 2000s, it started to become clear that other species also express a conserved functional homologue of hp53. Interestingly, it was shown that *Drosophila* genome encodes for a functional homologue of p53. However, unlike mammals, *Drosophila* genome contains only one p53 homologue—dp53, making it easier for studying its functions (Brodsky et al., 2000; Ollmann et al., 2000). Importantly, dp53 was also found to be important in mediating DNA damage induced cell death (Brodsky et al., 2000; Ollmann et al., 2000). Moreover, overexpression of mammalian p53 leads to cell death in *Drosophila*, suggesting a conservation of function (Brodsky et al., 2000). Expression analysis performed for dp53 regulated genes in response to irradiation led to the identification of ~27 Radiation Induced p53 dependent (RIPD) genes through two independent studies (Akdemir et al., 2007; Brodsky et al., 2004). Moreover, similar to mammals, dp53 was also found to transcriptionally upregulate a number of pro-apoptotic genes namely *rpr*, *hid* and *grim*. Interestingly, the binding of dp53 on *rpr* locus has been mapped and the sequence shows high similarity with mammalian p53 binding consensus sequence (Brodsky et al., 2000). These findings clearly suggest that *Drosophila* p53 (dp53) shares a conserved function with that of its mammalian homologue, thus making the fly a simple system to study p53 functions.

Surprisingly, the *dp53* null mutant does not show any defects or tumor; this is in marked contrast to the mammalian p53 mutants. However, dp53 has been shown to play a critical role in response to exposure to varied stimuli. In response to different kinds of cellular assaults, dp53 is known to engage in a variety of cellular functions which include cell death and cell proliferation via its crosstalk with other cellular signaling pathway (Mesquita et al., 2010; Shlevkov and Morata, 2012; Wells et al., 2006). Quite recently, *Drosophila* p53 has been implicated in programmed necrosis that occurs commonly in germ cells during spermatogenesis. This necrotic form of cell death has been found to be dependent on an atypical function of the initiator caspase, Dronc. Interestingly, p53 appears to be involved in a similar form of necrosis observed in mouse testes, suggesting an evolutionarily conserved function of p53 in this process (Napoletano et al., 2017).

2.3.2 p53-Independent Pathway of Cell Death in Response to DNA Damage

p53, even though believed to be a crucial player in acting as a mediator between DNA damage and death, can often be bypassed or not employed as a means to an end. Of the several strategies employed for this purpose, the first involves the p53 homologues namely p63 and p73 (Pietsch et al., 2008). The former requires the latter two to trigger apoptosis. Interestingly, p73 is pro-apoptotic, even in the absence of p53. It requires the transcriptional activation of PUMA, BAX-mitochondrial translocation and cytochrome *c* release; p73 is rarely mutated, but often overexpressed in tumors (Melino et al., 2002). p63 has the ability to suppress p73-mediated apoptosis, suggesting the existence of a more complex and entangled web of interactions between the three proteins.

Another avenue for triggering p53-independent apoptosis involves BCL-2 degradation, a classical sign detected in many p53-mutated contexts. The precise mechanism involved in this case is not clearly defined. The most widely believed notion is that phosphorylation of BCL-2 at Ser87 by the MAP kinases protects against its degradation (Breitschopf et al., 2000). p53 independent death can be accomplished by activating caspase 2 and the formation of piddosomes. This caspase happens to be the only reported member of its family to reside within the nucleus. It is important for both chemical- and UV-light induced apoptosis (Lassus, 2004; Zhivotovsky et al., 1999). Caspase 2 acts in two ways—by activating the BCl-2 family members and promoting the release of cytochrome *c*, independently of Bcl-2 (Guo et al., 2002; Robertson et al., 2002, 2004). However, what remains unclear is any evidence linking this caspase to a specifically defined DNA lesion.

In addition to apoptotic death, DNA damaging agents coupled with DDR have also been linked to necrosis. Using genetically defined cells, Zong et al. have shown that DNA-alkylating agents cause necrotic cell death, independent of p53 or Bax and Bak (Zong et al., 2004). Similar results were reported in human kidney proximal tubular (HK-2) cells, wherein it was shown that Doxorubicin induced DNA damage can result in necrosis via a Poly-ADP Ribose 1 (PARP1) dependent, but a p53-independent pathway. They could show that inhibition of p53 has no effect on Doxorubicin induced DNA damage, cell cycle changes and oxidative stress (Shin et al., 2015).

The role of p53-independent pathway of cell death in *Drosophila* has also been described in various contexts. One of the first few studies which identified this phenomena showed that the death which was triggered in the absence of p53 was dependent on the dose of the IR, with characteristics of an apoptotic response (Wichmann et al., 2006). This p53-independent cell death was found to be mediated by the pro-apoptotic gene *hid*, activated via JNK signaling (McNamee and Brodsky, 2009). In a separate study, another transcription factor E2F was shown to carry out cell death in the absence of p53, by inducing *hid* expression. It is interesting to note that the pro-apoptotic activity of E2F and p53 is independent of one another in the context of *Drosophila* development, but they appear to converge with regard to DNA damage response (Moon et al., 2008). Subsequent genome wide analysis of wild-type and *p53* mutant Drosophila larvae uncovered a role for p53 not just in the induction, but also in the repression of the transcripts of many genes. One of the top hits from such a microarray analysis was a gene called EF-1a, reduction in the levels of which led to a further enhancement of p53-independent apoptosis. This study did allude to the fact that there might be both positive and negative inputs which collaborate to induce p53 independent apoptosis (van Bergeijk et al., 2012).

2.4 DNA Damage Induced Death and Tissue Sculpting

A number of cancer therapies rely on IR and cisplatin-induced death to mediate its therapeutic benefits. These modes of treatment involve DNA damage of various kinds, which in turn result in tissue level changes. Various studies have attempted to uncover these defects using a number of in vitro and in vivo models. Apart from drug therapeutics, normal homeostatic development also employs DNA damage-induced apoptosis for the removal of unwanted cells, and subsequent tissue sculpting. This section will attempt to outline a few examples in this regard.

DNA damage response pathway playing a vital role in tissue sculpting, which has also been reported in various contexts. Mouse gastrulation is known to be associated with rapidly proliferating cells, which are hypersensitive to low dose radiation induced DNA damage, and known to undergo apoptosis without cell cycle arrest. Quite interestingly, it is only the embryonic cells but not the extra-embryonic ones which display hypersensitivity to genotoxic stress and undergo ATM- and p53-dependent apoptosis (Heyer et al., 2000). This suggests that the sensitivity to DNA damage induced apoptosis might be dictated by cell type or cell fate. In this context, it has also been shown that between implantation and gastrulation, mouse pluripotent epiblast cells expand enormously and depict an unusual hypersensitivity to DNA damage. Unlike other cells, upon low dose irradiation, these subsets of cells undergo mitotic arrest followed by p53-dependent apoptosis. This mechanism of protection is believed to ensure the elimination of every mutated cell before its expansion, thus ensuring greater tissue fitness (Laurent and Blasi, 2015). A number of phenotypes reported in both mouse models and human subjects lacking components of the DDR pathway can be attributed to defects in stem cell function. For instance, hematopoietic stem cells lacking ATM are known to display increased ROS levels coupled with an overall decrease in number and function over time owing to hematopoietic failure (Ito et al., 2004, 2006). ATM-deficient mice also harbor defects in germ cell development, with mutant animals exhibiting a progressive loss in germ stem cells with eventual infertility (Takubo et al., 2008). On the other hand, mutations in ATR also cause developmental defects, i.e., pre-gastrulation lethality in mice and Seckel syndrome in humans (Hakem, 2008; Seita et al., 2010).

From the point of view of cell and tissue-specificity in adult tissues, it was long speculated that there exists a functional relationship between the accumulation of DNA damage and apoptosis in the developing mouse brain. In an elegant set of complementary experiments which involved cell culture and in vivo studies, it was shown that chemical and radiation induced death in the brain requires p53 and caspase 9, whereas natural death did not (D'Sa-Eipper et al., 2001). Accumulation of DNA damage in the central nervous system may be more abundant than in other tissues, owing to the low DNA repair capacity of post-mitotic brain tissue. There is also evidence to suggest that the levels of telomerase activity decline dramatically during the time window when the death of Neural Precursor cells (NPC) occurs in the embryonic mouse cortex (Fu et al., 2000). There also exists evidence to suggest that a deficiency of DNA repair enzymes results in apoptosis of NPC in developing mouse embryos (Gilmore et al., 2000). Another set of stem cells which are particularly interesting are the intestinal stem cells. Contrary to popular beliefs that radiosensitivity is correlated with cell-cycle status, the apoptosis sensitivity of intestinal crypt cells is inversely correlated with their relative quiescence (Gudkov and Komarova, 2003). Some of the mechanisms proposed for this extreme sensitivity include an enhanced activation of the p53 pathway, lower expression of the anti-apoptotic protein Bcl-2 and general lack of DNA repair activity (Merritt et al., 1995; Potten et al., 2002).

The pivotal role played by different DNA repair mechanisms in the maintenance of tissue architecture and function is well elucidated by the severe clinical consequences reported in humans and mice harboring mutations for these genes (Hakem, 2008). Quite interestingly, a number of human conditions have been linked to defects in DNA-damage recognition and repair pathways, which includes Fanconi's Anemia, Bloom's or Werner's syndromes (both caused by mutations in DNA helicases), and a range of diseases associated with telomerase dysfunction and telomere instability (Kenyon and Gerson, 2007). The roles played by stem cells in mediating such tissue specific functions and the eventual clinical outcomes via their diverse DNA-damage recognition and DNA-repair mechanisms are only slowly starting to emerge (Kenyon and Gerson, 2007).

Apart from DNA damage accumulation and death, a number of studies have also showed how DNA repair enzymes can contribute to developmental death and regeneration. A mammalian DNA repair enzyme, namely Exonuclease1 (Exo1), plays a vital role in orchestrating cell death in fibroblasts. In fact, they could also show how this enzyme itself also happens to be a substrate for caspase 3, with the cleavage site mapped to Asp514 (Bolderson et al., 2009). Employing a more in vivo setting, Martin and Wong showed that the enforced expression of human 8-oxoguanine DNA glycosylase (hOGG1), a mtDNA repair enzyme and human apurinic-apyrimidinic endonuclease-1/hAPE1 in the mouse brain could rescue injury induced degeneration (Martin and Wong, 2017). In a different context, it has also been reported that in models of inflammation induced apoptosis, inhibiting hOGG1 is sufficient to rescue mtDNA integrity and apoptosis in chondrocytes exposed to TNF-alpha and IL1-alpha (Kang et al., 2010).

Cellular senescence is an irreversible condition exhibited by damaged cells which remain alive, but are unable to proliferate (Campisi and d'Adda di Fagagna, 2007). The robust activation of DDR pathways and the generation of DDR foci in senescent cells have been shown to act as major players in the establishment and maintenance of both replicative and oncogene-induced senescence (OIS). The senescent cells influence neighboring cells through the secretion of an array of diverse cytokines, chemokines, growth factors, and proteases known as the senescence-associated secretory phenotype (SASP) (Ghosh and Capell, 2016). A number of cytokines/factors released during senescence are known to have both beneficial and detrimental consequences, depending upon the context. For example, the SASP is associated with increased senescence of surrounding normal fibroblasts through non-cell autonomous mechanisms, thereby increasing the clearance of potentially tumorigenic cells (Kuilman et al., 2008; Neves et al., 2015). Conversely, the role of SASP has also been shown to have deleterious effects, impacting skin aging (through breakdown of MMP1-mediated collagen), osteoarthritis cardiovascular disease as well as diabetes (Demaria et al., 2014; Fyhrquist et al., 2013; Greene and Loeser, 2015; Muñoz-Espín and Serrano, 2014). SASP can also drive tumorigenesis through multi-factor effects. Many cancer studies have implicated the pleiotropic nature of the SASP influencing distinct steps of cancer progression: angiogenesis (Coppé et al., 2006), the induction of cancer stem cells (Cahu et al., 2012), also via triggering of the epithelial-mesenchymal transition (EMT) leading to changes in tumor microenvironment, cellular dedifferentiation, and enhanced metastatic capacity (Laberge et al., 2012). Finally, there is also accumulating evidence to suggest that nuclear DNA damage contributes to aging, with many studies showing a causative link between mitochondrial DNA damage and the major phenotypes associated with aging (Maynard et al., 2015). However, this matter is beyond the scope of this review.

In addition to deciding a cell fate between survival and death, the role of DDR machinery has also been shown to affect the overall organismal patterning and development in many different contexts. Classical examples being experiments conducted in chick embryos which show how increasing doses of radiation can affect limb bud formation (Wolpert et al., 1974). More recently, it has been shown that *C. elegans* lacking rad51 display a striking difference across the AP axis characterized by a loss of survival in the posterior half (Barghouth et al., 2018). In case of *Drosophila*, it is known that an IR-induced delay in developmental patterning in wing discs accompanies an animal-wide delay of the juvenile-adult transition, and both these delays require p53. The *p53* mutants display a lack of developmental delays and of damage resolution, which leads to aneuploidy and tissue defects, culminating in morphological abnormalities (Wells et al., 2006).

Another classical example of DNA damage response affecting patterning has been uncovered in the *Drosophila* female germline of a number of DNA repair mutants. It was shown that the constitutive activation of DDR interferes with EGFR signaling leading to the culmination of defects during embryo patterning (Abdu et al., 2002). There are also instances wherein tissues have developed various strategies to either cope with or prevent the transmission of damages arising from DNA damage stress. One example in this regard would be the *Drosophila* testis, wherein the interconnectivity of cells within a "cyst" makes the spermatogonia undergo a synchronized form of cell death, even when only a subset of them are exposed to DNA damage. This increased sensitivity of the germline to damage is believed to be a means to protect the integrity of gametes (Lu and Yamashita, 2017). In a different setting, a study conducted with overexpression of DNA repair genes found either negative or positive impacts on lifespan and stress resistance depending on the tissue of expression, stage of induction, sex and role of the gene in the DNA repair process (Shaposhnikov et al., 2015).

Studies in the fly midgut using cell type specific knockdowns for either ATM or ATR have shown that the loss of either protein leads to a decrease in the number of intestinal stem cells and oxidative-stress induced proliferation. Interestingly, this work suggested that ATR plays a more important role than ATM in the maintenance of tissue homeostasis in the adult gut (Park et al., 2015). On those lines, work done in the fly ovary has shown that DNA damage affects both germline stem cell self-renewal and progeny differentiation. The loss of Chk2 (lok) or its kinase activity can rescue the loss of GSCs and the defects in differentiation (Ma et al., 2016).

There is speculation that cellular decisions in the presence of DNA damage might also be influenced by regional signals that could involve crosstalk between tissues and organs. Apart from the well-studied cell autonomous effects of DDR, there is accumulating evidence which suggests that DDR signals can influence the surrounding cellular microenvironment, thus functioning in a paracrine manner. In addition to these well-studied cell-autonomous effects, emerging evidence now shows that the DDR signaling cascade can also function in a paracrine manner, thus influencing the biology of the surrounding cellular microenvironment. In this context, DDR plays an emerging role in shaping the damaged tumor microenvironment through the regulation of tissue repair and local immune responses, thereby providing a promising avenue for novel therapeutic interventions. Additionally, while DDR-mediated extracellular signals can convey information to surrounding undamaged cells, they can also feedback onto DNA-damaged cells to reinforce selected signaling pathways. Overall, these extracellular DDR signals can be subdivided into two time-specific waves: a rapid bystander effect occurring within a few hours of DNA damage; and a late, delayed, senescence-associated secretory phenotype generally requiring multiple days to establish.

3. ROLE OF CELL DEATH PLAYERS (CASPASES) IN CELLULAR FUNCTION AND IN DDR

The part played by cell death components and caspases in mediating the effects of DNA damage is quite well known, and has been described in detail in the previous sections. However, the putative roles played by caspases in affecting the DDR remain poorly understood. This would involve either an apoptotic or a non-apoptotic function of caspases. The next few sections and Fig. 4 attempt to summarizes the role of cell death players, in particular caspases, in either apoptotic or non-apoptotic functions with their corresponding effect on various cellular and physiological outcomes.

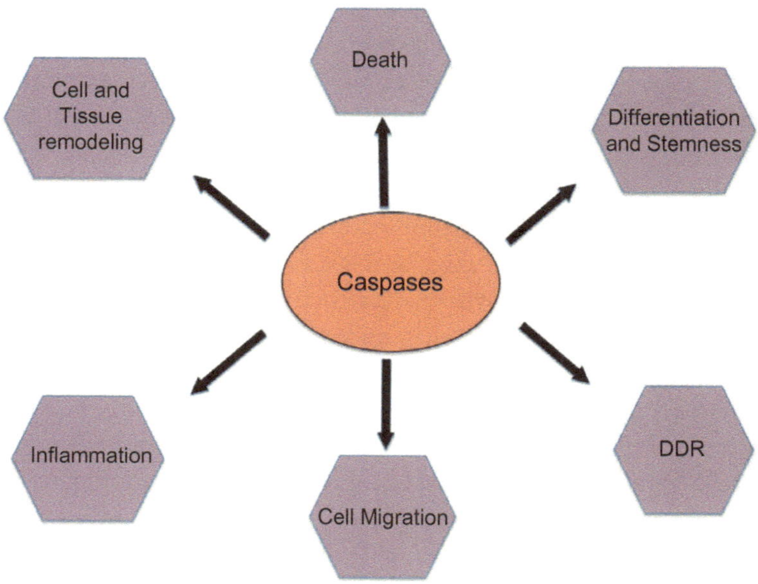

Fig. 4 Distinct cellular functions of caspases. Schematic showing the role of caspases in multiple cellular function apart from their classical role in cell death. These non-cell death functions of caspases are classified in the category of non-apoptotic role. These non-apoptotic functions include aspects of: cell migration, DDR, inflammation, differentiation, stemness.

3.1 Caspase Activation During Cell and Tissue Remodeling

Pruning of neurons is an important, yet poorly understood phenomenon. While it was always believed to be independent of caspases, work done in recent years suggests otherwise. A genetically encoded caspase activity sensor expressed only in pruning neurons revealed the involvement of these proteases in a non-apoptotic function (Williams et al., 2006). This activation of caspases is quite local and restricted to dendrites, with a more global activation resulting in the death of the neurons. Apparently, this spatially restricted activation of caspases is important for labeling the pruned dendrites for phagocytosis. This subcellular compartmentalization of caspase activity is also evident in other developmental processes like spermatid differentiation in flies, wherein caspase activity is found to be required for the removal of the bulk of cytoplasm during the process of sperm individualization (Arama et al., 2003). Other complex cellular remodeling processes like the terminal differentiation of various vertebrate cell types like lens cell, keratinocytes and erythrocytes also require a local activation of caspases to remove specific organelles within the cell (Ishizaki et al., 1998; Weil et al., 1999; Zermati et al., 2001). A classic example in this regard is the activation of caspase 3 during lens fiber differentiation, which is required for the processes of differentiation and enucleation in the equatorial epithelium (Ishizaki et al., 1998).

There is also evidence to suggest a more global involvement of caspases in processes of tissue morphogenesis, for example, *Drosophila* dorsal closure. Caspases are believed to play a role in sculpting the actomyosin cytoskeleton, thus influencing the progress and mechanics of the process (Muliyil and Narasimha, 2014). A similar phenomenon has been reported in mice which are deficient for caspase 8 in skin. In this context, the caspase 8 null skin is found to display epidermal hyperplasia arising from an aberrant immune signaling operating between epithelial, mesenchymal and leukocytic cells, and a recapitulation of multiple phases of a wound healing response (Lee et al., 2009). On a similar note, a genetic screen conducted in *C. elegans* to identify genes involved in global miRISC function uncovered CED-3 caspase to play a non-apoptotic function through proteolytic inactivation of a number miRNA targets, and thus influence developmental timing (Weaver et al., 2014). These examples clearly exemplify the diverse roles played by caspases in modulating tissue dynamics across evolution.

3.2 Caspases in Inflammation

The links between caspases and immunity happen to be quite historic. Caspase 1 or Interleukin beta converting enzyme (ICE) was first discovered as a unique cysteine protease that aids in the processing of IL1β. Since then, a number of caspases have been assigned roles in inflammatory responses. Infections or cellular stresses can trigger the formation of a canonical inflammasome via caspase 1 or a non-canonical inflammasome via caspase 11 (Kayagaki et al., 2011; Masters et al., 2010; Yu and Finlay, 2008). Other inflammatory caspases in case of mammals, namely, caspases 4, 5 and 11, have been shown to directly bind microbial pathogens. Interestingly, a number of them are activated by either bacterial infections or direct binding to lipopolysaccharides (Yang et al., 2015). Additionally, there is mounting evidence to suggest that caspase 8 contributes to the production of inflammatory cytokines (Henry and Martin, 2017). Contrary to past beliefs, there is evidence to postulate a potential cross-talk between the apoptotic and inflammatory functions of caspases, such that one feeds into the other.

3.3 Caspases Driving Cell Migration

The first ever reports implicating caspases in cell migration were based on experiments conducted in the Montell lab. A screen performed to identify genes involved in cell migration uncovered DIAP1 (*thread*) as a candidate that could significantly suppress the migration defects which exist in Rac dominant negative mutants. More specifically, they could show that DIAP1 is vital for the activation of Rac GTPAse dependent border cell migration in *Drosophila* ovaries (Geisbrecht and Montell, 2004). In the context of tumors, it has been reported that caspase 8 can promote cell motility via an optimal activation of calpains, Rac and lamellipodial assembly (Helfer et al., 2006). Subsequent studies showed that caspase 8 can act as part of a distinct biosensory complex which either promotes migration or death depending on the cellular context, thus hinting toward a possible role for caspase 8 in the progression of cancer and metastasis, which might be context dependent. This non-apoptotic function of caspase 8 is triggered by its interactions with components of both focal adhesions and early endosomes at the cellular periphery, enhancing focal adhesion turnover and promoting rapid integrin recycling to the cell surface (Graf et al., 2014).

3.4 Caspases in Differentiation or Maintenance of Stemness

There is enough evidence implicating caspases in either the maintenance of stemness or the induction of differentiation. One of the very first observations in this regard happens to be the fact that caspase 8 null mice exhibit early lethality owing to apoptotic flaws associated with a reduced expression of neurogenic markers and aberrant differentiation of neural precursors

(Sakamaki et al., 2002). At the same time, there is evidence to show that mouse embryonic stem cells (mESCs) mutant for caspase 3 and 9 have problems to differentiate (Fujita et al., 2008). This was attributed to their ability to cleave and process the stem cell factor, Nanog. On a similar note, mice lacking caspases 3 and 7 are known to have a reduced number of differentiated cardiomyocytes in the developing heart (Cardona et al., 2015). Interestingly, there is also evidence for the converse, i.e., an ectopic activation of caspase 3 can revert the differentiation of human fibroblasts to obtain iPSCs. In this context, it is quite important to note that the peak in expression of caspases 3 precedes the reprogramming process and seems to be dependent on Oct-4 (Li et al., 2010). In a different tissue context in mouse models, it has also been reported that the silencing of caspase 3 correlated with inappropriate levels of cell differentiation in peripheral tissues and a premature exit from quiescence of hematopoietic cells. At the molecular level, this was attributed to the inability of caspase 3 to regulate the Ras-MAPK pathway in response to specific cytokines, through unknown molecular mechanisms (Janzen et al., 2008). A similar observation has been made with caspase 2 deficient mice, wherein a deficiency in caspase 2 results in an increase in hematopoietic progenitors and skewed differentiation toward myeloid progenitors in aged mutant mice (Dawar et al., 2016).

Interestingly, in *Drosophila* the proteolytic activity of Dronc has been implicated in regulating the numbers of supernumerary sensory organ precursors via its ability to modulate GSK3β/shaggy function (Kanuka et al., 2005). Quite recently it has been shown that *crinkled*, the fly homologue of mammalian Non-muscle myosin 7A, acts as an adaptor protein bringing together Dronc and the inactive version of Shaggy/GSK3β to facilitate the proteolytic cleavage of the latter (Orme et al., 2016). There is also evidence suggesting the involvement of Dronc in regulating neuroblast stem cells via regulation of Numb. In this case, the regulation appears to be dependent on protein–protein interactions between Dronc and numb (Ouyang et al., 2011).

These observations taken together suggest a strong role for caspases in regulating the transition between pluripotency and stemness, potentially via the regulation of various substrates in a temporally defined fashion.

3.5 Caspases in DDR

In addition to their function as proteases during cellular apoptosis, caspases also play an important role in carrying out nuclear apoptosis by activating caspase activated DNAse (CADs), which in turn facilitates chromatin fragmentation and condensation (Kitazumi and Tsukahara, 2011; Sakahira et al., 1998). Studies conducted in Jurkat cells showed that Etoposide induced apoptosis relies upon the formation of γH2Ax, and the absence of H2ax phosphorylation was associated with impaired apoptosis. In a similar study, the role of caspase 3 was implicated in activating MST1 (mammalian STE20-like kinase 1), which phosphorylates H2Ax on ser-139 (Wen et al., 2010). Other studies using TRAIL (death receptor ligand) have also shown that induction of cell death can cause an activation of the DNA damage pathway. Although, some studies have also commented that γH2ax observed during cell death is markedly different from the one observed during DDR (Solier and Pommier, 2009). The γH2Ax formed upon an induction of cell death stimuli exhibits a pan nuclear-ring like morphology, whereas the DDR associated γH2Ax shows distinct foci (Solier and Pommier, 2009; Solier et al., 2009). In addition to their morphology, γH2Ax also differs in repair capacity, the ring like γH2Ax lacks functional MDC-1 and repair, whereas γH2Ax foci are pro-repair (Solier and Pommier, 2011). Moreover, caspases have been shown to dampen DDR directly by cleaving the apical kinase ATM to promote proper cell death over DNA repair (Smith et al., 1999).

Interestingly, our recent work in *Drosophila* has now shown a plausible link between caspases and components of the DDR machinery. It points toward a non-apoptotic role for initiator caspase Dronc in facilitating γH2Av formation and probably repair (Khan et al., 2017a,b). This study for the first time provides a model in which the cell death players might serve to promote DNA repair rather than cell death. In fact, it hints toward the possibility that this kind of dual function might be dictated by the extent of DNA damage in a cell and the choice between repair and death might be governed by levels of Dronc activation. In a separate study, Al-Khalaf and colleagues have identified the base excision repair component XRCC1 as a vital mediator of muscle differentiation. Interestingly, in their study Caspase-triggered XRCC1 repair foci formed within differentiating myonuclei and then dissolved with the progress of the maturation program (Al-Khalaf et al., 2016). Previous work from the same lab also elucidated how caspase mediated activation of caspase activated DNase (CAD) leads to DNA strand breaks which are important for the initiation of the terminal differentiation of skeletal muscle cells. A loss of caspase 3 resulted in a loss of strand break formation with a subsequent block in the myogenic program. They could show that this differentiation program relies on CAD targeting of the p21 promoter, such that a loss of caspase 3 meant a significant reduction in p21 expression (Larsen et al., 2010). All these studies allude toward the fact that both cell death and DDR might be intertwined and there might exist a cross-talk where the decision between repair or death could arise from feedback regulation between the two modules.

3.6 Mechanisms Driving the Non-apoptotic Functions of Caspases

The first ever caspase to be discovered was Ced-3 in *C. elegans* (Ellis and Horvitz, 1986; Yuan et al., 1993). The mammalian homologue for the same, namely Interleukin 1β converting enzyme (ICE), had already been implicated in immune functions (Cerretti et al., 1992; Thornberry et al., 1992). Thus, the first ever caspase to be discovered was known to function in both apoptotic and non-death related functions. Currently, there is enough evidence to suggest that caspases participate in a number of non-apoptotic functions ranging from inflammation, differentiation and even cell migration. The origins of these different functions have been attributed to various causes which include subcellular compartmentalization of caspase activity, temporal controls or a threshold based system of regulation. A number of different attempts have been made to ascertain the exact threshold of caspase activation through the development of new sensors and probes. One such example is that of a Forster Resonance energy transfer (FRET) probe that consists of a Caspase cleavage sequence (DEVD) flanked by two fluorescent proteins: ECFP and Venus, which enables a quantitative monitoring of caspase activity in vivo (Takemoto et al., 2003; Tyas et al., 2000). Another example is that of fluorescently labeled activity based probes (ABPs) that covalently label active caspases in vivo. These probes could help decipher the kinetics of caspase activation in tissues and organs of live mice (Blum et al., 2009). By measuring the buildup and decline of caspase activity using the above methods, one could potentially build profiles that fit either apoptotic versus non-apoptotic functions.

Majority of research points toward three possible ways in which non-apoptotic function of these cell death players can be executed: (1) A diverse substrate repertoire for caspases which could potentially influence the downstream functions that are regulated by this interesting family of proteases. (2) The other potential explanation could arise from a spatiotemporal regulation of these proteases, which helps them exert their varied yet distinct functions in a defined subcellular context. (3) Finally, the last but not the least explanation could be the extent of activation of caspases which leads to distinct physiological outcomes. The non-apoptotic roles played by caspases in DDR could fall into any one of the above categories. Future work directed toward understanding the exact cellular and molecular interplay between DDR players and caspases can help elucidate the detailed mechanisms at work.

4. ACTIVATION OF STRESS SIGNALING FOLLOWING DNA DAMAGE

As discussed in the previous section, cellular response to genomic damage is not only restricted to managing genomic integrity but also influences cellular physiology. There exists a complex crosstalk between the molecular players of DDR pathways and various stress signaling networks operational within a cell. In the first half, the known examples of cellular signaling pathways and their participation in DDR will be discussed. In the second half, we will try to highlight the importance of these signaling pathways in maintaining overall homeostatic balance, upon assaults to the genome.

4.1 Cellular Signaling Pathways Responding to DNA Damage

Various studies have identified the role of signaling pathways in regulating the cellular outcome of DNA damage and repair. This section discusses the key signaling pathways in mammals and underlying molecular mechanism of their function. So far, there are very few studies in *Drosophila* describing the role of cellular signaling; future work in this direction will prove quite enlightening.

4.1.1 NF-κB Signaling

The first hint of DNA damage leading to NF-κB activation came from studies conducted by Herrlich group in 1989, where they showed that exposure to UV results in the cytoplasmic accumulation of active NF-κB in human cells (Stein et al., 1989). Later, it was shown that the activation of NF-κB in response to UV-B exposure stems from the disruption of an interaction with its inhibitory partners (Simon et al., 1994). Mammalian NF-κB family consists of a group of five transcription factors (namely p65/RelA, RelB, c-Rel, p50, and p52), with their ability to bind to DNA as hetero- or homo-dimers. NF-κB in its inactive form remains sequestered in the cytoplasm, bound to its inhibitory partner, the IκB. NF-κB activating signals lead to the phosphorylation of IκB by IκB kinase (IKK)—a three-subunit complex, which is composed of two kinases (IKKα and IKKβ) and a regulatory subunit, the NF-κB essential modifier (NEMO). The IκB kinase (IKK) phosphorylates IκB leading to its ubiquitin-mediated proteasomal degradation, in turn releasing NF-κB dimers causing its translocation to the nucleus and activation of its target genes (Lawrence, 2009).

Earlier observations hinted toward an UV-induced activation of NF-κB signaling. Although, the precise mechanisms by which DNA damage occurring in the nucleus activates NF-κB in the cytoplasm have only recently been described (Janssens and Tschopp, 2006; Picco and Pagès, 2013). It was observed that *atm* null mutant mice were compromised in mounting DNA damage induced activation of NF-κB signaling. The reason for this was linked to the absence of the IKK activation by ATM kinase (Li et al., 2001). Thus, for the first time providing a possible mechanism of communication between nuclear DNA damage and signaling activation in cytoplasm. Later, other studies pieced different parts of the puzzle together, by clearly showing that NEMO serves as a carrier of the signal from the nucleus to the cytoplasm (Miyamoto, 2011; Wu et al., 2010). It is now known that in response to DNA damage, a small fraction of NEMO is transported into the nucleus. Once in the nucleus, NEMO is post-translationally modified by the addition of SUMO and PAR groups, resulting in increased nuclear accumulation of NEMO, as well as its interaction with ATM kinase (Stilmann et al., 2009; Yang et al., 2011). In addition to these initial modifications, NEMO is further ubiquitinated by the protein cIAP1 (an inhibitor of apoptosis), which is required for its nuclear export (Jin et al., 2009). Cytoplasmic export of NEMO delivers a signal of nuclear DNA damage causing activation of IκB kinase (IKK) via TAK1, resulting in the activation and nuclear translocation of NF-κB (Biton and Ashkenazi, 2011; Wu et al., 2010). Once in the nucleus, the NF-κB (p65/p50) heterodimer regulates the expression of its downstream targets in a sequence specific manner.

NF-κB signaling is often described as a pro-survival factor associated with an increased resistance to anticancer drugs. Seminal study performed by Wang and colleagues highlighted the central role played by NF-κB activation in abolishing cell death in response to cancer drug treatment, leading to chemotherapy-induced tumorigenesis (Wang et al., 1996). Since then, many papers have shown that the inhibition of NF-κB activation can potentially enhance the success rate of tumor therapy (Nakanishi and Toi, 2005). However, the celebrated role of NF-κB as a pro-survival factor has been challenged by multiple studies. These studies have shown that under certain cases of genotoxic stress, NF-κB can also contribute to apoptosis (Wang et al., 2017). This apparent duality of NF-κB signaling as a pro-survival or pro-apoptotic factor is suggested to originate from concomitant activation of other signaling pathways in the cell. One such example is the influence of p53 on NF-κB activation. It was shown that p53 counteracts NF-κB function by blocking its transcriptional activity either by competing for coactivators or by converting NF-κB members into repressors (Tergaonkar et al., 2002). These studies painted a picture, wherein NF-κB tumorigenic potential is tightly controlled by tumor suppressor genes such as p53 or JNK, such that during initial stages of tumor progression they keep a check on NF-κB activation by negatively regulating it.

The earliest reports linking NF-κB signaling to DNA damage in *Drosophila* came from studies which used a null mutant of Relish, a *Drosophila* NF-κB homologue. This study showed that the Relish null mutant larvae were more sensitive to UV-induced pigmentation, immunosuppression and apoptosis of hemocytes than wild-type. By utilizing gel mobility shift assay, they could demonstrate UV-induced NF-κB activation (Kim et al., 2004). Upon stimulation with specific peptidoglycans, Relish is converted to the N-terminal fragment (Relish-N) with transcriptional activity and the C-terminal fragment (Relish-C) harboring an ankyrin repeat domain that inhibits the nuclear translocation of Relish-N (Shibata et al., 2013). Interestingly, there is evidence to suggest that mutations in the ATM gene can cause neurodegeneration by activating Relish. This study could show that Relish mutations inhibited neurodegeneration in *ATM* mutants (Petersen et al., 2013). However, there is much to be understood about how this signaling is actually activated in various contexts, and the role of NF-κB signaling in regulating the cellular events following DNA damage.

4.1.2 Jun Kinase (JNK) Signaling

JNK belongs to the family of mitogen-activated protein kinase (MAPK) signal transduction pathway, consisting of JNK kinase kinases (JNKKKs), JNK kinases (JNKKs), JNK and the effectors jun and fos (Stronach, 2005). In response to numerous inputs, the Jun NH2-terminal kinase (JNK) signaling pathway regulates fundamental cell behaviors, such as proliferation, differentiation, morphogenesis, and apoptosis (Stronach, 2005). The complexity of JNK signaling arises in part due to the presence of multiple family members in each class, giving rise to diverse outputs (Morrison and Davis, 2003). This can be further complicated by the scaffolding proteins, which are required to assemble the multi-tiered module; the JNKK and JNK components most often constitute the core, whereas the JNKKK member might change depending on the scaffold (Morrison and Davis, 2003).

The first link between JNK signaling and DNA damage was suggested in 1994, from the observation that exposure of cells to UV leads to the activation of JNK1 (Drijard and Barrett, 1994). Later, studies conducted by Tan's group showed that JNK signaling was also elevated post gamma-irradiation. The activation of JNK upon exposure to both UV and gamma-radiation was thought to promote cell death in human cell lines (Chen et al., 1996). Another, interesting role of JNK to the genotoxic assaults was to promote γH2AX formation. As discussed earlier, phosphorylation of H2AX on serine S139 by ATM kinase is one of the early DNA damage signaling (Paull et al., 2000). Interestingly, blocking molecular players of JNK pathway with the help of either specific chemical inhibitors or by siRNAs against JNK1 and JNK2 leads to a reduction of H2AX phosphorylation in response to UV exposure. This suggests that, similar to ATM kinase, upon exposure to DNA damaging assaults, JNK signaling can also mediate phosphorylation of H2AX (Lu et al., 2006). The role of JNK dependent phosphorylation is described to promote apoptosis in cells facing extensive DNA damage. Moreover, JNK signaling has also been implicated in promoting DNA repair by several studies. One such example being that the lack of transcription factor c-Fos results in increased sensitivity to UV-C, owing to lack of efficient repair. Importantly, the cell death inducing role of JNK is shown to occur after prolonged DNA damage. This suggests an interesting possibility; JNK signaling can work in a biphasic model (1) during early phases: a short-term JNK activation can promote DNA repair; (2) prolonged phase: JNK serves a pro-apoptotic function (Biton and Ashkenazi, 2011; Tournier et al., 2000). Interestingly, a study by Biton and colleagues dissects out the two phases of DNA damage, initial phase/low DNA damage characterized by NF-κB activation, where NF-κB serves as a pro-survival signal; late phase/higher DNA damage leading to the activation of JNK signaling pathway accompanied by cellular apoptosis (Biton and Ashkenazi, 2011).

Various fly studies have elucidated the importance of JNK signaling in different cellular contexts, including cytoskeletal rearrangement during wound closure (Bosch et al., 2005; Ríos-Barrera and Riesgo-Escovar, 2013), elimination of damaged cells by injury-induced apoptosis (Moreno et al., 2002; Shlevkov and Morata, 2012), and even compensatory proliferation to replace damaged tissues (Bergantiños et al., 2010). Apart from its role in several cellular processes, JNK function has also been shown to be important in regulating the outcome of cells with damaged DNA. One study conducted in fly imaginal discs showed that JNK signaling plays a key role in promoting apoptosis in response to a diverse array of signals, including both the p53-dependent response to DNA damage and developmentally regulated apoptosis. In fact, they could show that when apoptosis is blocked, a transcriptional target of JNK, puckered inhibits tissue overgrowth by preventing unbridled JNK activation. This suggests that puckered might play an important role in deciding whether JNK signaling can promote apoptosis or proliferation in different cellular contexts (McEwen, 2005). Subsequent studies have shown that the transcription factors, Foxo and AP-1, are required downstream of JNK to trigger death in response to UV-induced DNA damage in the developing fly retina (Luo et al., 2007). Interestingly, both p53-dependent and independent cell death appear to be linked to JNK signaling. While there is evidence to suggest that p53 and JNK establish a positive feedback loop that amplifies the apoptotic stimulus, there is also data to indicate that p53-independent cell death also relies on JNK to induce cell death (McNamee and Brodsky, 2009; Shlevkov and Morata, 2012). This p53 independent cell death is used to limit aneuploidy in DNA damaged cells. What remains to be understood are the exact molecular mechanisms by which these two distinct forms of cells death are executed by the JNK pathway, i.e., whether tissue specific factors can compensate for one another.

4.1.3 EGFR Signaling

The epidermal growth factor receptor is a member of the ErbB family of receptors, a subfamily of four closely related receptor tyrosine kinases: EGFR (ErbB-1), HER2/neu (ErbB-2), Her3 (ErbB-3) and Her4 (ErbB-4). The activation of EGFR signaling can be brought out in either a ligand-dependent or ligand independent manner. In response to genotoxic stresses, such as ionizing radiation (IR), cisplatin, UV light, heat shock and H_2O_2, following its activation in the absence of a ligand EGFR translocates to the nucleus. Once in the nucleus, EGFR signaling regulates various aspects of DNA repair (Chen and Nirodi, 2007; Lu et al., 2011).

The first evidence that EGFR can directly promote or influence DNA repair came from studies showing the evidence of a physical interaction between EGFR and the DNA-PKcs (Bandyopadhyay et al., 1998). Since then, studies have shown that in non-irradiated cells, EGFR is normally present in the perinuclear space and upon exposure to radiation, EGFR translocates into the nucleus in a ligand-independent manner (Dittmann et al., 2005). Once in the nucleus, EGFR binds to DNA-PKcs, and forms EGFR-DNA-PK complex, which results in a rise in the activity of DNA-PKcs, just after few minutes post irradiation.

The role of EGFR in promoting DNA repair is not only restricted to the NHEJ pathway, but has also been shown to promote HR. It was shown that the inhibitors of ERK kinase also result in the suppression of ATM phosphorylation. Moreover, inhibitors of ATM kinase result in suppressing ERK1/2 phosphorylation. This suggests that ATM and ERK signaling might act in a feedback loop to promote DNA repair (Golding et al., 2007; Nyati et al., 2006). Interestingly, the role of EGFR has also been shown to promote HR by facilitating rad51 nucleofilament formation. This is evidenced by the fact that a blockage of EGFR in human breast cancer cells results in an attenuation of DNA damage-induced Rad51 foci (Li et al., 2008). Further studies will need to dissect out the finer cell type and context dependent interactions between different arms of the EGFR signaling and DDR. Additionally, more work perhaps in a more physiological setting as in flies will help to elucidate the complex interplay between various signaling pathways and their outcome upon DNA damage.

4.2 Integration of Cellular Response to Maintain Tissue and Organismal Homeostasis

In a multicellular organism, cellular stress needs to be integrated into a complex system, via cell autonomous and cell non-autonomous responses. The overall purpose of the system is to restore tissue and organismal homeostasis. One such classical example comes from a study published in 1977, where *Drosophila* wing imaginal discs were found to give rise to a full size adult wing, even after losing more than 50% of the cells to irradiation-induced damage (Haynie and Bryant, 1977). Under physiological conditions, tissue homeostasis plays an important role in serving as a quality check process during the course of development, as well as in maintaining tissue fitness (Neto-Silva et al., 2009). Tissue homeostasis often employs one or more of the following cellular mechanisms: removal of damaged cells via cell death, and/or replacing cell loss by increased proliferation of the neighboring tissue, and/or by protecting cells from untimely cell death. In addition to local tissue level adjustments, tissue homeostasis also involves systemic changes aimed at coordinating developmental timing with that of tissue repair and regeneration.

Compensatory proliferation (CP) and Apoptosis induced proliferation (AiP) provide interesting examples of tissue homeostasis, wherein the neighboring cells compensate for dying cells by undergoing an extra round of proliferation. In this context, the dying cells promote its neighbors to proliferate in a cell non-autonomous manner by secreting mitogens (Morata et al., 2011).

Pioneering studies have been performed in *Drosophila* imaginal discs using "undead cells," which were created by arresting dying cells prior to the execution of cell death via an overexpression of p35 (an inhibitor of effector caspase) (Lannan et al., 2007). These studies led to the isolation of numerous cell autonomous factors released by dying cells, such as mitogens Wg and Dpp (Morata et al., 2011; Pérez-Garijo et al., 2005; Perez-Garijo et al., 2009). Interestingly enough, it was also shown that the non-apoptotic functions of initiator caspase Dronc play a central role in both CP and AiP (Kondo et al., 2006; Ryoo et al., 2004). It is now known that in undead cells, Dronc activate p53 and JNK signaling, and the activated p53 and JNK are required to upregulate the expression of secreted mitogens Wg and Dpp (Pérez-Garijo et al., 2005; Perez-Garijo et al., 2009; Wells et al., 2006). The activation of JNK and p53 promotes death cell autonomously, which leads to cell proliferation in a non-autonomous manner. A recent study by Fogarty and colleagues nicely describes this in undead cells; Dronc induced extracellular ROS attracts activated macrophages, which in turn activates JNK signaling by secreting the TNFR ligand-eiger which in turn promotes AiP (Fogarty et al., 2016). A response similar to compensatory pathway called as "Phoenix Rising" has also been reported in mice post ionizing radiation induced cell death. The activity of Caspase 3 and 7 in dying cells has been shown to be important in mediating this by facilitating the release of prostaglandin E2 (Li et al., 2010). It is interesting to know that the phenomenon of compensatory proliferation is not only restricted to undead cells, but is also known to commonly occur in different kinds of mutations abrogating various cellular functions (Poulton et al., 2014; Slade and Staveley, 2015). A similar setting has also been provided by studies conducted using induced chromosomal instability, where it was shown that genomic instability can give rise to higher ROS, in turn leading to JNK activation that can cause cell death autonomously and cell proliferation in a non-autonomous manner (Clemente-Ruiz et al., 2016; Dekanty et al., 2015). Futuristically, it will be interesting to investigate that role of DDR players in regulating such tissue behaviors in response to DNA damage.

Another interesting phenomenon reported in IR studies has been the "radiation bystander effect," by which the irradiated cells make their neighbors more prone to the cell death. There is evidence for a soluble signal; media from irradiated cells can induce the bystander effect on naïve cells. Inhibitors of this so-called bystander effect include antioxidants (Mothersill and Seymour, 2006; Mothersill et al., 2000). Although studies in *Drosophila* have identified a pathway opposite to that of classical radiation-induced bystander effect, wherein the dying cells protect their neighbors from IR induced cytotoxicity. It was also shown that dying cells can cause an activation of anti-apoptotic bantam microRNA leading to increased radioprotection (Bilak et al., 2014). Further, studies await how the molecular players of DNA repair and signaling participate in different signaling pathways to mediate overall organismal homeostasis.

5. PERSPECTIVES: POSSIBLE ROLE OF DNA REPAIR PLAYERS IN PATTERNING TISSUE AND ORGANISM

Drosophila and *C. elegans* have been widely used as model organisms by many developmental biologists across the world. However, they are relatively under-studied in relation to DDR. It is only our work coupled with a few others, which have attempted to define the roles of various DDR players in functions other than DNA damage in *Drosophila*, with a special emphasis on developmental themes. A separate piece of work conducted by us also posits that *Drosophila* rad51 has functions which can regulate tissue patterning (Khan et al., 2017a,b). There is emerging evidence from work done in *C. elegans*, which shows that Notch can modulate DDR by binding to ATM and inactivates it. This mode of regulation of DDR by Notch appears to be independent of its transcriptional activity, and appears to be conserved in humans (Vermezovic et al., 2015). This is an example of how work done in nematodes has paved the way for further explorations in other organisms including humans. Interestingly, this study also shows how Notch appears to impair DDR signaling in the gonad cells of the nematode. Additionally, it also highlights the importance of using model organisms, to uncover tissue-specific relationships shared between complex signaling networks. This fact is especially important owing to different susceptibilities of various cells and tissues to DDR, which might be linked to their distinct cell cycle status during development (discussed in Section 2.4).

6. FUTURE QUESTIONS AND DIRECTIONS

The field of DNA damage and repair has grown extensively over the years. However, much needs to be still answered using the right techniques and a pertinent physiological setting. What remains almost unexplored is the ability of different types of lesions causing activation of distinct signaling pathways. Is there a common overlapping molecular toolkit in operation, or a preferential tissue bias to the response?

Interestingly, both p53-dependent and independent pathways, activated upon DNA damage, end up using the same downstream component, i.e., JNK signaling (Brodsky et al., 2004; McNamee and Brodsky, 2009). This points toward a plausible overlap in the downstream signaling loop operating downstream of a DNA damage. However, there is also evidence to suggest otherwise. Signaling pathways like NF-κB and JNK can decide to exert either pro-death or survival functions upon DDR. The cellular and molecular components that play a vital role in this decision-making process still remain fairly unknown. Again, there could be a tissue bias in place that tilts the choice in favor of one versus the other. On a similar note, it would be interesting to evaluate how different tissues exhibit diverse responses to the same DNA damage. Overall, it would be important to entangle matters of tissue sensitivity versus the molecular choice of different DDR components activating specific signaling pathways. While there is a wealth of knowledge for some tissues like the brain and stem cells, the same is lacking for many others, especially in the context of development and tissue homeostasis. Finally, it would be fascinating to explore whether DDR components can engage in wider unappreciated functions akin to various death regulators, i.e., caspases. While our work in conjunction with that of a few others is suggestive of this, much needs to be done in this direction.

With the resurgence of interdisciplinary science, coupled with new age tools and techniques, the field of DNA damage and cell signaling has witnessed an influx of novel concepts and themes. The fact that a number of mammalian mutants for DDR are early lethal had always made it difficult to investigate the developmental roles of various genes involved in this very pathway. The discovery of organoids can help fill this existing void, by lending cell and developmental biologists, alike, the ability to tweak levels of various DDR genes in a close to in vivo situation. On a similar note, the development of super-resolution microscopy that breaks the diffraction barrier can help researchers visualize nuclear events at a greater precision than before, thus paving the way for observations, which can better explain the signaling happening between the nucleus and the cytoplasm, especially under conditions of genotoxic stress. This is being increasingly used in various studies to recapitulate events that occur under UV induced DNA damage, by employing a multiphoton laser to create DNA lesions associated with UV stress (Daddysman and Fecko, 2013; Ferrando-May et al., 2013). Similarly, the rise of light sheet microscopy has enabled biologists to investigate tissue level changes in an in vivo setting with a never before precision, thus aiding in the documentation of phenotypic changes in cells and tissues, as and when they happen (Huisken et al., 2004). Photo-toxicity associated with such live imaging processes is known to be quite minimal (Kaufmann et al., 2012). This will greatly facilitate the visualization of various DNA damage related events and its outcomes, at a high temporal and spatial resolution, within tissues of organisms like zebrafish and Xenopus. To really make the best use of resources available, one can couple the use of light sheet microscopy with organoids to study various features of DDR signaling, in situations where the DDR mutants are early lethal. There are already studies which have started to explore the above routes as a means of testing the efficiency and reproducible outcomes of various DDR biosensors in live 3D spheroid models (Mondesert et al., 2015).

The use of high end mass spectrometry coupled with good precision sequencing can enable researchers to map proteome level changes post DNA damage. This will no doubt uncover novel players and interactions, which can potentially shed light on the complex interplay between DDR components and other signaling pathways. Already attempts of this kind have begun to be reported in various contexts. A report from the Schumacher lab has attempted to delineate a comprehensive map of the proteomic, phospho-proteomic and lipidomic changes that occur in *C. elegans*, in response to persistent DNA damage. More specifically, they could show that proteostasis shifts toward autophagy, with fatty acid metabolism being attenuated (Edifizi et al., 2017). Many studies of this kind are attempting to identify changes in abundance, interactions and modification of proteins in the context of genotoxic stress using various components of the proteomics toolkit.

Another set of important questions that begs some investigation are the distinct roles played by DDR players under genotoxic stress versus normal development. Much of the studies performed to uncover the roles played by DDR components in various contexts have been performed in the presence of various drug or radiation induced genotoxic stress. This tends to avoid investigation into functions performed by DDR players to maintain normal tissue homeostasis. Again, the CRISPR revolution might help in this respect. The creation of various endogenously tagged CRISPR lines would definitely help uncover the sub-cellular localization of various DDR proteins in different tissues and organisms under different contexts. This, in combination with techniques like the Auxin induced degron system, can also aid in creating an acute loss of DDR components in cells, thus enabling researchers in distinguishing phenotypes arising from an acute versus a chronic loss of these proteins (Nishimura et al., 2009). Additionally, there are other techniques like Trim21 mediated degradation, which can help uncover effects arising from transient versus chronic depletion of a protein of interest (Clift et al., 2017). This kind of analysis will also help uncover potential long-term adaptive effects that arise within tissues and organisms as a result of the loss of DDR proteins.

In conclusion, there are a lot of unopened questions waiting to be answered in the context of DDR and development. A complementary multidisciplinary approach that aims to uncover newer roles for DDR players in an in vivo setting during development will greatly benefit the field of DNA damage and repair. Counterintuitively, normal genome replication and even transcriptional events, per se, being entirely directional on the genome tract, are known to run into myriads of barriers along the way, thus giving rise to genomic stress that are efficiently repaired by the DDR players. Tuning such repair events in the context of varying levels of endogenous cellular stress must invariably involve cell-type specific signaling cascades, which remains an exciting new area of research. Current review is almost the first step in that direction keeping the mechanistic basis as the sole focus.

REFERENCES

Abdu, U., Brodsky, M., Schüpbach, T., 2002. Activation of a meiotic checkpoint during *Drosophila* oogenesis regulates the translation of Gurken through Chk2/Mnk. Curr. Biol. 12 (19), 1645–1651. https://doi.org/10.1016/S0960-9822(02)01165-X.

Abdu, U., et al., 2003. The Drosophila spn-D gene encodes a RAD51C-like protein that is required exclusively during meiosis. Genetics 165, 197–204.

Akdemir, F., et al., 2007. p53 directs focused genomic responses in Drosophila. Oncogene 26 (36), 5184–5193. https://doi.org/10.1038/sj.onc.1210328.

Alani, E., et al., 1997. Genetic and biochemical analysis of Msh2p-Msh6p: role of ATP hydrolysis and Msh2p-Msh6p subunit interactions in mismatch base pair recognition. Mol. Cell. Biol. 17 (5), 2436–2447. https://doi.org/10.1128/MCB.17.5.2436.

Al-Khalaf, M.H., et al., 2016. Temporal activation of XRCC1-mediated DNA repair is essential for muscle differentiation. Cell Discov. 2, 1–13. https://doi.org/10.1038/celldisc.2015.41.

Arama, E., Agapite, J., Steller, H., 2003. Caspase activity and a specific cytochrome C are required for sperm differentiation in *Drosophila*. Dev. Cell 4, 687–697.

Attardi, L.D., et al., 2000. PERP, an apoptosis-associated target of p53, in a novel member of the PMP-22/gas3 family. Genes Dev. 14 (6), 704–718. https://doi.org/10.1101/gad.14.6.704.

Bandyopadhyay, D., et al., 1998. Physical interaction between epidermal growth factor receptor and DNA-dependent protein kinase in mammalian cells. J. Biol. Chem. 273 (3), 1568–1573. https://doi.org/10.1074/jbc.273.3.1568.

Barbosa, V., Kimm, N., Lehmann, R., 2007. A maternal screen for genes regulating drosophila oocyte polarity uncovers new steps in meiotic progression. Genetics 176 (4), 1967–1977. https://doi.org/10.1534/genetics.106.069575.

Barghouth, P.G., Thiruvalluvan, M., LeGro, M., Oviedo, N.J., 2018. DNA damage and tissue repair: what we can learn from planaria. Semin. Cell Dev. Biol. 1084–9521. https://doi.org/10.1016/j.semcdb.2018.04.013.

Baumann, P., West, S.C., 1998. Role of the human RAD51 protein in homologous recombination and double-stranded-break repair. Trends Biochem. Sci. 23 (7), 247–251. https://doi.org/10.1016/S0968-0004(98)01232-8.

Bergantiños, C., Corominas, M., Serras, F., 2010. Cell death-induced regeneration in wing imaginal discs requires JNK signalling. Development (Cambridge, England) 137, 1169–1179. https://doi.org/10.1242/dev.045559.

Bi, X., Srikanta, D., et al., 2005a. *Drosophila* ATM and ATR checkpoint kinases control partially redundant pathways for telomere maintenance. Proc. Natl. Acad. Sci. U. S. A. 102 (42), 15167–15172. https://doi.org/10.1073/pnas.0504981102.

Bi, X., Gong, M., et al., 2005b. *Drosophila* ATM and Mre11 are essential for the G2/M checkpoint induced by low-dose irradiation. Genetics 171 (2), 845–847. https://doi.org/10.1534/genetics.105.047720.

Bilak, A., Uyetake, L., Su, T.T., 2014. Dying cells protect survivors from radiation-induced cell death in *Drosophila*. PLoS Genet. 10 (3), e1004220. https://doi.org/10.1371/journal.pgen.1004220.

Biton, S., Ashkenazi, A., 2011. NEMO and RIP1 control cell fate in response to extensive DNA damage via TNF-α feedforward signaling. Cell 145 (1), 92–103. https://doi.org/10.1016/j.cell.2011.02.023.

Blum, G., et al., 2009. Comparative assessment of substrates and activity based probes as tools for non-invasive optical imaging of cysteine protease activity. PLoS One 4 (7), e6374. https://doi.org/10.1371/journal.pone.0006374.

Bohgaki, T., Bohgaki, M., Hakem, R., 2010. DNA double-strand break signaling and human disorders. Genome Integr. 1 (1), 15. https://doi.org/10.1186/2041-9414-1-15.

Bolderson, E., et al., 2009. Involvement of Exo1b in DNA damage-induced apoptosis. Nucleic Acids Res. 37 (10), 3452–3463. https://doi.org/10.1093/nar/gkp194.

Bosch, M., et al., 2005. JNK signaling pathway required for wound healing in regenerating *Drosophila* wing imaginal discs. Dev. Biol. 280 (1), 73–86. https://doi.org/10.1016/j.ydbio.2005.01.002.

Breitschopf, K., et al., 2000. Posttranslational modification of Bcl-2 facilitates its proteasome-dependent degradation: molecular characterization of the involved signaling pathway. Mol. Cell. Biol. 20 (5), 1886–1896. https://doi.org/10.1128/MCB.20.5.1886-1896.2000.

Brodsky, M.H., et al., 2000. *Drosophila* p53 binds a damage response element at the reaper locus. Cell 101, 103–113. https://doi.org/10.1016/S0092-8674(00)80627-3.

Brodsky, M.H., et al., 2004. *Drosophila* melanogaster MNK/Chk2 and p53 regulate multiple DNA repair and apoptotic pathways following DNA damage. Mol. Cell. Biol. 24 (3), 1219–1231. https://doi.org/10.1128/MCB.24.3.1219.

Cadet, J., Douki, T., Ravanat, J.-L., 2008. Oxidatively generated damage to the guanine moiety of DNA: mechanistic aspects and formation in cells. Acc. Chem. Res. 41 (8), 1075–1083. https://doi.org/10.1021/ar700245e.

Cahu, J., Bustany, S., Sola, B., 2012. Senescence-associated secretory phenotype favors the emergence of cancer stem-like cells. Cell Death Dis. 3 (12), e446. https://doi.org/10.1038/cddis.2012.183.

Campisi, J., d'Adda di Fagagna, F., 2007. Cellular senescence: when bad things happen to good cells. Nat. Rev. Mol. Cell Biol. 8, 729. https://doi.org/10.1038/nrm2233.

Cardona, M., et al., 2015. Executioner caspase-3 and 7 deficiency reduces myocyte number in the developing mouse heart. PLoS One 10 (6), 1–21. https://doi.org/10.1371/journal.pone.0131411.

Cerretti, D.P., et al., 1992. Molecular cloning of the interleukin-1 beta converting enzyme. Science 256 (December 1991), 97–100. https://doi.org/10.1126/science.1373520.
Chai, J., et al., 2001. Crystal structure of a Procaspase-7 zymogen. Cell 107 (3), 399–407. https://doi.org/10.1016/S0092-8674(01)00544-X.
Chakraborty, R., et al., 2015. Corp regulates P53 in *Drosophila* melanogaster via a negative feedback loop. PLoS Genet. 11 (7), 1–25. https://doi.org/10.1371/journal.pgen.1005400.
Chen, D.J., Nirodi, C.S., 2007. The epidermal growth factor receptor: a role in repair of -radiation-induced DNA damage. Clin. Cancer Res. 13 (22), 6555–6560. https://doi.org/10.1158/1078-0432.CCR-07-1610.
Chen, Y., et al., 1996. The role of c-Jun N-terminal kinase (JNK) in apoptosis induced by ultraviolet C and radiation. Biochemistry 271 (50), 31929–31936. https://doi.org/10.1074/jbc.271.50.31929.
Chew, S.K., et al., 2004. The apical caspase dronc governs programmed and unprogrammed cell death in *Drosophila*. Dev. Cell 7, 897–907.
Ciccia, A., Elledge, S.J., 2010. The DNA damage response: making it safe to play with knives. Mol. Cell 40 (2), 179–204. https://doi.org/10.1016/j.molcel.2010.09.019.
Clemente-Ruiz, M., et al., 2016. Gene dosage imbalance contributes to chromosomal instability-induced tumorigenesis. Dev. Cell 36, 290–302. https://doi.org/10.1016/j.devcel.2016.01.008.
Clift, D., et al., 2017. A method for the acute and rapid degradation of endogenous proteins. Cell 171 (7), 1692–1706. https://doi.org/10.1016/j.cell.2017.10.033.
Coppé, J.-P., et al., 2006. Secretion of vascular endothelial growth factor by primary human fibroblasts at senescence. J. Biol. Chem. 281 (40), 29568–29574. https://doi.org/10.1074/jbc.M603307200.
Cory, S., Adams, J.M., 2002. The Bcl2 family: regulators of the cellular life-or-death switch. Nat. Rev. Cancer 2, 647. https://doi.org/10.1038/nrc883.
Curtin, N.J., 2012. DNA repair dysregulation from cancer driver to therapeutic target. Nat. Rev. Cancer 12 (12), 801–817. https://doi.org/10.1038/nrc3399.
Daddysman, M.K., Fecko, C.J., 2013. Quantifying the contribution of obstructions to anomalous diffusion in cell nuclei. Biophys. J. 104 (2), 341a. https://doi.org/10.1016/j.bpj.2012.11.1895.
Davis, A., Chen, D., 2013. DNA double strand break repair via non-homologous end-joining. Transl. Cancer Res. 2 (3), 130–143. https://doi.org/10.3978/j.issn.2218-676X.2013.04.02.DNA.
Dawar, S., et al., 2016. Impaired haematopoietic stem cell differentiation and enhanced skewing towards myeloid progenitors in aged caspase-2-deficient mice. Cell Death Dis. 7 (12). https://doi.org/10.1038/cddis.2016.406.
Dekanty, A., Barrio, L., Milán, M., 2015. Contributions of DNA repair, cell cycle checkpoints and cell death to suppressing the DNA damage-induced tumorigenic behavior of *Drosophila* epithelial cells. Oncogene 34 (8), 978–985. https://doi.org/10.1038/onc.2014.42.
Demaria, M., et al., 2014. An essential role for senescent cells in optimal wound healing through secretion of PDGF-AA. Dev. Cell 31 (6), 722–733. https://doi.org/10.1016/j.devcel.2014.11.012.
Dittmann, K., et al., 2005. Radiation-induced epidermal growth factor receptor nuclear import is linked to activation of DNA-dependent protein kinase. J. Biol. Chem. 280 (35), 31182–31189. https://doi.org/10.1074/jbc.M506591200.
Dixit, V.M., Ashkenazi, A., 1998. Death receptors: signaling and modulation. Science 281 (5381), 1305–1308.
Drijard, B., Barrett, T., 1994. JNKI: a protein kinase stimulated by UV light and Ha-Ras that binds and phosphorylates the c & n activation domain. Cell 76, 1025–1037.

D'Sa-Eipper, C., et al., 2001. DNA damage-induced neural precursor cell apoptosis requires p53 and caspase 9 but neither Bax nor caspase 3. Development 128 (1), 137–146.

Duncan, T., et al., 2002. Reversal of DNA alkylation damage by two human dioxygenases. Proc. Natl. Acad. Sci. U. S. A. 99 (26), 16660–16665. https://doi.org/10.1073/pnas.262589799.

Edifizi, D., et al., 2017. Multilayered reprogramming in response to persistent DNA damage in *C. elegans*. Cell Rep. 20 (9), 2026–2043. https://doi.org/10.1016/j.celrep.2017.08.028.

El-Deiry, W.S., et al., 1992. Definition of a consensus binding site for p53. Nat. Genet. 1, 45. https://doi.org/10.1038/ng0492-45.

Ellis, H.M., Horvitz, H.R., 1986. Genetic control of programmed cell death in the nematode *C. elegans*. Cell 44 (6), 817–829. https://doi.org/10.1016/0092-8674(86)90004-8.

Falck, J., Coates, J., Jackson, S.P., 2005. Conserved modes of recruitment of ATM, ATR and DNA-PKcs to sites of DNA damage. Nature 434, 605. https://doi.org/10.1038/nature03442.

Ferrando-May, E., et al., 2013. Highlighting the DNA damage response with ultrashort laser pulses in the near infrared and kinetic modeling. Front. Genet. 4, 135. https://doi.org/10.3389/fgene.2013.00135.

Flynn, R.L., Zou, L., 2011. ATR: a master conductor of cellular responses to DNA replication stress. Trends Biochem. Sci. 36 (3), 133–140. https://doi.org/10.1016/j.tibs.2010.09.005.

Fogarty, C.E., et al., 2016. Extracellular reactive oxygen species drive apoptosis-induced proliferation via *Drosophila* macrophages. Curr. Biol. 575–584. https://doi.org/10.1016/j.cub.2015.12.064.

Fousteri, M., Mullenders, L.H.F., 2008. Transcription-coupled nucleotide excision repair in mammalian cells: molecular mechanisms and biological effects. Cell Res. 18, 73. https://doi.org/10.1038/cr.2008.6.

Fu, W., et al., 2000. The catalytic subunit of telomerase is expressed in developing brain neurons and serves a cell survival-promoting function. J. Mol. Neurosci. 14, 3–15.

Fujita, J., et al., 2008. Caspase activity mediates the differentiation of embryonic stem cells. Cell Stem Cell 2 (6), 595–601. https://doi.org/10.1016/j.stem.2008.04.001.

Fyhrquist, F., Saijonmaa, O., Strandberg, T., 2013. The roles of senescence and telomere shortening in cardiovascular disease. Nat. Rev. Cardiol. 10, 274. https://doi.org/10.1038/nrcardio.2013.30.

Geisbrecht, E.R., Montell, D.J., 2004. A role for *Drosophila* IAP1-mediated caspase inhibition in Rac-dependent cell migration. 118, 111–125.

Genschel, J., et al., 1998. Isolation of MutSβ from human cells and comparison of the mismatch repair specificities of MutSβ and MutSα. J. Biol. Chem. 273 (31), 19895–19901. https://doi.org/10.1074/jbc.273.31.19895.

Ghabrial, A., Ray, R.P., Schüpbach, T., 1998. Okra and spindle-B encode components of the RAD52 DNA repair pathway and affect meiosis and patterning in *Drosophila* oogenesis. Genes Dev. 12 (17), 2711–2723. https://doi.org/10.1101/gad.12.17.2711.

Ghosh, K., Capell, B.C., 2016. The senescence-associated secretory phenotype: critical effector in skin cancer and aging. J. Invest. Dermatol. 136 (11), 2133–2139. https://doi.org/10.1016/j.jid.2016.06.621.

Gilmore, E.C., et al., 2000. Cell birth, cell death, cell diversity and DNA breaks: how do they all fit together? Trends Neurosci. 23 (3), 100–105. https://doi.org/10.1016/S0166-2236(99)01503-9.

Glassner, B.J., et al., 1999. DNA repair methyltransferase (Mgmt) knockout mice are sensitive to the lethal effects of chemotherapeutic alkylating agents. Mutagenesis 14 (3), 339–347. https://doi.org/10.1093/mutage/14.3.339.

Golding, S.E., et al., 2007. Extracellular signal-related kinase positively regulates ataxia telangiectasia mutated, homologous recombination repair, and the DNA damage response. Cancer Res. 67 (3), 1046–1053. https://doi.org/10.1158/0008-5472.CAN-06-2371.
Graf, R.P., Keller, N., Barbero, S., Stupack, D., 2014. Caspase-8 as a regulator of tumor cell motility. Curr. Mol. Med., 246–254. https://doi.org/10.2174/1566524014666140128111951.
Greene, M.A., Loeser, R.F., 2015. Aging-related inflammation in osteoarthritis. Osteoarthr. Cartil. 23 (11), 1966–1971. https://doi.org/10.1016/j.joca.2015.01.008.
Gu, Y., et al., 1997. Ku70-deficient embryonic stem cells have increased ionizing radiosensitivity, defective DNA end-binding activity, and inability to support V(D)J recombination. Proc. Natl. Acad. Sci. U. S. A. 94 (15), 8076–8081. https://doi.org/10.1073/pnas.94.15.8076.
Gudkov, A.V., Komarova, E.A., 2003. The role of p53 in determining sensitivity to radiotherapy. Nat. Rev. Cancer 3, 117. https://doi.org/10.1038/nrc992.
Guo, X.Z., et al., 2002. Effect of bax, bcl-2 and bcl-xL on regulating apoptosis in tissues of normal liver and hepatocellular carcinoma. World J. Gastroenterol. 8 (6), 1059–1062. https://doi.org/10.3748/wjg.v8.i6.1059.
Hakem, R., 2008. DNA-damage repair; the good, the bad, and the ugly. EMBO J. 27 (4), 589–605. https://doi.org/10.1038/emboj.2008.15.
Hari, K.L., et al., 1995. The mei-41 gene of D. melanogaster is a structural and functional homolog of the human ataxia telangiectasia gene. Cell 82 (5), 815–821. https://doi.org/10.1016/0092-8674(95)90478-6.
Harris, C.C., Hollstein, M., 1993. Clinical implications of the p53 tumor-suppressor gene. N. Engl. J. Med. 329 (18), 1318–1327. https://doi.org/10.1056/NEJM199310283291807.
Hawkins, C.J., et al., 2000. The *Drosophila* caspase DRONC cleaves following glutamate or aspartate and is regulated by DIAP1, HID, and GRIM. J. Biol. Chem. 275 (35), 27084–27093. https://doi.org/10.1074/jbc.M000869200.
Hay, B.A., Wolff, T., Rubin, G.M., 1994. Expression of baculovirus P35 prevents cell death in *Drosophila*. Development 120 (8), 2121–2129.
Haynie, J.L., Bryant, P.J., 1977. The effects of X-rays on the proliferation dynamics of cells in the imaginal wing disc of *Drosophila* melanogaster. Roux. Arch. Dev. Biol. 183 (2), 85–100. https://doi.org/10.1007/BF00848779.
Helfer, B., et al., 2006. Caspase-8 promotes cell motility and calpain activity under nonapoptotic conditions. Cancer Res. 66 (8), 4273–4278. https://doi.org/10.1158/0008-5472.CAN-05-4183.
Henry, C.M., Martin, S.J., 2017. Caspase-8 acts in a non-enzymatic role as a scaffold for assembly of a pro-inflammatory "FADDosome" complex upon TRAIL stimulation. Mol. Cell 65 (4). https://doi.org/10.1016/j.molcel.2017.01.022. 715–729.e5.
Heyer, B.S., et al., 2000. Hypersensitivity to DNA damage leads to increased apoptosis during early mouse development. Genes Dev. 14 (16), 2072–2084. http://www.ncbi.nlm.nih.gov/entrez/query.fcgi?cmd=Retrieve&db=PubMed&dopt=Citation&list_uids=10950870.
Heyer, W.D., et al., 2006. Rad54: the Swiss Army knife of homologous recombination? Nucleic Acids Res. 34 (15), 4115–4125. https://doi.org/10.1093/nar/gkl481.
Horn, H.F., Vousden, K.H., 2007. Coping with stress: multiple ways to activate p53. Oncogene 26 (9), 1306–1316. https://doi.org/10.1038/sj.onc.1210263.
Huisken, J., et al., 2004. Optical sectioning deep inside live embryos by selective plane illumination microscopy. Sci. Am. Assoc. Adv. Sci. 305 (5686), 1007–1009. https://doi.org/10.1126/science.1100035.
Ishizaki, Y., Jacobson, M.D., Raff, M.C., 1998. A role for caspases in lens fiber differentiation. J. Cell Biol. 140 (1), 153–158. https://doi.org/10.1083/jcb.140.1.153.

Ito, K., et al., 2004. Regulation of oxidative stress by ATM is required for self-renewal of haematopoietic stem cells. Nature 431, 997. https://doi.org/10.1038/nature02989.

Ito, K., et al., 2006. Reactive oxygen species act through p38 MAPK to limit the lifespan of hematopoietic stem cells. Nat. Med. 12, 446. https://doi.org/10.1038/nm1388.

Jackson, S.P., 2002. Sensing and repairing DNA double-strand breaks. Carcinogenesis 23 (5), 687–696.

Jackson, S.P., Bartek, J., 2009. The DNA-damage response in human biology and disease. Nature 461 (7267), 1071–1078. https://doi.org/10.1038/nature08467.

Janssens, S., Tschopp, J., 2006. Signals from within: the DNA-damage-induced NF-κB response. Cell Death Differ. 13 (5), 773–784. https://doi.org/10.1038/sj.cdd.4401843.

Janzen, V., et al., 2008. Hematopoietic stem cell responsiveness to exogenous signals is limited by caspase-3. Cell Stem Cell 2, 584–594.

Jhappan, C., et al., 1997. DNA-PKcs: a T-cell tumour suppressor encoded at the mouse scid locus. Nat. Genet. 17 (4), 483–486. https://doi.org/10.1038/ng1297-483.

Jin, H.S., et al., 2009. cIAP1, cIAP2, and xIAP act cooperatively via nonredundant pathways to regulate genotoxic stress-induced nuclear factor-κB activation. Cancer Res. 69 (5), 1782–1791. https://doi.org/10.1158/0008-5472.CAN-08-2256.

Jiricny, J., 2006. The multifaceted mismatch-repair system. Nat. Rev. Mol. Cell Biol 7, 335. https://doi.org/10.1038/nrm1907.

Kaina, B., 2003. DNA damage-triggered apoptosis: critical role of DNA repair, double-strand breaks, cell proliferation and signaling. Biochem. Pharmacol. 66 (8), 1547–1554. https://doi.org/10.1016/S0006-2952(03)00510-0.

Kang, E.H., et al., 2010. Adiponectin is a potential catabolic mediator in osteoarthritis cartilage. Arthritis Res. Ther. 12 (6), R231. https://doi.org/10.1186/ar3218.

Kanuka, H., et al., 2005. *Drosophila* caspase transduces shaggy/GSK-3β kinase activity in neural precursor development. EMBO J. 24 (21), 3793–3806. https://doi.org/10.1038/sj.emboj.7600822.

Kastenhuber, E.R., Lowe, S.W., 2017. Putting p53 in context. Cell 170 (6), 1062–1078. https://doi.org/10.1016/j.cell.2017.08.028.

Kaufmann, A., et al., 2012. Multilayer mounting enables long-term imaging of zebrafish development in a light sheet microscope. Development 139 (17), 3242–3247. https://doi.org/10.1242/dev.082586.

Kayagaki, N., et al., 2011. Non-canonical inflammasome activation targets caspase-11. Nature 479 (7371), 117–121. https://doi.org/10.1038/nature10558.

Kelekar, A., Thompson, C.B., 1998. Bcl-2-family proteins: the role of the BH3 domain in apoptosis. Trends Cell Biol. 8 (8), 324–330. https://doi.org/10.1016/S0962-8924(98)01321-X.

Kenyon, J., Gerson, S.L., 2007. The role of DNA damage repair in aging of adult stem cells. Nucleic Acids Res. 35 (22), 7557–7565. https://doi.org/10.1093/nar/gkm1064.

Khan, C., Muliyil, S., Ayyub, C., Rao, B.J., 2017a. Somatic developmental defects in dmrad51/spnA mutant show low temperature sensitivity with enhanced genomic damages and cell death. bioRixv https://doi.org/10.1101/180737.

Khan, C., Muliyil, S., Ayyub, C., Rao, B.J., 2017b. The initiator caspase Dronc plays a non-apoptotic role in promoting DNA damage signalling in D. melanogaster. J. Cell Sci. 130, 2984–2995. https://doi.org/10.1242/jcs.200782.

Kim, Y.-J., Wilson, D.M., 2012. Overview of base excision repair biochemistry. Curr. Mol. Pharmacol. 5 (1), 3–13.

Kim, Y.-S., et al., 2004. Role of relish, a *Drosophila* NF-κB protein, in response to UV irradiation. J. Korean Assoc. Cancer Prev. 9 (4), 237–243.

Kitazumi, I., Tsukahara, M., 2011. Regulation of DNA fragmentation: the role of caspases and phosphorylation. FEBS J. 278 (3), 427–441. https://doi.org/10.1111/j.1742-4658.2010.07975.x.

Klovstad, M., Abdu, U., Schüpbach, T., 2008. *Drosophila* brca2 is required for mitotic and meiotic DNA repair and efficient activation of the meiotic recombination checkpoint. PLoS Genet. 4 (2), e31. https://doi.org/10.1371/journal.pgen.0040031.

Kobayashi, J., et al., 2004. NBS1 and its functional role in the DNA damage response. DNA Repair 3 (8), 855–861. https://doi.org/10.1016/j.dnarep.2004.03.023.

Kondo, S., et al., 2006. DRONC coordinates cell death and compensatory proliferation. Mol. Cell. Biol. 26 (19), 7258–7268. https://doi.org/10.1128/MCB.00183-06.

Kuilman, T., et al., 2008. Oncogene-induced senescence relayed by an interleukin-dependent inflammatory network. Cell 133 (6), 1019–1031. https://doi.org/10.1016/j.cell.2008.03.039.

Kuwana, T., et al., 2002. Bid, Bax, and lipids cooperate to form supramolecular openings in the outer mitochondrial membrane. Cell 111 (3), 331–342. https://doi.org/10.1016/S0092-8674(02)01036-X.

Laberge, R.M., et al., 2012. Glucocorticoids suppress selected components of the senescence-associated secretory phenotype. Aging Cell 11 (4), 569–578. https://doi.org/10.1111/j.1474-9726.2012.00818.x.

Lamarche, B.J., Orazio, N.I., Weitzman, M.D., 2010. The MRN complex in double-strand break repair and telomere maintenance. FEBS Lett. 584 (17), 3682–3695. https://doi.org/10.1016/j.febslet.2010.07.029.

Lannan, E., Vandergaast, R., Friesen, P.D., 2007. Baculovirus caspase inhibitors P49 and P35 block virus-induced apoptosis downstream of effector caspase DrICE activation in *Drosophila* melanogaster cells. J. Virol. 81 (17), 9319–9330. https://doi.org/10.1128/JVI.00247-07.

LaRocque, J.R., et al., 2007. *Drosophila* ATR in double-strand break repair. Genetics 175 (3), 1023–1033. https://doi.org/10.1534/genetics.106.067330.

Larsen, B.D., et al., 2010. Caspase 3/caspase-activated DNase promote cell differentiation by inducing DNA strand breaks. Proc. Natl. Acad. Sci. U. S. A. 107 (9), 4230–4235. https://doi.org/10.1073/pnas.0913089107.

Lassus, P., 2004. Erratum: requirement for caspase-2 in stress-induced apoptosis before mitochondrial permeabilization. Science 306 (5702), 1683. https://www.scopus.com/inward/record.uri?eid=2-s2.0-10044224487&partnerID=40&md5=b7955072bfc7cb1c4970a7a537e48bf1.

Laurent, A., Blasi, F., 2015. Differential DNA damage signalling and apoptotic threshold correlate with mouse epiblast-specific hypersensitivity to radiation. Development 142 (21), 3675–3685. https://doi.org/10.1242/dev.125708.

Lawrence, T., 2009. The nuclear factor NF-kappaB pathway in inflammation. Cold Spring Harb. Perspect. Biol. 1 (6), 1–10. https://doi.org/10.1101/cshperspect.a001651.

Lee, Y., McKinnon, P.J., 2007. Responding to DNA double strand breaks in the nervous system. Neuroscience 145 (4), 1365–1374. https://doi.org/10.1016/j.neuroscience.2006.07.026.

Lee, P., et al., 2009. Dynamic expression of epidermal caspase 8 simulates a wound healing response. Nature 458 (7237), 519–523. https://doi.org/10.1038/nature07687.

Leibeling, D., Laspe, P., Emmert, S., 2006. Nucleotide excision repair and cancer. J. Mol. Histol. 37 (5), 225–238. https://doi.org/10.1007/s10735-006-9041-x.

Levine, A.J., Hu, W., Feng, Z., 2006. The P53 pathway: what questions remain to be explored? Cell Death Differ. 13 (6), 1027–1036. https://doi.org/10.1038/sj.cdd.4401910.

Li, G.-M., 2007. Mechanisms and functions of DNA mismatch repair. Cell Res. 18, 85. https://doi.org/10.1038/cr.2007.115.

Li, N., et al., 2001. ATM is required for IκB kinase (IKK) activation in response to DNA double strand breaks. J. Biol. Chem. 276 (12), 8898–8903. https://doi.org/10.1074/jbc.M009809200.

Li, L., et al., 2008. Erlotinib attenuates homologous recombinational repair of chromosomal breaks in human breast cancer cells. Cancer Res. 68 (22), 9141–9146. https://doi.org/10.1158/0008-5472.CAN-08-1127.

Li, F., et al., 2010. Apoptotic caspases regulate induction of iPSCs from human fibroblasts. Cell Stem Cell 7 (4), 508–520. https://doi.org/10.1016/j.stem.2010.09.003.

Li, Z., et al., 2011. Decision making of the p53 network: death by integration. J. Theor. Biol. 271 (1), 205–211. https://doi.org/10.1016/j.jtbi.2010.11.041.

Liu, Y., Kulesz-Martin, M., 2001. p53 protein at the hub of cellular DNA damage response pathways through sequence-specific and non-sequence-specific DNA binding. Carcinogenesis 22 (6), 851–860.

Lomonosova, E., Chinnadurai, G., 2009. BH3-only proteins in apoptosis and beyond: an overview. Oncogene 27, S2. https://doi.org/10.1038/onc.2009.39.

Lu, K.L., Yamashita, Y.M., 2017. Germ cell connectivity enhances cell death in response to DNA damage in the *Drosophila*. eLife 6, e27960. https://doi.org/10.7554/eLife.27960.

Lu, C., et al., 2006. Cell apoptosis: requirement of H2AX in DNA ladder formation, but not for the activation of Caspase-3. Mol. Cell 23 (1), 121–132. https://doi.org/10.1016/j.molcel.2006.05.023.

Lu, J., et al., 2011. Role of epidermal growth factor receptor in DNA damage repair. Chin. Sci. Bull. 56 (30), 3132–3137. https://doi.org/10.1007/s11434-011-4650-5.

Luo, X., et al., 2007. Foxo and Fos regulate the decision between cell death and survival in response to UV irradiation. EMBO J. 26 (2), 380–390. https://doi.org/10.1038/sj.emboj.7601484.

Ma, X., et al., 2016. DNA damage-induced Lok/CHK2 activation compromises germline stem cell self-renewal and lineage differentiation. Development 143 (23), 4312–4323. https://doi.org/10.1242/dev.141069.

Madigan, J.P., Chotkowski, H.L., Glaser, R.L., 2002. DNA double-strand break-induced phosphorylation of *Drosophila* histone variant H2Av helps prevent radiation-induced apoptosis. Nucleic Acids Res. 30 (17), 3698–3705. https://doi.org/10.1093/nar/gkf496.

Mao, Z., et al., 2008. Comparison of nonhomologous end joining and homologous recombination in human cells. DNA Repair (Amst.) 7, 1765–1771. https://doi.org/10.1016/j.dnarep.2008.06.018.

Mao, Z., et al., 2017. DNA repair by nonhomologous end joining and homologous recombination during cell cycle in human cells. Cell Cycle 7 (July), 2902–2906. https://doi.org/10.4161/cc.7.18.6679.

Marteijn, J.A., et al., 2014. Understanding nucleotide excision repair and its roles in cancer and ageing. Nat. Rev. Mol. Cell Biol. 15 (7), 465–481. https://doi.org/10.1038/nrm3822.

Martin, L.J., Wong, M., 2017. Enforced DNA repair enzymes rescue neurons from apoptosis induced by target deprivation and axotomy in mouse models of neurodegeneration. Mech. Ageing Dev. 161, 149–162. https://doi.org/10.1016/j.mad.2016.06.011.

Masters, S.L., et al., 2010. Activation of the NLRP3 inflammasome by islet amyloid polypeptide provides a mechanism for enhanced IL-1β 2 in type 2 diabetes. Nat. Immunol. 11 (10), 897–904. https://doi.org/10.1038/ni.1935.

Matos, J., West, S.C., 2014. Holliday junction resolution: regulation in space and time. DNA Repair 19, 176–181. https://doi.org/10.1016/j.dnarep.2014.03.013.

Maynard, S., et al., 2015. DNA damage, DNA repair, aging, and neurodegeneration. Cold Spring Harb. Perspect. Med. 5 (10). https://doi.org/10.1101/cshperspect.a025130.

McEwen, D.G., 2005. Puckered, a *Drosophila* MAPK phosphatase, ensures cell viability by antagonizing JNK-induced apoptosis. Development 132 (17), 3935–3946. https://doi.org/10.1242/dev.01949.

McNamee, L.M., Brodsky, M.H., 2009. P53-independent apoptosis limits DNA damage-induced aneuploidy. Genetics 182 (2), 423–435. https://doi.org/10.1534/genetics.109.102327.

Melino, G., De Laurenzi, V., Vousden, K.H., 2002. p73: friend or foe in tumorigenesis. Nat. Rev. Cancer 2, 605. https://doi.org/10.1038/nrc861.

Merritt, A.J., et al., 1995. Differential expression of bcl-2 in intestinal epithelia. Correlation with attenuation of apoptosis in colonic crypts and the incidence of colonic neoplasia. J. Cell Sci. 108, 2261–2271. http://www.ncbi.nlm.nih.gov/pubmed/7673346.

Mesquita, D., Dekanty, A., Milán, M., 2010. A dp53-dependent mechanism involved in coordinating tissue growth in *Drosophila*. PLoS Biol. 8 (12), e1000566. https://doi.org/10.1371/journal.pbio.1000566.

Miyamoto, S., 2011. Nuclear initiated NF-κB signaling: NEMO and ATM take center stage. Cell Res. 21 (1), 116–130. https://doi.org/10.1038/cr.2010.179.

Modrich, P., 1991. Mechanisms and biological effects of mismatch repair. Annu. Rev. Genet. 25 (1), 229–253. https://doi.org/10.1146/annurev.ge.25.120191.001305.

Mondesert, O., et al., 2015. Monitoring the activation of the DNA damage response pathway in a 3D spheroid model. PLoS One 10 (7), 1–12. https://doi.org/10.1371/journal.pone.0134411.

Moon, N., et al., 2008. E2F and p53 induce apoptosis independently during *Drosophila* development but intersect in the context of DNA damage. PLoS Genet. 4 (8), e1000153. https://doi.org/10.1371/journal.pgen.1000153.

Morata, G., Shlevkov, E., Pérez-Garijo, A., 2011. Mitogenic signaling from apoptotic cells in *Drosophila*. Dev. Growth Differ. 53 (2), 168–176. https://doi.org/10.1111/j.1440-169X.2010.01225.x.

Moreno, E., Yan, M., Basler, K., 2002. Evolution of TNF signaling mechanisms: JNK-dependent apoptosis triggered by Eiger, the *Drosophila* homolog of the TNF superfamily. Curr. Biol. 12 (14), 1263–1268. https://doi.org/10.1016/S0960-9822(02)00954-5.

Morrison, D.K., Davis, R.J., 2003. Regulation of MAP kinase signaling modules by scaffold proteins in mammals. Annu. Rev. Cell Dev. Biol. 19 (1), 91–118. https://doi.org/10.1146/annurev.cellbio.19.111401.091942.

Mothersill, C., Seymour, C., 2006. Radiation-induced bystander effects: evidence for an adaptive response to low dose exposures? Dose-Response 4 (4), 283–290. https://doi.org/10.2203/dose-response.06-111.Mothersill.

Mothersill, C., et al., 2000. Involvement of energy metabolism in the production of "bystander effects" by radiation. Br. J. Cancer 82 (10), 1740–1746. https://doi.org/10.1054/bjoc.2000.1109.

Muliyil, S., Narasimha, M., 2014. Mitochondrial ROS regulates cytoskeletal and mitochondrial remodeling to tune cell and tissue dynamics in a model for wound healing. Dev. Cell 28 (3), 239–252. https://doi.org/10.1016/j.devcel.2013.12.019.

Muller, P.A.J., Vousden, K.H., 2014. Mutant p53 in cancer: new functions and therapeutic opportunities. Cancer Cell 25 (3), 304–317. https://doi.org/10.1016/j.ccr.2014.01.021.

Müller, M., et al., 1998. p53 activates the CD95 (APO-1/Fas) gene in response to DNA damage by anticancer drugs. J. Exp. Med. 188 (11), 2033–2045. https://doi.org/10.1084/jem.188.11.2033.

Muñoz-Espín, D., Serrano, M., 2014. Cellular senescence: from physiology to pathology. Nat. Rev. Mol. Cell Biol. 15, 482. https://doi.org/10.1038/nrm3823.

Muzio, M., et al., 1998. An induced proximity model for caspase-8 activation. J. Biol. Chem. 273 (5), 2926–2930. https://doi.org/10.1074/jbc.273.5.2926.

Nagata, S., Golstein, P., 1995. The Fas death factor. Science 267 (5203), 1449–1456. https://doi.org/10.1126/science.7533326.

Nakanishi, C., Toi, M., 2005. Nuclear factor-κB inhibitors as sensitizers to anticancer drugs. Nat. Rev. Cancer 5, 297. https://doi.org/10.1038/nrc1588.

Napoletano, F., et al., 2017. p53-dependent programmed necrosis controls germ cell homeostasis during spermatogenesis. PLOS Genet. 13 (9), 1–21. https://doi.org/10.1371/journal.pgen.1007024.

Neto-Silva, R.M., Wells, B.S., Johnston, L.A., 2009. Mechanisms of growth and homeostasis in the *Drosophila* wing. Annu. Rev. Cell Dev. Biol. 25, 197–220. https://doi.org/10.1146/annurev.cellbio.24.110707.175242.

Neves, J., et al., 2015. Of flies, mice, and men: evolutionarily conserved tissue damage responses and aging. Dev. Cell 32 (1), 9–18. https://doi.org/10.1016/j.devcel.2014.11.028.

Nishimura, K., et al., 2009. An auxin-based degron system for the rapid depletion of proteins in nonplant cells. Nat. Methods 6, 917. https://doi.org/10.1038/nmeth.1401.

Nyati, M.K., et al., 2006. Ataxia telangiectasia mutated down-regulates phospho-extracellular signal-regulated kinase 1/2 via activation of MKP-1 in response to radiation. Cancer Res. 66 (24), 11554–11559. https://doi.org/10.1158/0008-5472.CAN-06-1935.

Ollmann, M., et al., 2000. *Drosophila* p53 is a structural and functional homolog of the tumor suppressor p53. Cell 101, 91–101.

Oren, M., 1999. Regulation of the p53 tumor suppressor protein. J. Biol. Chem. 274 (51), 36031–36034. https://doi.org/10.1074/jbc.274.51.36031.

Orii, K.E., et al., 2006. Selective utilization of nonhomologous end-joining and homologous recombination DNA repair pathways during nervous system development. Proc. Natl. Acad. Sci. U. S. A. 103 (26), 10017–10022. https://doi.org/10.1073/pnas.0602436103.

Orme, M.H., et al., 2016. The unconventional myosin CRINKLED and its mammalian orthologue MYO7A regulate caspases in their signalling roles. Nat. Commun. 7, 10972. https://doi.org/10.1038/ncomms10972.

Ouyang, Y., et al., 2011. Dronc caspase exerts a non-apoptotic function to restrain phospho-numb-induced ectopic neuroblast formation in *Drosophila*. Development 138 (11), 2185–2196. https://doi.org/10.1242/dev.058347.

Park, J.-S., et al., 2015. Requirement of ATR for maintenance of intestinal stem cells in aging *Drosophila*. Aging 7 (5), 307–318. https://doi.org/10.18632/aging.100743.

Paull, T.T., et al., 2000. A critical role for histone H2AX in recruitment of repair factors to nuclear foci after DNA damage. Curr. Biol. 10 (15), 886–895. https://doi.org/10.1016/S0960-9822(00)00610-2.

Pegg, A.E., Dolan, M.E., Moschel, R.C., 1995. Structure, function, and inhibition of O6-alkylguanine-DNA alkyltransferase. Prog. Nucleic Acid Res. Mol. Biol. 51, 167–223. https://doi.org/10.1016/S0079-6603(08)60879-X.

Pérez-Garijo, A., et al., 2005. Dpp signaling and the induction of neoplastic tumors by caspase-inhibited apoptotic cells in *Drosophila*. Proc. Natl. Acad. Sci. U. S. A. 102 (49), 17664–17669. https://doi.org/10.1073/pnas.0508966102.

Perez-Garijo, A., Shlevkov, E., Morata, G., 2009. The role of Dpp and Wg in compensatory proliferation and in the formation of hyperplastic overgrowths caused by apoptotic cells in the *Drosophila* wing disc. Development 136 (7), 1169–1177. https://doi.org/10.1242/dev.034017.

Petersen, A.J., Katzenberger, R.J., Wassarman, D.A., 2013. The innate immune response transcription factor relish is necessary for neurodegeneration in a *Drosophila* model of Ataxia-telangiectasia. Genetics 194 (1), 133–142. https://doi.org/10.1534/genetics.113.150854.

Picco, V., Pagès, G., 2013. Linking JNK activity to the DNA damage response. Genes Cancer 4 (9–10), 360–368. https://doi.org/10.1177/1947601913486347.

Pietsch, E.C., et al., 2008. The p53 family and programmed cell death. Oncogene 27 (50), 6507–6521. https://doi.org/10.1038/onc.2008.315.
Podhorecka, M., Skladanowski, A., Bozko, P., 2010. H2AX phosphorylation: its role in DNA damage response and cancer therapy. J. Nucleic Acids 2010, 920161. https://doi.org/10.4061/2010/920161.
Potten, C.S., Owen, G., Booth, D., 2002. Intestinal stem cells protect their genome by selective segregation of template DNA strands. J. Cell Sci. 115 (Pt. 11), 2381–2388. http://www.ncbi.nlm.nih.gov/pubmed/12006622.
Poulton, J.S., Cuningham, J.C., Peifer, M., 2014. Acentrosomal *Drosophila* epithelial cells exhibit abnormal cell division, leading to cell death and compensatory proliferation. Dev. Cell 30 (6), 731–745. https://doi.org/10.1016/j.devcel.2014.08.007.
Riley, T., et al., 2008. Transcriptional control of human p53-regulated genes. Nat. Rev. Mol. Cell Biol. 9, 402. https://doi.org/10.1038/nrm2395.
Ríos-Barrera, L.D., Riesgo-Escovar, J.R., 2013. Regulating cell morphogenesis: the *drosophila* Jun N-terminal kinase pathway. Genesis 51 (3), 147–162. https://doi.org/10.1002/dvg.22354.
Rivlin, N., et al., 2011. Mutations in the p53 tumor suppressor gene: important milestones at the various steps of tumorigenesis. Genes Cancer 2 (4), 466–474. https://doi.org/10.1177/1947601911408889.
Robertson, J.D., et al., 2002. Caspase-2 acts upstream of mitochondria to promote cytochrome c release during etoposide-induced apoptosis. J. Biol. Chem. 277 (33), 29803–29809. https://doi.org/10.1074/jbc.M204185200.
Robertson, J.D., et al., 2004. Processed caspase-2 can induce mitochondria-mediated apoptosis independently of its enzymatic activity. EMBO Rep. 5 (6), 643–648. https://doi.org/10.1038/sj.embor.7400153.
Roos, W.P., Kaina, B., 2006. DNA damage-induced cell death by apoptosis. Trends Mol. Med. 12 (9), 440–450. https://doi.org/10.1016/j.molmed.2006.07.007.
Ryoo, H.D., Gorenc, T., Steller, H., 2004. Apoptotic cells can induce compensatory cell proliferation through the JNK and the wingless signaling pathways. Dev. Cell 7 (4), 491–501. https://doi.org/10.1016/j.devcel.2004.08.019.
Sakahira, H., Enari, M., Nagata, S., 1998. Cleavage of CAD inhibtor in CAD activation and DNA degradation during apoptosis. Nature 391 (6662), 96–99.
Sakamaki, K., et al., 2002. Ex vivo whole-embryo culture of caspase-8-deficient embryos normalize their aberrant phenotypes in the developing neural tube and heart. Cell Death Differ. 9, 1196. https://doi.org/10.1038/sj.cdd.4401090.
Sancar, A., et al., 2004. Molecular mechanisms of mammalian DNA repair and the DNA damage checkpoints. Annu. Rev. Biochem. 73 (1), 39–85. https://doi.org/10.1146/annurev.biochem.73.011303.073723.
Sedgwick, B., et al., 2007. Repair of alkylated DNA: recent advances. DNA Repair 6 (4), 429–442. https://doi.org/10.1016/j.dnarep.2006.10.005.
Seita, J., Rossi, D.J., Weissman, I.L., 2010. Differential DNA damage response in stem and progenitor cells. Cell Stem Cell 7 (2), 145–147. https://doi.org/10.1016/j.stem.2010.07.006.
Sekelsky, J., 2017. DNA repair in *Drosophila*: mutagens, models, and missing genes. Genetics 205 (2), 471–490. https://doi.org/10.1534/genetics.116.186759.
Sekelsky, J.J., Brodsky, M.H., Burtis, K.C., 2000. DNA repair in *Drosophila*: insights from the *Drosophila* genome sequence. J. Cell Biol. 150 (2), 31–36. https://doi.org/10.1083/jcb.150.2.F31.
Shaposhnikov, M., et al., 2015. Lifespan and stress resistance in *Drosophila* with overexpressed DNA repair genes. Sci. Rep. 5, 15299. https://doi.org/10.1038/srep15299.
Shibata, T., et al., 2013. Transglutaminase-catalyzed protein-protein cross-linking suppresses the activity of the NF-κB like transcription factor relish. Sci. Signal. 6 (285), ra61. https://doi.org/10.1126/scisignal.2003970.

Shiloh, Y., 2006. The ATM-mediated DNA-damage response: taking shape. Trends Biochem. Sci. 31 (7), 402–410. https://doi.org/10.1016/j.tibs.2006.05.004.

Shiloh, Y., Ziv, Y., 2013. The ATM protein kinase: regulating the cellular response to genotoxic stress, and more. Nat. Rev. Mol. Cell Biol. 14, 197. https://doi.org/10.1038/nrm3546.

Shin, Y.J., et al., 2015. Nanobody-targeted E3-ubiquitin ligase complex degrades nuclear proteins. Sci. Rep. 5, 14269. https://doi.org/10.1038/srep14269.

Shlevkov, E., Morata, G., 2012. A dp53/JNK-dependant feedback amplification loop is essential for the apoptotic response to stress in *Drosophila*. Cell Death Differ. 19 (3), 451–460. https://doi.org/10.1038/cdd.2011.113.

Sibon, O.C.M., et al., 1999. The *Drosophila* ATM homologue Mei-41 has an essential checkpoint function at the midblastula transition. Curr. Biol. 9 (6), 302–312. https://doi.org/10.1016/S0960-9822(99)80138-9.

Sijmons, R.H., Hofstra, R.M.W., 2016. Clinical aspects of hereditary DNA mismatch repair gene mutations. DNA Repair 38, 155–162. https://doi.org/10.1016/j.dnarep.2015.11.018.

Simon, M.M., et al., 1994. UVB light induces nuclear factor κB (NFκB) activity independently from chromosomal DNA damage in cell-free cytosolic extracts. J. Invest. Dermatol. 102, 422–427. https://doi.org/10.1111/1523-1747.ep12372194.

Slade, J.D., Staveley, B.E., 2015. Compensatory growth in novel *Drosophila* Akt1 mutants. BMC. Res. Notes 8 (1), 77. https://doi.org/10.1186/s13104-015-1032-0.

Smith, G.C., et al., 1999. Cleavage and inactivation of ATM during apoptosis. Mol. Cell. Biol. 19 (9), 6076–6084. https://doi.org/10.1128/MCB.19.9.6076.

Solier, S., Pommier, Y., 2009. The apoptotic ring: a novel entity with phosphorylated histones H2AX and H2B and activated DNA damage response kinases. Cell Cycle 8 (12), 1853–1859. https://doi.org/10.4161/cc.8.12.8865.

Solier, S., Pommier, Y., 2011. MDC1 cleavage by caspase-3: a novel mechanism for inactivating the DNA damage response during apoptosis. Cancer Res. 71 (3), 906–913. https://doi.org/10.1158/0008-5472.CAN-10-3297.

Solier, S., et al., 2009. Death receptor-induced activation of the Chk2- and histone H2AX-associated DNA damage response pathways. Mol. Cell. Biol. 29 (1), 68–82. https://doi.org/10.1128/MCB.00581-08.

Staeva-Vieira, E., Yoo, S., Lehmann, R., 2003. An essential role of DmRad51/SpnA in DNA repair and meiotic checkpoint control. EMBO J. 22 (21), 5863–5874. https://doi.org/10.1093/emboj/cdg564.

Statistik, B.P., 2014. Base excision repair. CSHL Press XXXIII (2), 81–87. https://doi.org/10.1007/s13398-014-0173-7.2.

Stein, B., et al., 1989. UV-induced DNA damage is an intermediate step in UV-induced expression of human immunodeficiency virus type 1, collagenase, c-fos, and metallothionein. Mol. Cell. Biol. 9 (11), 5169–5181. https://doi.org/10.1128/MCB.9.11.5169.Updated.

Stilmann, M., et al., 2009. A nuclear poly(ADP-ribose)-dependent signalosome confers DNA damage-induced IκB kinase activation. Mol. Cell 36 (3), 365–378. https://doi.org/10.1016/j.molcel.2009.09.032.

Stronach, B., 2005. Dissecting JNK signaling, one KKKinase at a time. Dev. Dyn. 232 (3), 575–584. https://doi.org/10.1002/dvdy.20283.

Suwaki, N., Klare, K., Tarsounas, M., 2011. RAD51 paralogs: roles in DNA damage signalling, recombinational repair and tumorigenesis. Semin. Cell Dev. Biol. 22 (8), 898–905. https://doi.org/10.1016/j.semcdb.2011.07.019.

Suzuki, A., et al., 1997. Brca2 is required for embryonic cellular proliferation in the mouse. Genes Dev. 11 (10), 1242–1252. https://doi.org/10.1101/gad.11.10.1242.

Takemoto, K., et al., 2003. Spatio-temporal activation of caspase revealed by indicator that is insensitive to environmental effects. J. Cell Biol. 160 (2), 235–243. https://doi.org/10.1083/jcb.200207111.

Takubo, K., et al., 2008. Stem cell defects in ATM-deficient undifferentiated Spermatogonia through DNA damage-induced cell-cycle arrest. Cell Stem Cell 2 (2), 170–182. https://doi.org/10.1016/j.stem.2007.10.023.

Tearle, R.G., Nusslein-Volhard, C., 1987. Tubingen mutants and stock list. Drosoph. Inf. Serv. 209–269.

Tenev, T., et al., 2002. Jafrac2 is an IAP antagonist that promotes cell death by liberating Dronc from DIAP1. EMBO J. 21 (19), 5118–5129.

Tergaonkar, V., et al., 2002. p53 stabilization is decreased upon NFκB activation: a role for NFκB in acquisition of resistance to chemotherapy. Cancer Cell 1 (5), 493–503. https://doi.org/10.1016/S1535-6108(02)00068-5.

Thornberry, N.A., et al., 1992. A novel heterodimeric cysteine protease is required for interleukin-1β processing in monocytes. Nature 356, 768. https://doi.org/10.1038/356768a0.

Tournier, C., et al., 2000. Requirement of JNK for stress-induced activation of the cytochrome c-mediated death pathway. Science (New York, N.Y.) 288 (May), 870–874. https://doi.org/10.1126/science.288.5467.870.

Tsuzuki, T., et al., 1996. Targeted disruption of the Rad51 gene leads to lethality in embryonic mice. Proc. Natl. Acad. Sci. U. S. A. 93 (June), 6236–6240. https://doi.org/10.1073/pnas.93.13.6236.

Tyas, L., Brophy, V., Pope, A., 2000. Rapid caspase-3 activation during apoptosis revealed using fluorescence resonance energy transfer. EMBO J. 1 (3), 266–270. http://onlinelibrary.wiley.com/doi/10.1093/embo-reports/kvd050/full.

van Bergeijk, P., et al., 2012. Genome-wide expression analysis identifies a modulator of ionizing radiation-induced p53-independent apoptosis in *Drosophila* melanogaster. PLoS One 7 (5), e36539. https://doi.org/10.1371/journal.pone.0036539.

Vermezovic, J., et al., 2015. Notch is a direct negative regulator of the DNA-damage response. Nat. Struct. Mol. Biol. 22, 417. https://doi.org/10.1038/nsmb.3013.

Vucic, D., Kaiser, W.J., Miller, L.K., 1998. A mutational analysis of the Baculovirus inhibitor of apoptosis. J. Biol. Chem. 273 (51), 33915–33921.

Wang, C.-Y., Mayo, M.W., Baldwin, A.S., 1996. TNF- and cancer therapy-induced apoptosis: potentiation by inhibition of NF-κB. Science 274 (5288), 784–787. http://science.sciencemag.org/content/274/5288/784.abstract.

Wang, S.L., et al., 1999. The *Drosophila* caspase inhibitor DIAP1 is essential for cell survival and is negatively regulated by HID. Cell 98 (4), 453–463. https://doi.org/10.1016/S0092-8674(00)81974-1.

Wang, W., Mani, A.M., Wu, Z.-H., 2017. DNA damage-induced nuclear factor-kappa B activation and its roles in cancer progression. J. Cancer Metastasis Treat. 3 (3), 45. https://doi.org/10.20517/2394-4722.2017.03.

Weaver, B.P., et al., 2014. CED-3 caspase acts with miRNAs to regulate non-apoptotic gene expression dynamics for robust development in *C. elegans*. eLife 3 (November), 1–22. https://doi.org/10.7554/eLife.04265.

Weil, M., Raff, M.C., Braga, V.M., 1999. Caspase activation in the terminal differentiation of human epidermal keratinocytes. Curr. Biol. 9 (7), 361–364. https://doi.org/10.1016/S0960-9822(99)80162-6.

Wells, B.S., Yoshida, E., Johnston, L.A., 2006. Compensatory proliferation in *Drosophila* imaginal discs requires Dronc-dependent p53 activity. Curr. Biol. 16 (16), 1606–1615. https://doi.org/10.1016/j.cub.2006.07.046.

Wen, W., et al., 2010. MST1 promotes apoptosis through phosphorylation of histone H2AX. J. Biol. Chem. 285 (50), 39108–39116. https://doi.org/10.1074/jbc.M110.151753.

Wichmann, A., Jaklevic, B., Su, T.T., 2006. Ionizing radiation induces caspase-dependent but Chk2- and p53-independent cell death in *Drosophila* melanogaster. Proc. Natl. Acad. Sci. U. S. A. 103, 2–7.

Williams, D.W., et al., 2006. Local caspase activity directs engulfment of dendrites during pruning. Nat. Neurosci. 9 (10), 1234–1236. https://doi.org/10.1038/nn1774.

Wolpert, L., Hornbruch, A., Clarke, M.R.B., 1974. Positional information and positional signaling in Hydra. Am. Zool. 14 (2), 647–663.

Wu, G.S., et al., 1997. KILLER/DR5 is a DNA damage-inducible p53-regulated death receptor gene. Nat. Genet. 17, 141. https://doi.org/10.1038/ng1097-141.

Wu, Z.H., et al., 2010. ATM- and NEMO-dependent ELKS ubiquitination coordinates TAK1-mediated IKK activation in response to genotoxic stress. Mol. Cell 40 (1), 75–86. https://doi.org/10.1016/j.molcel.2010.09.010.

Yan, N., et al., 2004. Molecular mechanisms of DrICE inhibition by DIAP1 and removal of inhibition by reaper, hid and grim. Nat. Struct. Mol. Biol. 11 (5), 420–428. https://doi.org/10.1038/nsmb764.

Yang, Y., et al., 2011. A cytosolic ATM/NEMO/RIP1 complex recruits TAK1 to mediate the NF-κB and p38 mitogen-activated protein kinase (MAPK)/MAPK-activated protein 2 responses to DNA damage. Mol. Cell. Biol. 31 (14), 2774–2786. https://doi.org/10.1128/MCB.01139-10.

Yang, D., et al., 2015. Caspase-11 requires the pannexin-1 channel and the purinergic P2X7 pore to mediate pyroptosis and endotoxic shock. Immunity 43 (5), 923–932. https://doi.org/10.1016/j.immuni.2015.10.009.

Yu, H.B., Finlay, B.B., 2008. The caspase-1 inflammasome: a pilot of innate immune responses. Cell Host Microbe 4 (3), 198–208. https://doi.org/10.1016/j.chom.2008.08.007.

Yuan, J., et al., 1993. The *C. elegans* cell death gene ced-3 encodes a protein similar to mammalian interleukin-1β-converting enzyme. Cell 75 (4), 641–652. https://doi.org/10.1016/0092-8674(93)90485-9.

Zermati, Y., et al., 2001. Caspase activation is required for terminal erythroid differentiation. J. Exp. Med. 193 (2), 247–254. https://doi.org/10.1084/jem.193.2.247.

Zhivotovsky, B., et al., 1999. Caspases: their intracellular localization and translocation during apoptosis. Cell Death Differ. 6 (7), 644–651. https://doi.org/10.1038/sj.cdd.4400536.

Zhou, L., Steller, H., 2003. Distinct pathways mediate UV-induced apoptosis in *Drosophila* embryos. Dev. Cell 4 (4), 599–605. https://doi.org/10.1016/S1534-5807(03)00085-6.

Zong, W.X., et al., 2004. Alkylating DNA damage stimulates a regulated form of necrotic cell death. Genes Dev. 18 (11), 1272–1282. https://doi.org/10.1101/gad.1199904.

CHAPTER FIVE

Nucleic Acid Sensing in Mammals and Plants: Facts and Caveats

Martin Heil*, Isaac Vega-Muñoz
Departmento de Ingeniería Genética, CINVESTAV-Irapuato, Irapuato, Guanajuato, Mexico
*Corresponding author: e-mail address: mheil@ira.cinvestav.mx

Contents

International Review of Cell and Molecular Biology, Volume 345
ISSN 1937-6448
https://doi.org/10.1016/bs.ircmb.2018.10.003

Abstract

The accumulation of nucleic acids in aberrant compartments is a signal of danger: fragments of cytosolic or extracellular self-DNA indicate cellular dysfunctions or disruption, whereas cytosolic fragments of nonself-DNA or RNA indicate infections. Therefore, nucleic acids trigger immunity in mammals and plants. In mammals, endosomal Toll-like receptors (TLRs) sense single-stranded (ss) or double-stranded (ds) RNA or CpG-rich DNA, whereas various cytosolic receptors sense dsDNA. Although a self/nonself discrimination could favor targeted immune responses, no sequence-specific sensing of nucleic acids has been reported for mammals. Specific immune responses to extracellular self-DNA versus DNA from related species were recently reported for plants, but the underlying mechanism remains unknown. The subcellular localization of mammalian receptors can favor self/nonself discrimination based on the localization of DNA fragments. However, autoantibodies and diverse damage-associated molecular patterns (DAMPs) shuttle DNA through membranes, and most of the mammalian receptors share downstream signaling elements such as stimulator of interferon genes (STING) and the master transcription regulators, nuclear factor (NF)-κB, and interferon regulatory factor 3 (IRF3). The resulting type I interferon (IFN) response stimulates innate immunity against multiple threats—from infection to physical injury or endogenous DNA damage—all of which lead to the accumulation of eDNA or cytoplasmatic dsDNA. Therefore, no or only low selective pressures might have favored a strict self/nonself discrimination in nucleic acid sensing. We conclude that the discrimination between self- and nonself-DNA is likely to be less strict—and less important—than assumed originally.

ABBREVIATIONS

AIDS acquired immune deficiency syndrome
AIM2 absent in melanoma 2
APC antigen-presenting cell
ASC apoptosis-associated speck-like protein
CARD caspase recruitment domain
cGAMP cyclic GMP-AMP
cGAS cyclic GMP-AMP synthase
DAI DNA-dependent activator of IFN regulatory factors
DAMP damage-associated molecular pattern
DC dendritic cell
dsDNA double-stranded DNA
dsRNA double-stranded RNA
eATP extracellular ATP
eDNA extracellular DNA
eIF2α eukaryotic initiation factor 2α
HAMP herbivore-associated molecular pattern
HDPs host defense peptides
HIV-1 human immunodeficiency virus 1
HMGB high-mobility group box protein 1
HSP70 heat shock protein 70

IFI16	interferon-inducible 16
IFN	interferon
IL	interleukin
IRF	IFN regulatory factor
JA	jasmonic acid
JAZ	jasmonate-ZIM-domain
LGP2	laboratory of genetics and physiology 2
LPS	lipopolysaccharides
LRR	leucine-rich repeat
MAPK	mitogen-activated protein kinase
MDA5	melanoma differentiation-associated gene-5
mtDNA	mitochondrial DNA
MyD 88	myeloid differentiation primary response 88
NF-κB	nuclear factor kappa beta
NLRP3	NOD-like, LRR- and PYD-containing 3
NOD	nucleotide-binding oligomerization domain
PAMP	pathogen-associated molecular pattern
pDCs	plasmacytoid dendritic cells
PKR	protein kinase R
PRR	pattern-recognition receptor
PYD	pyrin domain
RAGE	receptor for advanced glycation end products
RIG	retinoic acid-inducible gene
RLR	retinoic acid-inducible gene-like receptors
ROS	reactive oxygen species
SA	salicylic acid
SLE	systemic lupus erythematosus
ssDNA	single-stranded DNA
ssRNA	single-stranded RNA
STING	stimulator of interferon genes
TLR	Toll-like receptor
TREX1	three prime repair exonuclease 1
ZBP1	*Z*-DNA binding protein 1

1. INTRODUCTION

Organisms from the entire tree of life—including bacteria, fungi, plants, and animals—have evolved a plethora of mechanisms aimed at maintaining homeostasis in spite of endogenous and exogenous damage. Since most factors that cause damage are unpredictable, successful immunity begins with the detection of the current problems, followed by signaling cascades that coordinate a timely and directed response aimed at clearance (at least in the case of infection) and repair. Immunology originally was defined as the "science of self/nonself discrimination" (Golub and Green, 1991), hence, it was proposed that "the immune system evolved to discriminate infectious nonself from non-infectious self" (Janeway, 1992). However, since then, multiple lines of evidence revealed that successful immunity critically depends on a continuous surveillance for integrity or, vice-versa, damage. Therefore, Mat zinger's danger model states that the first line of any successful immune response depends on the recognition of endogenous "danger signals" (Matzinger, 1994, 2002; Yatim et al., 2017).

These danger signals, also known as "damage-associated molecular patterns" (DAMPs), are evolutionarily conserved molecules of host origin that allow organisms to recognize the "damaged self" (Heil, 2009) when they appear in aberrant compartments such as the extracellular space or—in the case of molecules of mitochondrial or nuclear origin—the cytoplasm (Bianchi, 2007; Chen and Nuñez, 2010; Choi and Klessig, 2016; Heil and Land, 2014; Lotze et al., 2007; Ranf, 2017; Yatim et al., 2017). Thus, the same molecules are "invisible" to the immune system as long as they fulfill their normal physiological role during homeostasis, but they become "visible"—i.e., immunogenic—in situations of danger. Situations that result in the release of DAMPs comprise mechanical disruption, infection, and diverse types of cellular dysfunction that cause the loss of the integrity of nuclei, mitochondria, or entire cells (Kolb et al., 2017; O'Neill, 2013; Shrivastava et al., 2016). Although programmed cell death (i.e., apoptosis) was originally considered to be non-immunogenic, certain types of programmed cell death, such as pyroptosis and necroptosis, are characterized by the release of DAMPs and therefore become immunogenic (Galluzzi et al., 2018b; Kaczmarek et al., 2013; Kolb et al., 2017; Orzalli and Kagan, 2017).

Nucleic acids of host (self) origin—including nuclear or mitochondrial DNA (mtDNA)—represent frequently reported mammalian DAMPs that have been associated with various types of disease (Gallucci and Maffei, 2017; Galluzzi et al., 2018b; Garg et al., 2015; Land et al., 2016a; Mills et al., 2017; Yatim et al., 2017). For example, enhanced levels of mtDNA have been reported from the plasma of patients suffering rheumatoid arthritis or *Systemic lupus erythematosus* (SLE), various cancers, hypertension, and in HIV-1 infected individuals (Arshad et al., 2018; Barrat et al., 2005; Cossarizza et al., 2011; Liu et al., 2016; McCarthy et al., 2015; Pinti et al., 2012; Shin et al., 2013), and high levels of cell-free DNA of nuclear origin have been observed in the sera of Sjögren's syndrome patients and were related to pro-inflammatory processes that characterize this autoimmune disease (Vakrakou et al., 2018). Enhanced plasma levels of cell-free mtDNA or nuclear DNA were also observed after severe injury, i.e., trauma (Aswani et al., 2018; Simmons et al., 2013). In this case, serum DNA can occur also in histone-DNA complexes, and elevated levels of these complexes were correlated with poor clinical outcome (Russell et al., 2018). Finally, first reports indicate an immunogenic role of self-dsDNA also for plants (Barbero et al., 2016; Duran-Flores and Heil, 2018; Lee et al., 2016; Niehl et al., 2016; Park et al., 2016; Yakushiji et al., 2009).

However, fragments of dsDNA or dsRNA are also "one of the few molecular patterns that can be used to detect viruses" (Deguine, 2017) and then function as indicators of the "infectious nonself," i.e., pathogen-associated molecular patterns (PAMPs) (Krapp et al., 2018). In fact, nonself nucleic acids represent the first reported triggers of antiviral immunity (Isaacs et al., 1963; Rotem et al., 1963). Since then, and in particular since the outbreak of the AIDS pandemic, significant breakthroughs have been achieved in our understanding of how the immune system senses nucleic acids to mount a successful immune response to viruses (Altfeld and Gale, 2015; Chen et al., 2016; Gallucci and Maffei, 2017).

In short, fragments of nucleic acids trigger immunity in organisms from different kingdoms. Whereas immune responses to nonself nucleic acids usually should be beneficial, responses to self-DNA are increasingly being identified as causal reasons of autoimmune diseases (Ablasser, 2018; Dhanwani et al., 2018). These factors make nucleic acids particularly interesting, universal modulators of immunity. Multiple reviews document the rapidly evolving knowledge on nucleic acid sensing in mammals (Barber, 2015; Chow et al., 2018; Gallucci and Maffei, 2017; Hartmann, 2017; Luecke and Paludan, 2017; Lupfer and Anand, 2016; Pelka et al., 2016; Qiao et al., 2017; Roers et al., 2016; Schlee and Hartmann, 2016; Wu and Chen, 2014; Xiao, 2015; Zevini et al., 2017). Therefore, the aim of this chapter is to give a short, comparative overview on the role of nucleic acids as DAMPs in mammals and plants. In Section 4, we discuss the mammalian DNA sensors Toll-like receptor 9 (TLR9), *Z*-DNA binding protein 1 (ZBP1), or DNA-dependent activator of IFN regulatory factors (DAI), receptor for advanced glycation end products (RAGE), cyclic GMP-AMP synthase (cGAS), and the inflammasome-forming receptors absent in melanoma 2 (AIM2), interferon-inducible 16 (IFI16), and the members of the nucleotide-binding oligomerization domain (NOD)-like, leucine-rich repeat (LRR)- and pyrin domain (PYD)-containing (NLRP) family, such as NLRP3. In Section 5, we discuss the mammalian RNA receptors from the family of retinoic acid-inducible gene (RIG)-like receptors

(RLRs) RIG1, melanoma differentiation-associated gene 5 (MDA5), and laboratory of genetics and physiology 2 (LGP2), the TLRs 3, 7, 8, and 10, and the dsRNA binding kinase, PKR. The little knowledge available on immunogenic effects of nucleic acids in plants is summarized in Section 6.

We should state clearly that the distinction between RNA receptors and DNA receptors is partially arbitrary. For example, RAGE senses RNA and DNA, TLRs and cGAS have been reported to sense RNA:DNA hybrids, and the dsRNA sensor, PKR, can also activate dsDNA-dependent PRRs such as AIM2 and NLRP3. Therefore, Section 7 focuses on receptor specificity in the context of self/nonself discrimination. We discuss the potential relevance of the species-specific immunogenic effects of eDNA in plants and of the trafficking of eDNA through membranes that is facilitated by certain DAMPs or the receptor RAGE. This trafficking of extracellular nucleic acids can lead to their exposure to intracellular receptors and might strongly enhance the immunogenic effects of self-eDNA. Finally, we take an evolutionary perspective to discuss whether and to which degree a self/nonself discrimination is required for a successful nucleic acid-driven immunity and suggest that the concentration at which nucleic acid fragments accumulate might represent an underestimated element that facilitates targeted immune responses to numerous threats. We hope that the very incomplete list of articles that we review in this chapter, and our suggestions and speculations, will motivate others to contemplate the multiple roles of nucleic acids in immunity, and that this chapter helps to close the gap between the plant sciences and the immunological sciences.

2. DAMPs ARE ENDOGENOUS DANGER SIGNALS

Definitions of the term DAMPs differ in whether or not they include molecules that are synthesized de novo after damage (Bianchi, 2007; Choi and Klessig, 2016; Gallo and Gallucci, 2013; Garg et al., 2015; Krysko et al., 2012; Land et al., 2016b; Ramadan et al., 2017). Nevertheless, the most classical DAMPs remain those molecules that exist already before damage happens and that "have a physiological role inside the cell but acquire additional functions when they are exposed to the extracellular environment" (Vénéreau et al., 2015). More recently, the term "constitutive DAMPs" has been suggested for the same group of molecules (Yatim et al., 2017). Of note, DAMPs are not necessarily released passively from dying cells but can also be actively secreted by stressed cells and then facilitate an early response that might counteract the stress before cells are irreversibly damaged (Gardella et al., 2002). Upon sensing by a diverse set of germline-encoded pattern-recognition receptors (PRRs), DAMPs trigger early immunity-related signals, particularly Ca^{2+} influx, the formation of reactive oxygen species (ROS) and the activation of mitogen-activated protein kinases (MAPKs), in organisms from plants to mammals (Acevedo et al., 2015; Bianchi, 2007; Choi and Klessig, 2016; Duran-Flores and Heil, 2016; Heil and Land, 2014; Tanaka et al., 2014; Tang et al., 2012).

In mammals, DAMPs contribute to all central steps of successful immunity: DAMPs (i) are essential for the detection of damage (Bianchi, 2007; Gallucci and Matzinger, 2001; Garg et al., 2015; Heil and Land, 2014), (ii) serve as cytokines, i.e., as signals to other elements of innate immunity (Andersson et al., 2018a; Kang et al., 2014; Magna and Pisetsky, 2014; Yang et al., 2015), (iii) contribute to the clearance of invading microorganisms (Allam et al., 2014; Bording-Jorgensen et al., 2017; Brinkmann et al., 2004; Hancock et al., 2016; Ingelsson et al., 2018; Mollerherm et al., 2016), and (iv) even are crucial for repair (Hancock et al., 2016; Hernández-Oñate and Herrera-Estrella, 2015; Pandolfi et al., 2016).

A pivotal role of DAMPs is the priming of multiple elements of innate immunity. Hereinafter, we use "priming" to denominate the preparation of the immune system for a faster or stronger response to a given stress (Martinez-Medina et al., 2016). In mammals, DAMPs prime the immune system via at least three consecutively acting mechanisms. First, as cytokines, they trigger the maturation of dendritic cells (DCs) to antigen-presenting cells (APCs) (Heil and Land, 2014; Yatim et al., 2017) or recruit T-lymphocytes to the site of injury (Chen and Nuñez, 2010; Heil and Land, 2014; Matzinger, 2002; Yatim et al., 2017). Second, the sensing of DAMPs by TLRs and diverse inflammasomes in immune cells such as DCs, APCs, and macrophages, triggers the release of transcription factors such as nuclear factor (NF)-κB, type I interferon (IFN I), and of pro-inflammatory cytokines such as interleukin (IL)-1β and IL-18 (Gougeon, 2017; Lugrin and Martinon, 2018; Mathur et al., 2018; Mills et al., 2017; Paludan and Bowie, 2013; Trigg and Ferguson, 2015). Ultimately, DAMPs can prime the expression of components of the inflammasome, in particular NLRP3 (Abderrazak et al., 2015; Carlos et al., 2017; Dib et al., 2015).

Examples of mammalian DAMPs are soluble or crystallized uric acid, fragments of hyaluronan and other molecules of the extracellular matrix, HMGB1, histones, small host defense peptides (HDPs) which are formed upon cellular damage by proteolysis of existing proteins, eATP, and delocalized DNA fragments of host origin (self-DNA), including mtDNA and nuclear DNA (Boyapati et al., 2016; Braga et al., 2017; Chen and Nuñez, 2010; Garg et al., 2015; Gougeon, 2017; Heil and Land, 2014; Ingelsson et al., 2018; Ishii et al., 2001; Mills et al., 2017; Nakahira et al., 2015; Silk et al., 2017; Vénéreau et al., 2015; Yatim et al., 2017). For plants, examples include oligogalacturonides and other cell wall fragments that are sensed via cell wall-associated kinases (Bacete et al., 2018; Benedetti et al., 2015, 2018; De Lorenzo et al., 2011; Ferrari et al., 2013; Savatin et al., 2014), HMGB3 (Choi et al., 2016), systemin and other peptides which—similar to mammalian HDPs—are released from larger precursor proteins upon damage (Bartels et al., 2013; Flury et al., 2013; Klauser et al., 2015; Shinya et al., 2018), eATP (Choi and Klessig, 2016; Duran-Flores and Heil, 2016; Heil et al., 2012; Tanaka et al., 2014), and—most recently—eDNA (Barbero et al., 2016; Duran-Flores and Heil, 2015, 2018). Thus, DAMPs in mammals and plants resemble each other at the functional level or are even identical. However, it remains an open question whether these similarities represent analogies (i.e., the results of an independent parallel evolution) or homologies (i.e., traits that share a common evolutionary origin) (Heil et al., 2016).

3. EXTRACELLULAR ATP (eATP)

Although ATP is a nucleotide, rather than a nucleic acid, we shortly discuss the role of eATP in this section, because eATP represents a paramount example of a classical or "constitutive DAMP" sensu (Yatim et al., 2017). ATP represents the universal source of energy for metabolic processes and, thus, can be expected to occur in all living cells. However, as stated by Tanaka et al. (2014), "once cellular ATP appears in the apoplast, ATP becomes a signal." As a DAMP, eATP was reported to trigger immune responses in algae (Torres et al., 2008), plants (Demidchik et al., 2003; Feng et al., 2015; Tanaka et al., 2014; Tripathi et al., 2018), fungi (Hernández-Oñate and Herrera-Estrella, 2015; Medina-Castellanos et al., 2014), insects (Moreno-García et al., 2014), fish (Kawate et al., 2009), and mammals (Chen and Nuñez, 2010; Ding and Tan, 2016; Faas et al., 2017; Gombault et al., 2013; Séror et al., 2011; Zeiser et al., 2011). Early immunity-related responses to eATP that are shared among plants, fungi and mammals comprise the enhancement of intracellular Ca^{2+} concentrations, the activation of MAPKs, and the formation of ROS (Heil and Land, 2014; Kim et al., 2006; Medina-Castellanos et al., 2014; Roux and Steinebrunner, 2007; Shimada et al., 2012; Torres et al., 2008).

3.1 eATP Sensing in Mammals

Concentrations of eATP can reach the 100 μM range in inflammatory tissues, as compared to the low nanomolar range in non-inflammatory tissues (Di Virgilio et al., 2017). Such elevated concentrations of eATP can result from tissue disruption or lytic forms of cells death, such as pyroptosis (Garg et al., 2015; Graziano et al., 2015; Kalogeris et al., 2012; Kolb et al., 2017; Land et al., 2016b), but ATP can also be secreted actively by stressed cells (Faas et al., 2017). In higher mammals, the perception of eATP depends mainly on two classes of receptors: metabotropic P2Y receptors, which activate intracellular signaling pathways through G proteins, and ligand (ATP)-gated ionotropic P2X receptors, which are associated with pores that open upon eATP binding and consecutively allow for Ca^{2+} influx and K^{+} efflux (Di Virgilio et al., 2017). Downstream to Ca^{2+} signaling, eATP sensing triggers cytokine production and promotes the migration of leukocytes—effects that can be suppressed by the application of an ATPase that hydrolyses eATP (McDonald et al., 2010)—and ultimately favors pro-inflammatory innate and/or adaptive immune responses against pathogens (Di Virgilio et al., 2017).

Besides stimulating the immunity to ongoing threats, eATP even influences the generation of immunological memory: following the fate of co-adoptively transferred wild-type and *P2rx7*-deficient (*P2rx7*$^{-/-}$) CD8+ T cells after the infection of mice with acute lymphocytic choriomeningitis virus, da Silva et al. (2018) observed a progressive defect in the maintenance of *P2rx7*$^{-/-}$ cells. Consecutive experiments revealed that the activation of P2RX7 by eATP not only triggered the acute immune response but also promoted the survival of memory CD8+ T cells (da Silva et al., 2018).

Furthermore, eATP can exert its pro-inflammatory functions indirectly via the activation of inflammasomes, such as the NLRP3 inflammasome or the (NACHT-, LRR-, and PYD-containing) (NALP3) inflammasome (Abderrazak et al., 2015; Lupfer and Anand, 2016; Savage et al., 2012). Inflammasomes are large protein complexes that control the caspase1-mediated release of inflammatory cytokines such as IL-1β and IL-18 (Chen and Nuñez, 2010), and evidence is accumulating that supports eATP as an activator of various types of inflammasomes (Carlos et al., 2017; McDonald et al., 2010; Savage et al., 2012; Shimada et al., 2012; Yatim et al., 2017). For example, eATP activated the NLRP3 inflammasome in murine macrophages in a ROS-dependent manner and enhanced bacterial killing (Bording-Jorgensen et al., 2017), and eATP released from cancer cells that die via autophagy activated the NLRP3 inflammasome in the macrophages that engulf these cancer cells (Gombault et al., 2013). Furthermore, eATP can induce mitochondrial dysfunction and apoptosis, which results in the release of oxidized mtDNA: another DAMP that activates the NLRP3 inflammasome (Shimada et al., 2012) (see Section 4.5). Therefore, eATP has been described as "one of the best-characterized activators of the NLRP3 inflammasome" (McDonald et al., 2010).

A particularly interesting example of the function of eATP as a DAMP is provided by its contribution to the distinction between pathogenic and non-pathogenic bacteria by intestinal immune cells. Since both groups of micro-organisms share most surface molecules, immune responses to common bacterial "microbe-associated molecular patterns" (MAMPs) such as flagellin or lipopolysaccharides (LPS) would lead to chronic inflammation in the intestine. However, Ivison et al. (2011) observed that epithelial cells responded to flagellin only when co-exposed to increased concentrations of eATP which, as a DAMP, indicates the presence of damaged cells (Ivison et al., 2011). In other words, eATP triggers the perception of an "unproblematic" MAMP as a "potentially problematic" PAMP.

3.2 eATP Sensing in Plants

Some 15 years ago, Demidchik and colleagues still asked "Is ATP a signalling agent in plants?" (Demidchik et al., 2003). A few years later, sufficient evidence had accumulated to report for eATP "an unexpected role as a signaller in plants" (Roux and Steinebrunner, 2007). For example, eATP caused membrane depolarization in the root hairs of *Arabidopsis thaliana* (Lew and Dearnaley, 2000), triggered Ca^{2+} influxes (Roux and Steinebrunner, 2007), and the formation of ROS in *Medicago truncatula* (Kim et al., 2006) and the alga, *Dasycladus vermicularis* (Torres et al., 2008), and the synthesis of jasmonic acid (JA) in and lima bean (*Phaseolus lunatus*) (Heil et al., 2012) and Arabidopsis (Tripathi et al., 2018): a plant hormone that controls immunity to chewing herbivores and necrotrophic pathogens (Choi and Klessig, 2016; Feng et al., 2015; Tanaka et al., 2014). At the phenotypic level, eATP induced immunity to viral and bacterial pathogens in tobacco (Chivasa et al., 2009), to herbivores in lima bean (Heil et al., 2012), and to the bacterial pathogen, *Pseudomonas syringae*, in Arabidopsis (Chen et al., 2017). The immunity to biotrophic pathogens can include the hypersensitive response, a programmed form of cell death that plays a similar role as apoptosis or necroptosis in the mammalian immune system (Feng et al., 2015). Finally, eATP even affected stomatal aperture or developmental processes (Chen et al., 2017; Clark et al., 2014; Lew and Dearnaley, 2000; Lim et al., 2014; Tanaka et al., 2010, 2014).

The effects of eATP on ROS formation in plants are mediated by membrane-bound NADPH oxidase (Demidchik et al., 2009), whereas the induction of JA and of JA-dependent responses by eATP has recently been linked to a destabilizing effect of eATP on jasmonate ZIM-domain (JAZ) proteins (Tripathi et al., 2018). The JAZ proteins are negative key regulators of JA-dependent responses: upon JA-perception by a ubiquitin ligase, JAZ proteins become hydrolyzed and liberate JA-responsive genes for transcription (Staswick, 2008; Thines et al., 2007; Zhang et al., 2017). Finally, the effects of eATP on cytosolic levels of Ca^{2+} (Roux and Steinebrunner, 2007) found their explanation with the discovery of the eATP receptor DOes not Respond to Nucleotides 1 (DORN1), a lectin receptor kinase that is required for ATP-induced Ca^{2+} influxes, and MAPK-dependent signaling in Arabidopsis (Choi et al., 2014).

At the latest with the discovery of a specific receptor, eATP can be considered an accepted DAMP for the plant kingdom (Tanaka et al., 2014; Van Holle and Van Damme, 2018). Interestingly, DORN1 shows no sequence homologies to the well-studied purinergic eATP receptors of higher mammals (Choi et al., 2014; Di Virgilio et al., 2017; Tanaka et al., 2014; Van Holle and Van Damme, 2018). Apparently, evolution has "invented" the same "use" of eATP as a DAMP at least two times independently. This analogy nicely illustrates the selective benefits of using endogenous alarm signals to mount and shape immune responses (Choi et al., 2014; Di Virgilio et al., 2017; Gombault et al., 2013; Hernández-Oñate and Herrera-Estrella, 2015; Tanaka et al., 2014).

4. MAMMALIAN DNA SENSORS

DNA is a macromolecule that is well known for storing the genetic information and encoding for hundreds of thousands of proteins (Venter et al., 2001). Due to the biological importance of DNA, maintaining its stability and integrity is crucial (Hartmann, 2017). Factors that usually secure the integrity of DNA are its compartmentalization in the nucleus, a site where the nuclear envelope helps to protect the DNA from detrimental factors, and its packaging into chromatin, i.e., the intimate binding of dsDNA to proteins such as histones and chaperones (Giglia-Mari et al., 2011). However, numerous endogenous and exogenous factors—from severe metabolic dysfunctions, cancer, or mutations in genes that are involved in DNA damage repair, to infections by intracellular pathogens—can cause the release of dsDNA fragments from the nucleus or—in cases of an uncontrolled disruption of cells or lytic forms of programmed cell death—even result in the appearance eDNA (Galluzzi et al., 2018b; Kaczmarek et al., 2013; Kolb et al., 2017; Orzalli and Kagan, 2017; Shrivastava et al., 2016; Yatim et al., 2017).

According to this highly diverse set of dysfunctions that result in the appearance of fragmented dsDNA in aberrant compartments, dsDNA triggers a plethora of immunity-related responses that in many, but not all, cases, involve IFN I-dependent signaling (Altfeld and Gale, 2015; Carlos et al., 2017; Gallucci and Maffei, 2017; Gao et al., 2015; Hartmann, 2017; Li and Chen, 2018; Lugrin and Martinon, 2018; Margolis et al., 2017; Paludan and Bowie, 2013; Qiao et al., 2017; Saxena et al., 2017; Schlee and Hartmann, 2016; Surpris and Poltorak, 2016; Tao et al., 2016; Wang et al., 2017). Genomic DNA can also induce the maturation of APCs (Ishii et al., 2001). Of note, immune cells might even secrete DNA—as a pure molecule or bound in chromatin—to the extracellular space, where it binds bacteria and thereby directly contributes to pathogen clearance (Brinkmann et al., 2004; Ingelsson et al., 2018).

4.1 Toll-Like Receptor 9

Toll-like receptors form an evolutionarily conserved family of LRR receptors which, besides the extracellular (or endosomal) LRR domain, consist of a single transmembrane domain and a cytoplasmatic signaling domain termed the Toll/IL-1 receptor domain (Thompson and Locarnini, 2007). TLRs sense highly diverse set of DAMPs and PAMPs (Andersson et al., 2018b; Bianchi, 2007; Heil and Land, 2014; Kirschning and Bauer, 2001; Lotze et al., 2007; Paludan and Bowie, 2013; Portou et al., 2015). A central player in the sensing of microbial and viral dsDNA by mammalian immune cells such as B cells, dendritic cells, and macrophages is TLR9, which recognizes unmethylated CpG DNA (Ahmad-Nejad et al., 2002; Bauer et al., 2001; Hemmi et al., 2000; Rutz et al., 2004). The CpG motif, in which an unmethylated cytosine residue is followed by a guanosine residue (Bauer et al., 2001; Hemmi et al., 2000; Rutz et al., 2004), is more frequent in viral and prokaryotic DNA than in eukaryotic DNA and, thus, it represents a molecular signature that mammalian immune cells can use to identify microbial nonself-DNA (Krieg, 2002; Pisetsky, 2012). The activation of TLR9 triggers a signaling cascade that involves the myeloid differentiation (MyD) primary response protein MyD88 and results a type I IFN response (Brown et al., 2011; Rutz et al., 2004).

Due to its preference for unmethylated CpG-rich DNA, TLR9 originally was assumed to discriminate between self- and nonself-DNA (Ahmad-Nejad et al., 2002, Bauer et al., 2001, Hemmi et al., 2000, Rutz et al., 2004). Eukaryotic DNA also contains CpG motifs that could potentially be sensed by TLR9 but normally, they are masked by methylation (Cross et al., 1997; Krieg, 2002; Rutz et al., 2004). Moreover, TLR9 is located to endosomal membranes and thus, only senses DNA that appears in the endosome, whereas under normal physiological conditions, self-DNA is compartmentalized in the nucleus and in the mitochondria. Even when genomic DNA is released from necroptotic or pyroptotic cells, it should appear in the extracellular space, and nucleases such as three prime repair exonuclease 1 (TREX1) aid to degrade dsDNA fragments that happen to appear in different compartments (Gilliet et al., 2008). Therefore, under most conditions, cytosolic dsDNA stems from bacteria or viruses that have been engulfed by macrophages for their digestion, or from intracellular bacteria and viruses (Ahmad-Nejad et al., 2002). The importance of the localization of TLR9 for a successful maintenance of homeostasis is illustrated by the observation that mice expressing a mutant TLR9 that mis-localizes to the cell surface displayed spontaneous fatal inflammatory disease, due to the overactivation of TLR9 by endogenous DNA (Mouchess et al., 2011). Therefore, the localization of TLR9 on endolysosomal membranes favors the recognition of bacterial/viral DNA while avoiding self-DNA autoimmunity, under most physiological conditions (Barton et al., 2006; Crowl et al., 2017; Gilliet et al., 2008; Lamphier et al., 2006; Pelka et al., 2016).

Nevertheless, CpG-rich DNA has been observed since a long time in anti-DNA antibody immune complexes that were isolated from the plasma of patients with SLE (Sano and Morimoto, 1982), and TLR9-sensing of self-DNA can be considered a widely accepted factor in SLE by now (Barrat et al., 2005; Boulé et al., 2004; Gilliet et al., 2008; Lamphier et al., 2006; Tian et al., 2007). Polymorphisms in TLR9 were associated with differences in the progress of disease in HIV-1 infected individuals (Bochud et al., 2007), and TLR9-dependent pro-inflammatory effects of self-DNA have also been associated with hypertension, cancer, and even with inflammation-related complications during pregnancy (Goldfarb et al., 2018; Lamphier et al., 2006; Tuomela et al., 2013).

At the first glance, the stimulation of the intracellular TLR9 by extracellular self-DNA appears to be counterintuitive. In this context, mainly two factors become important. First, CpG motifs do not allow a fully accurate discrimination between self- and nonself-derived DNA (Roers et al., 2016) because mtDNA—due to the evolutionary history of mitochondria—resembles bacterial DNA in being circular and containing unmethylated CpG motifs (Ries et al., 2013). When cellular disruption harms the integrity of mitochondria, mtDNA is released as intracellular DAMP which can leave the cell under certain circumstances. For example, mitochondrial DAMPs were observed to accumulate in the circulation after severe injury caused by trauma (Simmons et al., 2013), and among them, mtDNA can stimulate TLR9 (McCarthy et al., 2015; Nakahira et al., 2015; Pinti et al., 2012; Zhang et al., 2010, 2014). Second, even genomic DNA was found to be a potent agonist for TLR9 that induced IFN-Iα production by plasmacytoid dendritic cells (pDCs) (Barrat et al., 2005; Gilliet et al., 2008). Since pDCs are a major source of type I IFN, aberrant self-DNA sensing is now a widely accepted factor in SLE (Barrat et al., 2005; Bode et al., 2016; Gilliet et al., 2008; Lamphier et al., 2006).

In addition, recent evidence suggests that compartmentalization is not as efficient as originally assumed. For example, IL-26 formed complexes with self-DNA released by dying host cells that triggered the production of type I IFN by pDCs via an activation of TLR9 (Meller et al., 2015) (see Section 7.4 for details on the DAMPs assisted trafficking of eDNA through membranes). In conclusion, the sensing of self-DNA by TLR9 is increasingly being discussed in the context of several pathological scenarios (Farrugia and Baron, 2017).

4.2 ZBP1/DAI

Whereas TLR9 is bound to the endosomal compartment, DAI functions as a sensor for cytosolic DNA (Takaoka et al., 2007; Wang et al., 2008). Upon DNA binding, DAI associates with TANK-binding kinase 1 (TBK1) and interferon regulatory factor 3 (IRF3) to activate the master transcription regulators, IRF3 and NF-κB, and trigger a type I IFN response (Ishii et al., 2008; Kuriakose and Kanneganti, 2018; Lin et al., 2016; Takaoka et al., 2007). The corresponding gene—termed originally DLM-1—was identified using a differential display approach as a tumor-associated and IFN-inducible gene expressed in macrophages around tumor cells (Fu et al., 1999). L929 cells responded to the treatment with poly (dA:dT) of increasing length with a stronger IFN-1 induction, an observation related to the fact that DAI requires three DNA binding domains (Zα, Zβ, and D3) to become fully active (Wang et al., 2008), whereas DAI has "no strict preference for the sequence and configuration of DNA" (Wang et al., 2008). That is, since its original discovery, DAI was related to the sensing of aberrant self-DNA (Takaoka et al., 2007, Wang et al., 2008). Later studies revealed that DAI only binds to *Z*-form (i.e., left-handed helix) DNA, for which reason most authors now prefer the term ZBP1 (Kuriakose and Kanneganti, 2018). This specificity is not entirely exclusive, as ZBP1 also binds synthetic B-form DNA—poly (dA:dT)—and even Z-form RNA (Assil and Paludan, 2017; Maelfait et al., 2017). Nevertheless, the role of ZBP1/DAI in DNA sensing was debated, since DAI-deficient mice still showed innate or adaptive immune responses to B-DNA or DNA vaccination (Ishii et al., 2008). In fact, several authors still suggest "no role or redundant roles for ZBP1 in DNA-mediated responses" (Kuriakose and Kanneganti, 2018).

Further studies provided evidence for a critical role of ZBP1 in the pro-inflammatory effects of lymphocyte-derived "self-apoptotic DNA" (Zhang et al., 2013). For example, inflammatory processes caused by smoking were related to the sensing of mtDNA by ZBP1 (Szczesny et al., 2018). Moreover, the NLRP3-inflammasome-mediated necroptosis or pyroptosis of influenza virus-infected cells was found to be regulated by ZBP1 (Kuriakose et al., 2016).

4.3 RAGE

Whereas TLR9 is bound to the endosome and ZBP1/DAI is localized in the cytoplasm, RAGE is a transmembrane protein that senses molecules outside of the cell. Pro-inflammatory processes of diverse origins were linked to RAGE expression in endothelial cells, T and B cells, monocytes, and DCs (Kierdorf and Fritz, 2013), and RAGE expression on the activated endothelium can trigger the adhesion and transmigration of leukocytes (Kierdorf and Fritz, 2013). Although RAGE had been discovered as a receptor of advanced glycation end products (i.e., in the context of metabolic disorders), it acts as a multi-ligand receptor that binds to diverse molecules including DAMPs such as HMGB1, S100 proteins, or HSP70 (Bierhaus et al., 2005; Hudson and Lippman, 2018; Kierdorf and Fritz, 2013; Lee and Park, 2013; Rojas et al., 2017). In particular the activation of RAGE by HMGB1 has been related to sterile inflammation and to cancer, diabetes, and Alzheimer's disease (Sims et al., 2010).

RAGE binds DNA and RNA in a sequence-independent manner via an interaction with the charged sugar-phosphate backbones of nucleic acids

(Bierhaus et al., 2005), and recent evidence demonstrated a role for RAGE in DNA damage repair (Kumar et al., 2017). A further mechanism via which RAGE acts in synergy with other receptors of nucleic acids is the sequestering of eDNA or eRNA: once bound to RAGE, extracellular nucleic acids can be taken up into the cell and, thereby, become accessible to intracellular sensors (Bertheloot et al., 2016; Sirois et al., 2013) (see Section 7.4).

After agonist recognition, RAGE activates the pro-inflammatory transcription factor NF-κB which, upon translocation of the active form to the nucleus, induces the transcription of cytokines and of RAGE itself. A particularly important feature in this context is the prolonged activation of NF-κB by RAGE, which apparently overcomes several endogenous negative feedback control mechanisms (Bierhaus et al., 2005). Moreover, RAGE triggers a NADPH-dependent formation of ROS within the cell (Rojas et al., 2017) and thereby also activates downstream sensors such as the NLRP3 inflammasome. This mechanism might explain the multiple links between the NLRP3 inflammasome and metabolic disorders such as diabetes that recently have been discovered (Haneklaus and O'Neill, 2015; Liston and Masters, 2017).

4.4 The cGAS-cGAMP-STING Pathway

Multiple ground-breaking studies published in 2013 linked cGAS and its product—the cyclic dinucleotide GMP-AMP (cGAMP)—to the sensing of cytoplasmatic dsDNA (Ablasser et al., 2013; Civril et al., 2013; Diner et al., 2013; Gao et al., 2013; Li et al., 2013; Sun et al., 2013). Upon dsDNA sensing, cGAS catalyzes the synthesis of cGAMP, a second messenger that activates the stimulator of interferon genes (STING), which then migrates to the Golgi apparatus where it activates IRF3 and NF-κB (Burdette and Vance, 2013; Cai et al., 2014; Gallucci and Maffei, 2017; Galluzzi et al., 2018a; Kato et al., 2017; Li and Chen, 2018; Margolis et al., 2017; Tao et al., 2016; Wang et al., 2016; Xiao and Fitzgerald, 2013). Due to the antiviral, antimicrobial, antitumor, and pro-inflammatory effects of the resulting type I IFN response, the cGAS-cGAMP-STING pathway plays an essential role in the protection from exogenous and endogenous stressors, including endogenous DNA damage (Cai et al., 2016; Gallucci and Maffei, 2017; Gao et al., 2015; Hartmann, 2017; Hornung et al., 2014; Li and Chen, 2018; Liu and Wang, 2016), the infection with intracellular bacteria such as *Leisteria monocytogenes*, *Francisella*, or *Mycobacterium tuberculosis* (Hansen et al., 2014; Storek et al., 2015; Watson et al., 2015), and—evidently—viruses, including retroviruses such as HIV-1 (Altfeld and Gale, 2015; Cai et al., 2014; Gantier, 2017; Gao et al., 2013; Krapp et al., 2018; Nasi et al., 2015).

cGAS discriminates B-form dsDNA from A-form dsDNA and dsRNA, while B-form dsDNA activates cGAS in a sequence-independent manner, because cGAS binds to its sugar-phosphate backbone (Andreeva et al., 2017; Gantier, 2017; Mertens, 2015). Therefore, cGAS does not discriminate between self- and nonself-DNA (Li et al., 2017; Surpris and Poltorak, 2016). For example, dsDNA released from dying cardiomyocytes triggered cGAS-STING-dependent type I IFN signaling in phagocytic monocytes and or/macrophages, which ultimately caused an increased inflammation related to myocardial infarction (King et al., 2017). The stimulation of cGAS by self-DNA (Gao et al., 2015), including mtDNA, is now considered one of the mechanisms that connect DNA damage to several autoimmune diseases and to cancer (Barber, 2015; Chen et al., 2016; Dhanwani et al., 2018; Galluzzi et al., 2018a; Krapp et al., 2018; Kretschmer and Lee-Kirsch, 2017; Li and Chen, 2018; Liu et al., 2016; Ng et al., 2018; Szczesny et al., 2018).

Moreover, the lack of a specificity for any sequence motifs also explains the counterintuitive fact that cGAS initiates type I IFN responses against RNA viruses (van Montfoort et al., 2014). In this case, cGAS senses host mtDNA that is released upon mitochondrial damage as it occurs, e.g., during an infection with dengue virus (Aguirre and Fernandez-Sesma, 2017; Sun et al., 2017). Most intriguingly, dengue virus can block this danger signal by promoting cGAS degradation via the viral protease cofactor N2SB, thereby minimizing the translocation of cGAMP to neighboring cells (Aguirre et al., 2017). Similarly, the gammaherpes virus-specific protein, ORF52, was found to directly inhibit cGAS enzymatic activity, thereby allowing Kaposi's sarcoma-associated herpesvirus to evade its recognition by the DNA sensors of the infected host cells (Wu et al., 2015).

Rather than sequence, the length of dsDNA fragments is important for their cGAS-mediated immunogenic properties (Gantier, 2017). In principle, cGAS is efficiently activated by dsDNA fragments of at least 24–40 bp length (Chen et al., 2016; Roers et al., 2016). However, at low concentrations, dsDNA induced IFN I secretion in a length-dependent manner over a wide length-span of DNA, ranging from the minimal stimulatory length to several kilobases (Luecke et al., 2017; Luecke and Paludan, 2017). Most recently, Du and Chen (2018) demonstrated that in the presence of dsDNA, cGAS formed liquid-like droplets, which fused with each other or allowed rapid exchange of the cGAS substrates, ATP and GTP, thereby functioning as micro reactors (Du and Chen, 2018). The formation of these micro reactors could be observed only at concentrations of >30 nM of both cGAS and dsDNA. Furthermore, more and larger droplet was formed with longer dsDNA fragments, and the synthesis of cGAMP correlated with the degree of phase separation (Du and Chen, 2018). Based on these observations, it was suggested that "phase separation focuses DNA sensing" and thereby benefits cGAS-dependent responses to nonself-DNA (Ablasser, 2018).

4.5 Inflammasome-Associated DNA Sensors (AIM2, IFI16, NLRP3)

As mentioned in Section 1, several DNA sensors can form inflammasomes: large protein complexes in which PRRs such as AIM2, IFI16, or NLRP3, assemble with the adaptor protein apoptosis-associated speck-like and caspase recruitment domain (CARD) containing protein (ASC) to form macromolecular complexes (ASC specks) that liberate active caspase-1 from pro-caspase and thereby trigger the massive production of IL-1β and IL-18 (Braga et al., 2017; Chen and Núñez, 2011; Franklin et al., 2018; Gougeon, 2017; Mathur et al., 2018; Mills et al., 2017; Saxena et al., 2017; Shrivastava et al., 2016; Verma et al., 2016). In many cells, the full activity of the NLRP3 inflammasome requires a first, priming step in which expression of NLRP3 is induced by microbial TLR ligands or endogenous molecules such as IL-1β (Broz and Dixit, 2016; He et al., 2016). Once primed, the full activation of the NLRP3 inflammasome can be triggered by a highly diverse set of molecules including both PAMPs and DAMPs (Broz and Dixit, 2016).

Priming of the NLRP3 inflammasome by mtDNA has been related to age-related macular degeneration (Carlos et al., 2017; Dib et al., 2015; Martinon, 2012; Shimada et al., 2012). A systemic priming of NLRP3 was reported for peripheral blood mononuclear cells from patients with primary Sjögren's syndrome and was related to enhanced serum concentrations of cell-free DNA, IL-18, and ASCs, as well as to an accumulation of cytosolic dsDNA in these cells (Vakrakou et al., 2018). The broad range of molecules that have been reported to activate the NLRP3 inflammasome makes their direct sensing by NLRP3 questionable. It appears more likely that NLRP3 plays a role as central integrator of multiple cellular signaling cascades (He et al., 2016) which are involved in the general sensing of disturbed homeostasis (Liston and Masters, 2017). For example, inflammasome activation can trigger the release of HMGB1 (Lamkanfi et al., 2010).

Whereas discussion continues concerning the role NLRP3 as a DNA sensor, several studies confirmed AIM2 and IFI16 as sensors of self-dsDNA (Chen and Nuñez, 2010; Lugrin and Martinon, 2018; Lupfer and Anand, 2016;

Man et al., 2016; Mathur et al., 2018; Shrivastava et al., 2016; Unterholzner et al., 2010). The HIN domain—which characterizes the AIM2 and IFI16 receptors (Unterholzner et al., 2010)—binds directly to the sugar-phosphate backbone of DNA and, thus, exerts no specificity toward specific sequences (Jin et al., 2012; Lugrin and Martinon, 2018; Ni et al., 2016). Therefore, AIM2 can detect bacterial, viral, and self-dsDNA in the cytoplasm (Di Micco et al., 2016; Jones et al., 2010; Rathinam et al., 2010; Reinholz et al., 2013), although dsDNA fragments must have a minimum size of ~80 bp for optimal AIM2-mediated IL-1β induction (Jin et al., 2012). Upon binding of cytosolic dsDNA, AIM2 associates with ASCs to form an inflammasome (Hornung et al., 2009). Because activated AIM2 also triggers the formation of the ASC pyroptosome 4, sensing of dsDNA via AIM2 triggers pyroptosis (Fernandes-Alnemri et al., 2009).

The sensing of self-dsDNA by AIM2 has been reported in several studies (Fernandes-Alnemri et al., 2009; Hornung et al., 2009), making the AIM2 inflammasome an important sensor of exogenous as well as endogenous cellular damage (Lugrin and Martinon, 2018; Man et al., 2016). During tubular necrosis, the release of dsDNA and its consecutive uptake into macrophages and AIM2-mediated recognition has been suggested as a key factor during necroinflammation (Komada et al., 2018). Similarly, the antiretroviral drug, Nelfinavir, has been reported to impair the integrity of the nuclear envelope; the resulting release of self-dsDNA fragments from the nucleus triggered pyroptosis in an AIM2-dependent manner (Di Micco et al., 2016). Several authors also suggested that AIM2 can sense self-dsDNA delivered by exosomes, which are endosome-derived organelles that contain proteins, lipids, RNA, DNA, among other molecules, and that mediate intercellular communication and regulate immune signaling (Lee et al., 2011). Exosomes can protect self-DNA from extracellular nucleases and facilitate its delivery into the cytosol of immune cells (Asada et al., 2018; Lian et al., 2017; Lugrin and Martinon, 2018). For example, self-DNA released in an exosome-dependent manner from chemotherapy-damaged intestinal cells activated the AIM2 inflammasome in immune cells, leading to IL-1β and IL-18 secretion (Lian et al., 2017).

5. MAMMALIAN RNA SENSORS

Like DNA, RNA is a potent immunogenic molecule—both as ssRNA and as dsRNA. The presence of dsRNA in endosomes usually results from the phagocytosis of viruses or virus-infected cells, whereas cytoplasmatic dsRNA is massively produced by replicating viruses. Thus, dsDNA arguably is a particularly potent virus-associated PAMP. In mammals, viral RNA can induce pro-inflammatory cytokines and type I IFN, bacterial mRNA can induce the production of IL-2, and 16S ribosomal RNA sensing leads to IFN-α production (Chow et al., 2018; Eberle et al., 2009; Gitlin et al., 2006). In fact, RNA fragments were among the first "foreign nucleic acids" for which Isaacs, Cox, and Rotem described an antiviral and IFN-inducing activity (Isaacs et al., 1963; Rotem et al., 1963). That is, early research focused on RNAs as PAMPs, which form a central component of the mammalian interferon system and act as efficient activators of antiviral responses. Cytoplasmic dsRNA is sensed by RLRs (Chow et al., 2018; Gitlin et al., 2006; Gürtler and Bowie, 2013; Kato et al., 2006, 2008; Lässig and Hopfner, 2017; Luecke and Paludan, 2017) whereas endosomal ssRNA or dsRNA are sensed by TLRs (Thompson and Locarnini, 2007). However, RNA sensors such as TLR3 and 9 can also sense circulating endogenous dsRNA as a DAMP: an effect that explains the role of these receptors in immune responses to damage caused by UV radiation but that can also cause autoimmunity (Amarante and Watanabe, 2010; Bernard et al., 2012; Chen et al., 2014; Lamphier et al., 2006; Nellimarla and Mossman, 2014). In plants, as to the best of our knowledge, no RNA receptor has been described, although evidence exists for a role of RNA as a PAMP in plants (Bhat and Ryu, 2016; da Silva de Faria et al., 2013; Lee et al., 2016; Niehl et al., 2016).

5.1 Toll-Like Receptors 3, 7, 8, and 10

While TLR9 senses DNA, most other TLRs exert a preference for RNAs (Thompson and Locarnini, 2007). Specifically, TLR3 and TLR10 sense dsRNA and small interfering RNAs (Alexopoulou et al., 2001; Amarante and Watanabe, 2010; Lee et al., 2018) and signal via the Toll-IL-1 receptor domain containing adaptor-inducing IFN-β (TRIF) to activate NF-κB and IRF3 (Chen et al., 2014; Gürtler and Bowie, 2013; Kumar et al., 2006). At least for TLR3, the cytokine production in response to poly(I:C) was shown to depend on MyD88 (Alexopoulou et al., 2001). Similarly, TLR7 and TLR8 recognize viral ssRNA and bacterial RNAs (mRNA, rRNA, small RNA) and trigger innate immune responses signal through MyD88-dependent pathways (Botos et al., 2011; Campos et al., 2017; Lee et al., 2017; Mancuso et al., 2009). A feature that TLR3 shares with TLR9 is its translocation from the endoplasmic reticulum to the endosome, thereby allowing to sense PAMPs that stem from phagocytosed pathogens (Murshid et al., 2015). TLR7 and TLR8 share the property to sense mainly ssRNA, and both TLRs are implicated in the recognition of RNA viruses, phagosomal bacteria, and bacterial spores (Chow et al., 2018). However, TLR7 and TLR8 differ with respect to the types of cells in which they are expressed and in the mechanisms of cytokine induction. TLR7 is mainly expressed in pDCs, whereas TLR8 is primarily expressed in macrophages and myeloid DCs (Cervantes et al., 2012; Gilliet et al., 2008). Moreover, TLR7 stimulates the production of type I IFNs in a MyD88-dependent manner (Cao et al., 2009). Therefore, MyD88-deficient pDCs have defective responses against a wide range viruses, because both TLR7 and TLR9 depend on MyD88 to trigger antiviral responses (Gilliet et al., 2008). By contrast, TLR8 activation induces pro-inflammatory cytokine production (Vollmer et al., 2005). TLR8 recognizes ssRNA via two distinct sites, one that is specific to uridine and the other to a short oligonucleotide (Tanji et al., 2015). TLR8 has shown to be upregulated by phagocytosis of *Mycobacterium tuberculosis* and *Helicobacter pylori* in THP-1 cells (Davila et al., 2008; Gantier et al., 2010). IFN Iβ induction upon *Borrelia burgdorferi* phagocytosis has shown to be entirely dependent on TLR8 and through IRF7 (Cervantes et al., 2011). Due to these differences, the activation of both TLR7 and TLR8 may be important for subsequent immune responses (Jung et al., 2015).

In a similar way to the other nucleic acid sensing TLRs, TLR10 is localized in the endosome in order to prevent aberrant activation by self-RNA. Activation of TLR10 by poly(I:C) is followed by the recruitment of MyD88 and ultimately triggers a type I IFN response (Lee et al., 2014). For example, TLR10 has been described to be upregulated in human macrophages during *Influenza virus* confrontation, possibly implying that TLR10 may be activated by RNA-protein complexes (Lee et al., 2014). Although TLR10 has clearly been demonstrated to respond to infection (Nagashima et al., 2015; Regan et al., 2013), others suggested a mainly inhibitory role of TLR10 in pro-inflammatory processes (Oosting et al., 2014). Intriguingly, no generally accepted ligand could be identified for TLR10, although recent in vitro studies demonstrated that TLR10 can bind dsRNA at endosomal pH (Lee et al., 2018). However, it has also been suggested that TLR10 can sequester dsRNA from TLR3, thus inhibiting TLR3 signaling (Lee et al., 2018).

Toll-like receptors have mainly been studied in the context of PAMP detection. However, besides being activated by dsRNA and small interfering RNAs, at least TLR3 also responds to self-RNA (Amarante and Watanabe, 2010; Bernard et al., 2012; Kumar et al., 2006; Takahashi et al., 2006; Takemura et al., 2014; Zhang et al., 2007). Similarly, TLR7 can sense self-RNA rich in uridine or guanosine, small nuclear ribonucleoproteins, and self-RNA-autoantibodies and, thus, might trigger type I IFN responses in pDCs and B cells upon sensing of DAMPs (Gilliet et al., 2008; Lau et al., 2005; Vollmer et al., 2005).

5.2 RIG-Like Receptors

The family of RLRs comprises, besides RIG1, its homolog, MDA5, and the negative regulator of RIG1, LGP2 (Gitlin et al., 2006; Lässig and Hopfner, 2017; Rothenfusser et al., 2005; Yoneyama and Fujita, 2012). All three receptors are expressed in the cytoplasm, respond to virus infection and share highly conserved domains, including a DExD/H-box RNA helicase domain. Therefore, RLRs sense dsRNA—including synthetic molecules such as poly(I:C) (Feng et al., 2012; Kato et al., 2006, 2008; ; Lässig and Hopfner, 2017)—although at different levels of efficiency (Kato et al., 2006). In addition to dsRNA, RIG1 also senses ssRNA (Pichlmair et al., 2006), whereas MDA5 can also sense ssRNA:dsRNAs web-like structures that are formed during virus replication/transcription (Pichlmair et al., 2009). RIG1 and MDA5 possess a CARD, which mediates the production of IL-1β and other pro-inflammatory and antiviral cytokines, for which reason RIG1 and MDA5 are considered key players in the type I IFN response to viral infections (Gack et al., 2007; Pham et al., 2016; Yoneyama et al., 2005).

Upon RNA recognition, RLRs translocate to the mitochondria, where they activate TNF receptor-associated factor (TRAF) proteins: a family of ubiquitin ligases that control IRF3 and NF-κB (Nistal-Villán et al., 2015; Parisien et al., 2018). By contrast, LGP2 lacks a CARD and was reported to inhibit RIG1-mediated responses to viruses (Rothenfusser et al., 2005). Two explanations have been suggested for this function of LGP2 as a negative regulator of antiviral responses. First, due to its ability to bind dsRNA, LGP2 could sequester dsRNA and thereby prevent its detection by RIG1 (Rothenfusser et al., 2005). More recently, a direct interference of LGP2 with diverse TRAF proteins was proposed to inhibit their ubiquitin ligase activity (Parisien et al., 2018).

Both RIG1 and MDA5 respond to synthetic dsRNA and viruses like Reovirus, Dengue virus, and West Nile virus (Wu and Chen, 2014).

Nevertheless, both receptors seem to sense partly different spectra of RNA viruses. For example, RIG1 is activated by several viruses in the Paramyxoviridae (Sendai, Newcastle disease viruses), Flaviviridae (Japanese encephalitus and HCV viruses), and Rhabdoviridae (vesicular stomatitis virus), whereas MDA5 is activated by viruses in the Picornaviridae family (Polio virus and EMCV) (Kato et al., 2006). In fact, the receptors differ in their response to RNAs of different lengths. Short RNA fragments (~300 bp) are recognized by RIG1, whereas longer fragments (>4 kb) are recognized by MDA5 (Kato et al., 2008; Lässig and Hopfner, 2017). Furthermore, it has been suggested that MDA5 senses RNA-webs rather than simply long dsRNA molecules (Pichlmair et al., 2009). In addition to length, the discrimination between self- and nonself-RNA by RIG1 appears to depend on the presence of a triphosphate group coupled with a short double-stranded RNA 5′ end, a motif that is absent from mammalian mRNAs (Hornung et al., 2006; Lässig and Hopfner, 2017; Pichlmair et al., 2006; Schlee et al., 2009; Schmidt et al., 2009). Although RIG1 and MDA5 recognize different dsRNA motifs, they signal via a shared downstream pathway. Upon dsRNA sensing, RIG1 and MDA5 bind to IFN-β promoter stimulator 1 via direct CARD-CARD interactions and thereby stimulate its aggregation on the mitochondrial membrane, which ultimately leads to the activation of transcription factors such as NF-κB and diverse IRFs and thereby induces a type I IFN response (Pham et al., 2016). In short, while experimental evidence strongly suggests that RLRs acts as PAMPs receptors, it remains an open question whether these receptors can sense self-RNA. However, several type I interferonopathies but also SLE have been linked to mutations in members of the RLRs, particularly MDA5 (Lässig and Hopfner, 2017).

5.3 Protein Kinase R (PKR)

The activation of the IFN I inducible serine/threonine protein kinase PKR by dsRNA is known since >20 years (Davis and Watson, 1996). Upon activation, PKR undergoes autophosphorylation reactions that activate the kinase domain, which consecutively phosphorylates eukaryotic initiation factor 2α (eIF2α); a translation factor that acts on cellular as well as viral RNA (Radetskyy et al., 2018). An important step in the activation of PKR is the dsRNA-mediated formation of dimers. Since only dsRNA with a minimal length of 30 bp can bind two PKR monomers, PKR activation depends on the fragment size of dsRNA molecules (Lemaire et al., 2008). Although these considerations supported only dsRNA as an agonist of

PKR, recent studies reported that also ssRNA can activate PKR via binding to the basic region adjacent to the N-terminus of the kinase domain (Mayo and Cole, 2017).

Furthermore, PKR can directly interact with inflammasome-forming PRRs such as NLRP3 and AIM2 via at least two independent pathways which do, or do not, depend on the kinase activity of PKR (Hett et al., 2013). Using macrophages from *Pkr*$^{-/-}$ and *Pkr*$^{+/+}$ mice, Lu et al. (2012) detected a significantly inhibited IL-1β production in *Pk*$^{-/-}$ macrophages in response to ATP or other stimulating agents. Interestingly, the expression of NLRP3 and pro-IL-1β did not differ significantly between *Pkr*$^{-/-}$ and *Pkr*$^{+/+}$ macrophages, but the secretion of active IL-1β in response to live *E. coli* was significantly inhibited in *Pkr*$^{-/-}$ macrophages (Lu et al., 2012). In a later study using primary bone marrow-derived macrophages obtained from p58^{IPK} knockout mice, the chaperone p58^{IPK} was identified as a negative regulator of PKR and, thereby, of PKR-mediated activation of the NLRP3 inflammasome (Boriushkin et al., 2016).

6. EFFECTS OF NUCLEIC ACIDS ON PLANT IMMUNITY

Plants, like mammals, possess PRRs that trigger immunity upon the sensing of DAMPs and PAMPs and—as a plant-specific group of molecules—of herbivore-associated molecular patterns (HAMPs) (Jones et al., 2016; Jones and Dangl, 2006). Early signaling steps comprise the influx of extracellular Ca^{2+} into the cytosol, membrane depolarization, the production of ROS, and the activation of MAPKs (Couto and Zipfel, 2016; Wu and Baldwin, 2010; Zipfel, 2014). Thus, early immune responses to DAMPs and PAMPs appear to be conserved among plants and animals (Bigeard et al., 2015; Duran-Flores and Heil, 2016; Heil and Land, 2014). Downstream to these earliest signals, plant immunity is orchestrated mainly by two hormones: besides JA (see Section 3.2), salicylic acid (SA) controls the resistance to sap-sucking herbivores and biotrophic pathogens (Nguyen et al., 2016; Thaler et al., 2012; Wu and Baldwin, 2010).

Anecdotic evidence published since the 1970s linked plant resistance to DNA damage (recently reviewed in Hadwiger and Tanaka, 2018), demonstrated that eDNA can be taken up into plant roots and function as a signaling molecule (Paungfoo-Lonhienne et al., 2010), or suggested a direct role of eDNA in the immunity of plant root tips (Hawes et al., 2011; Tran et al., 2016; Wen et al., 2009), similarly to neutrophil extracellular traps (Brinkmann et al., 2004). However, in spite of these observations—and in an astonishing parallel to the immunological sciences—plant scientists focused on the perception of the "nonself," whereas plant "damaged-self recognition" was considered only recently (Heil, 2009; Lotze et al., 2007). In particular, the immunogenic properties of nucleic acids have received surprisingly little attention in the plant sciences. As to the best of our knowledge, Yakushiji et al. (2009) were the first to report that bacterial DNA can induce the formation of ROS in *Arabidopsis thaliana*. Even more recently, immunogenic properties in plants were also shown for bacterial RNA (Lee et al., 2016), for dsRNA (Niehl et al., 2016), and for self-eDNA (Barbero et al., 2016; Duran-Flores and Heil, 2018).

6.1 Nonself Nucleic Acids Induce Plant Immunity

Using *Eco*RI-digested plasmid DNA, Yakushiji et al. (2009) demonstrated that bacterial DNA induces the formation of ROS and callose deposition in *Arabidopsis thaliana*. A stimulation of plant immunity was then also reported for bacterial RNAs and was shown to involve ROS, Ca^{2+}, and MAPK signaling and both major plant defense hormones, JA and SA (Lee et al., 2016). Endocytosis inhibitors significantly reduced the immunity-inducing activity of bacterial eDNA in Arabidopsis (Yakushiji et al., 2009), an observation that supported intracellular sensors for nonself-DNA. In addition, *Sma*I- and HapII-digested plasmid DNA or EcoRI-digested herring DNA was inactive, demonstrating that the recognition of bacterial eDNA in Arabidopsis requires the same CpG motif that is sensed by mammalian TLR9 (Yakushiji et al., 2009). However, TLRs have not been described for plants (Couto and Zipfel, 2016). In plants, PAMPs and DAMPs are mainly recognized via receptor-like kinases (Bacete et al., 2018; Saijo et al., 2018; Van Holle and Van Damme, 2018). Leucine-rich repeat (LRR)-containing PRRs preferentially bind proteins or peptides such as bacterial flagellin (a PAMP) or endogenous "AtPep peptides," which in analogy to mammalian HDPs are released from pre-existing endogenous proteins and, thus, represent DAMPs (Boutrot and Zipfel, 2017; Couto and Zipfel, 2016; Van Holle and Van Damme, 2018; Zipfel, 2014). However, the nucleotide-binding LRR protein (Rx NLR) of potato also binds nucleic acids, with similar preferences for single-stranded and double-stranded DNA (Fenyk et al., 2016). Further studies demonstrated that in vitro-generated dsRNAs and the dsRNA analogue, poly(I:C), induce plant immunity via a pathway that depends on the membrane-bound co-receptor Somatic Embryogenesis Receptor-Like Kinase 1 (Niehl et al., 2016).

6.2 Self-DNA-Induced Plant Immunity

Whereas the above-mentioned studies focused on the perception of nonself nucleic acids in plants, comparative studies using the non-host resistance of pea endocarp tissue against *Fusarium solani* f. sp. *Phaseoli*, a fungal pathogen of bean, reported an induction of pathogenesis-related genes, the deposition of callose, and enhanced activity of phenylalanine-ammonia lyase, after the application of DNA dyes or other DNA-intercalating compounds (Hadwiger and Schwochau, 1970a, 1971a), UV radiation (Hadwiger and Schwochau, 1971b), or DNA-damaging enzymes such as DNase (Hadwiger, 2015; Hadwiger and Tanaka, 2017). Furthermore, histones (Hadwiger and Schwochau, 1970b) and HMG protein (Klosterman et al., 2003) triggered plant immunity in the same model system, and Arabidopsis-HMGB3 induced innate immunity in this plant via a SA-dependent pathway (Choi et al., 2016). Based on this evidence for links between DNA damage and pathogen resistance in plants (Hadwiger and Tanaka, 2018), DNA has been suggested to act as a DAMP in plants (Duran-Flores and Heil, 2015, 2016; Gust et al., 2017) and some authors even discussed the use of DNA as a "bio-stimulant" or "vaccine" for plants (Quintana-Rodriguez et al., 2018; Vargas-Hernandez et al., 2017).

Two recent studies provided more direct evidence for the immunogenic function of self-eDNA in plants (Barbero et al., 2016, Duran-Flores and Heil, 2018). The application of fragmented self-DNA triggered membrane depolarization and Ca^{2+} signaling in lima bean and maize (*Zea mays*) (Barbero et al., 2016) as well as the generation of ROS and the activation of MAPKs in common bean (*Phaseolus vulgaris*) (Duran-Flores and Heil, 2018). In lima bean, self-eDNA also triggered the secretion of extrafloral nectar (EFN), a common indirect defense against herbivores (Heil, 2015), and the resistance to a bacterial pathogen, *Pseudomonas syringae* (Duran-Flores and Heil, 2018). Intriguingly, the effects of eDNA depended on its biological source. Lima bean DNA—but not maize DNA—triggered membrane depolarization in lima bean, and vice-versa (Barbero et al., 2016). In common bean, the formation of ROS, EFN secretion, and resistance to *P. syringae* responded more strongly to genomic self-eDNA (obtained from plants in the same species) than to nonself-eDNA (obtained from *Phaseolus lunatus* and *Acacia farnesiana*) and MAPKs were only activated by self-eDNA, but not by nonself-eDNA (Duran-Flores and Heil, 2018). The species used as sources of the nonself-DNA belong to the same family and—in the case of *P. lunatus*—even the same genus as the receiver plant. Consecutive analyses revealed that nuclear eDNA had stronger immunogenic properties than mtDNA or plastid-derived DNA and showed concentration-dependent effects of self-eDNA on the synthesis of the defense hormones, JA and SA (Duran-Flores and Heil, unpublished data).

The species-specific and concentration-dependent effects of self-eDNA on immunity-related traits in bean and maize plants (Barbero et al., 2016, Duran-Flores and Heil, 2018) resembled its concentration-dependent inhibitory effect on seed germination and the growth of plants (Duran-Flores and Heil, 2018; Mazzoleni et al., 2015a) as well as on the growth of each one bacterium, fungus, protozoon, and insect species (Mazzoleni et al., 2015b). The causal mechanisms behind this inhibitory effect of self-eDNA remain elusive. For example, the application of random fragments that represent basically the entire genome might cause inhibition by interference with mRNA molecules (Carteni et al., 2016), similarly as known for small interfering DNA (Croset et al., 2013). Intriguingly, a recent study confirmed that the overexpression of small sequences of genomic DNA in mouse cell lines generates immunogenic cytoplasmatic self-dsDNA fragments, which acted as "natural antisense" molecules that suppressed the complementary RNA by direct interference (Asada et al., 2018).

While direct interference represents an elegant explanation of the inhibitory effects of eDNA on growth, it is less intriguing how it could explain the specific induction of immunity-related traits. Alternatively, the inhibitory effects of self-eDNA might represent the consequences of the costs of the induced resistance. Induced plant immunity causes significant metabolic costs, which frequently become phenotypically apparent as an—at least transient—arrestment of growth (for reviews, see Cipollini and Heil, 2010, Heil and Baldwin, 2002, Vos et al., 2013, Walters et al., 2011, Walters and Heil, 2007), somewhat similar to fatigue in infected humans. In fact, reports on the growth-inhibitory effects of self-eDNA (but not nonself-eDNA) (Mazzoleni et al., 2015a) had motivated the hypothesis that self-eDNA might function as a DAMP in plants (Duran-Flores and Heil, 2015).

How can eDNA cause species-specific effects when we apply fragmented genomic DNA of two congeneric plant species that share >98% sequence identity? As to the best of our knowledge, no DNA receptor has been reported from plants, and none of the receptors that are known from mammals would allow for this distinction when assuming a sequence-dependent sensing mechanism. A recent study linked the immunogenic effects of eDNA in lettuce seedlings to changes in the methylation pattern of genomic CpG DNA, although—in apparent contrast to the earlier observations—self- and nonself-DNA elicited similar responses in this system (Vega-Muñoz et al., 2018).

Signal interfering DNA, short dsDNA molecules, may exert immunogenic effects via an inhibition of DNA damage repair (Croset et al., 2013; Quanz et al., 2009) and multiple lines of evidence link the DNA damage response to innate immunity, at least in mammals (Dhanwani et al., 2018; Krapp et al., 2018; Li and Chen, 2018; Lugrin and Martinon, 2018; Nakad and Schumacher, 2016). Even in plants, DNA-damaging factors exert immunogenic effects (Hadwiger and Tanaka, 2018). Thus, it is tempting to speculate that the application of fragmented genomic DNA could drive responses that are similar to the intrinsic DNA damage response.

7. SPECIFICITY IN NUCLEIC ACID SENSING

7.1 Consequences of Aberrant Self-DNA Sensing

In spite of the above-mentioned mechanisms that favor the distinction between self- and nonself-dsDNA, evidence is quickly accumulating that identifies DNA damage, an impaired DNA damage response, or an impaired cleavage of DNA fragments, as factors that drive cancer as well as numerous inflammatory and autoimmune diseases (Ahn et al., 2012; Barber, 2015; Chen et al., 2016; Gall et al., 2012; Gallucci and Maffei, 2017; Gao et al., 2015; Jakobs et al., 2015; Li and Chen, 2018; Lugrin and Martinon, 2018; Nakad and Schumacher, 2016; Roers et al., 2016; Toussaint et al., 2017; Trigg and Ferguson, 2015). Since DNA damage can drive chronic inflammation and vice-versa, the underlying mechanisms are prone to form vicious feed-forward loops (Nakad and Schumacher, 2016). For example, self-eDNA can stimulate the maturation of APCs (Ishii et al., 2001) and AIM2-dependent pyroptosis (Fernandes-Alnemri et al., 2009; Hornung et al., 2009). Pyroptosis is well known to lead to the release of pro-inflammatory DAMPs, including self-eDNA, which can trigger further inflammation via AIM2-dependent or -independent mechanisms. For example, besides AIM2, also IFI16 has been related to various autoimmune diseases (Krapp et al., 2018), and RAGE activates the pro-inflammatory transcription factor NF-κB, which induces the transcription of RAGE itself, opening a further avenue toward positive feed-forward effects (Bierhaus et al., 2005). Thereby, "pyroptosis can breed more pyroptosis" (Doitsh and Greene, 2016).

Under healthy, physiological conditions, the accumulation of self-DNA in the cytoplasm or the extracellular space is prevented by the activity of DNases, such as endosomal DNase II, cytoplasmatic TREX1, and extracellular exDNase (Hawes et al., 2015; Lan et al., 2014). Numerous studies linked inflammation caused by deficient activity of DNases to the cGAS-STING pathway (Dhanwani et al., 2018; Galluzzi et al., 2018a) and correspondingly, mice deficient in cGAS or STING also exhibited a reduced type I IFN response (King et al., 2017). Mice lacking DNase II suffer polyarthritis-like diseases inflammatory disease because self-DNA that accumulated outside the nucleus or phagocytosed DNA from apoptotic cells is not digested and becomes accessible to intracellular DNA sensors (Ahn et al., 2012; Jakobs et al., 2015; Lan et al., 2014). Similarly, TREX1-deficient mouse DCs and macrophages exhibited significant leakage of nuclear DNA into the cytoplasm which—likely mediated by STING—caused a constitutive activation of APCs in these mice (Ahn et al., 2014). *Trex1*$^{-/-}$ mice exhibited a spontaneous IFN I response even when treated with high doses of antiretroviral drugs, an observation that excluded endogenous retroelements as the causal reason of the observed inflammatory processes (Achleitner et al., 2017), and biallelic loss-of-function mutations in TREX1 or MDA5 cause the lupus-like Aicardi-Goutières syndrome, in which an endogenous DNA substrate of TREX1 triggers a cGAS-dependent type I IFN response (Achleitner et al., 2017; Gall et al., 2012; Pichlmair et al., 2009).

Even in humans, mutations causing DNase II deficiency could be linked to severe IFN-1-mediated autoinflammation (Rodero et al., 2017). In general terms, aberrant self-dsDNA sensing has been suggested as a factor in a diverse set of "Type I interferonopathies" (Lee-Kirsch et al., 2015) such as arthritis (Man et al., 2016), atherosclerosis (Döring et al., 2012), cardiac inflammation (Lavine and Mann, 2017), chronic neurodegenerative diseases (Gao et al., 2015; Ioannidou et al., 2016), inflammatory processes induced by chemotherapy (Lian et al., 2017), "inflamm-aging" (Franceschi et al., 2000; Ioannidou et al., 2016), pancreatitis (Nakamura et al., 2012), psoriasis (Lande et al., 2007; Man et al., 2016), rhinovirus-induced type-2 allergic asthma exacerbation (Toussaint et al., 2017), and SLE (Bai et al., 2018; Barrat et al., 2005; Gilliet et al., 2008; Shin et al., 2013; Zhang et al., 2013). Recent studies even suggest the therapeutic use of nucleic acid-binding polymers (NABPs) as scavengers that capture circulating nucleic acids and thereby avoid their pro-inflammatory TLR-mediated effects in the context of SLE (Naqvi et al., 2018).

In short, mutations in multiple elements of nucleic acid sensing—or their activation in aberrant physiological conditions—can cause a detrimental activation of innate immunity by self-DNA. Moreover, the same elements represent attractive targets for pathogens that evolved to suppress or overcome their hosts' immunity. For example, the HIV-1 protein Tat sequesters dsRNA and also binds directly to PKR, thereby inhibiting the detection of viral dsRNA by PKR and probably other dsRNA sensors (Radetskyy et al., 2018).

7.2 Sequence-Related Motifs vs Localization

The diversity and severity of pathological situations that can be created by aberrant nucleic acid sensing has motivated a strong research focus on mechanisms that favor self/nonself discrimination in nucleic acid sensing. As discussed in Sections 4.1 and 5.1, TLRs are central players in the recognition of RNA and DNA in mammals, and both sequence-dependent and -independent mechanisms contribute to the specificity in the recognition process (Lässig and Hopfner, 2017; Schlee and Hartmann, 2016). The unmethylated CpG motif was originally considered as an element that allows for a specific recognition of viral or bacterial DNA by TLR9 (Ahmad-Nejad et al., 2002; Bauer et al., 2001; Krieg, 2002; Ohto et al., 2015), and some authors even suggested this recognition process to be "species-specific" (Bauer et al., 2001). The same motif was proposed to be required for the specific recognition of bacterial eDNA in plants (Yakushiji et al., 2009). However, TLR9 senses mtDNA as well (McCarthy et al., 2015; Pinti et al., 2012; Ries et al., 2013; Zhang et al., 2014), and also eukaryotic genomic DNA contains CpG motifs (Ashikawa, 2001; Cross et al., 1997), which could be sensed by TLR9 as long as they are not masked by methylation (Gilliet et al., 2008). In fact, recent evidence supports a recognition of self-DNA from dying human cancer cells or murine fetal DNA by TLR9 (Goldfarb et al., 2018; Tuomela et al., 2013).

Similarly, cGAS was originally coined as "an innate immune sensor of HIV and other retroviruses" (Gao et al., 2013) or a PRR that recognizes "viral reverse transcriptase products" (Altfeld and Gale, 2015), and was assumed to be subject to a "sequence-specific activation" by DNA structures as they are commonly found in primary HIV-1 cDNA (Herzner et al., 2015). By contrast, O'Neill—immediately upon its discovery—connected the cGAS-STING-mediated recognition of cytosolic dsDNA to inflammatory and autoimmune responses (O'Neill, 2013), an interpretation that since then has been widely accepted (Andreeva et al., 2017; Gallucci and Maffei, 2017; Gantier, 2017; Kato et al., 2017; King et al., 2017; Kretschmer and Lee-Kirsch, 2017; Li and Chen, 2018; Liu et al., 2016; Qiao et al., 2017; Zevini et al., 2017). Intriguingly, the CpG motif was found to be enriched in DNA/anti-DNA complexes that were isolated from the sera of patients with SLE >30 years ago (Sano and Morimoto, 1982). In summary, the known mammalian dsDNA sensors have only a limited capacity to discriminate microbial from endogenous DNA (Roers et al., 2016).

Nevertheless, nuclear DNA or mtDNA that is released from pyroptotic cells should show up in the extracellular space, whereas the known receptors of DNA—with the exception of RAGE—are localized within the cell (Luecke and Paludan, 2017; Paludan and Bowie, 2013). Therefore, self-eDNA should not be immunogenic. In support of this assumption, genomic DNA and other exogenously applied oligonucleotides must be transfected into the cells in order to become immunogenic, in most experimental setups. In short, the localization of dsDNA fragments is now considered a key element that allows self/nonself discrimination in nucleic acid sensing (Gallucci and Maffei, 2017; Gilliet et al., 2008; Luecke and Paludan, 2017; Schlee and Hartmann, 2016).

In spite of the above-mentioned mechanisms that favor self/nonself discrimination in DNA sensing "various DNAs, whether pathogen-derived or host-derived, can activate the immune system" (Wang et al., 2008). In the following two sections, we discuss two non-exclusive explanations of this phenomenon: central hubs that integrate upstream signals from different PRRs and DAMPs or autoantibody-assisted trafficking of eDNA into cells.

7.3 Central Hubs in Nucleic Acid Sensing

Although TLR3 senses dsRNA, while TLRs7 and TLR8 sense ssRNA and TLR9 senses CpG-rich DNA, all four TLRs trigger downstream pro-inflammatory responses in a MyD88-dependent pathway (Alexopoulou et al., 2001; Botos et al., 2011; Hemmi et al., 2002; Mancuso et al., 2009; Wagner, 2004). Moreover, the DNA sensors cGAS and ZBP1/DAI and the RNA sensors RIG1 and MDA5, all trigger a type I IFN response via the activation of IRF3 and NF-κB (Cai et al., 2014; Gallucci and Maffei, 2017; Kato et al., 2017; Li and Chen, 2018; Margolis et al., 2017; Tao et al., 2016; Wang et al., 2016; Xiao and Fitzgerald, 2013). In a recent review, Zevini and colleagues summarize the mechanisms and consequences of the crosstalk between RNA-sensing and dsDNA-sensing pathways (Zevini et al., 2017). Moreover, due to its localization in the cell membrane, RAGE does have access to eDNA. Since RAGE activates NF-κB which in turn triggers the expression of RAGE, the resulting feed-forward mechanism can cause a particularly prolonged activation of NF-κB (Bierhaus et al., 2005) and consecutively the induction of multiple pro-inflammatory effects by eDNA.

Inflammasomes form a further central hub that can mediate a type I IFN response to multiple ligands. For example, PKR triggered the formation of NLRP3- and AIM2-dependent inflammasomes (Lu et al., 2012), thereby allowing for an RNA-dependent activation of caspase 1. Since inflammasome-mediated cell death (pyroptosis) causes the release of pro-inflammatory DAMPs such as HMGB1 (Lamkanfi et al., 2010) and self-nucleic acids, inflammasome activation is prone to positive feed-forward scenarios, as exemplified by the AIM2-mediated pyroptotic cell death caused by exogenously applied genomic DNA or by drugs that damage the nuclear envelope (Di Micco et al., 2016; Fernandes-Alnemri et al., 2009). In the light of these scenarios, the controversial observations concerning a role of ZBP1/DAI as a DNA sensor (see Section 4.2). ZBP1/DAI and review by Kuriakose and Kanneganti (2018) might find their mechanistic explanation in the role of ZBP1/DAI as a potent inducer of necroptosis (Kuriakose et al., 2016; Maelfait et al., 2017): if ZBP1/DAI triggered inflammatory responses to self-eDNA that cause necroptosis, it would favor the release of even more self-eDNA which, in turn, can trigger immunity via ZBP1/DAI-independent pathways. In short, most sensors of nucleic acids share downstream signaling elements that cause a type I IFN responses independently of the upstream receptor: ligand interactions (Trigg and Ferguson, 2015).

Similarly, in plants, the specificity that can be observed at the level of receptor-ligand binding is not reflected at the level of hormonal and phenotypic responses: independently of the activated sensors, most plant immune responses to damage (including DNA damage by, e.g., UV light) or to infection channel into JA- and SA-dependent pathways. Phenotypic responses include cell wall strengthening, lignification, and the induction of phytoalexins and other secondary compounds and of so-called pathogenesis-related proteins. Intriguingly, many secondary compounds are anti-oxidants or radical scavengers and, thus, protect from damaging radiation as well as from biological enemies, and common pathogenesis-related proteins have chitinase or β-glucanase activity and, thus, can attack the cell walls of diverse fungal and bacterial pathogens (Jones and Dangl, 2006).

7.4 DAMPs Mediated DNA Trafficking

The counterintuitive finding of a TLR9 stimulation by eDNA found its mechanistic explanation with the observation of immune complexes containing DNA and autoantibodies Intriguingly, the CpG motif was found enriched in DNA/anti-DNA complexes from the sera of patients with SLE (Sano and Morimoto, 1982). In complexes with autoantibodies (IgGs), eDNA is protected from the lysis by extracellular DNases and moreover, such complexes facilitate DNA uptake into immune cells, thereby exposing self-eDNA to RAGE or TLR9 in B cells and pDCs (Barrat et al., 2005;

Boulé et al., 2004; Gilliet et al., 2008; Leadbetter et al., 2002) or to the NLRP3 inflammasome in human monocytes (Shin et al., 2013).

Besides autoantibodies, DNA binding proteins can stabilize extracellular and cytoplasmatic dsDNA and condense self-DNA into particles that can be endocytosed (Magna and Pisetsky, 2016). Frequently reported DAMPs in this context comprise HMGB1, histones, and several HDPs (Barrat et al., 2005; Chamilos et al., 2012; Gilliet et al., 2008; Gougeon, 2017; Lande et al., 2007, 2015; Leadbetter et al., 2002; Lövgren et al., 2004; Magna and Pisetsky, 2016; McGlasson et al., 2017; Sandgren et al., 2004; Westman et al., 2015). Thus, HMGB1/DNA complexes are being discussed in the context of diverse autoimmune diseases (Kang et al., 2014; Magna and Pisetsky, 2014). In particular SLE is characterized by the presence of immune stimulatory complexes containing DNA- and RNA-associated proteins such as HMGB1 (Bai et al., 2018; Lamphier et al., 2006; Lee-Kirsch et al., 2015). For example, HMGB1/DNA immune complexes stimulated cytokine production through a TLR9-MyD88 pathway that also involved the multi-ligand receptor RAGE (Tian et al., 2007) and activated the AIM2-inflammasome (Liu et al., 2014).

Similarly, extracellular histones enhanced the TLR9-mediated immunogenic properties of self-eDNA in a murine model of hepatic ischemia/reperfusion injury (Huang et al., 2011). In human trauma patients, enhanced serum levels of histone-DNA complexes were correlated with poor clinical outcome (Russell et al., 2018). Of note, histone H4 can bind to TLR4 on monocytes and then triggered a CXCL10-dependent chemoattraction of leucocytes that was independent of eDNA (Westman et al., 2015).

Furthermore, iDAMPs sensu (Yatim et al., 2017) like the HDP, LL-37, form complexes with human or microbial eDNA that can be taken up by human pDCs and monocytes (Chamilos et al., 2012; Döring et al., 2012; Lande et al., 2007; Sandgren et al., 2004). For example, LL-37 enhanced the pro-inflammatory effects of self-DNA on keratinocytes, and effect that was related to invasive cholesteatoma (Chi et al., 2014). As a second, perhaps additional mechanism, LL-37 can favor the recognition of CpG DNA by TLR9 (Nakagawa and Gallo, 2015). Similarly, IL-26, which is overexpressed in many chronic inflammatory diseases, is released from Th17 cells in complexes with DNA of host or bacterial origin, and these complexes triggered a TLR9-dependent type I IFN response in pDCs (Meller et al., 2015). IL-26 was also reported to form immunogenic complexes with mtDNA or nuclear self-DNA bound in neutrophil extracellular traps, which can be taken up into monocytes to trigger STING- and inflammasome-dependent cytokine secretion (Liu et al., 2016; Poli et al., 2017). Finally, RAGE can facilitate the uptake of extracellular nucleic acids into endosomes (Bertheloot et al., 2016; Paz et al., 2017; Sirois et al., 2013).

Synergistic effects have also been reported for self-eRNA, which strongly enhanced the pro-inflammatory response of human macrophages to a TLR2 agonist (Noll et al., 2017) and for the plant defense peptide OsPep3, which enhanced the immune stimulatory effect of a HAMP (Shinya et al., 2018). Unfortunately, these last two studies apparently did not consider DNA-complex forming as a potential mechanism for the observed synergistic effects. In summary, "intracellular DAMP and PAMP receptors may be exposed to and activated by mediators that would never reach them without the helpful HMGB1 transport" (Yang et al., 2015) and DAMPs "act as adjuvants that further increase the host response in addition to the initial cause of inflammation" (Westman et al., 2015).

8. HOW CAN WE GAIN A BROADER UNDERSTANDING OF NUCLEIC ACID SENSING?

Most reports on new receptors tended to overestimate the level of specificity at which these PRR sense, e.g., foreign nucleic acids. As mentioned above, TLR9 was first considered as sensing only bacterial and viral CpG-rich DNA, cGAS was coined as HIV-specific receptor or—at least—receptor of retroviral dsDNA, and ZBP1/DAI was considered to sense exclusively *Z*-form dsDNA. However, it then became clear that TLR9 also senses mtDNA, that cGAS exerts no sequence-related specificity at all, and that ZBP1/DAI, at the very least, also senses Z-RNA. Consecutively, most researchers focused on the localization of the accumulating nucleic acid fragments as a factor that allows immune cells to distinguish self- from nonself-DNA.

However, even the activity of RNases and DNases such as TREX1 and the restriction of most nucleic acid sensors to the cytoplasm or the endosome, are not sufficient to guarantee a clear self/nonself discrimination in DNA sensing. In particular, the above-mentioned trafficking of eDNA into immune cells clearly demonstrates that and how intracellular PRRs can get access to self-eDNA that had been released from dying cells into the extracellular space. Do all these observations reveal "problems" in the sensing of nucleic acids by the immune system, i.e., potentially detrimental side-effects that natural selection has failed to eliminate? In the following sections, we present some evolutionary considerations concerning the selective benefits of a strict self/nonself discrimination in nucleic acid sensing and suggest further research lines.

8.1 Evolutionary Considerations

Does self-DNA-driven autoimmunity indicate that natural selection failed to optimize nucleic acid sensing? In fact, most observations on detrimental effects of self-nucleic acid sensing were made using mutated mice or human diseases that are related to mutations in relevant genes. The fitness of experimentally generated knockout mice arguably would be very low under natural conditions, and it appears likely that "problematic" alleles that cause SLE and other autoinflammatory diseases also would be counter-selected. Moreover, diabetes and other "contemporary" metabolic diseases occur when the mammalian metabolism and immune system have to function under a surplus of energy-rich food: conditions to which mammals never were exposed during their evolution (Hotamisligil, 2017). Similarly, we now are exposed to many artificial compounds with oxidative and pro-inflammatory properties that did not exist before. Thus, a part of the above-mentioned diseases can certainly be related to a sub-optimal self/nonself discrimination of nucleic acids that becomes particularly evident when the immune system has to function under conditions under which it did not evolve.

However, the "DAMPs assisted trafficking" that provides immunogenic properties to self-eDNA and the signaling hubs that integrate signals from diverse receptors favor an alternative, although non-exclusive explanation. Perhaps, a specific self/nonself discrimination simply is less important than most of us have always thought. Many researchers might have based their work on the a priori assumption that the immune system must recognize the foreign and avoid immune responses to endogenous molecules. However, fragments of dsDNA or dsRNA that accumulate in the cytoplasm or extracellular space are reliable indicators of cellular damage, independently of whether they originate from intracellular pathogens or the host (Bierhaus et al., 2005; Desmet and Ishii, 2012; O'Neill, 2013; Wang et al., 2008). Moreover, multiple observations link infection, metabolic disorders, and DNA damage/DNA repair to each other: all these pathological scenarios trigger shared elements of the innate and even the adaptive arm of the immune system (Haneklaus and O'Neill, 2015; Nakad and Schumacher, 2016; Yatim et al., 2017). For example, the activation of cGAS-STING by genomic DNA links DNA damage to cellular senescence and to adequate immune responses to tumor-derived eDNA (Li and Chen, 2018; Liu and Wang, 2016; Qiao et al., 2017; Tao et al., 2016; Zevini et al., 2017). Overwhelming evidence has accumulated showing that the cGAS-cGAMP-STING-mediated type I IFN response allows for efficient immunity to, e.g., HIV-1 infection AND to cancer (recently reviewed in Krapp et al., 2018; Ng et al., 2018, respectively). Similarly, PKR has been reported as a tumor suppressor besides its well-known role in antiviral immunity (Davis and Watson, 1996; Pham et al., 2016; Radetskyy et al., 2018), and also AIM2 triggers adequate immune responses to cancer as well as to infection (Corrales et al., 2016; Lian et al., 2017; Man et al., 2016; Pham et al., 2016; Radetskyy et al., 2018),

Apparently, it is time to re-think the relevance of self/nonself discrimination in nucleic acid sensing. Why SHOULD the innate immune system evolve mechanisms that allow for a strict discrimination, when the type I IFN response is "a nearly universal defensive response to pathogens and cell or tissue damage" (Dhanwani et al., 2018), i.e., against a broad spectrum of threats that share as a common element the accumulation of eDNA or cytoplasmatic dsDNA? Similarly, the evolutionary arms-race between plants and their enemies has resulted in a highly specified recognition-evasion network, whereas downstream pathways converge in the induction of broad-spectrum responses that act against diverse pathogens, herbivores, and even abiotic forms of stress (Jones and Dangl, 2006).

8.2 Open Questions and Research Perspectives

As mentioned in Section 6.2, we and others detected recently that in plants, self-eDNA (applied as fragmented genomic DNA) can exert much stronger immunogenic effects than nonself-eDNA, even when the latter stems from a closely related species (Barbero et al., 2016; Duran-Flores and Heil, 2018). These observations indicate a role of eDNA as a DAMP in plants. However, to support this hypothesis, future studies should (i) show that natural damage leads to the release of eDNA from plant cells, (ii) characterize the nature of these eDNA fragments (e.g., single-stranded or double-stranded, fragment length, concentrations that are reached in the extracellular space), and (iii) identify the mechanism via which eDNA induces immunity-related traits in a species-specific way. Moreover, it remains to be elucidated whether and to which degree eDNA can exert similar effects on human cells. However, at the very least, these observations indicate that eDNA can trigger eukaryote immunity via as-yet unknown mechanisms. Considering that "receptor-specificity" seemingly represented a general goal in the research into nucleic acid sensing, it appears surprising that many empirical studies relied on the use of artificial polynucleotides such as poly (dA:dT) or poly (I:C). Thus, a significant part of the empirical research is intrinsically based on the inability of innate immune receptors to discriminate between microbial and artificial oligonucleotides.

Moreover, the examples compiled in Section 7.4 show that and how IgGs and DAMPs can provide immunogenic properties to extracellular nucleic acids. Since mechanical tissue damage and lytic forms of programmed cell death release nuclear DNA, mtDNA, and other DAMPs, it appears likely that these synergistic effects occur under most "natural situations," but they will be overlooked in all experimental setups that use purified nucleic acids. Finally, the concentration of nucleic acid fragments is a factor that requires more attention. Membranes are no absolute barriers, and high concentrations of eDNA fragments arguably favor their diffusion through membranes. Quantitative arguments show up in frequently used phrases, such as "excessive DNA endocytosis," "self-DNA accumulated in the cytosol" or, even more explicitly, "DNAs, …, can activate the immune system when exposed *en masse* to the cytosol" (Wang et al., 2008). However, as to the best of our knowledge, relatively few studies have focused explicitly on concentration effects in mammalian DNA sensing. For example Fernandes-Alnemri et al. reported a partial conversion of AIM2-dimers to oligomers at 0.5 μg dsDNA and a complete conversion at 1 μg dsDNA (Fernandes-Alnemri et al., 2009). The length-specific effects of dsDNA on cGAS became apparent only at very low concentrations of dsDNA—concentrations—which the authors argued to be more likely to resemble the natural situation in infected tissues (Gantier, 2017; Luecke et al., 2017)—and the recently discovered cGAS micro reactors were formed only at concentrations >30 nM of both dsDNA and cGAS (Du and Chen, 2018). Additional examples of studies that considered concentration-dependent effects are summarized in Table 1.

We suggest that concentration represents an overlooked element that might favor targeted immune responses to dsDNA from different sources, independently of the classical paradigm of self/nonself discrimination. Enhanced concentrations of foreign dsDNA indicate infection by certain bacteria, DNA viruses, or retroviruses, whereas self-dsDNA indicates a loss of the integrity of the nucleus (Di Micco et al., 2016) or other forms of severe cellular dysfunction. Thus, in most scenarios, the concentration of dsDNA in the cytoplasm or in the extracellular space should be positively correlated with the degree of danger, which means that a concentration-dependent

Table 1 Concentration-Dependent Effects of DNA Fragments in Nucleic Acid Sensing

Organism/Cell Type	Sensor	DNA Fragment(s)	Quantity/ Concentration	Method	Response	Reference
In vitro	AIM2	64-mer dsDNA	0.5 and 1 μg dsDNA	Immunoblotting	AIM2-oligomerization partial at 0.5 and complete at 1 μg dsDNA, AIM2-dimer disappears at 1 μg	Fernandes-Alnemri et al. (2009)
Mouse macrophages (LPS-primed)	AIM2	Poly (dA:dT), mouse genomic DNA, *E. coli* DNA	$1\,\mu g\,mL^{-1}$ DNA per 2×10^6 cells	Transfection	Caspase-1 activation	Fernandes-Alnemri et al. (2009)
THP1-cells	AIM2	Poly (dA:dT)	0, 0.6, and 1.5 μg poly (dA:dT)	Transfection	Formation of ASC pyroptosomes	Fernandes-Alnemri et al. (2009)
RAW 264.7 macrophages		Modified immuno-stimulatory DNA (15–45 bp), poly (dA:dT)	10, 100, $1000\,ng\,mL^{-1}$ poly(dA:dT) at 0.25, 1.0, and $2.0\,\mu g\,mL^{-1}$	Transfection	IFN-β production in a concentration-dependent manner, being ISD (45 bp) slightly more potent	Karayel et al. (2009)
Macrophages, dendritic cells	AIM2	Poly (dA:dT)	0.75–3 μg per 1×10^6 cells	Transfection	Caspase-1 cleavage, IL-1β production, and IL-18 AIM2-dependent production	Rathinam et al. (2010)
AIM2 −/ macrophages and dendritic cells	AIM2	Tumor-derived DNA	$0.1–1000\,ng\,mL^{-1}$	Stimulation	IFN-β expression is enhanced in AIM2-deficient cells peaking at 0.1 and $1\,\mu g\,mL^{-1}$ DNA, leading to the activation of the cGAS-STING pathway	Corrales et al. (2016)

THP-1, fibroblasts, human peripheral blood mononuclear cells	cGAS	PCR-derived and restriction-digested DNA (88–4570 bp)	0.0167, 0.033, and 0.167 $\mu g\,mL^{-1}$	Transfection	At low concentrations (0.167 $\mu g\,mL^{-1}$), long fragments (2027, 4003 bp) strongly stimulated type I IFN	Luecke et al. (2017)
HEK293T, SLK	cGAS	Interferon stimulatory DNA + ORF52	2–30 $\mu g\,mL^{-1}$	Transfection	Low—intermediate concentrations of DNA in the presence of OR52 inhibited cGAS	Wu et al. (2015)
In vitro, and cGAS-deficient MEF cells with human FL-cGAS	cGAS	HT DNA and synthetic oligonucleotides	<10–900 nM	In vitro	Phase separation leading to the formation of micro reactors with active cGAS	Du and Chen (2018)
THP-1, HEK293, macrophages	IFI16	Poly(dA:dT), viral DNA, mammalian DNA, and bacterial DNA	0.5–5 $\mu g\,mL^{-1}$	Transfection	HEK293 *IFNB* promoter stimulation only by poly(dA:dT) while in THP-1 all dsDNAs-induced IFN-β mRNAs	Unterholzner et al. (2010)
Macrophages, dendritic cells	NALP3	Adenoviral DNA, ssDNA, poly (dA:dT), *E. coli* DNA, CpG oligonucleotides, plasmid DNA, and salmon sperm DNA	1 μg	Incubation, transfection	Transfected but not incubated adenoviral DNA-activated IL-1β Inflammasome also responded to non-adenoviral DNA but in a NALP3-independent manner, relying on ASC and caspase-1	Muruve et al. (2008)

Continued

Table 1 Concentration-Dependent Effects of DNA Fragments in Nucleic Acid Sensing—cont'd

Organism/Cell Type	Sensor	DNA Fragment(s)	Quantity/ Concentration	Method	Response	Reference
AIM2$^{-/-}$ macrophages	NALP3	Mouse mtDNA	1 μg	Transfection	Caspase-1-dependent maturation of IL-1β and IL-18	Nakahira et al. (2010)
In vitro macrophages, in vivo rats	TLR9	Mouse nuclear DNA and mtDNA	10 μg mL^{-1}	Transfection, intravenous injection	mtDNA-induced TNF-α, IL-6, and IL-10 production and upregulating *NF-κB*, *IκB-α*, and TLR9 mRNAs expression in vivo and in vitro	Zhang et al. (2014)
Lymphocytes	TLR9	Genomic DNA residue in EV71 vaccine purification (20–50 bp)	TLR9 expression assay: 0.38, 3.8, and 38 ng mL^{-1} EV71 neutralizing antibodies assay: 1.5, 6, and 24 ng mL^{-1}	Incubation, injection	Residual DNA in EV71 vaccines upregulates in a concentration-dependent manner TLR9 mRNA expression and promotes EV71 neutralizing antibodies	Shao et al. (2016)
MDA-MB-231, D54MG, T47-D	TLR9	DNA-LL37 complex	1 μg DNA-10 μg LL37	Matrigel invasion assay	Cancer death cell DNA uptake by surviving cancer cells inducing TLR9	Tuomela et al. (2013)
RAW 264.7 macrophages	TLR9	ODNs, telomer ODNs	ODNs—1–20 μg mL^{-1}	Incubation	IL-6 release stimulation by ODNs being 5 μg mL^{-1} the concentration with maximum effect. Telomer ODNs failed to induce IL-6 release	Goldfarb et al. (2018)

RAW 264.7 macrophages	TLR9	Fetal and placental gDNA, fragmented fetal, and placental gDNA (200–700 bp), Telomer-depleted genomic DNA	DNA 0–5 $\mu g\,mL^{-1}$	Incubation	Fetal and placental genomic DNA stimulated IL-6 production, telomer-depleted DNA produced a threefold increase in IL-6 production, while fragmented DNA with telomere presence failed to induce IL-6	Goldfarb et al. (2018)
RAW 264.7 macrophages	TLR9	gDNA + DOTAP complex	DNA: 15 $\mu g\,mL^{-1}$ DOTAP: 5–120 $\mu g\,mL^{-1}$	Incubation	High DNA-DOTAP complex failed to induce IL-6 while greater DOTAP-DNA ratios increased IL-6 secretion levels	Goldfarb et al. (2018)

(but source-independent) type I IFN response to eDNA or cytosolic dsDNA might be sufficient to guarantee an adequate immune response.

In vitro testing of this hypothesis could compare the effects of different concentrations of eDNA from different biological origins on established cellular model systems such as PBMCs, macrophages, or dendritic cells. Readouts ideally would comprise early signaling steps such as Ca^{2+} fluxes, ROS formation, and MAPK activation, intermediate steps such as the expression of DNA sensors, and late steps such as the formation of inflammasomes and the release of IFN I or cytokines. Ideally, such experiments would also consider synergistic effects that become evident only when cells are simultaneously exposed to eDNA and those DAMPs that facilitate DNA trafficking.

9. CONCLUSIONS

Self/nonself discrimination in the context of nucleic acid sensing appears to be less strict and less important than it would be expected under Janeway's paradigm, and high concentrations of nucleic acid fragments and their DAMPs assisted trafficking might act in synergy to relativize the importance of the (original) localization of dsDNA fragments. We conclude that most of the evidence that we compiled in this section supports Matzinger's danger model and suggests the concentration of nucleic acid fragments as a danger signal of underestimated relevance. Recent observations in plants reveal the existence of an as-yet unknown mechanism via which genomic eDNA can trigger immunity-related effects at a surprisingly high level of specificity. If confirmed for mammalian immune cells, this mechanism would strongly relativize the suitability of artificial oligonucleotides if we aim to detect all potential effects of nucleic acids on the immune system. Furthermore, experiments that use purified oligonucleotides will overlook synergistic effects that are mediated by DAMPs. We suggest that the immunological sciences would benefit from an intensified communication across disciplines and that the (re-)integration of traditional screening designs, as they had been applied in the early days of immunology (Isaacs et al., 1963), could help to gain a more complete understanding of the full spectrum of scenarios in which nucleic acids become immunogenic and their multiple effects on the immunity of plants and mammals, in health and disease.

ACKNOWLEDGMENT

M.H. acknowledges financial support from CONACYT de México (grant 258119).

REFERENCES

Abderrazak, A., Syrovets, T., Couchie, D., et al., 2015. NLRP3 inflammasome: from a danger signal sensor to a regulatory node of oxidative stress and inflammatory diseases. Redox Biol. 4, 296–307.

Ablasser, A., 2018. Phase separation focuses DNA sensing. Science 361, 646–647.

Ablasser, A., Goldeck, M., Cavlar, T., Deimling, T., Witte, G., 2013. cGAS produces a 2′-5′-linked cyclic dinucleotide second messenger that activates STING. Nature 498, 380–384.

Acevedo, F.E., Rivera-Vega, L.J., Chung, S.H., Ray, S., Felton, G.W., 2015. Cues from chewing insects—the intersection of DAMPs, HAMPs, MAMPs and effectors. Curr. Opin. Plant Biol. 26, 80–86.

Achleitner, M., Kleefisch, M., Hennig, A., et al., 2017. Lack of Trex1 causes systemic autoimmunity despite the presence of antiretroviral drugs. J. Immunol. 199, 2261–2269.

Aguirre, S., Fernandez-Sesma, A., 2017. Collateral damage during dengue virus infection: making sense of DNA by cGAS. J. Virol. 91, e01081-16.

Aguirre, S., Luthra, P., Sanchez-Aparicio, M. T., et al. (2017). Dengue virus NS2B protein targets cGAS for degradation and prevents mitochondrial DNA sensing during infection. Nat. Microbiol. 2, 17037.

Ahmad-Nejad, P., Hacker, H., Rutz, M., Bauer, S., Vabulas, R.M., Wagner, H., 2002. Bacterial CpG-DNA and lipopolysaccharides activate Toll-like receptors at distinct cellular compartments. Eur. J. Immunol. 32, 1958–1968.

Ahn, J., Gutman, D., Saijo, S., Barber, G., 2012. STING manifests self DNA-dependent inflammatory disease. Proc. Natl. Acad. Sci. U. S. A. 109, 19386–19391.

Ahn, J., Ruiz, P., Barber, G.N., 2014. Intrinsic self-DNA triggers inflammatory disease dependent on STING. J. Immunol. 193, 4634–4642.

Alexopoulou, L., Holt, A.C., Medzhitov, R., Flavell, R.A., 2001. Recognition of double-stranded RNA and activation of NF-kappa B by Toll-like receptor 3. Nature 413, 732–738.

Allam, R., Kumar, S.R., Darisipudi, M., Anders, H.-J., 2014. Extracellular histones in tissue injury and inflammation. J. Mol. Med. 92, 465–472.

Altfeld, M., Gale, M., 2015. Innate immunity against HIV-1 infection. Nat. Immunol. 16, 554–562.

Amarante, M.K., Watanabe, M.A.E., 2010. Toll-like receptor 3: involvement with exogenous and endogenous RNA. Int. Rev. Immunol. 29, 557–573.

Andersson, U., Yang, H., Harris, H., 2018a. Extracellular HMGB1 as a therapeutic target in inflammatory diseases. Expert Opin. Ther. Targets 22, 263–277.

Andersson, U., Yang, H., Harris, H., 2018b. High-mobility group box 1 protein (HMGB1) operates as an alarmin outside as well as inside cells. Semin. Immunol. 9. https://doi.org/10.1016/j.smim.2018.02.011. publ. online March 2018.

Andreeva, L., Hiller, B., Kostrewa, D., et al., 2017. cGAS senses long and HMGB/TFAM-bound U-turn DNA by forming protein-DNA ladders. Nature 549, 394–398.

Arshad, O., Gadawska, I., Sattha, B., Côté, H.C.F., Hsieh, A.Y.Y., 2018. Elevated cell-free mitochondrial DNA in filtered plasma is associated with HIV infection and inflammation. J. Acquir. Immune Defic. Syndr. 78, 111–118.

Asada, K., Ito, K., Yui, D., Tagaya, H., and Yokota, T. (2018). Cytosolic genomic DNA functions as a natural antisense. Sci. Rep. 8, 8551.

Ashikawa, I., 2001. Gene-associated CpG islands in plants as revealed by analyses of genomic sequences. Plant J. 26, 617–625.

Assil, S., Paludan, S.R., 2017. Live and let die: ZBP1 senses viral and cellular RNAs to trigger necroptosis. EMBO J. 36, 2470–2472.

Aswani, A., Manson, J., Itagaki, K., et al. (2018). Scavenging circulating mitochondrial DNA as a potential therapeutic option for multiple organ dysfunction in trauma hemorrhage. Front. Immunol. 9, 891.

Bacete, L., Mélida, H., Miedes, E., Molina, A., 2018. Plant cell wall-mediated immunity: cell wall changes trigger disease resistance responses. Plant J. 93, 614–636.

Bai, Y., Tong, Y., Liu, Y., Hu, H., 2018. Self-dsDNA in the pathogenesis of systemic lupus erythematosus. Clin. Exp. Immunol. 191, 1–10.

Barber, G.N., 2015. STING: infection, inflammation and cancer. Nat. Rev. Immunol. 15, 760–770.

Barbero, F., Guglielmotto, M., Capuzzo, A., and Maffei, M. E. (2016). Extracellular self-DNA (esDNA), but not heterologous plant or insect DNA (etDNA), induces plasma membrane depolarization and calcium signaling in Lima bean (*Phaseolus lunatus*) and maize (*Zea mays*). Int. J. Mol. Sci. 17, 1659.

Barrat, F.J., Meeker, T., Gregorio, J., et al., 2005. Nucleic acids of mammalian origin can act as endogenous ligands for Toll-like receptors and may promote systemic lupus erythematosus. J. Exp. Med. 202, 1131–1139.

Bartels, S., Lori, M., Mbengue, M., et al., 2013. The family of peps and their precursors in Arabidopsis: differential expression and localization but similar induction of pattern-triggered immune responses. J. Exp. Bot. 64, 5309–5321.

Barton, G.M., Kagan, J.C., Medzhitov, R., 2006. Intracellular localization of Toll-like receptor 9 prevents recognition of self DNA but facilitates access to viral DNA. Nat. Immunol. 7, 49–56.

Bauer, S., Kirschning, C.J., Häcker, H., et al., 2001. Human TLR9 confers responsiveness to bacterial DNA via species-specific CpG motif recognition. Proc. Natl. Acad. Sci. U. S. A. 98, 9237–9242.

Benedetti, M., Pontiggia, D., Raggi, S., et al., 2015. Plant immunity triggered by engineered in vivo release of oligogalacturonides, damage-associated molecular patterns. Proc. Natl. Acad. Sci. U. S. A. 112, 5533–5538.

Benedetti, M., Verrascina, I., Pontiggia, D., et al., 2018. Four Arabidopsis berberine bridge enzyme-like proteins are specific oxidases that inactivate the elicitor-active oligogalacturonides. Plant J. 94, 260–273.

Bernard, J.J., Cowing-Zitron, C., Nakatsuji, T., et al., 2012. Ultraviolet radiation damages self noncoding RNA and is detected by TLR3. Nat. Med. 18, 1286–1290.

Bertheloot, D., Naumovski, A.L., Langhoff, P., et al., 2016. RAGE enhances TLR responses through binding and internalization of RNA. J. Immunol. 197, 4118–4126.

Bhat, A., Ryu, C.-M., 2016. Plant perceptions of extracellular DNA and RNA. Mol. Plant 9, 956–958.

Bianchi, M.E., 2007. DAMPs, PAMPs and alarmins: all we need to know about danger. J. Leukoc. Biol. 81, 1–5.

Bierhaus, A., Humpert, P.M., Morcos, M., et al., 2005. Understanding RAGE, the receptor for advanced glycation end products. J. Mol. Med. 83, 876–886.

Bigeard, J., Colcombet, J., Hirt, H., 2015. Signaling mechanisms in pattern-triggered immunity (PTI). Mol. Plant 8, 521–539.

Bochud, P.-Y., Hersberger, M., Taffé, P., et al., 2007. Polymorphisms in Toll-like receptor 9 influence the clinical course of HIV-1 infection. AIDS 21, 441–446.

Bode, C., Fox, M., Tewary, P., et al., 2016. Human plasmacytoid dentritic cells elicit a type I interferon response by sensing DNA via the cGAS-STING signaling pathway. Eur. J. Immunol. 46, 1615–1621.

Bording-Jorgensen, M., Alipour, M., Danesh, G., Wine, E., 2017. Inflammasome activation by ATP enhances *Citrobacter rodentium* clearance through ROS generation. Cell. Physiol. Biochem. 41, 193–204.

Boriushkin, E., Wang, J.J., Li, J.H., Bhatta, M., Zhang, S.X., 2016. p58(IPK) suppresses NLRP3 inflammasome activation and IL-1 beta production via inhibition of PKR in macrophages. Sci. Rep. 6, 25013.

Botos, I., Segal, D., Davies, D., 2011. The structural biology of Toll-like receptors. Structure 19, 447–459.
Boulé, M.W., Broughton, C., Mackay, F., Akira, S., Marshak-Rothstein, A., Rifkin, I.R., 2004. Toll-like receptor 9-dependent and -independent dendritic cell activation by chromatin–immunoglobulin G complexes. J. Exp. Med. 199, 1631–1640.
Boutrot, F., Zipfel, C., 2017. Function, discovery, and exploitation of plant pattern recognition receptors for broad-spectrum disease resistance. Annu. Rev. Phytopathol. 55, 257–286.
Boyapati, R.K., Rossi, A.G., Satsangi, J., Ho, G.T., 2016. Gut mucosal DAMPs in IBD: from mechanisms to therapeutic implications. Mucosal Immunol. 9, 567–582.
Braga, T.T., Forni, M.F., Correa-Costa, M., et al., 2017. Soluble uric acid activates the NLRP3 inflammasome. Sci. Rep. 7, 39884.
Brinkmann, V., Reichard, U., Goosmann, C., et al., 2004. Neutrophil extracellular traps kill bacteria. Science 303, 1532–1535.
Brown, J., Wang, H., Hajishengallis, G.N., Martin, M., 2011. TLR-signaling networks: an integration of adaptor molecules, kinases, and cross-talk. J. Dent. Res. (formerly Crit. Rev. Oral Biol.) 90, 417–427.
Broz, P., Dixit, V.M., 2016. Inflammasomes: mechanism of assembly, regulation and signalling. Nat. Rev. Immunol. 16, 407–420.
Burdette, D.L., Vance, R.E., 2013. STING and the innate immune response to nucleic acids in the cytosol. Nat. Immunol. 14, 19–26.
Cai, X., Chiu, Y.H., Chen, Z.J.J., 2014. The cGAS-cGAMP-STING pathway of cytosolic DNA sensing and signaling. Mol. Cell 54, 289–296.
Cai, R., Liu, L., Luo, B., et al., 2016. Caspase-1 activity in CD4 T cells is downregulated following antiretroviral therapy for HIV-1 infection. AIDS Res. Hum. Retroviruses 33, 164–171.
Campos, P.C., Gomes, M.T.R., Guimarïes, E.S., Guimarïes, G., Oliveira, S.C., 2017. TLR7 and TLR3 sense *Brucella abortus* RNA to induce proinflammatory cytokine production but they are dispensable for host control of infection. Front. Immunol. 8, 28.
Cao, W., Bover, L., Cho, M., et al., 2009. Regulation of TLR7/9 responses in plasmacytoid dendritic cells by BST2 and ILT7 receptor interaction. J. Exp. Med. 206, 1603–1614.
Carlos, D., Costa, F.R.C., Pereira, C.A., et al., 2017. Mitochondrial DNA activates the NLRP3 inflammasome and predisposes to type 1 diabetes in murine model. Front. Immunol. 8, 164.
Carteni, F., Bonanomi, G., Giannino, F., et al., 2016. Self-DNA inhibitory effects: underlying mechanisms and ecological implications. Plant Signal. Behav. 11, e1158381.
Cervantes, J.L., Dunham-Ems, S.M., La Vake, C.J., et al., 2011. Phagosomal signaling by *Borrelia burgdorferi* in human monocytes involves toll-like receptor (TLR) 2 and TLR8 cooperativity and TLR8-mediated induction of IFN. Proc. Natl. Acad. Sci. U. S. A. 108, 3683–3688.
Cervantes, J.L., Weinerman, B., Basole, C., Salazar, J.C., 2012. TLR8: the forgotten relative revindicated. Cell. Mol. Immunol. 9, 434–438.
Chamilos, G., Gregorio, J., Meller, S., et al., 2012. Cytosolic sensing of extracellular self-DNA transported into monocytes by the antimicrobial peptide LL37. Blood 120, 3699–3707.
Chen, G.Y., Nuñez, G., 2010. Sterile inflammation: sensing and reacting to damage. Nat. Rev. Immunol. 10, 826–837.
Chen, G.Y., Núñez, G., 2011. Inflammasomes in intestinal inflammation and cancer. Gastroenterology 141, 1986–1999.
Chen, C., Feng, Y., Zou, L., et al., 2014. Role of extracellular RNA and TLR3-Trif signaling in myocardial ischemia-reperfusion injury. J. Am. Heart Assoc. 3, e000683.

Chen, Q., Sun, L.J., Chen, Z.J.J., 2016. Regulation and function of the cGAS-STING pathway of cytosolic DNA sensing. Nat. Immunol. 17, 1142–1149.
Chen, D., Cao, Y., Li, H., et al., 2017. Extracellular ATP elicits DORN1-mediated RBOHD phosphorylation to regulate stomatal aperture. Nat. Commun. 8, 2265.
Chi, Z., Wang, Z., Wang, K., Zhu, Y., Qin, S., 2014. Cathelicidin antimicrobial peptide LL-37 in cholesteatoma enables keratinocyte reactivity with cytosolic DNA. Scand. J. Immunol. 79, 214–221.
Chivasa, S., Murphy, A.M., Hamilton, J.M., Lindsey, K., Carr, J.P., Slabas, A.R., 2009. Extracellular ATP is a regulator of pathogen defence in plants. Plant J. 60, 436–448.
Choi, H.W., Klessig, D.F., 2016. DAMPs, MAMPs, and NAMPs in plant innate immunity. BMC Plant Biol. 16, 232.
Choi, J., Tanaka, K., Cao, Y., et al., 2014. Identification of a plant receptor for extracellular ATP. Science 343, 290–294.
Choi, H.W., Manohar, M., Manosalva, P., Tian, M.Y., Moreau, M., Klessig, D.F., 2016. Activation of plant innate immunity by extracellular high mobility group box 3 and its inhibition by salicylic acid. PLoS Pathog. 12, 1005518.
Chow, K.T., Gale, M., Loo, Y.-M., 2018. RIG-I and other RNA sensors in antiviral immunity. Annu. Rev. Immunol. 36, 667–694.
Cipollini, D., Heil, M., 2010. Costs and benefits of induced resistance to herbivores and pathogens in plants. CAB Rev. Perspect. Agric. Vet. Sci. Nutr. Nat. Resour. 10, 005.
Civril, F., Deimling, T., de Oliveira Mann, C.C., Ablasser, A., Moldt, M., 2013. Structural mechanism of cytosolic DNA sensing by cGAS. Nature 498, 332–337.
Clark, G.B., Morgan, R.O., Fernandez, M.P., Salmi, M.L., Roux, S.J., 2014. Breakthroughs spotlighting roles for extracellular nucleotides and apyrases in stress responses and growth and development. Plant Sci. 225, 107–116.
Corrales, L., Woo, S.R., Williams, J.B., McWhirter, S.M., Dubensky, T.W., Gajewski, T.F., 2016. Antagonism of the STING pathway via activation of the AIM2 inflammasome by intracellular DNA. J. Immunol. 196, 3191–3198.
Cossarizza, A., Pinti, M., Nasi, M., et al., 2011. Increased plasma levels of extracellular mitochondrial DNA during HIV infection: a new role for mitochondrial damage-associated molecular patterns during inflammation. Mitochondrion 11, 750–755.
Couto, D., Zipfel, C., 2016. Regulation of pattern recognition receptor signalling in plants. Nat. Rev. Immunol. 16, 537–552.
Croset, A., Cordelières, F.P., Berthault, N., et al., 2013. Inhibition of DNA damage repair by artificial activation of PARP with siDNA. Nucleic Acids Res. 41, 7344–7355.
Cross, S.H., Lee, M., Clark, V.H., Craig, J.M., Bird, A.P., Bickmore, W.A., 1997. The chromosomal distribution of CpG islands in the mouse: evidence for genome scrambling in the rodent lineage. Genomics 40, 454–461.
Crowl, J.T., Gray, E.E., Pestal, K., Volkman, H.E., Stetson, D.B., 2017. Intracellular nucleic acid detection in autoimmunity. Annu. Rev. Immunol. 35, 313–336.
da Silva de Faria, I.J., Proveti Olmo, R., Guimarães Silva, E., Trindade Marques, J., 2013. dsRNA sensing during viral infection: lessons from plants, worms, insects, and mammals. J. Interferon Cytokine Res. 33, 239–253.
da Silva, H.B., Beura, L.K., Wang, H.G., et al., 2018. The purinergic receptor P2RX7 directs metabolic fitness of long-lived memory CD8(+) T cells. Nature 559, 264–268.
Davila, S., Hibberd, M.L., Dass, R.H., et al., 2008. Genetic association and expression studies indicate a role of Toll-like receptor 8 in pulmonary tuberculosis. PLoS Genet. 4, e1000218.
Davis, S., Watson, J.C., 1996. In vitro activation of the interferon-induced, double-stranded RNA-dependent protein kinase PKR by RNA from the 3′ untranslated regions of human alpha-tropomyosin. Proc. Natl. Acad. Sci. U. S. A. 93, 508–513.

De Lorenzo, G., Brutus, A., Savatin, D.V., Sicilia, F., Cervone, F., 2011. Engineering plant resistance by constructing chimeric receptors that recognize damage-associated molecular patterns (DAMPs). FEBS Lett. 585, 1521–1528.

Deguine, J., 2017. Common Senses. Trends Immunol. 38, 691–692.

Demidchik, V., Nichols, C., Oliynyk, M., Dark, A., Glover, B.J., Davies, J.M., 2003. Is ATP a signaling agent in plants? Plant Physiol. 133, 456–461.

Demidchik, V., Shang, Z.L., Shin, R., et al., 2009. Plant extracellular ATP signalling by plasma membrane NADPH oxidase and Ca2+ channels. Plant J. 58, 903–913.

Desmet, C.J., Ishii, K.J., 2012. Nucleic acid sensing at the interface between innate and adaptive immunity in vaccination. Nat. Rev. Immunol. 12, 479–491.

Dhanwani, R., Takahashi, M., Sharma, S., 2018. Cytosolic sensing of immuno-stimulatory DNA, the enemy within. Curr. Opin. Immunol. 50, 82–87.

Di Micco, A., Frera, G., Lugrin, J., et al., 2016. AIM2 inflammasome is activated by pharmacological disruption of nuclear envelope integrity. Proc. Natl. Acad. Sci. U. S. A. 113, E4671–E4680.

Di Virgilio, F., Dal Ben, D., Sarti, A.C., Giuliani, A.L., Falzoni, S., 2017. The P2X7 receptor in infection and inflammation. Immunity 47, 15–31.

Dib, B., Lin, H.J., Maidana, D.E., et al., 2015. Mitochondrial DNA has a pro-inflammatory role in AMD. BBA Mol. Cell. Res. 1853, 2897–2906.

Diner, E.J., Burdette, D.L., Wilson, S.C., Monroe, K.M., Kellenberger, C.A., 2013. The innate immune DNA sensor cGAS produces a noncanonical cyclic dinucleotide that activates human STING. Cell Rep. 3, 1355–1361.

Ding, Q.F., Tan, K.S., 2016. The danger signal extracellular ATP is an inducer of *Fusobacterium nucleatum* biofilm dispersal. Front. Cell. Infect. Microbiol. 6, 155.

Doitsh, G., Greene, W.C., 2016. Dissecting how CD4 T cells are lost during HIV infection. Cell Host Microbe 19, 280–291.

Döring, Y., Manthey, H.D., Drechsler, M., et al., 2012. Auto-antigenic protein-DNA complexes stimulate plasmacytoid dendritic cells to promote atherosclerosis. Circulation 125, 1673–1683.

Du, M., Chen, Z.J., 2018. DNA-induced liquid phase condensation of cGAS activates innate immune signaling. Science 361, 704–709. publ. online 5 July 2018. https://doi.org/10.1126/science.aat1022.

Duran-Flores, D., Heil, M., 2015. Growth inhibition by self-DNA: a phenomenon and its multiple explanations. New Phytol. 207, 482–485.

Duran-Flores, D., Heil, M., 2016. Sources of specificity in plant damaged-self recognition. Curr. Opin. Plant Biol. 32, 77–87.

Duran-Flores, D., Heil, M., 2018. Extracellular self-DNA as a damage-associated molecular pattern (DAMP) that triggers self-specific immunity induction in plants. Brain Behav. Immun. 72, 78–88.

Eberle, F., Sirin, M., Binder, M., Dalpke, A.H., 2009. Bacterial RNA is recognized by different sets of immunoreceptors. Eur. J. Immunol. 39, 2537–2547.

Faas, M.M., Sáez, T., de Vos, P., 2017. Extracellular ATP and adenosine: the Yin and Yang in immune responses? Mol. Aspects Med. 55, 9–19.

Farrugia, M., Baron, B., 2017. The role of Toll-like receptors in autoimmune diseases through failure of the self-recognition mechanism. Int. J. Inflamm. 8391230.

Feng, Q., Hato, S.V., Langereis, M.A., et al., 2012. MDA5 detects the double-stranded RNA replicative form in picornavirus-infected cells. Cell Rep. 2, 1187–1196.

Feng, H., Guan, D., Bai, J., Sun, K., Jia, L., 2015. Extracellular ATP: a potential regulator of plant cell death. Mol. Plant Pathol. 16, 633–639.

Fenyk, S., Dixon, C.H., Gittens, W.H., et al., 2016. The tomato nucleotide-binding leucine-rich repeat immune receptor I-2 couples DNA-binding to nucleotide-binding domain nucleotide exchange. J. Biol. Chem. 291, 1137–1147.

Fernandes-Alnemri, T., Yu, J.-W., Datta, P., Wu, J., Alnemri, E.S., 2009. AIM2 activates the inflammasome and cell death in response to cytoplasmic DNA. Nature 458, 509–513.

Ferrari, S., Savatin, D., Sicilia, F., Gramegna, G., Cervone, F., De Lorenzo, G., 2013. Oligogalacturonides: plant damage-associated molecular patterns and regulators of growth and development. Front. Plant Sci. 4, 49.

Flury, P., Klauser, D., Schulze, B., Boller, T., Bartels, S., 2013. The anticipation of danger: microbe-associated molecular pattern perception enhances AtPep-triggered oxidative burst. Plant Physiol. 161, 2023–2035.

Franceschi, C., Bonafe, M., Valensin, S., et al., 2000. Inflamm-aging—an evolutionary perspective on immunosenescence. In: Toussaint, O., Osiewacz, H.D., Lithgow, G.J., Brack, C. (Eds.), Molecular and Cellular Gerontology. In: vol. 908. New York Acad Sciences, New York, pp. 244–254.

Franklin, B.S., Latz, E., Schmidt, F.I., 2018. The intra- and extracellular functions of ASC specks. Immunol. Rev. 281, 74–87.

Fu, Y., Comella, N., Tognazzi, K., Brown, L.F., Dvorak, H.F., Kocher, O., 1999. Cloning of DLM-1, a novel gene that is up-regulated in activated macrophages, using RNA differential display. Gene 240, 157–163.

Gack, M.U., Shin, Y.C., Joo, C.H., et al., 2007. TRIM25 RING-finger E3 ubiquitin ligase is essential for RIG-I-mediated antiviral activity. Nature 446, 916–920.

Gall, A., Treuting, P., Elkon, K.B., et al., 2012. Autoimmunity initiates in nonhematopoietic cells and progresses via lymphocytes in an interferon-dependent autoimmune disease. Immunity 36, 120–131.

Gallo, P.M., Gallucci, S., 2013. The dendritic cell response to classic, emerging, and homeostatic danger signals. Implications for autoimmunity. Front. Immunol. 4, 138.

Gallucci, S., Maffei, M., 2017. DNA sensing across the tree of life. Trends Immunol. 38, 719–732.

Gallucci, S., Matzinger, P., 2001. Danger signals: SOS to the immune system. Curr. Opin. Immunol. 13, 114–119.

Galluzzi, L., Vanpouille-Box, C., Bakhoum, S.F., Demaria, S., 2018a. SnapShot: cGAS-STING signaling. Cell 173, 276.

Galluzzi, L., Vitale, I., Aaronson, S.A., et al., 2018b. Molecular mechanisms of cell death: recommendations of the nomenclature committee on cell death 2018. Cell Death Differ. 25, 486–541.

Gantier, M.P., 2017. Length does matter for cGAS. EMBO Rep. 18, 1675–1676.

Gantier, M.P., Irving, A.T., Kaparakis-Liaskos, M., et al., 2010. Genetic modulation of TLR8 response following bacterial phagocytosis. Hum. Mutat. 31, 1069–1079.

Gao, D., Wu, J., Wu, Y.-T., et al., 2013. Cyclic GMP-AMP synthase is an innate immune sensor of HIV and other retroviruses. Science 341, 903–906.

Gao, D.X., Li, T., Li, X.D., et al., 2015. Activation of cyclic GMP-AMP synthase by self-DNA causes autoimmune diseases. Proc. Natl. Acad. Sci. U. S. A. 112, E5699–E5705.

Gardella, S., Andrei, C., Ferrera, D., et al., 2002. The nuclear protein HMGB1 is secreted by monocytes via a non-classical, vesicle-mediated secretory pathway. EMBO Rep. 3, 995–1001.

Garg, A.D., Galluzzi, L., Apetoh, L., et al., 2015. Molecular and translational classifications of DAMPs in immunogenic cell death. Front. Immunol. 6, 588.

Giglia-Mari, G., Zotter, A., Vermeulen, W., 2011. DNA damage response. Cold Spring Harb. Perspect. Biol. 3, a000745.

Gilliet, M., Cao, W., Liu, Y.-J., 2008. Plasmacytoid dendritic cells: sensing nucleic acids in viral infection and autoimmune diseases. Nat. Rev. Immunol. 8, 594–606.

Gitlin, L., Barchet, W., Gilfillan, S., et al., 2006. Essential role of mda-5 in type I IFN responses to polyriboinosinic: polyribocytidylic acid and encephalomyocarditis picornavirus. Proc. Natl. Acad. Sci. U. S. A. 103, 8459–8464.

Goldfarb, I.T., Adeli, S., Berk, T., Phillippe, M., 2018. Fetal and placental DNA stimulation of TLR9: a mechanism possibly contributing to the pro-inflammatory events during parturition. Reprod. Sci. 25, 788–796.

Golub, E.S., Green, D.R., 1991. Immunology: A Synthesis. Sunderland, MA, Sinauer.

Gombault, A., Baron, L., Couilllin, I., 2013. ATP release and purinergic signaling in NLRP3 inflammasome activation. Front. Immunol. 3, 414.

Gougeon, M.L., 2017. Alarmins and central nervous system inflammation in HIV-associated neurological disorders. J. Intern. Med. 281, 433–447.

Graziano, F., Desdouits, M., Garzetti, L., et al., 2015. Extracellular ATP induces the rapid release of HIV-1 from virus containing compartments of human macrophages. Proc. Natl. Acad. Sci. U. S. A. 112, E3265–E3273.

Gürtler, C., Bowie, A.G., 2013. Innate immune detection of microbial nucleic acids. Trends Microbiol. 21, 413–420.

Gust, A.A., Pruitt, R., Nürnberger, T., 2017. Sensing danger: key to activating plant immunity. Trends Plant Sci. 22, 779–791.

Hadwiger, L.A., 2015. Nonhost resistance: self-inflicted DNA damage by fungal DNase accumulation is a major factor in terminating fungal growth in the pea-*Fusarium solani* f.sp *phaseoli* interaction. Physiol. Mol. Plant Pathol. 92, 79–87.

Hadwiger, L.A., Schwochau, M.E., 1970a. Role of dyes and other DNA intercalating compounds in activating genes for phytoalexin production. Phytopathology 60, 1294.

Hadwiger, L.A., Schwochau, W.E., 1970b. Induction of phenylalanine ammonia lyase and pisatin in pea pods by poly-lysine, spermidine or histone fractions. Biochem. Biophys. Res. Commun. 38, 683–691.

Hadwiger, L.A., Schwochau, M.E., 1971a. Specificity of deoxyribonucleic acid intercalating compounds in the control of phenylalanine ammonia lyase and pisatin levels. Plant Physiol. 47, 346–351.

Hadwiger, L.A., Schwochau, M.E., 1971b. Ultraviolet light-induced formation of pisatin and phenylalanine ammonia lyase. Plant Physiol. 47, 588–590.

Hadwiger, L.A., Tanaka, K., 2017. Non-host resistance: DNA damage is associated with SA signaling for induction of PR genes and contributes to the growth suppression of a pea pathogen on pea endocarp tissue. Front. Plant Sci. 8, 446.

Hadwiger, L.A., Tanaka, K., 2018. DNA damage and chromatin conformation changes confer nonhost resistance: a hypothesis based on effects of anti-cancer agents on plant defense responses. Front. Plant Sci. 9, 1056.

Hancock, R.E.W., Haney, E.F., Gill, E.E., 2016. The immunology of host defence peptides: beyond antimicrobial activity. Nat. Rev. Immunol. 16, 321–334.

Haneklaus, M., O'Neill, L.A.J., 2015. NLRP3 at the interface of metabolism and inflammation. Immunol. Rev. 265, 53–62.

Hansen, K., Prabakaran, T., Laustsen, A., et al., 2014. Listeria monocytogenes induces IFNβ expression through an IFI16-, cGAS- and STING-dependent pathway. EMBO J. 33, 1654–1666.

Hartmann, G., 2017. Nucleic acid immunity. In: Alt, F.W. (Ed.), Advances in Immunology. In: vol. 133. Academic Press, Burlington, pp. 121–169. Chapter Four.

Hawes, M.C., Curlango-Rivera, G., Wen, F., White, G.J., VanEtten, H.D., Xiong, Z., 2011. Extracellular DNA: the tip of root defenses? Plant Sci. 180, 741–745.

Hawes, M.C., Wen, F., Elquza, E., 2015. Extracellular DNA: a bridge to cancer. Cancer Res. 75, 4260–4264.

He, Y., Hara, H., Nunez, G., 2016. Mechanism and regulation of NLRP3 inflammasome activation. Trends Biochem. Sci. 41, 1012–1021.

Heil, M., 2009. Damaged-self recognition in plant herbivore defence. Trends Plant Sci. 14, 356–363.

Heil, M., 2015. Extrafloral nectar at the plant-insect interface: a spotlight on chemical ecology, phenotypic plasticity, and food webs. Annu. Rev. Entomol. 60, 213–232.
Heil, M., Baldwin, I.T., 2002. Fitness costs of induced resistance: emerging experimental support for a slippery concept. Trends Plant Sci. 7, 61–67.
Heil, M., Land, W.G., 2014. Danger signals—damaged-self recognition across the tree of life. Front. Plant Sci. 5, 579.
Heil, M., Ibarra-Laclette, E., Adame-Álvarez, R.M., et al., 2012. How plants sense wounds: damaged-self recognition is based on plant-derived elicitors and induces octadecanoid signaling. PLoS One 7, e30537.
Heil, M., Land, W., Tör, M., 2016. Editorial: wound recognition across the tree of life. Front. Plant Sci. 7, 1319.
Hemmi, H., Takeuchi, O., Kawai, T., et al., 2000. A Toll-like receptor recognizes bacterial DNA. Nature 408, 740–745.
Hemmi, H., Kaisho, T., Takeuchi, O., et al., 2002. Small anti-viral compounds activate immune cells via the TLR7 MyD88-dependent signaling pathway. Nat. Immunol. 3, 196–200.
Hernández-Oñate, M.A., Herrera-Estrella, A., 2015. Damage response involves mechanisms conserved across plants, animals and fungi. Curr. Genet. 61, 359–372.
Herzner, A.-M., Hagmann, C.A., Goldeck, M., et al., 2015. Sequence-specific activation of the DNA sensor cGAS by Y-form DNA structures as found in primary HIV-1 cDNA. Nat. Immunol. 16, 1025–1033.
Hett, E.C., Slater, L.H., Mark, K.G., et al., 2013. Chemical genetics reveals a kinase-independent role for protein kinase R in pyroptosis. Nat. Chem. Biol. 9, 398–405.
Hornung, V., Ellegast, J., Kim, S., Brzózka, K., Jung, A., 2006. 5′-triphosphate RNA is the ligand for RIG-I. Science 314, 994–997.
Hornung, V., Ablasser, A., Charrel-Dennis, M., Bauernfeind, F., Horvath, G., 2009. AIM2 recognizes cytosolic dsDNA and forms a caspase-1-activating inflammasome with ASC. Nature 458, 514–518.
Hornung, V., Hartmann, R., Ablasser, A., Hopfner, K.-P., 2014. OAS proteins and cGAS: unifying concepts in sensing and responding to cytosolic nucleic acids. Nat. Rev. Immunol. 14, 521–528.
Hotamisligil, G.S., 2017. Inflammation, metaflammation and immunometabolic disorders. Nature 542, 177–185.
Huang, H., Evankovich, J., Yan, W., et al., 2011. Endogenous histones function as alarmins in sterile inflammatory liver injury through Toll-like receptor 9 in mice. Hepatology 54, 999–1008.
Hudson, B.I., Lippman, M.E., 2018. Targeting RAGE signaling in inflammatory disease. Annu. Rev. Med. 69, 349–364.
Ingelsson, B., Soderberg, D., Strid, T., et al., 2018. Lymphocytes eject interferogenic mitochondrial DNA webs in response to CpG and non-CpG oligodeoxynucleotides of class C. Proc. Natl. Acad. Sci. U. S. A. 115, E478–E487.
Ioannidou, A., Goulielmaki, E., Garinis, G.A., 2016. DNA damage: from chronic inflammation to age-related deterioration. Front. Genet. 7, 187.
Isaacs, A., Rotem, Z., Cox, R.A., 1963. Foreign nucleic acids as stimulus to make interferon. Lancet 2, 113–116.
Ishii, K.J., Suzuki, K., Coban, C., et al., 2001. Genomic DNA released by dying cells induces the maturation of APCs. J. Immunol. 167, 2602–2607.
Ishii, K.J., Kawagoe, T., Koyama, S., et al., 2008. TANK-binding kinase-1 delineates innate and adaptive immune responses to DNA vaccines. Nature 451, 725–729.
Ivison, S.M., Himmel, M.E., Mayer, M., et al., 2011. The stress signal extracellular ATP modulates antiflagellin immune responses in intestinal epithelial cells. Inflamm. Bowel Dis. 17, 319–333.

Jakobs, C., Perner, S., Hornung, V., 2015. AIM2 drives joint inflammation in a self-DNA triggered model of chronic polyarthritis. PLoS One 10, e0131702.

Janeway, C.A., 1992. The immune system evolved to discriminate infectious nonself from noninfectious self. Immunol. Today 13, 11–16.

Jin, T.C., Perry, A., Jiang, J.S., et al., 2012. Structures of the HIN domain: DNA complexes reveal ligand binding and activation mechanisms of the AIM2 inflammasome and IFI16 receptor. Immunity 36, 561–571.

Jones, J.D.G., Dangl, J.L., 2006. The plant immune system. Nature 444, 323–329.

Jones, J.W., Kayagaki, N., Broz, P., et al., 2010. Absent in melanoma 2 is required for innate immune recognition of *Francisella tularensis*. Proc. Natl. Acad. Sci. U. S. A. 107, 9771–9776.

Jones, J.D.G., Vance, R.E., Dangl, J.L., 2016. Intracellular innate immune surveillance devices in plants and animals. Science 354, aaf6395.

Jung, S., Von Thülen, T., Laukemper, V., et al., 2015. A single naturally occurring 2′-O-methylation converts a TLR7- and TLR8-activating RNA into a TLR8-specific ligand. PLoS One 10, e0120498.

Kaczmarek, A., Vandenabeele, P., Krysko, D.V., 2013. Necroptosis: the release of damage-associated molecular patterns and its physiological relevance. Immunity 38, 209–223.

Kalogeris, T., Baines, C.P., Krenz, M., Korthuis, R.J., 2012. Cell biology of ischemia/reperfusion injury. Int. Rev. Cell Mol. Biol. 298, 229–317.

Kang, R., Chen, R., Zhang, Q., et al., 2014. HMGB1 in health and disease. Mol. Aspects Med. 40, 1–116.

Karayel, E., Burckstummer, T., Bilban, M., et al., 2009. The TLR-independent DNA recognition pathway in murine macrophages: ligand features and molecular signature. Eur. J. Immunol. 39, 1929–1936.

Kato, H., Takeuchi, O., Sato, S., et al., 2006. Differential roles of MDA5 and RIG-I helicases in the recognition of RNA viruses. Nature 441, 101–105.

Kato, H., Takeuchi, O., Mikamo-Satoh, E., et al., 2008. Length-dependent recognition of double-stranded ribonucleic acids by retinoic acid-inducible gene-I and melanoma differentiation-associated gene 5. J. Exp. Med. 205, 1601–1610.

Kato, K., Omura, H., Ishitani, R., Nureki, O., 2017. Cyclic GMP–AMP as an endogenous second messenger in innate immune signaling by cytosolic DNA. Annu. Rev. Biochem. 86, 541–566.

Kawate, T., Michel, J.C., Birdsong, W.T., Gouaux, E., 2009. Crystal structure of the ATP-gated $P2X_4$ ion channel in the closed state. Nature 460, 592–598.

Kierdorf, K., Fritz, G., 2013. RAGE regulation and signaling in inflammation and beyond. J. Leukoc. Biol. 94, 55–68.

Kim, S.Y., Sivaguru, M., Stacey, G., 2006. Extracellular ATP in plants. Visualization, localization, and analysis of physiological significance in growth and signaling. Plant Physiol. 142, 984–992.

King, K.R., Aguirre, A.D., Ye, Y.X., et al., 2017. IRF3 and type I interferons fuel a fatal response to myocardial infarction. Nat. Med. 23, 1481–1487.

Kirschning, C.J., Bauer, S., 2001. Toll-like receptors: cellular signal transducers for exogenous molecular patterns causing immune responses. Int. J. Med. Microbiol. 291, 251–260.

Klauser, D., Desurmont, G.A., Glauser, G., et al., 2015. The *Arabidopsis* pep-PEPR system is induced by herbivore feeding and contributes to JA-mediated plant defence against herbivory. J. Exp. Bot. 66, 5327–5336.

Klosterman, S.J., Choi, J.J., Hadwiger, L.A., 2003. Analysis of pea HMG-I/Y expression suggests a role in defence gene regulation. Mol. Plant Pathol. 4, 249–258.

Kolb, J.P., Oguin, T.H., Oberst, A., Martinez, J., 2017. Programmed cell death and inflammation: winter is coming. Trends Immunol. 38, 705–718.

Komada, T., Chung, H., Lau, A., et al., 2018. Macrophage uptake of necrotic cell DNA activates the AIM2 inflammasome to regulate a proinflammatory phenotype in CKD. J. Am. Soc. Nephrol. 29, 1165–1181.

Krapp, C., Jønsson, K., Jakobsen, M.R., 2018. STING dependent sensing—does HIV actually care? Cytokine Growth Factor Rev. 40, 68–76.

Kretschmer, S., Lee-Kirsch, M.A., 2017. Type I interferon-mediated autoinflammation and autoimmunity. Curr. Opin. Immunol. 49, 96–102.

Krieg, A.M., 2002. CpG motifs in bacterial DNA and their immune effects. Annu. Rev. Immunol. 20, 709–760.

Krysko, D.V., Garg, A.D., Kaczmarek, A., Krysko, O., Agostinis, P., Vandenabeele, P., 2012. Immunogenic cell death and DAMPs in cancer therapy. Nat. Rev. Cancer 12, 860–875.

Kumar, A., Zhang, J., Yu, F.-S.X., 2006. Toll-like receptor 3 agonist poly(I:C)-induced antiviral response in human corneal epithelial cells. Immunology 117, 11–21.

Kumar, V., Fleming, T., Terjung, S., et al., 2017. Homeostatic nuclear RAGE-ATM interaction is essential for efficient DNA repair. Nucleic Acids Res. 45, 10595–10613.

Kuriakose, T., Kanneganti, T.D., 2018. ZBP1: innate sensor regulating cell death and inflammation. Trends Immunol. 39, 123–134.

Kuriakose, T., Man, S.M., Malireddi, R.K.S., et al., 2016. ZBP1/DAI is an innate sensor of influenza virus triggering the NLRP3 inflammasome and programmed cell death pathways. Sci. Immunol. 1, aag2045.

Lamkanfi, M., Sarkar, A., Vande Walle, L., et al., 2010. Inflammasome-dependent release of the alarmin HMGB1 in endotoxemia. J. Immunol. 185, 4385–4392.

Lamphier, M.S., Sirois, C.M., Verma, A., Golenbock, D.T., Latz, E., 2006. TLR9 and the recognition of self and non-self nucleic acids. Ann. N. Y. Acad. Sci. 1082, 31–43.

Lan, Y.Y., Londono, D., Bouley, R., Rooney, M.S., Hacohen, N., 2014. Dnase2a deficiency uncovers lysosomal clearance of damaged nuclear DNA via autophagy. Cell Rep. 9, 180–192.

Land, W.G., Agostinis, P., Gasser, S., Garg, A.D., Linkermann, A., 2016a. DAMP-induced allograft and tumor rejection: the circle is closing. Am. J. Transplant. 16, 3322–3337.

Land, W.G., Agostinis, P., Gasser, S., Garg, A.D., Linkermann, A., 2016b. Transplantation and damage-associated molecular patterns (DAMPs). Am. J. Transplant. 16, 3338–3361.

Lande, R., Gregorio, J., Facchinetti, V., et al., 2007. Plasmacytoid dendritic cells sense self-DNA coupled with antimicrobial peptide. Nature 449, 564–569.

Lande, R., Chamilos, G., Ganguly, D., et al., 2015. Cationic antimicrobial peptides in psoriatic skin cooperate to break innate tolerance to self-DNA. Eur. J. Immunol. 45, 203–213.

Lässig, C., Hopfner, K.-P., 2017. Discrimination of cytosolic self and non-self RNA by RIG-I-like receptors. J. Biol. Chem. 292, 9000–9009.

Lau, C.M., Broughton, C., Tabor, A.S., et al., 2005. RNA-associated autoantigens activate B cells by combined B cell antigen receptor/Toll-like receptor 7 engagement. J. Exp. Med. 202, 1171–1177.

Lavine, K.J., Mann, D.L., 2017. Recognition of self-DNA drives cardiac inflammation: why broken hearts fail. Nat. Med. 23, 4455.

Leadbetter, E.A., Rifkin, I.R., Hohlbaum, A.M., Beaudette, B.C., Shlomchik, M.J., Marshak-Rothstein, A., 2002. Chromatin–IgG complexes activate B cells by dual engagement of IgM and Toll-like receptors. Nature 416, 603–607.

Lee, E.J., Park, J.H., 2013. Receptor for advanced glycation endproducts (RAGE), its ligands, and soluble RAGE: potential biomarkers for diagnosis and therapeutic targets for human renal diseases. Genomics Inform. 11, 224–229.

Lee, T.H., D'Asti, E., Magnus, N., Al-Nedawi, K., Meehan, B., Rak, J., 2011. Microvesicles as mediators of intercellular communication in cancer-the emerging science of cellular 'debris'. Semin. Immunopathol. 33, 455–467.

Lee, S.M.Y., Kok, K.-H., Jaume, M., et al., 2014. Toll-like receptor 10 is involved in induction of innate immune responses to influenza virus infection. Proc. Natl. Acad. Sci. U. S. A. 111, 3793–3798.

Lee, B., Park, Y.S., Lee, S., Song, G.C., Ryu, C.M., 2016. Bacterial RNAs activate innate immunity in Arabidopsis. New Phytol. 209, 785–797.

Lee, E.K., Chung, K.W., Kim, Y.R., et al., 2017. Small RNAs induce the activation of the pro-inflammatory TLR7 signaling pathway in aged rat kidney. Aging Cell 16, 1026–1034.

Lee, S.M.Y., Yip, T.F., Yan, S., et al., 2018. Recognition of double-stranded RNA and regulation of interferon pathway by toll-like receptor 10. Front. Immunol. 9, 00516.

Lee-Kirsch, M.A., Wolf, C., Kretschmer, S., Roers, A., 2015. Type I interferonopathies-an expanding disease spectrum of immunodysregulation. Semin. Immunopathol. 37, 349–357.

Lemaire, P.A., Anderson, E., Lary, J., Cole, J.L., 2008. Mechanism of PKR activation by dsRNA. J. Mol. Biol. 381, 351–360.

Lew, R.R., Dearnaley, J.D.W., 2000. Extracellular nucleotide effects on the electrical properties of growing *Arabidopsis thaliana* root hairs. Plant Sci. 153, 1–6.

Li, T., Chen, Z.J., 2018. The cGAS–cGAMP–STING pathway connects DNA damage to inflammation, senescence, and cancer. J. Exp. Med. 215, 1287–1299.

Li, X.-D., Wu, J., Gao, D., Wang, H., Sun, L., Chen, Z.J., 2013. Pivotal roles of cGAS-cGAMP signaling in antiviral defense and immune adjuvant effects. Science 341, 1390–1394.

Li, Y., Wilson, H.L., Kiss-Toth, E., 2017. Regulating STING in health and disease. J. Inflamm. 14, 11.

Lian, Q., Xu, J., Yan, S., et al., 2017. Chemotherapy-induced intestinal inflammatory responses are mediated by exosome secretion of double-strand DNA via AIM2 inflammasome activation. Cell Res. 27, 784–800.

Lim, M.H., Wu, J., Clark, G., Roux, S.J., 2014. Apyrases (NTPDases) and extracellular nucleotides regulate plant responses to biotic and abiotic stresses. Purinergic Signal 10, 712.

Lin, J., Kumari, S., Kim, C., et al., 2016. RIPK1 counteracts ZBP1-mediated necroptosis to inhibit inflammation. Nature 540, 124–128.

Liston, A., Masters, S.L., 2017. Homeostasis-altering molecular processes as mechanisms of inflammasome activation. Nat. Rev. Immunol. 17, 208–214.

Liu, X., Wang, C., 2016. The emerging roles of the STING adaptor protein in immunity and diseases. Immunology 147, 285–291.

Liu, L., Yang, M., Kang, R., et al., 2014. HMGB1–DNA complex-induced autophagy limits AIM2 inflammasome activation through RAGE. Biochem. Biophys. Res. Commun. 450, 851–856.

Liu, S., Feng, M., Guan, W.X., 2016. Mitochondrial DNA sensing by STING signaling participates in inflammation, cancer and beyond. Int. J. Cancer 139, 736–741.

Lotze, M.T., Zeh, H.J., Rubartelli, A., et al., 2007. The grateful dead: damage-associated molecular pattern molecules and reduction/oxidation regulate immunity. Immunol. Rev. 220, 60–81.

Lövgren, T., Eloranta, M.L., Bave, U., Alm, G.V., Ronnblom, L., 2004. Induction of interferon-a production in plasmacytoid dendritic cells by immune complexes containing nucleic acid released by necrotic or late anoptotic cells and lupus IgG. Arthritis Rheum. 50, 1861–1872.

Lu, B., Nakamura, T., Inouye, K., et al., 2012. Novel role of PKR in inflammasome activation and HMGB1 release. Nature 488, 670–674.

Luecke, S., Paludan, S.R., 2017. Molecular requirements for sensing of intracellular microbial nucleic acids by the innate immune system. Cytokine 98, 4–14.

Luecke, S., Holleufer, A., Christensen, M.H., et al., 2017. cGAS is activated by DNA in a length-dependent manner. EMBO Rep. 18, 1707–1715.

Lugrin, J., Martinon, F., 2018. The AIM2 inflammasome: sensor of pathogens and cellular perturbations. Immunol. Rev. 281, 99–114.

Lupfer, C., Anand, P.K., 2016. Integrating inflammasome signaling in sexually transmitted infections. Trends Immunol. 37, 703–714.

Maelfait, J., Liverpool, L., Bridgeman, A., Ragan, K.B., Upton, J.W., Rehwinkel, J., 2017. Sensing of viral and endogenous RNA by ZBP1/DAI induces necroptosis. EMBO J. 36, 2529–2543.

Magna, M., Pisetsky, D.S., 2014. The role of HMGB1 in the pathogenesis of inflammatory and autoimmune diseases. Mol. Med. 20, 138–146.

Magna, M., Pisetsky, D.S., 2016. The alarmin properties of DNA and DNA-associated nuclear proteins. Clin. Ther. 38, 1029–1041.

Man, S.M., Karki, R., Kanneganti, T.D., 2016. AIM2 inflammasome in infection, cancer, and autoimmunity: role in DNA sensing, inflammation, and innate immunity. Eur. J. Immunol. 46, 269–280.

Mancuso, G., Gambuzza, M., Midiri, A., et al., 2009. Bacterial recognition by TLR7 in the lysosomes of conventional dendritic cells. Nat. Immunol. 10, 587–594.

Margolis, S.R., Wilson, S.C., Vance, R.E., 2017. Evolutionary origins of cGAS-STING signaling. Trends Immunol. 38, 733–743.

Martinez-Medina, A., Flors, V., Heil, M., et al., 2016. Recognizing plant defense priming. Trends Plant Sci. 21, 818–822.

Martinon, F., 2012. Dangerous liaisons: mitochondrial DNA meets the NLRP3 inflammasome. Immunity 36, 313–315.

Mathur, A., Hayward, J.A., Man, S.M., 2018. Molecular mechanisms of inflammasome signaling. J. Leukoc. Biol. 103, 233–257.

Matzinger, P., 1994. Tolerance, danger, and the extended family. Annu. Rev. Immunol. 12, 991–1045.

Matzinger, P., 2002. The danger model: a renewed sense of self. Science 296, 301–305.

Mayo, C.B., Cole, J.L., 2017. Interaction of PKR with single-stranded RNA. Sci. Rep. 7, 3335.

Mazzoleni, S., Bonanomi, G., Incerti, G., et al., 2015a. Inhibitory and toxic effects of extracellular self-DNA in litter: a mechanism for negative plant–soil feedbacks? New Phytol. 205, 1195–1210.

Mazzoleni, S., Cartenì, F., Bonanomi, G., et al., 2015b. Inhibitory effects of extracellular self-DNA: a general biological process? New Phytol. 206, 127–132.

McCarthy, C.G., Wenceslau, C.F., Goulopoulou, S., et al., 2015. Circulating mitochondrial DNA and Toll-like receptor 9 are associated with vascular dysfunction in spontaneously hypertensive rats. Cardiovasc. Res. 107, 119–130.

McDonald, B., Pittman, K., Menezes, G.B., et al., 2010. Intravascular danger signals guide neutrophils to sites of sterile inflammation. Science 330, 362–366.

McGlasson, S.L., Semple, F., MacPherson, H., Gray, M., Davidson, D.J., Dorin, J.R., 2017. Human-defensin 3 increases the TLR9-dependent response to bacterial DNA. Eur. J. Immunol. 47, 658–664.

Medina-Castellanos, E., Esquivel-Naranjo, E.U., Heil, M., Herrera-Estrella, A., 2014. Extracellular ATP activates MAPK and ROS signaling during injury response in the fungus *Trichoderma atroviride*. Front. Plant Sci. 5, 659.

Meller, S., Di Domizio, J., Voo, K.S., et al., 2015. T(H)17 cells promote microbial killing and innate immune sensing of DNA via interleukin 26. Nat. Immunol. 16, 970–979.

Mertens, C., 2015. Chemical modifications of DNA activate the cGAS/STING signaling pathway even in the presence of the cytosolic exonuclease TREX1. Mathematisch-Naturwissenschaftliche Fakultät. Rheinische Friedrich-Wilhelms-Universität Bonn, Bonn.

Mills, E.L., Kelly, B., O'Neill, L.A.J., 2017. Mitochondria are the powerhouses of immunity. Nat. Immunol. 18, 488–498.
Mollerherm, H., von Kockritz-Blickwede, M., Branitzki-Heinemann, K., 2016. Antimicrobial activity of mast cells: role and relevance of extracellular DNA traps. Front. Immunol. 7, 265.
Moreno-García, M., Recio-Tótoro, B., Claudio-Piedras, F., Lanz-Mendoza, H., 2014. Injury and immune response: applying the danger theory to mosquitoes. Front. Plant Sci. 5, 451.
Mouchess, M.L., Arpaia, N., Souza, G., et al., 2011. Transmembrane mutations in Toll-like receptor 9 bypass the requirement for Ectodomain proteolysis and induce fatal inflammation. Immunity 35, 721–732.
Murshid, A., Gong, J., Ahmad, R., Borges, T.J., Calderwood, S.K., 2015. Scavenger receptor SREC-I promotes double stranded RNA-mediated TLR3 activation in human monocytes. Immunobiology 220, 823–832.
Muruve, D.A., Petrilli, V., Zaiss, A.K., et al., 2008. The inflammasome recognizes cytosolic microbial and host DNA and triggers an innate immune response. Nature 452, 103–108.
Nagashima, H., Iwatani, S., Cruz, M., et al., 2015. Toll-like receptor 10 in *Helicobacter pylori* infection. J. Infect. Dis. 212, 1666–1676.
Nakad, R., Schumacher, B., 2016. DNA damage response and immune defense: links and mechanisms. Front. Genet. 7, 147.
Nakagawa, Y., Gallo, R.L., 2015. Endogenous intracellular cathelicidin enhances TLR9 activation in dendritic cells and macrophages. J. Immunol. 194, 1274–1284.
Nakahira, K., Haspel, J.A., Rathinam, V.A.K., et al., 2010. Autophagy proteins regulate innate immune responses by inhibiting the release of mitochondrial DNA mediated by the NALP3 inflammasome. Nat. Immunol. 12, 222–230.
Nakahira, K., Hisata, S., Choi, A.M.K., 2015. The roles of mitochondrial damage-associated molecular patterns in diseases. Antioxid. Redox Signal. 23, 1329–1350.
Nakamura, T., Ito, T., Igarashi, H., et al., 2012. Cytosolic double-stranded DNA as a damage-associated molecular pattern induces the inflammatory response in rat pancreatic stellate cells: a plausible mechanism for tissue injury-associated pancreatitis. Int. J. Inflamm. 2012, 504128.
Naqvi, I., Gunaratne, R., McDade, J.E., et al., 2018. Polymer-mediated inhibition of pro-invasive nucleic acid DAMPs and microvesicles limits pancreatic cancer metastasis. Mol. Ther. 26, 1020–1031.
Nasi, M., De Biasi, S., Bianchini, E., et al., 2015. Analysis of inflammasomes and antiviral sensing components reveals decreased expression of NLRX1 in HIV-positive patients assuming efficient antiretroviral therapy. AIDS 29, 1937–1941.
Nellimarla, S., Mossman, K.L., 2014. Extracellular dsRNA: its function and mechanism of cellular uptake. J. Interferon Cytokine Res. 34, 419–426.
Ng, K.W., Marshall, E.A., Bell, J.C., Lam, W.L., 2018. cGAS–STING and cancer: dichotomous roles in tumor immunity and development. Trends Immunol. 39, 44–54.
Nguyen, D., Rieu, I., Mariani, C., van Dam, N.M., 2016. How plants handle multiple stresses: hormonal interactions underlying responses to abiotic stress and insect herbivory. Plant Mol. Biol. 91, 727–740.
Ni, X.M., Ru, H., Ma, F., et al., 2016. New insights into the structural basis of DNA recognition by HINa and HINb domains of IFI16. J. Mol. Cell Biol. 8, 51–61.
Niehl, A., Wyrsch, I., Boller, T., Heinlein, M., 2016. Double-stranded RNAs induce a pattern-triggered immune signaling pathway in plants. New Phytol. 211, 1008–1019.
Nistal-Villán, E., Rodríguez-Garciá, E., Di Scala, M., et al., 2015. A RIG-I 2CARD-MAVS200 chimeric protein reconstitutes IFN-β induction and antiviral response in models deficient in type I IFN response. J. Innate Immun. 7, 466–481.

Noll, F., Behnke, J., Leiting, S., et al., 2017. Self-extracellular RNA acts in synergy with exogenous danger signals to promote inflammation. PLoS One 12, e0190002.

Ohto, U., Shibata, T., Tanji, H., et al., 2015. Structural basis of CpG and inhibitory DNA recognition by Toll-like receptor 9. Nature 520, 702–705.

O'Neill, L.A.J., 2013. Sensing the dark side of DNA. Science 339, 763–764.

Oosting, M., Cheng, S.C., Bolscher, J.M., et al., 2014. Human TLR10 is an anti-inflammatory pattern-recognition receptor. Proc. Natl. Acad. Sci. U. S. A. 111, E4478–E4484.

Orzalli, M.H., Kagan, J.C., 2017. Apoptosis and necroptosis as host defense strategies to prevent viral infection. Trends Cell Biol. 27, 800–809.

Paludan, S.R., Bowie, A.G., 2013. Immune sensing of DNA. Immunity 38, 870–880.

Pandolfi, F., Altamura, S., Frosali, S., Conti, P., 2016. Key role of DAMP in inflammation, cancer, and tissue repair. Clin. Ther. 38, 1017–1028.

Parisien, J.P., Lenoir, J.J., Mandhana, R., et al., 2018. RNA sensor LGP2 inhibits TRAF ubiquitin ligase to negatively regulate innate immune signaling. EMBO Rep. 19, e45176.

Park, Y.S., Lee, B., Ryu, C.M., 2016. Getting to PTI of bacterial RNAs: triggering plant innate immunity by extracellular RNAs from bacteria. Plant Signal. Behav. 11, e1198866.

Paungfoo-Lonhienne, C., Lonhienne, T.G.A., Mudge, S.R., et al., 2010. DNA is taken up by root hairs and pollen, and stimulates root and pollen tube growth. Plant Physiol. 153, 799–805.

Paz, S., Hsiao, J., Cauntay, P., et al., 2017. The distinct and cooperative roles of Toll-like receptor 9 and receptor for advanced glycation end products in modulating in vivo inflammatory responses to select CpG and non-CpG oligonucleotides. Nucleic Acid Ther. 27, 272–284.

Pelka, K., Shibata, T., Miyake, K., Latz, E., 2016. Nucleic acid-sensing TLRs and autoimmunity: novel insights from structural and cell biology. Immunol. Rev. 269, 60–75.

Pham, A.M., Santa Maria, F.G., Lahiri, T., Friedman, E., Marie, I.J., Levy, D.E., 2016. PKR transduces MDA5-dependent signals for type I IFN induction. PLoS Pathog. 12, e1005489.

Pichlmair, A., Schulz, O., Tan, C.P., et al., 2006. RIG-I-mediated antiviral responses to single-stranded RNA bearing 5′-phosphates. Science 314, 997–1001.

Pichlmair, A., Schulz, O., Tan, C.-P., et al., 2009. Activation of MDA5 requires higher-order RNA structures generated during virus infection. J. Virol. 83, 10761–10769.

Pinti, M., Mussini, C., Cossarizza, A., 2012. Mitochondrial DNA: a proinflammatory 'enemy from within' during HIV infection? Cell Death Dis. 3, e307.

Pisetsky, D.S., 2012. The origin and properties of extracellular DNA: from PAMP to DAMP. Clin. Immunol. 144, 32–40.

Poli, C.L., Augusto, J.F., Dauve, J., et al., 2017. IL-26 confers proinflammatory properties to extracellular DNA. J. Immunol. 198, 3650–3661.

Portou, M.J., Baker, D., Abraham, D., Tsui, J., 2015. The innate immune system, toll-like receptors and dermal wound healing: a review. Vascul. Pharmacol. 71, 31–36.

Qiao, J., Tang, H.D., Fu, Y.X., 2017. DNA sensing and immune responses in cancer therapy. Curr. Opin. Immunol. 45, 16–20.

Quanz, M., Chassoux, D., Berthault, N., Agrario, C., Sun, J.-S., Dutreix, M., 2009. Hyperactivation of DNA-PK by double-strand break mimicking molecules disorganizes DNA damage response. PLoS One 4, e6298.

Quintana-Rodriguez, E., Duran-Flores, D., Heil, M., Camacho-Coronel, X., 2018. Damage-associated molecular patterns (DAMPs) as future plant vaccines that protect crops from pests. Sci. Hortic. 237, 207–220.

Radetskyy, R., Daher, A., Gatignol, A., 2018. ADAR1 and PKR, interferon stimulated genes with clashing effects on HIV-1 replication. Cytokine Growth Factor Rev. 40, 48–58.
Ramadan, A., Land, W.G., Paczesny, S., 2017. Editorial: danger signals triggering immune response and inflammation. Front. Immunol. 8, 979.
Ranf, S., 2017. Sensing of molecular patterns through cell surface immune receptors. Curr. Opin. Plant Biol. 38, 68–77.
Rathinam, V.A.K., Jiang, Z.Z., Waggoner, S.N., et al., 2010. The AIM2 inflammasome is essential for host defense against cytosolic bacteria and DNA viruses. Nat. Immunol. 11, 395–403.
Regan, T., Nally, K., Carmody, R., et al., 2013. Identification of TLR10 as a key mediator of the inflammatory response to *Listeria* monocytogenes in intestinal epithelial cells and macrophages. J. Immunol. 191, 6084–6092.
Reinholz, M., Kawakami, Y., Salzer, S., et al., 2013. HPV16 activates the AIM2 inflammasome in keratinocytes. Arch. Dermatol. Res. 305, 723–732.
Ries, M., Schuster, P., Thomann, S., Donhauser, N., Vollmer, J., Schmidt, B., 2013. Identification of novel oligonucleotides from mitochondrial DNA that spontaneously induce plasmacytoid dendritic cell activation. J. Leukoc. Biol. 94, 123–135.
Rodero, M.P., Tesser, A., Bartok, E., et al., 2017. Type I interferon-mediated autoinflammation due to DNase II deficiency. Nat. Commun. 8, 2176.
Roers, A., Hiller, B., Hornung, V., 2016. Recognition of endogenous nucleic acids by the innate immune system. Immunity 44, 739–754.
Rojas, A., Morales, M.A., Araya, P., González, I., 2017. RAGE—the receptor of advanced glycation end products. eLS. https://doi.org/10.1002/9780470015902.a0027298.
Rotem, Z., Cox, R.A., Isaacs, A., 1963. Inhibition of virus multiplication by foreign nucleic acid. Nature 197, 564–566.
Rothenfusser, S., Goutagny, N., DiPerna, G., et al., 2005. The RNA helicase Lgp2 inhibits TLR-independent sensing of viral replication by retinoic acid-inducible gene-I. J. Immunol. 175, 5260–5268.
Roux, S.J., Steinebrunner, I., 2007. Extracellular ATP: an unexpected role as a signaler in plants. Trends Plant Sci. 12, 522–527.
Russell, R.T., Christiaans, S.C., Nice, T.R., et al., 2018. Histone-complexed DNA fragments levels are associated with coagulopathy, endothelial cell damage, and increased mortality after severe pediatric trauma. Shock 49, 44–52.
Rutz, M., Metzger, J., Gellert, T., et al., 2004. Toll-like receptor 9 binds single-stranded CpG-DNA in a sequence- and pH-dependent manner. Eur. J. Immunol. 34, 2541–2550.
Saijo, Y., Loo, E.P.-i., Yasuda, S., 2018. Pattern recognition receptors and signaling in plant–microbe interactions. Plant J. 93, 592–613.
Sandgren, S., Wittrup, A., Cheng, F., et al., 2004. The human antimicrobial peptide LL-37 transfers extracellular DNA plasmid to the nuclear compartment of mammalian cells via lipid rafts and proteoglycan-dependent endocytosis. J. Biol. Chem. 279, 17951–17956.
Sano, H., Morimoto, C., 1982. DNA isolated from anti-DNA antibody immune-complexes in systemic *Lupus erythematosus* is rich in guanine-cytosine content. J. Immunol. 128, 1341–1345.
Savage, C., Lopez-Castejon, G., Denes, A., Brough, D., 2012. NLRP3-Inflammasome activating DAMPs stimulate an inflammatory response in glia in the absence of priming which contributes to brain inflammation after injury. Front. Immunol. 3, 288.
Savatin, D. V., Gramegna, G., Modesti, V., and Cervone, F. (2014). Wounding in the plant tissue: the defense of a dangerous passage. Front. Plant Sci. 5, 470.
Saxena, A.R., Gao, L.Y., Srivatsa, S., et al., 2017. Oxidized and degraded mitochondrial polynucleotides (DeMPs), especially RNA, are potent immunogenic regulators in primary mouse macrophages. Free Radic. Biol. Med. 104, 371–379.

Schlee, M., Hartmann, G., 2016. Discriminating self from non-self in nucleic acid sensing. Nat. Rev. Immunol. 16, 566–580.

Schlee, M., Roth, A., Hornung, V., et al., 2009. Recognition of 5′ triphosphate by RIG-I helicase requires short blunt double-stranded RNA as contained in panhandle of negative-strand virus. Immunity 31, 25–34.

Schmidt, A., Schwerd, T., Hamm, W., et al., 2009. 5′-triphosphate RNA requires base-paired structures to activate antiviral signaling via RIG-I. Proc. Natl. Acad. Sci. U. S. A. 106, 12067–12072.

Séror, C., Melki, M.T., Subra, F., et al., 2011. Extracellular ATP acts on P2Y2 purinergic receptors to facilitate HIV-1 infection. J. Exp. Med. 208, 1823–1834.

Shao, J., Gao, F., Lin, H.J., et al., 2016. Short-fragment DNA residue from vaccine purification processes promotes immune response to the new inactivated EV71 vaccine by upregulating TLR9 mRNA. Plos One 11, e0153867.

Shimada, K., Crother, T.R., Karlin, J., et al., 2012. Oxidized mitochondrial DNA activates the NLRP3 inflammasome during apoptosis. Immunity 36, 401–414.

Shin, M.S., Kang, Y., Lee, N., et al., 2013. Self dsDNA induces IL-1β production from human monocytes by activating NLRP3 inflammasome in the presence of anti-dsDNA antibodies (P4083). J. Immunol. 190, 1407–1415.

Shinya, T., Yasuda, S., Hyodo, K., et al., 2018. Integration of danger peptide signals with herbivore-associated molecular pattern signaling amplifies anti-herbivore defense responses in rice. Plant J. 94, 626–637.

Shrivastava, G., León-Juárez, M., García-Cordero, J., Meza-Sánchez, D.E., Cedillo-Barrón, L., 2016. Inflammasomes and its importance in viral infections. Immunol. Res. 64, 1101–1117.

Silk, E., Zhao, H.L., Weng, H., Ma, D.Q., 2017. The role of extracellular histone in organ injury. Cell Death Dis. 8, e2812.

Simmons, J.D., Lee, Y.-L., Mulekar, S., et al., 2013. Elevated levels of plasma mitochondrial DNA DAMPs are linked to clinical outcome in severely injured human subjects. Ann. Surg. 258, 591–598.

Sims, G.P., Rowe, D.C., Rietdijk, S.T., Herbst, R., Coyle, A.J., 2010. HMGB1 and RAGE in inflammation and cancer. Annu. Rev. Immunol. 28, 367–388.

Sirois, C.M., Jin, T.C., Miller, A.L., et al., 2013. RAGE is a nucleic acid receptor that promotes inflammatory responses to DNA. J. Exp. Med. 210, 2447–2463.

Staswick, P.E., 2008. JAZing up jasmonate signaling. Trends Plant Sci. 13, 66–71.

Storek, K.M., Gertsvolf, N.A., Ohlson, M.B., Monack, D.M., 2015. cGAS and Ifi204 cooperate to produce type I IFNs in response to *Francisella* infection. J. Immunol. 194, 3236–3245.

Sun, L., Wu, J., Du, F., Chen, X., Chen, Z.J., 2013. Cyclic GMP-AMP synthase is a cytosolic DNA sensor that activates the type I interferon pathway. Science 339, 786–791.

Sun, B., Sundström, K.B., Chew, J.J., et al., 2017. Dengue virus activates cGAS through the release of mitochondrial DNA. Sci. Rep. 7, 3594.

Surpris, G., Poltorak, A., 2016. The expanding regulatory network of STING-mediated signaling. Curr. Opin. Microbiol. 32, 144–150.

Szczesny, B., Marcatti, M., Ahmad, A., et al., 2018. Mitochondrial DNA damage and subsequent activation of Z-DNA binding protein 1 links oxidative stress to inflammation in epithelial cells. Sci. Rep. 8, 914.

Takahashi, N., Yamada, T., Narita, N., Fujieda, S., 2006. Double-stranded RNA induces production of RANTES and IL-8 by human nasal fibroblasts. Clin. Immunol. 118, 51–58.

Takaoka, A., Wang, Z., Choi, M.K., et al., 2007. DAI (DLM-1/ZBP1) is a cytosolic DNA sensor and an activator of innate immune response. Nature 448, 501–505.

Takemura, N., Kawasaki, T., Kunisawa, J., et al., 2014. Blockade of TLR3 protects mice from lethal radiation-induced gastrointestinal syndrome. Nat. Commun. 5, 3492.
Tanaka, K., Swanson, S.J., Gilroy, S., Stacey, G., 2010. Extracellular nucleotides elicit cytosolic free calcium oscillations in Arabidopsis. Plant Physiol. 154, 705–719.
Tanaka, K., Choi, J., Cao, Y., Stacey, G., 2014. Extracellular ATP acts as a damage-associated molecular pattern (DAMP) signal in plants. Front. Plant Sci. 5, 446.
Tang, D., Kang, R., Coyne, C.B., Zeh, H.J., Lotze, M.T., 2012. PAMPs and DAMPs: signal 0s that spur autophagy and immunity. Immunol. Rev. 249, 158–175.
Tanji, H., Ohto, U., Shibata, T., et al., 2015. Toll-like receptor 8 senses degradation products of single-stranded RNA. Nat. Struct. Biol. 22, 109–115.
Tao, J.L., Zhou, X., Jiang, Z.F., 2016. cGAS-cGAMP-STING: the three musketeers of cytosolic DNA sensing and signaling. IUBMB Life 68, 858–870.
Thaler, J.S., Humphrey, P.T., Whiteman, N.K., 2012. Evolution of jasmonate and salicylate signal crosstalk. Trends Plant Sci. 17, 260–270.
Thines, B., Katsir, L., Melotto, M., et al., 2007. JAZ repressor proteins are targets of the SCF^{COI1} complex during jasmonate signalling. Nature 448, 661–666.
Thompson, A.J., Locarnini, S.A., 2007. Toll-like receptors, RIG-I-like RNA helicases and the antiviral innate immune response. Immunol. Cell Biol. 85, 435–445.
Tian, J., Avalos, A.M., Mao, S.Y., et al., 2007. Toll-like receptor 9-dependent activation by DNA-containing immune complexes is mediated by HMGB1 and RAGE. Nat. Immunol. 8, 487–496.
Torres, J., Rivera, A., Clark, G., Roux, S.J., 2008. Participation of extracellular nucleotides in the wound response of *Dasycladus vermicularis* and *Acetabularia acetabulum* (Dasycladales, Chlorophyta). J. Phycol. 44, 1504–1511.
Toussaint, M., Jackson, D.J., Swieboda, D., et al., 2017. Host DNA released by NETosis promotes rhinovirus-induced type-2 allergic asthma exacerbation. Nat. Med. 23, 681–691.
Tran, T.M., MacIntyre, A., Hawes, M., Allen, C., 2016. Escaping underground nets: extracellular DNases degrade plant extracellular traps and contribute to virulence of the plant pathogenic bacterium *Ralstonia solanacearum*. PLoS Pathog. 12, e1005686.
Trigg, B.J., Ferguson, B.J., 2015. Functions of DNA damage machinery in the innate immune response to DNA virus infection. Curr. Opin. Virol. 15, 56–62.
Tripathi, D., Zhang, T., Koo, A.J., Stacey, G., Tanaka, K., 2018. Extracellular ATP acts on jasmonate signaling to reinforce plant defense. Plant Physiol. 176, 511–523.
Tuomela, J., Sandholm, J., Kaakinen, M., et al., 2013. DNA from dead cancer cells induces TLR9-mediated invasion and inflammation in living cancer cells. Breast Cancer Res. Treat. 142, 477–487.
Unterholzner, L., Keating, S.E., Baran, M., et al., 2010. IFI16 is an innate immune sensor for intracellular DNA. Nat. Immunol. 11, 997–1004.
Vakrakou, A.G., Boiu, S., Ziakas, P.D., Xingi, E., Boleti, H., Manoussakis, M.N., 2018. Systemic activation of NLRP3 inflammasome in patients with severe primary Sjögren's syndrome fueled by inflammagenic DNA accumulations. J. Autoimmun. 91, 23–33.
Van Holle, S., Van Damme, E.J.M., 2018. Signaling through plant lectins: modulation of plant immunity and beyond. Biochem. Soc. Trans. 46, 217–233.
van Montfoort, N., Olagnier, D., Hiscott, J., 2014. Unmasking immune sensing of retroviruses: interplay between innate sensors and host effectors. Cytokine Growth Factor Rev. 25, 657–668.
Vargas-Hernandez, M., Macias-Bobadilla, I., Guevara-Gonzalez, R.G., et al., 2017. Plant hormesis management with biostimulants of biotic origin in agriculture. Front. Plant Sci. 8, 01762.

Vega-Muñoz, I., Feregrino-Pérez, A.A., Torres-Pacheco, I., Guevara-González, R.G., 2018. Exogenous fragmented DNA acts as a damage-associated molecular pattern (DAMP) inducing changes in CpG DNA methylation and defence-related responses in *Lactuca sativa*. Funct. Plant Biol. 45, 1065–1072.

Vénéreau, E., Ceriotti, C., Bianchi, M.E., 2015. DAMPs from death to new life. Front. Immunol. 6, 422.

Venter, J.C., Adams, M.D., Myers, E.W., et al., 2001. The sequence of the human genome. Science 291, 1304–1351.

Verma, V., Dhanda, R.S., Møller, N.F., Yadav, M., 2016. Inflammasomes and their role in innate immunity of sexually transmitted infections. Front. Immunol. 7, 540.

Vollmer, J., Tluk, S., Schmitz, C., et al., 2005. Immune stimulation mediated by autoantigen binding sites within small nuclear RNAs involves toll-like receptors 7 and 8. J. Exp. Med. 202, 1575–1585.

Vos, I.A., Pieterse, C.M.J., van Wees, S.C.M., 2013. Costs and benefits of hormone-regulated plant defences. Plant Pathol. 62, 43–55.

Wagner, H., 2004. The immunobiology of the TLR9 subfamily. Trends Immunol. 25, 381–386.

Walters, D., Heil, M., 2007. Costs and trade-offs associated with induced resistance. Physiol. Mol. Plant Pathol. 71, 3–17.

Walters, D.R., Havis, N.D., Sablou, C., Walsh, D.J., 2011. Possible trade-off associated with the use of a combination of resistance elicitors. Physiol. Mol. Plant Pathol. 75, 188–192.

Wang, Z.C., Choi, M.K., Ban, T., et al., 2008. Regulation of innate immune responses by DAI (DLM-1/ZBP1) and other DNA-sensing molecules. Proc. Natl. Acad. Sci. U. S. A. 105, 5477–5482.

Wang, F., Alain, T., Szretter, K.J., et al., 2016. S6K-STING interaction regulates cytosolic DNA-mediated activation of the transcription factor IRF3. Nat. Immunol. 17, 514–522.

Wang, Y., Ning, X., Gao, P., et al., 2017. Inflammasome activation triggers caspase-1-mediated cleavage of cGAS to regulate responses to DNA virus infection. Immunity 46, 393–404.

Watson, R.O., Bell, S.L., MacDuff, D.A., et al., 2015. The cytosolic sensor cGAS detects *Mycobacterium tuberculosis* DNA to induce type I interferons and activate autophagy. Cell Host Microbe 17, 811–819.

Wen, F.S., White, G.J., VanEtten, H.D., Xiong, Z.G., Hawes, M.C., 2009. Extracellular DNA is required for root tip resistance to fungal infection. Plant Physiol. 151, 820–829.

Westman, J., Papareddy, P., Dahlgren, M.W., et al., 2015. Extracellular histones induce chemokine production in whole blood ex vivo and leukocyte recruitment in vivo. PLoS Pathog. 11, e1005319.

Wu, J.Q., Baldwin, I.T., 2010. New insights into plant responses to the attack from insect herbivores. Annu. Rev. Genet. 44, 1–24.

Wu, J., Chen, Z.J., 2014. Innate immune sensing and signaling of cytosolic nucleic acids. Annu. Rev. Immunol. 32, 461–488.

Wu, J.J., Li, W.W., Shao, Y.M., et al., 2015. Inhibition of cGAS DNA sensing by a herpesvirus virion protein. Cell Host Microbe 18, 333–344.

Xiao, T.S., 2015. The nucleic acid-sensing inflammasomes. Immunol. Rev. 265, 103–111.

Xiao, T.S., Fitzgerald, K.A., 2013. The cGAS-STING pathway for DNA sensing. Mol. Cell 51, 135–139.

Yakushiji, S., Ishiga, Y., Inagaki, Y., Toyoda, K., Shiraishi, T., Ichinose, Y., 2009. Bacterial DNA activates immunity in *Arabidopsis thaliana*. J. Gen. Plant Pathol. 75, 227–234.

Yang, H., Wang, H., Chavan, S.S., Andersson, U., 2015. High mobility group box protein 1 (HMGB1): the prototypical endogenous danger molecule. Mol. Med. 21, S6–S12.

Yatim, N., Cullen, S., Albert, M.L., 2017. Dying cells actively regulate adaptive immune responses. Nat. Rev. Immunol. 17, 262–275.

Yoneyama, M., Fujita, T., 2012. Cytoplasmic sensing of viral double-stranded RNA and activation of innate immunity by RIG-I-like receptors. In: Wang, R. (Ed.), Innate Immune Regulation and Cancer Immunotherapy. Springer New York, New York, NY, pp. 51–60.

Yoneyama, M., Kikuchi, M., Matsumoto, K., et al., 2005. Shared and unique functions of the DExD/H-box helicases RIG-I, MDA5, and LGP2 in antiviral innate immunity. J. Immunol. 175, 2851–2858.

Zeiser, R., Penack, O., Holler, E., Idzko, M., 2011. Danger signals activating innate immunity in graft-versus-host disease. J. Mol. Med. 89, 833–845.

Zevini, A., Olagnier, D., Hiscottt, J., 2017. Crosstalk between cytoplasmic RIG-I and STING sensing pathways. Trends Immunol. 38, 194–205.

Zhang, S.Y., Jouanguy, E., Ugolini, S., et al., 2007. TLR3 deficiency in patients with herpes simplex encephalitis. Science 317, 1522–1527.

Zhang, Q., Raoof, M., Chen, Y., et al., 2010. Circulating mitochondrial DAMPs cause inflammatory responses to injury. Nature 464, 104–108.

Zhang, W.J., Zhou, Q., Xu, W., et al., 2013. DNA-dependent activator of interferon-regulatory factors (DAI) promotes lupus nephritis by activating the calcium pathway. J. Biol. Chem. 288, 13534–13550.

Zhang, J.Z., Liu, Z., Liu, J., Ren, J.X., Sun, T.S., 2014. Mitochondrial DNA induces inflammation and increases TLR9/NF-kappa B expression in lung tissue. Int. J. Mol. Med. 33, 817–824.

Zhang, L., Zhang, F., Melotto, M., Yao, J., He, S.Y., 2017. Jasmonate signaling and manipulation by pathogens and insects. J. Exp. Bot. 68, 1371–1385.

Zipfel, C., 2014. Plant pattern-recognition receptors. Trends Immunol. 35, 345–351.

CHAPTER SIX

Nucleic Acid Sensing in Invertebrate Antiviral Immunity

Pei-Hui Wang[*,†,1], **Jian-Guo He**[‡,§,1]
[*]Advanced Medical Research Institute, Shandong University, Jinan, China
[†]School of Biomedical Sciences, The University of Hong Kong, Pokfulam, Hong Kong
[‡]MOE Key Laboratory of Aquatic Product Safety/State Key Laboratory of Biocontrol, School of Life Sciences, Sun Yat-Sen University, Guangzhou, China
[§]School of Marine Sciences, Sun Yat-Sen University, Guangzhou, China
[1]Corresponding authors: e-mail address: pei-hui.wang@connect.hku.hk; lsshjg@mail.sysu.edu.cn

Contents

Abstract

Innate immunity is an ancient and conserved defense mechanism against infectious agents. It is activated after pathogen-associated molecular pattern sensing by germline-encoded pattern-recognition receptors, including specialized nucleic acid sensors. In vertebrates, nucleic acid sensing activates the dominant antiviral pathway that induces interferon (IFN) response and enhances antigen-specific adaptive immunity. Although canonical IFN system is absent in invertebrates, nucleic acid stimulation, and viral infection trigger an inducible non-specific antiviral response that exhibits multiple similarities to vertebrate IFN system. Invertebrates lack the adaptive immunity that provides long-term antigen-specific protection from pathogens. Meanwhile, the RNA interference (RNAi) pathway senses

International Review of Cell and Molecular Biology, Volume 345
ISSN 1937-6448
https://doi.org/10.1016/bs.ircmb.2018.11.002

viral nucleic acids and triggers the sequence-specific degradation of viral RNAs, thereby representing a specific antiviral mechanism of invertebrates. RNAi provides the primary antiviral response in some invertebrates but plays minimal or no role in vertebrate antiviral immunity. This review summarizes the nucleic acid sensing-mediated antiviral immunity in invertebrates.

1. INTRODUCTION

Innate and adaptive immunities are two complementary systems of vertebrates in sensing and limiting invading pathogens. Innate immunity is activated within hours to prevent early phase infection, whereas adaptive immunity requires days to weeks to provide pathogen-specific clearance (Kawai and Akira, 2011). Innate immunity controls infections before the initial adaptive immunity takes effect. Invertebrates lack a canonical adaptive immune system because of the absence of B cells, T cells, major histocompatibility complex, and antibodies. Innate immunity is activated by pathogen-associated molecular pattern (PAMP) sensing through a limited number of germline-encoded pattern-recognition receptors (PRRs) (Takeuchi and Akira, 2010). PAMPs are common microbial components that are essential for the host's survival and difficult to change, thereby rendering them ideal targets for immune sensing. Most PAMPs, such as peptidoglycan (PGN), lipopolysaccharides, single- and double-stranded RNA (ssRNA and dsRNA, respectively), ssDNA, and dsDNA, effectively activate innate immunity from lower invertebrates to mammals (Wang et al., 2015b; Wu and Chen, 2014). The immune detection of viruses relies almost exclusively on viral nucleic acid sensing. Nucleic acids serving as PAMPs significantly expand the repertoire of microbes detectable by the immune system (Wu and Chen, 2014). Bacterial secondary messengers cyclic dinucleotides (CDNs), including *c*-di-AMP, *c*-di-GMP, and 3′3′-cGAMP, constitute a novel class of PAMPs that bind with the stimulator of interferon (IFN) genes (STING), which is an endoplasmic reticulum-localized adaptor in DNA-sensing pathways, and induce IFN production (Danilchanka and Mekalanos, 2013; Tan et al., 2018).

In mammals, PRRs for nucleic acid sensing are generally divided into two groups according to the localization of immune sensing (Tan et al., 2018; Wu and Chen, 2014). Group one includes Toll-like receptors (TLRs) localized on intracellular compartments, such as endosomes and lysosomes. TLRs sense various forms of nucleic acids from the bacteria and viruses and function mostly in immune cells. TLRs are type I integral membrane glycoproteins characterized by extracellular domains that contain varying numbers of leucine-rich repeat (LRR) motifs and a cytosolic signaling domain called Toll/interleukin 1 receptor (IL-1R) homology (TIR) domain, which is homologous to that of IL-1R. LRR motifs mediate the recognition of a wide range of PAMPs, and the TIR domain interacts with adapter proteins, including TRIF and MyD88, for signal transduction and subsequent innate immune activation. A total of 10 human and 12 mouse TLRs have been reported to date. TLR10 exists in humans but is absent in mice, while TLR11, TLR12, and TLR13 are present in mice but not in humans. Among TLRs, TLR3, TLR7, TLR8, and TLR13 are involved in pathogen-derived RNA recognition, whereas TLR9 recognizes unmethylated CpG-DNA (Tan et al., 2018; Wu and Chen, 2014). Group two includes cytoplasm-restricted RNA and DNA sensors expressed in almost all cell types. Retinoic acid-inducible gene-I (RIG-I)-like receptors (RLRs) are major cytoplasmic RNA sensors. RLRs have three types, namely, RIG-I, melanoma differentiation-associated gene 5 (MDA5), and Laboratory of Genetics and Physiology 2 (LGP2), all of which contain a central DExD/H-box helicase domain for RNA binding (Tan et al., 2018; Wu and Chen, 2014). RIG-I recognizes relatively short dsRNA, and the presence of a 5′ triphosphate end significantly enhances its IFN-inducing activity. Self-RNA is modified by adding a 7-methylguanosine cap at the 5′ ends or by removing 5′ triphosphate before transport into the cytoplasm. Thus, RIG-I can discriminate between viral and self-RNAs. MDA5 recognizes relatively long dsRNA (Tan et al., 2018). RIG-I and MDA5 contain two N-terminal caspase recruitment domains (CARDs) involved in signal transduction but not LGP2. Without CARDs, LGP2 is expected to inhibit the RLR pathway, but genetic knockout and related functional assays indicated that LGP2 facilitates viral RNA recognition by RIG-I and MDA5 (Wu and Chen, 2014). After dsRNA sensing, RIG-I/MDA5 binds with and activates mitochondrial antiviral signaling (MAVS) protein via CARD–CARD-mediated interactions. MAVS interacts with TRAF3 and subsequently activates TBK1 and IκB kinase (IKK) complex, thereby activating the IFN-regulatory factor 3 (IRF3) and NF-κB pathways (Tan et al., 2018). DExD/H-box helicase family proteins that are distantly related to RLRs, such as DDX3, DHX9, DDX60, and DDX1–DDX21–DHX36 complex, also function as cytoplasmic RNA sensors and activate immune responses in certain cell types, such as dendritic cells and macrophages (Goubau et al., 2013). Cyclic GMP-AMP synthase (cGAS) and absent in melanoma 2 (AIM2) are the main cytoplasmic dsDNA sensors (Tan et al., 2018).

cGAS senses cytoplasmic dsDNA and produces 2′3′-cGAMP from the substrates, that is, ATP and GTP (Wu et al., 2013). cGAMP, which is structurally similar to bacterial CDNs, binds to STING, and activates the IRF3 and NF-κB pathways. AIM2 contains an N-terminal pyrin domain (PYD) and a C-terminal HIN domain responsible for dsDNA binding. AIM2 senses cytoplasmic dsDNA and activates caspase-1 inflammasomes, thereby controlling the cleavage of pro-inflammatory cytokines pro-IL-1β and pro-IL-18 and the subsequent secretion of their mature forms (Wang et al., 2018c). In addition to cGAS that functions as a nonredundant and general cytoplasmic DNA sensor in IFN production via STING, several proteins are also candidates of cytoplasmic DNA sensors; these proteins include the DNA-dependent activator of IRFs, IFN-gamma-inducible protein 16 (IFI16), DDX41, DNA-dependent protein kinase, and DNA damage sensor, meiotic recombination 11 homolog A (Wu and Chen, 2014). IFI16 is an AIM2-like receptor protein that contains one N-terminal PYD and two C-terminal dsDNA-binding domains (i.e., HINa and HINb). IFI16 predominantly localizes to the nucleus but translocates into the cytoplasm after viral infection in macrophages and lymphocytes to act as cytoplasmic dsDNA sensor and activate the IRF3 pathway by interacting with STING (Wang et al., 2018c). In mouse dendritic cells, DDX41 not only directly binds with dsDNA through its DEADc domain but also with bacterial CDNs and then activates the IRF3 pathway by forming a DDX41–STING–TBK1 complex (Wu and Chen, 2014). Therefore, TLRs, RLRs, and cGAS are essential nucleic acid sensors for host defense against viral infection (Goubau et al., 2013; Wu and Chen, 2014).

Upon nucleic acid sensing, PRRs activate immune signaling pathways through various adapter proteins, such as MyD88 and TRIF in the TLR pathway, MAVS in the RLR pathway, and STING in the DNA-sensing pathway, as well as through kinases and transcription factors, such as IRFs and NF-κB (Goubau et al., 2013; Takeuchi and Akira, 2010). This phenomenon upregulates the production of IFNs and pro-inflammatory cytokines. IFNs induce apoptotic cell death in virally infected cells, thereby rendering cells resistant to viral infection and activating adaptive immunity (Takeuchi and Akira, 2010). IFNs also induce hundreds of IFN-stimulated genes (ISGs) through the Janus kinase (JAK)/signal transducer and activator of transcription (STAT) pathway. ISGs inhibit viral replication by targeting almost every step of the viral life cycle and exhibit combined antiviral effects (Wu and Chen, 2014). Pro-inflammatory cytokines are pleiotropic proteins that regulate the cell death of inflammatory tissues, modify vascular endothelial permeability, recruit blood cells to inflamed tissues, and induce acute-phase production (Takeuchi and Akira, 2010; Wu and Chen, 2014). The hallmark of innate immune activation is IFN and pro-inflammatory cytokine induction. Nucleic acid sensing by PRRs also initiates cell-mediated immunity through the induction of phagocytosis, autophagy, and cell death (Takeuchi and Akira, 2010; Wu and Chen, 2014).

Invertebrates and vertebrates shared a common invertebrate ancestor; thus, invertebrates have similar core elements of innate immune systems to that of vertebrates. PRR-dependent immune activation is evolutionarily conserved by activating the NF-κB and JAK/STAT pathways in some invertebrates, including arthropods and mollusks (Green and Speck, 2018). Viral nucleic acid sensing induces both specific and non-specific antiviral immunity in invertebrates. Some proteins homologous to nucleic acid sensors described above are encoded by invertebrate genomes. Meanwhile, insect Dicer-2 and oyster RIG-I-1 have been well studied as cytoplasmic RNA sensors to initiate antiviral siRNA pathway and IFN-like responses, respectively (Green and Speck, 2018; Guo et al., 2018). However, the function of most proteins remains uncharacterized, especially in nucleic acid sensing. Thus, this paper summarizes the recent advances in understanding nucleic acid sensing-induced antiviral immunity in invertebrates.

2. SPECIFIC ANTIVIRAL IMMUNITY VIA RNA SILENCING

2.1 RNA Silencing

2.1.1 Discovery of RNAi

RNA interference (RNAi) is a post-transcriptional regulatory mechanism of gene silencing and is dependent on small noncoding RNAs (ncRNAs) with 21–31 nucleotides (nts). RNAi was first described in the nematode *Caenorhabditis elegans*, whose sequence-specific gene silencing can be achieved by injecting long dsRNA (Fire et al., 1998). Since then, dsRNA-initiated gene silencing has been observed in plants, protozoans, insects, and vertebrates. Thus, RNAi is an evolutionarily conserved mechanism that mediates sequence-specific gene silencing across the plant and animal kingdoms (Ding, 2010; Skalsky and Cullen, 2010).

2.1.2 RNA Silencing Pathways

RNA silencing acts as a nucleic acid sensing system that can be activated after RNA sensing. DNA also triggers RNA silencing through RNA intermediates for being processed into small ncRNAs (Guo et al., 2018; Skalsky and Cullen, 2010). The nature of nucleic acids determines the type of small ncRNA generated: small interfering RNAs (siRNAs), microRNAs (miRNAs), and p-element induced wimpy testis (PIWI)-interacting RNAs (piRNAs), which correspond to three distinct RNA silencing pathways (Ding, 2010; Guo et al., 2018; Skalsky and Cullen, 2010). These small ncRNAs are usually defined as to their origin, size, interaction with specific Ago proteins, and functions. Perfectly base-paired long dsRNAs are the precursors of siRNA populations, whereas a single miRNA is typically transcribed and processed from a miRNA-encoding gene that contains an imperfect intramolecular stem-loop (Skalsky and Cullen, 2010). In contrast to siRNAs and miRNAs with approximately 21–24 nts, piRNAs typically contain 24–31 nts, and its processing is independent of Dicer (Mussabekova et al., 2017). Similar to siRNAs, miRNAs and piRNAs can guide target mRNA slicing when they are bound to a catalytically active Ago protein and have near-perfect sequences that are complementary to their respective target RNAs (Fig. 1) (Mussabekova et al., 2017).

Dicer-2 senses and cleaves *endo-* and exogenous dsRNAs from cellular transcripts and viral replicative intermediates, respectively, to produce siRNAs (Ding, 2010; Guo et al., 2018). These siRNAs work with Ago proteins to search and destroy cellular or viral ssRNAs in a sequence-specific manner. By mismatching with target mRNAs outside the seed region, miRNAs mostly function as translation repressors of mRNA targets instead of slicing inducers. miRNAs regulate important biological processes. When miRNA pathway deficiency occurs, severe developmental defects form (Skalsky and Cullen, 2010). piRNAs protect genomes against transposon mobilization and expression (Iwasaki et al., 2015). During a viral infection, virus-derived ncRNAs (vncRNAs) are produced, which share features with host endogenous siRNAs, miRNAs, or piRNAs. Similar to host ncRNAs, vncRNAs induce RNA silencing by targeting viral RNAs, thereby initiating specific antiviral immunity (Ding, 2010). The antiviral immunity of invertebrates most relies on the siRNA pathway, although miRNA and piRNA pathways participate in antiviral response as well (Guo et al., 2018). In most conditions, RNAi and RNA silencing are defined as the same thing, referring to RNA-mediated regulation of gene expression (Blair and Olson, 2015). In this review, we use the definition proposed as followings: RNAi is defined as gene silencing by siRNAs and dsRNA, whereas RNA silencing is described as an Ago-mediated silencing mechanism directed by siRNAs, miRNAs, or piRNAs (Ding, 2010).

Studies on RNA silencing have mostly focused on *Drosophila melanogaster*, employing widely available genetic tools; mosquitoes as vectors for

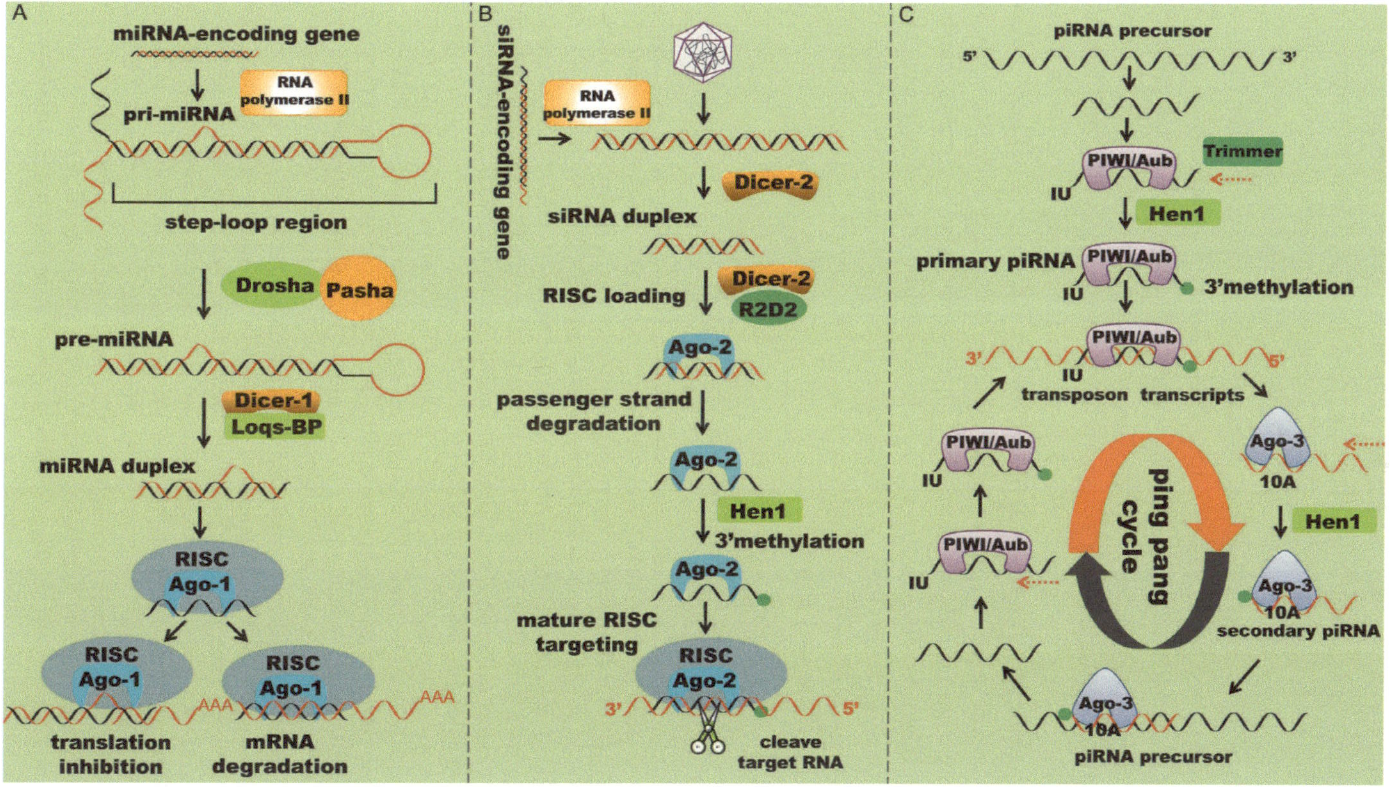

Fig. 1 See figure legend on next page.

transmitting arboviruses, which are arthropod-borne viruses or a group of viruses that are transmitted to humans from arthropod vectors, such as Zika virus (ZIKV), West Nile virus (WNV), dengue virus serotypes 1–4 (DENV 1–4), Japanese encephalitis virus (JEV), yellow fever virus (YFV), and chikungunya virus (CHIKV); and honeybees and silkworms with high ecological and economic values (Fung et al., 2018). The three small ncRNA pathways, which are also highly conserved in mammals, are well defined in *D. melanogaster* (Ding, 2010) (Fig. 1). *Drosophila* genome encodes two Dicers, five Agos, and other core components required for siRNA, miRNA, and piRNA pathways. Dicer-1 and Dicer-2 initiate miRNA and siRNA pathways, respectively, and the piRNA pathway is independent ofDicer. dsRNA activates the siRNA pathway that involves Dicer-2 and Ago-2 and plays a major role in antiviral immunity. Dicer-1 and Ago-1 are implicated in the miRNA pathway that modulates the expression of cellular genes. Three Ago family proteins, namely, PIWI, aubergine (Aub), and Ago-3, participate in the piRNA pathway that regulates mobile genetic elements in germ line cells (Fung et al., 2018).

Fig. 1 *Drosophila* siRNA, miRNA, and piRNA pathways that are highly conserved in vertebrates. (A) Primary miRNAs (pri-miRNAs) are transcribed by RNA polymerase and then recognized and processed by Drosha and Pasha (called DGCR8 in vertebrates) to generate pre-miRNAs in the nucleus. After exporting onto the cytoplasm, Dicer-1 associated with Loqs-PB (called PACT in vertebrates) will process pre-miRNAs into a mature miRNA duplex that is loaded onto Ago-1 and form miRNA-programmed RISC to mediate translation inhibition or degradation of the target mRNA. (B) dsRNAs from both endogenous and exogenous resources are sensed by Dicer-2 and cleaved into a 19–22 nt siRNA duplex. Subsequently, siRNA duplexes are loaded onto Ago-2, and the pre-RNA-induced silencing complex (RISC) is assembled with R2D2. Ago-2 removes the passenger strand of the siRNA duplex and leaves only a guide strand that will be methylated at the 3′ end by Hen1 to increase siRNA stability and promote RISC maturation. With the used of the siRNA guide strand, Ago-2 specifically cleaves the complementary target of the viral RNA or cellular mRNA, thereby leading to viral RNA degradation or gene knockdown. (C) piRNAs are generated from piRNA clusters via an unclear mechanism independent of Dicer. piRNAs are mostly antisense (shown in black) to transposon transcripts. PIWI and Aub participate in the biogenesis of primary piRNAs that have 1U bias at the 5′ end. After binding with PIWI/Aub, piRNA intermediates are trimmed and methylated at the 3′ end to generate mature PIWI-piRISC, which will cleave complementary transcripts and produce sense piRNA (shown in red) with adenosine at the 10th position 10 (10A). Ago-3 binds to the low-abundance sense piRNAs and collaborates with PIWI/Aub to amplify piRNAs through ping-pong cycle.

2.2 siRNAs in Arthropod Antiviral Immunity

2.2.1 siRNA Biogenesis

The siRNA pathway is activated by dsRNAs from viral replicative intermediates or cellular transcripts with secondary structures, which are recognized by the RNase III enzyme Dicer-2 and cleaved into 19–22 nt duplex siRNAs (Ding, 2010; Skalsky and Cullen, 2010). A common method for assembling a pre-RNA-induced silencing complex (RISC) is the loading of siRNA on a slicer Ago-2, an RNase H-type endonuclease, with dsRNA-binding cofactor R2D2 and RNA helicase Armitage. Ago-2 is the effector of RISC that binds and unwinds siRNAs, cleaves, and removes the passenger strand of a duplex, thereby leaving the guide strand (Guo et al., 2018). The stability of RISC is enhanced, and RISC maturation is facilitated through the methylation of the remaining guide strand of siRNA via 2′-*O*-methylation by Hen1 at the 3′ end (Ding, 2010). Ago-2 uses the bound guide strand to form a perfect Watson–Crick base-paired duplex with a complementary target of viral ssRNAs or cellular mRNA transcripts, which are specifically cleaved for degradation at the center of the duplex by Ago-2 endonuclease activity. The cleavage process results in virus inhibition or cellular gene downregulation (Fig. 1) (Ding, 2010; Skalsky and Cullen, 2010). dsRNA is recognized as an important PAMP of viruses that can trigger the robust degradation of viral RNA via the siRNA pathway.

2.2.2 siRNA-Mediated Antiviral Immunity in Flies

Virus-derived siRNAs (vsiRNAs) are widely detected in *C. elegans*, *D. melanogaster*, mosquitoes, silkworms, and vertebrates (Ding, 2010; Guo et al., 2018). vsiRNAs are also involved in specific antiviral immunity and commonly detected in cells or animals infected by viruses, including positive strand RNA, negative strand RNA, dsRNA, ssDNA, and dsDNA viruses (Bronkhorst et al., 2012; Kemp et al., 2013; Mueller et al., 2010). These RNAs provide the footprints of antiviral response processes and useful information for the sensing of viral nucleic acids. The first experimental evidence showing vncRNA targets virus for degradation was obtained in *D. melanogaster* S2 cells (Li et al., 2002). vncRNAs detected in flock house virus (FHV)-infected S2 cells contain 21–22 nts and are similar to siRNAs processed from long dsRNAs by Dicer-2. Similar to siRNAs derived from dsRNA, FHV vncRNA sequences correspond to any region of the viral genome and have either negative or positive polarity. They are distinct from the subsequently discovered virus-derived miRNAs (vmiRNAs), which are processed from specific sites and polarity of the viral genome. Thus, these FHV vncRNAs are classified as siRNAs (Ding, 2010). FHV infection causes the rapid accumulation of FHV-derived siRNAs from positive and negative strand polarities. vsiRNAs are detected in FHV-infected cells and flies. Given that *Dicer-2* mutant abolishes vsiRNA production; an intact siRNA pathway is required (Kemp and Imler, 2009). Flies that lack *Dicer-2* or *Ago-2* cannot effectively clear viral RNAs, and this condition results in uncontrolled viral proliferation; therefore, such flies are hypersensitive to FHV infection (Nayak et al., 2013). Moreover, FHV-encoded B2 protein is a viral silencing repressor (VSR) targeting the siRNA pathway that promotes virus replication and is essential for FHV infection. FHV B2 is dispensable to viral replication in *Ago-2* depleted cells. When deprived of FHV B2, vsiRNA abundance increases considerably. At the later stages of virus infection, vsiRNA accumulation significantly declines (Li et al., 2002). vsiRNAs co-immunoprecipitate with Ago-2 rather than with Ago-1 (Czech et al., 2008; Kemp and Imler, 2009). Flies carrying the defective mutants of the siRNA pathway components *Dicer-2*, *Ago-2*, or *R2D2* are hypersensitive to FHV infection (Nayak et al., 2013).

The siRNA pathway, especially its core components Dicer-2, R2D2, and Ago-2, play a major role in *Drosophila* antiviral immunity. vsiRNAs are produced in a Dicer-2-dependent manner in flies or cells infected with various types of RNA and DNA viruses (Nayak et al., 2013). When infected with RNA viruses, including *Drosophila* C virus (DCV) (Galiana-Arnoux et al., 2006), Cricket paralysis virus (CrPV) (van Rij et al., 2006), FHV (Wang et al., 2006), or Sindbis virus (SINV) (Galiana-Arnoux et al., 2006; Kemp and Imler, 2009), flies with *Dicer-2*, *Ago-2*, or *R2D2* mutants succumb more rapidly than wild-type animals (Galiana-Arnoux et al., 2006; Kemp and Imler, 2009). Moreover, flies with *Dicer-2* or *Ago-2* mutants are more susceptible to West Nile virus (WNV) infection than wild-type controls (Chotkowski et al., 2008). The knockdown of *Ago-2* increases the susceptibility of *Drosophila* S2 cells to WNV infection, and this finding suggests that WNV-homologous small RNAs produced in infected cells are vsiRNAs involved in antiviral response (Chotkowski et al., 2008). Vesicular stomatitis virus (VSV) replicates efficiently in *Dicer-2*, *Ago-2*, or *R2D2* mutant flies (Mueller et al., 2010). *Drosophila Ars2* depletion increases susceptibility to a diverse set of RNA viruses, including FHV, DCV, SINV, and VSV (Sabin et al., 2009). Flies that lack siRNA components, such as *Ago-2*, *R2D2*, or *Armitage*, become increasingly susceptible to the dsRNA virus *Drosophila* X virus (DXV), FHV, DCV, SINV, and VSV

(Galiana-Arnoux et al., 2006; Mueller et al., 2010; Nayak et al., 2013; Sabin et al., 2009; Zambon et al., 2006). The siRNA pathway protects *D. melanogaster* against invertebrate iridescent virus 6 (IIV-6), a dsDNA virus (Bronkhorst et al., 2012, 2014). Compared with wild-type animals, flies with *Dicer-2* or *Ago-2* mutant are more susceptible to IIV-6 infection (Bronkhorst et al., 2012; Kemp et al., 2013). DCV and CrPV (*Dicistroviridae*), DXV (*Birnaviridae*), FHV (*Nodaviridae*), VSV (*Rhabdoviridae*), WNV (*Flaviviridae*), SINV (*Alphaviridae*), and IIV-6 (*Betairidovirinae*) belong to different families of viruses. Thus, the siRNA pathway has a broad spectrum of antiviral activity in *D. melanogaster* (Nayak et al., 2013).

2.2.3 siRNA-Mediated Antiviral Immunity in Mosquitoes

Although studies on the siRNA pathway that is mainly performed in *Drosophila* have provided primary protection against viral infection, important observations in other insects, specifically mosquitoes, also reinforce the relevance of this pathway in antiviral immunity. Mosquitoes have the orthologs of all the RNA silencing components of *Drosophila* (Samuel et al., 2018). In mosquitoes and cell lines, vsiRNAs processed by Dicer-2 are readily detected during an arbovirus infection, and a direct antiviral response that is essential to a pathogenic outcome is observed, indicating that the RNAi pathway targets invading viruses and suppresses virus replications (Keene et al., 2004; Leger et al., 2013; Samuel et al., 2018; Sanchez-Vargas et al., 2009). Small ncRNA profiles indicate that vsiRNAs are produced during viral infection, primarily showing the typical feature of 21 nts processed by Dicer-2 and asymmetrical distribution along the length of the viral genome (Brackney et al., 2009). The introduction of dsRNAs targeting viral genomes remarkably inhibits viral replication. The knockdown of the siRNA pathway components *Ago-2*, *Dicer-2*, or *R2D2* increases viral load and virus dissemination in mosquitoes (Cheng et al., 2016; Keene et al., 2004; Samuel et al., 2018; Sanchez-Vargas et al., 2009). The viral loads of SINV and YFV in *Dicer-2*-knocked out *Aedes aegypti* are considerably higher than those in wild-type mosquitoes. Viruses expressing VSRs replicate to a greater extent and spread better throughout hosts than those that do not express VSRs (Samuel et al., 2016). The disease phenotypes of RNAi-null mutant flies infected with wild-type viruses is similarly enhanced. The ectopic expression of FHV B2 in SINV and O′ nyong-nyong virus (ONNV, *Togaviridae*) significantly inhibits vsiRNA production (Cheng et al., 2016; Myles et al., 2008; Samuel et al., 2018). siRNA-mediated antiviral immunity is also involved in several virus–mosquito models, such as WNV/*Culex quinquefasciatus*, ONNV/*Anopheles gambiae*, and SINV–DENV–YFV/*A. aegypti* (Prasad et al., 2013). Thus, the RNAi pathway is a critical defense of mosquito antiviral immunity.

2.2.4 siRNA-Mediated Antiviral Immunity in Honeybees

All of the core elements, such as Dicer-like, Ago-2, and R2D2, of the siRNA pathway are found in honeybees (Brutscher et al., 2015). dsRNA intermediates that are generated during the replication of ssRNA virus and secondary RNA structures within a genome serve as the substrates of honeybee Dicer-like proteins and activate the siRNA pathway, thereby inducing vsiRNA production. The delivery of viral-specific dsRNAs provides honeybees with high protection against acute viral infections (McMenamin et al., 2018), and honeybees fed with combinations of Israeli acute paralysis virus (IAPV) and IAPV-specific dsRNA have reduced IAPV levels compared with those fed with IAPV only. Moreover, IAPV-specific vsiRNAs are detected in honeybees fed with IAPV-specific dsRNA, indicating that dsRNA is cleaved and processed by Dicer-like protein and Ago-2 (Brutscher et al., 2015). Honeybees fed with dsRNA targeting deformed wing virus (DWV) exhibit reduced mortality, viral load, and deformed wing symptoms (McMenamin et al., 2018). In naturally infected honeybees, vsiRNAs from IAPV, Kashmir bee virus (KBV), and DWV are documented and characterized, indicating that the siRNA pathway plays a critical role in honeybee antiviral defense against various viruses (Brutscher and Flenniken, 2015). siRNA pathway components of honeybees, such as *Dicer-like* and *Ago-2*, can be greatly upregulated by viral infection (Brutscher et al., 2017; Galbraith et al., 2015). Similar regulatory mechanisms of the siRNA pathway have also been observed in crustaceans (see Section 2.2.9) but not yet in *Drosophila* (Chen et al., 2011; Labreuche et al., 2010; McMenamin et al., 2018). This finding suggests that the regulation of the siRNA pathway may differ in arthropods.

2.2.5 Antiviral siRNA Pathway in Other Insects

Studies on the siRNA pathway, including deep viral small ncRNA profiling, have mainly focused on flies and mosquitoes. Studies on other insects demonstrated that the siRNA pathway is the major antiviral strategy in combating viral infection. For example, transgenic silkworms expressing dsRNAs of *Bombyx mori* nucleopolyhedrovirus (BmNPV) show enhanced antiviral capacity against this virus (Santos et al., 2018). In *B. mori*, the overexpression of Dicer-2 or Ago-2 enhances siRNA silencing effects, reduces CrPV-induced cell mortality, and strengthens antiviral effects, whereas the knockdown of *Dicer-2* or *Ago-2* elicits opposite effects (Santos et al., 2018). The presence of vncRNAs from *B. mori* cytoplasmic polyhedrosis virus (BmCPV) indicates that the siRNA pathway is operative in *B. mori* against BmCPV infection, and *B. mori* Dicer-2 predominantly targets viral dsRNA and produces 20-nt vsiRNAs (Zografidis et al., 2015). These 20-nt vsiRNAs are also found in *Helicoverpa armigera* infected with single NPV and in lepidopteran Sf9 cells infected with baculovirus (Jayachandran et al., 2012; Mehrabadi et al., 2015). The cultured cells from *Culicoides sonorensis* infected with either Bluetongue virus or Schmallenberg virus produce 21-nt vsiRNAs with a typical feature of production by a Dicer-like nuclease (Marques and Imler, 2016). An increasing number of insect species have a functional siRNA pathway in antiviral immunity, and this pathway is recognized as the major antiviral strategy in most insects and even in noninsect arthropods, such as ticks and crustaceans. The core genes of the siRNA pathway, such as *Dicer* and *Ago* genes, are identified in the tick *Ixodes scapularis*. dsRNA-induced gene silencing is also found in tick cell lines. *I. scapularis* cells infected with Langat virus (*Flaviviridae*) produce vsiRNAs with an average length of 22 nts (Blair and Olson, 2015).

2.2.6 Viral dsRNAs as PAMP of the siRNA Pathway

dsRNAs are undetectable in healthy cells but are generated during the replication cycle of nearly all viruses (Bronkhorst and van Rij, 2014). The accumulated viral dsRNA-activating mammalian immune pathways, such as RLR pathways, originates from the viral replication intermediates of ssRNA viruses, viral genomes of dsRNA viruses, structured RNA elements in viral ssRNA genomes or viral transcripts, and dsRNAs formed by the hybridization of the overlapping transcripts of DNA viruses (Ding, 2010; Tan et al., 2018). These viral dsRNAs may be used in the siRNA machinery for the restriction of virus replication. Insect Dicer-2 senses and cleaves viral dsRNAs and produces vsiRNAs. In the process, one strand of vsiRNA is loaded into RISC. The vsiRNA guides the RISC to complementary viral mRNAs and genomic RNAs, thereby facilitating cleavage and degradation and inhibiting viral replication (Fig. 1) (Bronkhorst and van Rij, 2014). The viral resources used for insect Dicer-2 sensing and processing are identified through RNA-seq in infected cells or animals (Kemp and Imler, 2009). The discovered vsiRNAs are used for mapping corresponding viral genomes; meanwhile, viral dsRNAs used as Dicer-2 substrates for vsiRNA production can be classified as dsRNA virus genomes, ssRNA virus replication intermediates, overlapping viral transcripts that can form dsRNA, and ssRNA-structured elements from genomes or transcripts

(Gammon and 7Mello, 2015). During ssRNA virus replication, dsRNA replication intermediates serve as major Dicer-2 substrates for vsiRNA production. During dsRNA virus infection, the genomic dsRNA itself acts as the main source of Dicer-2 cleavage. vsiRNAs derived from dsDNA viruses are mostly mapped in a viral genome region and predicted to produce structured transcripts and convergent overlapping transcriptions, which are ideal resources for dsRNA generation (Gammon and Mello, 2015). Thus, viral dsRNAs function as PAMPs for insect Dicer-2 sensing and induce specific antiviral immunity.

2.2.7 Viral Suppressors of Antiviral siRNA Pathway

Given the major role of the siRNA pathway in insect antiviral immunity, viruses encode VSRs for targeting siRNA pathway (Gammon and Mello, 2015). The importance of VSRs in antagonizing siRNA-mediated antiviral immunity and facilitating viral infection has been extensively studied, and the FHV B2 protein is mostly used as a model. The FHV genome consists of RNA1 and RNA2, which encode viral RNA-dependent RNA polymerase (RdRP) and viral capsid protein, respectively. FHV, unlike B2 mutant FHV, is highly pathogenic to wild-type flies. B2 mutant FHV can restore its virulence when infected with *Dicer-2*- or *Ago-2*-depleted flies. FHV B2 mechanically binds to long dsRNAs and siRNAs, preventing FHV dsRNA from Dicer-2 sensing, processing, and siRNA loading into RISC (Gammon and Mello, 2015; Li et al., 2002). FHV B2 directly interacts and inhibits Dicer-2, and B2 proteins from other *Nodaviridae* viruses, such as Wuhan nodavirus and Nodamura virus, unitize a similar mechanism to antagonize the siRNA pathway. B2 proteins not only bind to long dsRNAs and siRNAs but also interact with and inhibit Dicer-2. Hence, they can target multiple processes, including Dicer-2-mediated siRNA biogenesis and siRNA incorporation into Ago2-containing RISC involved in the siRNA pathway (Nayak et al., 2013). The B2 proteins of a mosquito-specific nodavirus mosinovirus (MoNV) suppress long dsRNA-induced RNAi but not siRNA-induced RNAi. MoNV-infected mosquito cells resistant to RNAi are triggered by siRNA, indicating that another viral factor may repress RNAi after siRNA biogenesis occurs (Schuster et al., 2014). In addition to nodaviruses, other virus families encode dsRNA-binding VSR proteins, such as the 1A protein of *Dicistroviridae* family DCV (van Mierlo et al., 2012; van Rij et al., 2006), VP3 proteins of *Birnaviridae* family Culex Y virus (CYV) and DXV (van Cleef et al., 2014). The 340R protein of *Iridoviridae* family IIV-6 binds to long dsRNA and inhibits Dicer-2 sensing and processing of viral dsRNAs (Bronkhorst et al., 2014). The VP3 proteins of CYV and DXV and the 340R protein of IIV-6 bind to siRNAs and block their loading into RISC (Bronkhorst et al., 2014; van Cleef et al., 2014). Moreover, orf27 from *Heliothis virescens ascovirus* (HvAV) is an RNase III enzyme that can directly degrade siRNA duplexes or compete with Dicer-2 for dsRNA substrates to prevent their incorporation into Ago-2 (Gammon and Mello, 2015; Hussain et al., 2010). Although VSRs have great diversities, non-specific dsRNA binding appears to be a common theme (Samuel et al., 2018). In contrast to dsRNA binding, some VSRs, such as 1A protein of CrPV and VP1 proteins of Nora virus and Nora-like virus, can physically interact with Ago-2 and block Ago-2-mediated RNA cleavage (van Mierlo et al., 2012, 2014). VP1 protein from a divergent Nora virus isolated from *D. immigrans* interacts with and suppresses *D. immigrans* Ago-2 but not *D. melanogaster* Ago-2. Thus, viral suppressors have host-specific activities. The antiviral siRNA pathway in *Drosophila* undergoes rapid adaptive evolution (Obbard et al., 2006; van Mierlo et al., 2014). In *Drosophila*, Dicer-2, Ago-2, and R2D2 exert a strong selective pressure to escape targeting. They are among the fastest evolving 3% of all *Drosophila* genes, and they evolve significantly faster than other immune-related and housekeeping genes (Obbard et al., 2006). This phenomenon provides an excellent example for the coevolution of host siRNA machineries and viral suppression mechanisms.

Although the B2 protein from MoNV, a putative arbovirus, functions as a VSR, most arboviruses, including some well-characterized arboviruses, do not encode a VSR to avoid strong pathology and insect vector death, which may cause a reduced virus transmission efficiency (Gammon and Mello, 2015). YFV capsid protein was shown to inhibit siRNA pathway in *A. aegypti* by binding to long dsRNAs and interfering with the production of vsiRNAs by Dicer. Moreover, the VSR activities of the flavivirus capsid proteins appear to be broadly conserved (Samuel et al., 2016). However, other studies cannot observe the VSR activities of these arboviruses (Leggewie and Schnettler, 2018; Varjak et al., 2017). SINV carrying FHV B2 shows an enhanced viral pathogenicity in mosquitoes (Gammon and Mello, 2015). However, the flaviviruses WNV and DENV produce large amounts of sub-genomic flavivirus RNA (sfRNA) containing extensive secondary structures that compete with other dsRNA substrates for Dicer sensing and processing, thus disrupting viral dsRNA processing (Gammon and Mello, 2015; Goertz et al., 2016; Pompon et al., 2017). WNV sfRNA inhibits dsRNA cleavage by Dicer in vitro. Therefore, sfRNA may act as a nucleic acid-based regulator of

RNA silencing pathways, which may be conserved among flaviviruses (Schnettler et al., 2012). The DENV NS4B protein displays VSR activity and blocks dsRNA cleavage by human Dicer through an unknown mechanism without binding to dsRNA (Kakumani et al., 2013). The genomes of the honeybee viruses IAPV, KBV, and acute bee paralysis virus all encode highly conserved DvExNPGP motifs that are responsible for VSR function. Naturally IAPV-infected bees fed with siRNAs targeting the DvExNPGP motif have lower IAPV load than control bees fed with siRNAs targeting the sequence of the IAPV internal ribosome entry site (Brutscher and Flenniken, 2015). Therefore, these honeybee-infecting viruses may encode a VSR to counteract siRNA-mediated antiviral defense, which builds up the importance of the siRNA pathway in honeybee antiviral defense. Overall, viruses have multifaceted mechanisms to inhibit antiviral siRNA pathway, highlighting the pivotal role of RNAi in antiviral immunity.

2.2.8 dsRNA Uptake and Systemic RNAi Response

In addition to cell-intrinsic immunity, systemic responses that mediate antiviral signals spreading from infection sites to distant uninfected tissues are also important because they can alert uninfected cells and amplify antiviral immune responses (Nayak et al., 2013). In the nematode *C. elegans*, Dicer-processed siRNAs are amplified to secondary siRNAs by endogenous RdRPs, which not only facilitate the spread of antiviral siRNAs but also amplify and sustain antiviral responses. *C. elegans* systemic RNAi-defective proteins (SIDs) play an essential role in this procedure by passively transporting dsRNA into cells (Nayak et al., 2013). *Drosophila* has no RdRP or SID orthologs, although systemic RNAi plays a critical role in its antiviral response (Ding, 2010). dsRNAs released by infected cells may trigger RNAi in distant cells upon internalization via the dsRNA uptake pathway. In *Drosophila*, dsRNA uptake is associated with vesicles and scavenger receptors, proteins involved in phagocytosis and endocytosis, and other proteins directly or indirectly implicated in endocytosis (Kingsolver et al., 2013; Saleh et al., 2006; Ulvila et al., 2006). The underlying mechanism has been recently clarified (Goic et al., 2013; Tassetto et al., 2017). In *Drosophila*, dsRNAs are taken up by macrophage-like hemocytes from infected cells and reverse transcribed into virus-derived complementary DNAs by endogenous transposon reverse transcriptase, which can act as templates for the de novo synthesis of secondary vsiRNAs. In naive flies, these vsiRNAs are secreted in exosome-like vesicles and confer passive protection against virus challenge through systemic RNAi-mediated antiviral immunity

(Goic et al., 2013; Tassetto et al., 2017). Although an SID ortholog is absent in *Drosophila*, it has been found in other insects, such as *B. mori* with three SID orthologs and *Spodoptera exigua* with one SID ortholog. However, whether these SID orthologs participate in systemic RNAi responses remains unknown (Xu et al., 2016a).

2.2.9 Antiviral siRNA Pathway in Crustaceans

Crustaceans, such as *Daphnia*, crab, and shrimp, have great ecological and agricultural values (He et al., 2015b). Crustaceans have an RNAi system that is similar to that of insects. Nearly all RNAi elements are found in shrimp. Shrimp Dicer-1 is highly similar to the Dicer-1 proteins of most invertebrates, whereas shrimp Dicer-2 is highly similar to insect Dicer-2 (Chen et al., 2011; He et al., 2015b). An increasing number of RNAi-related proteins, such as Ago family proteins, Ars2, TRBP, Drosha, Pasha, and SID-1, have been found in shrimps (He et al., 2015b). Some RNAi components can be upregulated by dsRNA injection and viral infection (Chen et al., 2011; Labreuche et al., 2010). Shrimp miRNAs, such as miR-100, are also involved in defense against viral infection. To date, no piRNA has been described from crustaceans or related viruses (Liu et al., 2016).

The introduction of dsRNA or siRNA targeting viral genes into shrimps significantly inhibits the replication of viruses, including white spot syndrome virus (WSSV), Taura syndrome virus (TSV), infectious myonecrosis virus (IMNV), yellow head virus (YHV), and *Penaeus monodon* densovirus (pmDNV) (He et al., 2015b). The injection of dsRNA of a TSV-specific sequence, not WSSV vp19-dsRNA, provides effective protection against TSV infection (Robalino et al., 2005). The Pacific white shrimp *Litopenaeus vannamei* injected with dsRNA corresponding to the ORF1 of IMNV (ORF1-dsRNA) exhibits robust and long-lasting antiviral response and 100% protection from disease and mortality caused by IMNV. *L. vannamei* treated with IMNV ORF1-dsRNA survives from the initial infection of IMNV and develops resistance against subsequent infection with a 100-fold dose of virus after more than 50 days (Loy et al., 2012). Furthermore, the coinjection of dsRNAs corresponding to the ORF1 and ORF2 of PmDNV efficiently suppresses PmDNV replication (Ho et al., 2011). The black tiger shrimp *Penaeus monodon* treated with YHV protease-dsRNA but not with GFP-dsRNA has a low YHV replication level and decreased mortality after YHV challenge (Tirasophon et al., 2007). In shrimp primary cells, low YHV replication is observed when cells are transfected with dsRNA targeting viral nonstructural (NS) genes (Liu et al., 2009).

The injection of dsRNA corresponding to the NS1 and structural protein gene (VP) of PmDNV can effectively inhibit viral propagation in naturally pre-infected shrimp. In heavily infected shrimp, the potency of the coinjection of NS1-dsRNA and VP-dsRNA for viral clearance is greater than that of the injection of only NS1-dsRNA (Attasart et al., 2011).

The injection of the synthesized vp28-specific siRNA of WSSV (vp28-siRNA) strongly inhibits viral infection in the kuruma shrimp *Marsupenaeus japonicus*. The continuous injection of vp28-siRNA, but not the mutant vp28-siRNA, at different times post-challenge completely eradicates WSSV infection (Xu et al., 2007). Thus, vp28-siRNA-mediated antiviral immunity against WSSV is highly specific and sequence dependent (Wu et al., 2007). The injection of dsRNA corresponding to WSSV vp28 generates a highly focused population of 22-nt vp28-siRNA and induces the antiviral response against WSSV infection (Nilsen et al., 2017). Meanwhile, vp28-siRNA is produced naturally during WSSV infection and detected by small RNA sequencing in WSSV-infected shrimp tissues. Thus, vp28-siRNA is a virus-derived siRNA generated from WSSV transcripts during infection (Huang and Zhang, 2013; Sabin and Cherry, 2013). The knockdown of *Dicer-2*, but not *Dicer-1* or *Drosha*, abolishes vp28-siRNA accumulation, suggesting that the production of vp28-siRNA depends on the siRNA pathway but not on the miRNA pathway. The number of WSSV genomic copies highly increases when *Dicer-2* and vp28-siRNA are depleted, and this increase renders the shrimp highly susceptible to WSSV infection. Hence, the biogenesis step of the shrimp RNAi pathway is fundamentally required for siRNA-mediated WSSV suppression in vivo (Huang and Zhang, 2013; Sabin and Cherry, 2013). The binding of vp28-siRNA to Ago-2 rather than to Ago-1 indicates that vp28-siRNAs produced during WSSV infection are incorporated into an Ago-2-containing RISC. vp28-siRNA restrains WSSV replication using antiviral RISC targeting viral mRNAs for degradation, thereby inhibiting viral gene expression at the post-transcriptional level. Whether this antiviral response is executed at the post-transcriptional level can be furtherly determined by the comparison of the changes in viral gene abundance between cells or animals with intact and deficient siRNA pathway (Sabin and Cherry, 2013).

The injection and oral consumption of dsRNA trigger system-wide gene silencing, and thus shrimps may possess systemic RNAi similar to nematodes and insects. In contrast to *Drosophila* lacking SID, shrimps have at least one SID ortholog. In *C. elegans*, SID-1 (CeSID-1) is responsible for systemic RNAi, and SID-2 is required for dsRNA uptake. *L. vannamei SID-1* ortholog (*LvSID-1*) encodes 11 transmembrane domains similar to CeSID-1, and it is strongly upregulated after dsRNA stimulation (Maruekawong et al., 2018). The knockdown of *LvSID-1* increases mortality after WSSV infection (Labreuche et al., 2010). When *LvSID-1* is upregulated, the uptake of Cy3-labeled dsRNA is synchronously improved, thereby enhancing gene silencing. In invertebrates, in addition to the transmembrane channel mechanism of dsRNA uptake through SIDs, extracellular dsRNA uptake occurs through an endocytosis-mediated mechanism, which requires receptor recognition and an endocytosis pathway (Xu et al., 2016a).

Given the pivotal role of siRNA-mediated antiviral immunity in shrimp, various methods for generating dsRNA and subsequent delivery into animals have been developed as therapeutic strategies for controlling diseases affecting farmed shrimps (Itsathitphaisarn et al., 2017). For example, the delivery of vp28-siRNAs expressed in bacteria or encapsulated by β-1,3-D-glucan into shrimp inhibits WSSV replication and is thus a potential mechanism for treatment development (He et al., 2015b). Feeding shrimp with a formulated diet containing mixed inactivated bacteria harboring dsRNAs corresponding to PmDNV NS1 and VP significantly eliminates PmDNV replication and reduces shrimp mortality, showing potential for disease control applications in commercial aquaculture (Chimwai et al., 2016). Further studies on the characterization of RNAi-related genes and antiviral mechanisms can revolutionize RNAi technology as a therapeutic approach and efficient strategy for viral disease control in crustacean aquaculture.

2.3 miRNAs in Arthropod Antiviral Immunity

2.3.1 miRNA Biogenesis

miRNAs, which are 19–24 nt in length and typically bind to the complementary sequences of the 3′-UTRs of target mRNAs, modulate gene expression at the post-transcriptional levels. miRNAs are initially identified in *C. elegans* and represent one of the most abundant gene families widely expressed in animals, plants, protists, and viruses (Lee et al., 1993; Skalsky and Cullen, 2010). In an miRbase, over 2588 miRNAs are found in humans, 434 miRNAs in *C. elegans*, and 466 miRNAs in *D. melanogaster* (Bronkhorst et al., 2014). *Drosophila* miRNAs participate in the regulation of up to half of all its coding genes. In mammals, miRNAs regulate more than 60% of all protein-coding genes, and a single miRNA can even target over 300 different transcripts (Skalsky and Cullen, 2010). miRNAs are involved in the regulation of various cellular and physiological processes, such as development, stress response, apoptosis, and immune response (Skalsky and Cullen, 2010).

Similar to protein-coding mRNAs, the biogenesis of miRNAs starts with the transcription of long primary miRNAs (pri-miRNAs) in the nucleus (Fig. 1) (Skalsky and Cullen, 2010). Pri-miRNAs, encoded by polycistronic transcripts or embedded within the introns of protein-coding genes, are regulated by independent promoters. Pri-miRNAs are typically capped at the 5′ end, are polyadenylated, and are composed of one or more 80-nt hairpins individually consisting of approximately 32-nt imperfect stem and a large terminal loop. This structure is recognized and processed by a microprocessor complex composed of the ribonuclease enzyme Drosha and dsRNA-binding protein DGCR8, producing approximately 60-nt precursor miRNAs (pre-miRNAs) (Trobaugh and Klimstra, 2017). Pre-miRNAs are exported into the cytoplasm by the exportin-5 and Ran-GTP complex and then processed by RNase III enzyme Dicer-1 associated with TAR RNA-binding protein (TRBP) with the help of a dsRNA-binding cofactor Loqs-PB, generating mature 22-nt RNA duplexes. The 22-nt duplex miRNA is loaded onto Ago-1 and forms the miRNA-programmed RISC (Bernier and Sagan, 2018; Skalsky and Cullen, 2010). Mature miRNAs typically bind to complementary sequences of the 3′ UTRs of target mRNAs and induce initial translation inhibition and mRNA degradation through slicing or decay (Skalsky and Cullen, 2010; Trobaugh and Klimstra, 2017). The fate of a target mRNA is determined by the degree of complementarity. Perfect complementarity generally results in endonucleolytic cleavage and target mRNA degradation, which is commonly reported in plants. Imperfect complementarity modulates translation and mRNA destabilization, which is mostly observed in animals and viral miRNA targets (Skalsky and Cullen, 2010).

2.3.2 miRNAs in Insect Antiviral Immunity

Viruses also encode miRNAs (vmiRNAs), which regulate viral RNAs or host targets or both. Conversely, host miRNAs target and disrupt viral nucleic acids (Mussabekova et al., 2017). Such interactions are important for virus pathogenesis because they determine the degree to which hosts restrict viral infection. The interactions are modified over evolutionary time as the genomes of both host and pathogen evolve (Gottwein and Cullen, 2008). vmiRNA biogenesis appears to be mediated solely by the host miRNA pathway, and no viral protein involved in miRNA processing has been identified so far. vmiRNAs exploit a conserved gene regulatory mechanism within host cells to modulate their own and cellular genes, thereby for exacerbating viral infection (Mussabekova et al., 2017;

Skalsky and Cullen, 2010). vmiRNAs regulate the fundamental cellular processes in immunity, apoptosis, and key steps in transition from latent to lytic infection of viruses. Unlike viral proteins, vmiRNAs regulate host and viral genes and avoid exposure to host immune forces (Mussabekova et al., 2017; Skalsky and Cullen, 2010). miRNAs play important roles in host antiviral defense and viral pathogenesis, making them ideal targets for therapeutic intervention. The emerging interface between viruses and miRNAs may uncover novel pathways that can promote or limit virus replication and contribute to drug design (Skalsky and Cullen, 2010).

Viral infection changes host miRNA profiles. Differentially expressed miRNAs may potentially target viral or cellular genes as part of an antiviral response. In turn, viruses may subvert the antiviral effects of cellular and viral miRNAs and benefit their own infection (Skalsky and Cullen, 2010). Hz-miR-24, which is upregulated by the viral infection in *Heliothis zea*, possesses predictable antiviral ability by targeting HvAV RNA polymerase (Hussain et al., 2008). *A. albopictus* miR-252 (alb-miR-252), an abundant cellular miRNA upregulated by the viral infection, participates in antiviral immunity in mosquitoes by targeting DENV-2 envelope protein (Yan et al., 2014). The inhibition of alb-miR-252 significantly enhances viral replication, whereas the oversupply of alb-miR-252 mimics reduces the rate of viral genomic RNA accumulation (Yan et al., 2014). *B. mori* miR-390 targets and effectively downregulates BmNPV *cg30* (Kang et al., 2018).

Host miRNAs modulated by viral infection target cellular genes and facilitate viral infection. In *A. albopictus*, *metalloprotease m41 FtsH* encodes a protein that enhances WNV replication. Upon WNV infection, aae-miR-2940 which positively targets *metalloprotease m41 FtsH* is downregulated, thus inhibiting viral replication (Slonchak et al., 2014). In *B. mori*, BmCPV infection highly upregulates *insulin-related peptide binding protein 2* (*IBP2*), a target of host miR-278-3p, which is involved in *B. mori* immune response. miR-278-3p overexpression downregulates *IBP2* expression. Synchronously, the transcription level of BmCPV RNA increases, indicating that miR-278-3p participates in defense against BmCPV infection (Wu et al., 2016). miR-956-3p level substantially decreased after DCV infection, whereas the cellular target of miR-956-3p, *Ect4*, is upregulated. In *Ect4* knocked down flies, DCV-induced mortality occurs quickly and virus accumulation increases (Monsanto-Hearne et al., 2017).

vmiRNAs target cellular genes, thereby disrupting miRNA biogenesis, cell proliferation and survival, and antiviral responses and facilitating viral replication. vmiRNAs also target viral genes, which regulate viral replication, thus facilitating the switching between lytic and latent viral phases. vmiRNAs target viral genes and regulate viral infection, especially during late-stage infection (Asgari, 2015). HvAV-miR-1, which originates from a stem-loop of the major capsid protein sequence of HvAV, targets and directs the transcriptional degradation of viral DNA polymerase I gene for the modulation of HvAV replication (Hussain et al., 2008). BmNPV-miR-3 targets and regulates the expression of viral *P6.9* and other late genes. The regulation of BmNPV late genes by BmNPV-miR-3 at the early stage of infection helps BmNPV escape host early immune response (Singh et al., 2014). AcMNPV-miR-1, which is encoded by *Autographa californica* NPV, restrains viral infection in cells, although it facilitates viral infection in larvae by targeting multiviral genes, such as *ac18*, *ac94*, and *ac95* (Zhu et al., 2016). hv-miR-246 and hv-miR-2959 are generated from a noncoding *persistency-associated gene 1* (*pag1*) of *Helicoverpa zea* nudivirus-1 (HzNV-1), which is the only viral transcript detectable during viral latency. These miRNAs target viral major early gene *hhi1*, which is responsible for the establishment of productive viral infection. hv-miR-246 and hv-miR-2959 directly induce latency in cells infected with a *pag1*-deleted latency-deficient virus, suggesting that vmiRNAs contribute to the establishment of viral latency (Wu et al., 2011). In the late phase of infection, vmiRNAs evade viral overreplication and compromise host survival. This condition facilitates viral dissemination and continuous virion production. In addition to targeting viral genes, cellular genes are targeted by vmiRNAs to facilitate viral infection. bmnpv-miR-1 from BmNPV targets the GTP-binding nuclear protein Ran, which plays an essential role in the export of pre-miRNA from the nucleus to the cytoplasm by forming a complex with exportin-5 (Asgari, 2015; Singh et al., 2012). bmnpv-miR-1 overexpression downregulates Ran expression, leading to a global reduction of mature miRNAs. Consequently, BmNPV viral load dramatically increases. The blocking of bmnpv-miR-1 upregulates Ran expression and prevents BmNPV replication (Asgari, 2015; Singh et al., 2012). Thus, vmiRNAs counter host antiviral systems by targeting cellular genes for effective proliferation. Overall, the regulation of viral replication by vmiRNA through targeting viral and/or cellular genes may be a general phenomenon of successful viral infection. No vmiRNA has been identified in cytoplasmic DNA viruses that infect neither vertebrates nor invertebrates (Asgari, 2015).

Most RNA viruses replicate in the cytoplasm, where they cannot access the nuclear protein Drosha. These viruses unlikely produce vmiRNAs that destroy viral genome or transcripts via gene silencing. But studies show that

Drosha shuttles between the nucleus and the cytoplasm during viral infection (Asgari, 2015; Skalsky and Cullen, 2010). vmiRNAs are usually produced from laboratory-engineered RNA viruses, including influenza virus, SINV, and VSV, without negatively affecting viral genome (Asgari, 2015). The first RNA virus that was reported to encode miRNA is HIV-1 (Whisnant et al., 2013). Since then, miRNAs encoded by invertebrate RNA viruses have also been identified. For example, the Kunjin strain of WNV, a cytoplasmic RNA virus of the *Flaviviridae* family, encodes an miRNA-like RNA named KUN-miR-1 from the terminal stem-loop structure (Hussain et al., 2012). In mosquito cells, the 3′ stem-loop of WNV transcript either from a plasmid or from a *Semliki Forest virus* (SFV) RNA replicon can produce mature KUN-miR-1. The silencing of *Dicer-1* rather than *Dicer-2* reduces KUN-miR-1 levels, and the knockdown of KUN-miR-1 significantly reduces viral RNA replication in mosquito cells. KUN-miR-1 positively regulates its target *GATA4*, which promotes WNV infection. Moreover, *GATA4* induction by KUN-miR-1 promotes viral replication (Hussain et al., 2012). An miRNA-like RNA produced from the 3′ UTR of DENV genomes (DENV-vsRNA-5) targets viral NS1 and regulates viral replication. DENV-vsRNA-5 exhibits similar miRNA characteristics, and its expression levels are Dicer-1 and Ago-1, but not Dicer-2 dependent (Hussain and Asgari, 2014). The presence of WNV- and DENV-derived miRNAs demonstrates that cytoplasmic RNA viruses also encode miRNAs through a cellular miRNA processing machinery for the modulation of viral infection.

2.3.3 miRNAs in Crustacean Antiviral Immunity

miRNAs from other invertebrates and related viruses are involved in antiviral defense. WSSV infection alters shrimp miRNA expression profiles (Kaewkascholkul et al., 2016; Yang et al., 2012), and WSSV itself encodes vmiRNAs to modulate host–virus interactions (He and Zhang, 2012; Huang et al., 2014; Liu et al., 2016). The high-throughput sequencing of small RNAs from WSSV-infected shrimp reveals that 25 are upregulated and 6 are downregulated. Most miRNAs are expressed at the early stages of WSSV infection, and miRNA biogenesis machineries, such as Drosha, Dicer-1, and Ago-1, are required for successful miRNA production (Verbruggen et al., 2016). These miRNAs are predicted to target cellular genes involved in innate immune response, including phagocytosis, apoptosis, autophagy, endocytosis, RNAi, Toll pathway, antimicrobial humoral response, and prophenoloxidase system (Huang et al., 2012; Huang and Zhang, 2012; Yang et al., 2012).

Shrimp miRNAs not only participate in innate immune regulation but also directly target viral genes. Shrimp miR-7 targets the 3′ UTR of the early gene wsv477, which is involved in early DNA replication and virus proliferation (Huang and Zhang, 2012). The injection of synthesized shrimp miR-7 strongly decreases the number of WSSV genomic copies. Crab miR-7 enhances WSSV replication by inhibiting Myd88, an adaptor protein in the Toll pathway (Huang et al., 2018b). Crab miR-217 targets another member of this pathway called tube and promotes WSSV infection (Huang et al., 2017d). Shrimp miR-965 upregulated by WSSV infection inhibits viral replication by targeting *wsv240* and phagocytosis-related gene *autophagy-related protein 5* (*ATG5*) (Shu et al., 2016). miR-10a, the most highly upregulated shrimp miRNA during WSSV infection, promotes viral replication by targeting the 5′ UTR of at least three viral genes, namely, *vp26*, *vp28*, and *wssv102*, to improve their translation. In miR-10a-inhibited shrimp, the number of WSSV genomic copies decreases (Huang et al., 2017b). Shrimp miR-12 simultaneously induces antiviral phagocytosis and apoptosis to inhibit viral proliferation by synchronously targeting and degrading cellular *PTEN*, *BI-1*, and *wsv024* (Shu and Zhang, 2017). The overexpression of shrimp miR-34, which is upregulated during WSSV infection, considerably decreases the number of WSSV genomic copies and shrimp mortality, and thus miR-34 possibly plays a positive role in antiviral defense (Cui et al., 2017). miR-71 targets *cap-1* and miR-13b targeted *knickkopf* to simultaneously regulate viral infection and autophagy. In miR-71- and miR-13b-knockdown shrimp, autophagy cannot be induced by WSSV infection, although under this condition autophagy activation unlikely influences viral replication, indicating that miR-71 and miR-13b are essential for WSSV infection and autophagy (He et al., 2017b). miR-100 from *M. japonicus* functions as an anti-apoptosis miRNA by targeting *trypsin*. miR-100 inhibition increases the apoptotic rate of shrimp hemocytes and prevents WSSV infection (Yang et al., 2014). miR-100 not only targets *p53* to inhibit apoptosis but also affects phagocytosis, phenoloxidase activity, superoxide dismutase activity, and total hemocyte count. The knockdown of miR-100 reduces the mortality of WSSV-infected shrimp (Wang and Zhu, 2017). Shrimp miRNA-S5, which is upregulated by WSSV infection, facilitates viral replication by regulating the phagocytosis and apoptosis of hemocytes (Wang and Zhu, 2018). Shrimp miRNA-9041 and miRNA-9850 targets *STAT* to suppress the induction of *dynamin* via the JAK/STAT pathway, increasing virus replication (Huang et al., 2016). Shrimp miRNA-1959 targets the 3′ UTR of the IκB ortholog *Cactus* and induces its degradation, thereby enhancing the activation of NF-κB-like transcription factor Dorsal (Zuo et al., 2016), which is an activator of some WSSV genes (Wang et al., 2011, 2013a). Thus, miRNA-1959 promotes WSSV infection by upregulating viral gene expression and enhancing WSSV replication (Xu et al., 2016c). In summary, crustacean miRNAs function in antiviral responses by targeting host and viral genes.

WSSV expresses more than 100 vmiRNAs, and most of them are detected at the early stages of infection (He and Zhang, 2012). Drosha, Dicer-1, and Ago-1 are required for successful vmiRNA production. WSSV-vmiRNAs regulate the balance between the promotion of viral infection and its latency. WSSV-vmiRNAs target *wsv094*, *wsv177*, *wsv248*, and *wsv309*, which are related to latency. WSSV-miR-66 targets *wsv094* and *wsv177*, whereas WSSV-miR-68 targets *wsv248* and *wsv309* to promote WSSV infection (He et al., 2014). In shrimp, both WSSV-miR-N13 and WSSV-miR-N23 target *Dorsal* to suppress Toll pathway-induced antiviral immune response, thereby exacerbating WSSV infection (Ren et al., 2017). WSSV-miR-N24 targets *caspase 8* and represses shrimp antiviral apoptosis, thereby increasing the number of WSSV genomic copies (Huang et al., 2014). WSSV-miR-22 promotes WSSV infection by targeting *STAT* to block the activation of the antiviral JAK/STAT pathway (Ren et al., 2015). WSSV-miR-N12 targets *wsv399* and induces virus latency (Cui et al., 2015). WSSV-miR-N32 can synchronously target two viral targets, *wsv459* and *wsv322*, remarkably reducing WSSV proliferation (He et al., 2017a). Overall, WSSV-vmiRNAs function as regulatory factors involved in virus life cycle and host antiviral immunity.

2.3.4 Immune Sensing of miRNAs by PRRs

Some miRNAs can be detected as PAMPs by innate immune system, which is independent of their conventional role in repressing gene expression. Tumor-secreted miR-21 and miR-29a bind to human TLR8 or murine TLR7 as ligands to trigger a TLR-dependent NF-κB activation and promote the secretion of pro-inflammatory and pro-metastatic cytokines, such as IL-6 and TNF-α (Fabbri et al., 2012). The sequence of let-7b contains a core GU-rich motif that is also present in HIV ssRNA40, a known ligand of TLR7. Extracellular let-7b acts as a potent activator of TLR7 signaling in mouse neurons. Synthetic let-7b induces the release of TNF-α from microglia and macrophages by directly activating TLR7 signaling. Under pathological conditions, let-7b is released to stimulate TLR7, thereby sending a danger signal to neurons and accelerating their decay. The GU-rich motif seems to be essential for the sensing of miRNAs by TLRs

(Fabbri et al., 2012; Lehmann et al., 2012). However, whether other miRNAs and vmiRNAs can be sensed as PAMPs and activate immune response remains unclear.

2.4 piRNAs in Arthropod Antiviral Immunity

2.4.1 piRNA Biogenesis

piRNAs contain 24–31 nts and are thus slightly longer than miRNAs and siRNAs, possess 2′-O-methyl modification sites at the 3′ end, and exhibit strong bias for uridine at the 5′ end (1U) (Iwasaki et al., 2015). piRNAs are generated in distinct genomic regions called piRNA clusters via a Dicer-independent mechanism (Huang et al., 2017c) (Fig. 1). piRNAs bind to the PIWI subfamily proteins of the Ago family and guide the proteins to target RNAs through base pairing. Similar to other Ago family proteins, PIWI proteins exert an endonucleolytic slicer activity that induces post-transcriptional silencing through the cleavage of complementary RNAs (Huang et al., 2017c; Miesen et al., 2015). piRNAs are generated via two pathways: primary processing and ping-pong amplification (Iwasaki et al., 2015). Long piRNA precursors are transcribed from piRNA clusters with the help of specific transcription factors. piRNA precursors are exported to the cytoplasm and possibly processed into intermediates, but the detailed processes remain unknown (Huang et al., 2017c). PIWI and Aub are mainly involved in primary piRNA biogenesis. Aub- and PIWI-bound primary piRNAs are antisense and have a bias toward having uridine (U) at their 5′ end (1U bias) (Huang et al., 2017c; Iwasaki et al., 2015). Once bound, piRNA intermediates are trimmed and 2′-O methylated at the 3′ end, forming mature PIWI-piRNA-induced silencing complexes that are subsequently transported to the nucleus. These complexes then cleave complementary transcripts (Wang et al., 2018e). This process generates a new sense piRNA with adenosine at position 10 (10A). After being loaded onto Ago-3 and 2′-O-methylated at the 3′ end, the secondary piRNAs generate Aub-bound piRNAs through the same mechanism. This piRNA amplification loop is referred to as the "ping-pong" cycle (Fig. 1) (Iwasaki et al., 2015). PIWI proteins and their associated piRNAs are highly enriched in the germline tissues of animals. The canonical function of the piRNA pathway is to protect genome integrity in animal germ cells by silencing transposons and selfish genetic elements with the ability to randomly integrate into a host's genome (Lewis et al., 2018).

2.4.2 piRNAs in Insect Antiviral Immunity

The piRNA pathway also participates in insect antiviral immunity. A class of virus-derived piRNAs (vpiRNA) are identified from *A. aegypti*, *A. Albopictus*, and mosquito cells infected with viruses (Iwasaki et al., 2015; Miesen et al., 2015). vpiRNAs derived from the integrated endogenous viral elements of the vertebrate genomes are also detected (Guo et al., 2018). vpiRNA biogenesis is independent of the siRNA and miRNA pathways. Thus, vpiRNAs possibly constitute a novel line of defense against arboviruses in invertebrates.

The initial evidence of vpiRNA production was obtained in *Drosophila* ovarian somatic sheet (OSS) cells that were persistently infected with several RNA viruses, including dsRNA viruses, such as DXV, American nodavirus, *Drosophila* birnavirus, and *Drosophila* tetravirus, as well as positive sense ssRNA viruses, such as DCV and norovirus, suggesting that the piRNA pathway may be involved in antiviral response (Wu et al., 2010). OSS cells express PIWI proteins but not Aub and Ago-3. Accordingly, vpiRNAs produced in OSS cells show a characteristic 5′ 1U bias but have no 10A bias, indicating the absence of the secondary amplification of vpiRNAs (Wu et al., 2010). vpiRNAs are absent in adult flies (Lewis et al., 2018; Petit et al., 2016). Although PIWI proteins are abundantly expressed in *Drosophila* germ cells, no detectable vpiRNAs are found in the cells naturally infected with *D. melanogaster* Sigma virus (MelSV). *Drosophila* PIWI proteins are largely confined to gonads and somatic cells lacking the necessary piRNA pathway elements, and the piRNA pathway may not involve in *Drosophila* antiviral immunity (Wang et al., 2018e).

The PIWI subfamily proteins are greatly expanded in mosquitoes and widely expressed in somatic tissues, in contrast to germline-specific expression in *Drosophila*, raising the possibility that the piRNA pathway has expanded functions in mosquitoes (Miesen et al., 2016b). The piRNA pathway participates in arbovirus replication and contributes to antiviral immunity. vpiRNAs are expressed in the cells and somatic tissues of mosquitoes infected with various arboviruses, including family members of *Togaviridae*, *Flaviviridae*, *Bunyaviridae*, and *Reoviridae* (Leggewie and Schnettler, 2018; Miesen et al., 2016b). vpiRNAs not only exhibit the typical size distribution of small ncRNAs (around 24–30 nts) but also show signatures of 5′ 1U and 10A biases, suggesting that both primary and secondary piRNAs are produced when mosquitoes are virally infected (Leggewie and Schnettler, 2018; Miesen et al., 2016b). vpiRNAs from *Flaviviridae*, such as DENV and *Togaviridae*, including SINV, show 2′-*O*-methyl modification at its 3′

end (Miesen et al., 2015). This modification is present in all PIWI-loaded mature piRNAs. vpiRNAs from *Togaviridae* arboviruses are directly associated with PIWI proteins and produced in a PIWI-dependent manner. vpiRNAs are identified in C6/36 cells infected with cell-fusing agent virus (*Flaviviridae*), SINV, or La Crosse virus (*Bunyaviridae*) (Brackney et al., 2010; Nayak et al., 2013). In Aag2 cells, vpiRNA production from SINV and DENV is mainly dependent on PIWI-5 and Ago-3, and the silencing of *PIWI-5* and *Ago-3* causes a decrease in vpiRNA production (Miesen et al., 2016a). PIWI-5 binds to antisense vpiRNAs with 1U-bias, and Ago-3 binds sense vpiRNAs with 10A-bias. PIWI-5 cleaves the sense strand of SINV and generates secondary sense-strand piRNAs, which are subsequently loaded into Ago-3. Thus, ping-pong amplification is initiated by PIWI-5-bound primary piRNAs from the SINV antisense strand (Miesen et al., 2015). In mosquitoes *A. aegypti* and *A. albopictus*, CHIKV infection induces the production of vpiRNAs, in addition to vsiRNAs, indicating that both pathways target CHIKV in somatic cells (Morazzani et al., 2012). Ping-pong-amplified vpiRNAs are present 9 days after CHIKV infection in *A. albopictus*. vsiRNAs and vpiRNAs are both detectable in *A. albopictus* infected with DENV-2. vpiRNAs are the second most abundant viral small ncRNAs after vsiRNAs (Wang et al., 2018e). vpiRNAs are primarily detected after the first wave of vsiRNAs. Moreover, vpiRNAs vastly outnumber vsiRNAs at 72h post-infection of Rift Valley fever virus (*Bunyaviridae*). In mosquitoes, post-DENV-2 infection, vpiRNAs are the most dominant vncRNAs on day 2, whereas vsiRNAs reach the highest level on day 9 (Miesen et al., 2016b). The differential accumulations of vsiRNAs and vpiRNAs during infection may shape antiviral response in mosquitoes. The proportions of vsiRNAs and vpiRNAs are essential to the transition from acute to persistent infection. Nevertheless, the exact and distinct roles of vsiRNAS and vpiRNAs in viral infection and antiviral immunity require further investigation (Miesen et al., 2016b).

The antiviral function of the piRNA pathway is demonstrated in arbovirus-infected *A. albopictus* cell line C6/36, an ideal model for piRNA studies on its *Dicer-2* deficiency (Brackney et al., 2010; Hess et al., 2011; Nayak et al., 2013). In C6/36 cells, siRNA-mediated antiviral immunity is defective, whereas vpiRNAs can mount antiviral responses. The knockdown of *PIWI-4* in *A. aegypti* cells increases SFV viral loads. Similarly, the knockdown of *Ago-3* in *Anopheles* mosquitoes increases ONNV titers, suggesting that the piRNA pathway plays a role in virus clearance (Miesen et al., 2016b; Nayak et al., 2013). Many virus-encoded VSRs target siRNA pathway; however, no convictive data show the suppression of the piRNA pathway by arboviruses, suggesting that the siRNA pathway still plays a major role in mosquito antiviral defense (Miesen et al., 2016b; Nayak et al., 2013). In contrast to with wild-type CHIKV, CHIKV carrying FHV B2 induces a slight decrease in piRNA levels during infection (Morazzani et al., 2012). However, whether vpiRNA downregulation is caused by directly targeting the piRNA pathway, indirect effects, effects of RNA abundance, or effects of accessibility remains unclear (Morazzani et al., 2012). Furthermore, the possibility that arboviruses have mechanisms for evading or disrupting the piRNA pathway is interesting and can feature the importance of the piRNA pathway in mosquito antiviral immunity. Up to date, studies performed in mosquitoes have not yet directly demonstrated the antiviral ability of vpiRNAs in vivo (Miesen et al., 2015). Therefore, novel hypotheses and designs of experiments may help to clarify the function of vpiRNAs (Guo et al., 2018).

2.4.2.1 piRNAs in Other Arthropods

PIWI proteins are frequently and highly expressed in the germline tissues of vertebrates. Their somatic expression levels in neurons, cancer cells, and stem cells are known, but whether these somatically expressed PIWI proteins target viral RNAs and contribute to vpiRNA production remains elusive (Iwasaki et al., 2015). Studies of the invertebrate piRNA pathway are mostly performed in flies, mosquitoes, and nematodes. The duplication of the somatic expression of PIWI proteins is reported in the pea aphid *Acyrthosiphon pisum* suggesting that a functional piRNA pathway may be involved in vpiRNA biogenesis in insects other than mosquitoes (Miesen et al., 2016b). One recent study reported that the germline expressions of piRNAs were detected in all 20 arthropods examined, whereas piRNAs were only readily found in somatic tissues of 16 arthropod species, including three chelicerates *Limulus polyphemus*, *Centruroides sculpturatus*, and *Parasteatoda tepidariorum*, one myriapod *Strigamia maritima*, and 12 insect species (Lewis et al., 2018). vpiRNAs with 25–30 nts and 1U bias derived from RNA viruses and DNA viruses are found in four arthropod species, including *C. sculpturatus*, *Tribolium castaneum*, *Diabrotica virgifera*, and *Plutella xylostella*. However, no ping-pong amplification of vpiRNAs was detected in any of these species (Lewis et al., 2018). In crustaceans, no piRNA has been reported. WSSV encodes numerous small RNAs with 21–23 nts and with the feature of Dicer products. These RNAs do not show the typical characteristics of 1U or 10A bias of piRNAs and may not be piRNAs (Liu et al., 2016). Although the piRNA pathway may involve in antiviral immunity against some viruses in certain arthropod species, the siRNA pathway seems to play a major role in defense against most viruses across the arthropods. The ancestral role of the piRNA pathway was to protect the germline; and somatic piRNAs targeting transposable elements and mRNAs are common among arthropods (Lewis et al., 2018).

2.4.3 Substrate for vpiRNA Biogenesis

In the siRNA pathway, cytoplasmic dsRNAs serve as danger signals during viral infection, act as PAMPs for Dicer-2 sensing to generate vsiRNAs and activate antiviral immunity in insects. In the piRNA pathway, PIWI proteins sense viral ssRNAs as substrates for vpiRNA biogenesis (Huang et al., 2017c). However, the mechanism by which PIWI proteins distinguishes viral RNA from host RNAs and determines which viral RNA to use for vpiRNA biogenesis remains elusive (Miesen et al., 2016b). Vertebrate PRRs recognize viral RNAs from self-RNA through non-self-RNA features (Tan et al., 2018; Wu and Chen, 2014). Nevertheless, studies have yet to determine whether PIWI proteins act as immune sensors to specifically recognize viral RNAs or are recruited to non-self-RNA features by other helper proteins in the same manner as some PRRs in mammals (Miesen et al., 2016b). In contrast to vsiRNAs produced at equal levels of the arbovirus genomes, the target selection of viral genomes by PIWI proteins indicates that vpiRNAs are mostly confined to specific hotspot regions (Miesen et al., 2015). vpiRNAs are predominantly produced from a subgenomic ssRNA transcribed from internal promoters, such as those in some alphaviruses SINV, CHIKV, and SFV (Miesen et al., 2015). This observation is explained by the increased expression of subgenomic RNA relative to genomic RNA. For example, a subgenomic promoter yields more subgenomic ssRNA than full-length genomic RNA during SINV infection (Miesen et al., 2016b). Another possibility is that subgenomic ssRNA is more accessible for the piRNA machinery because it is required for the translation of structural proteins at later stages of infection (Miesen et al., 2016b).

2.5 Antiviral RNA Silencing Pathways in Nematodes

RNAi, which was first described in *C. elegans*, operates similarly to a defense mechanism of insects against viral infection (Fire et al., 1998; Nayak et al., 2013). *C. elegans* genome encodes a single *Dicer-1* (*CeDicer-1*) and 27 *Ago* family genes (de Faria et al., 2013). CeDicer-1 senses and processes dsRNAs for the siRNA and miRNA pathways. Primary siRNAs processed by

CeDicer-1 are predominantly 23 nts bound by RNA interference-deficient 1 (RDE-1), which is Ago-2 ortholog. RRF-1, a *C. elegans* RdRp required for RNAi amplification, amplifies primary siRNAs independent of Dicer-1 processing and generates secondary siRNAs incorporated directly onto WAGOs, which are nematode-specific Ago family proteins (de Faria et al., 2013). *C. elegans* secondary siRNAs are 22-nt ssRNAs with guanidine (G) in the 5′ end and antisense to a target mRNA, often referred to as 22G RNAs and thought to be the downstream effectors of RNAi pathways (Guo et al., 2013). Virus-derived small RNAs are detected in *C. elegans* infected with FHV and VSV, which are not their natural pathogens. Increased viral titers are observed in *C. elegans* carrying mutant RNAi pathway components (Guo et al., 2018). *C. elegans* cells defective for *RDE-1*, *RDE-4* (R2D2 ortholog), or *RRF-1* is hypersensitive to VSV infection (Schott et al., 2005; Wilkins et al., 2005). Moreover, *C. elegans* lacking *RDE-1* supports the efficient replication of B2-deficient FHV (Lu et al., 2005). vsiRNAs in FHV-infected *C. elegans* are derived throughout the full genome with an antisense: sense ratio of 2:1 likely because of the preferential generation of secondary siRNAs from an antisense strand by RRF-1 (de Faria et al., 2013). vsiRNAs are severely reduced in *RDE-1* mutant, and they even disappear in *RDE-4* or *RFF-1* mutants, suggesting the important roles of these siRNA components in siRNA biogenesis or stability (de Faria et al., 2013). Orsay virus, a natural pathogen of *C. elegans*, readily infects *C. elegans* carrying *RDE-1* or *RDE-4* mutants and yields higher levels of viral RNAs and infection symptoms than wild-type strains (Felix et al., 2011). A small RNA library from *C. elegans* infected with the Orsay virus shows that sense vsiRNAs have a broad size distribution and have no 5′ nucleotide bias, which may represent Dicer cleavage products or viral RNA degradation intermediates. By contrast, most antisense viral small RNAs have a clear 22-nt peak in the size distribution and a G preference at the first base of the 5′ end and serve as the signature of 22G RNAs (Felix et al., 2011). *C. elegans* has three Dicer-related helicases (DRHs), which are absent in insects but are implicated in the siRNA pathway (Fig. 4). The helicase domains of the DRHs are highly homologous to those of mammalian RIG-I receptors. DRH-1 is essential to antiviral defense against FHV. DRH-1 can facilitate the biogenesis of siRNAs by catalyzing the translocation of the Dicer complex on viral long dsRNA precursors (Coffman et al., 2017). siRNAs induced either by exogenous dsRNA or by viral replicon are enhanced in *DRH-2 C. elegans* mutants, suggesting that DRH-2 functions as a negative regulator of antiviral immunity. DRH-3, sharing a similar domain structure with DRH-1 (Fig. 4), is required for the biogenesis of secondary but not primary siRNAs, suggesting that DRH-1 and DRH-3 contribute to siRNA-mediated antiviral immunity through distinct mechanisms (Guo et al., 2013).

miRNAs are initially identified in *C. elegans*, however, no vmiRNA has been identified in nematodes (Lee et al., 1993). A pool of piRNAs precisely with 21 nts and 5′ U called 21U-RNAs is generated in *C. elegans*. Unlike insects and vertebrates, *C. elegans* cannot produce secondary piRNAs via the ping-pong amplification loop. Although piRNA pathway is extensively studied in *C. elegans*, no vpiRNA has been detected in this animal (Iwasaki et al., 2015).

2.6 Antiviral RNA Silencing Pathways in Mollusks

Oysters of the phylum Mollusca are common invertebrates in aquatic ecology and have considerable economic value in aquaculture worldwide. The complete genome of the Pacific oyster *Crassostrea gigas* has been sequenced and assembled (Green and Speck, 2018). The *C. gigas* genome encodes the core elements of antiviral RNAi machinery, such as Dicer and Ago-2. Injection of dsRNAs can knockdown the expression of endogenous genes in a sequence-specific manner, thereby indicating that the siRNA-mediated gene silencing pathway is clearly functional in mollusks. However, whether the mollusk siRNA pathway is involved in host defense against viral infection remains unknown to date (Green et al., 2015; Green and Speck, 2018).

C. gigas small ncRNA libraries from hemocytes after bacterial infection have been sequenced. miRNA profiles are changed after immune stimulation, some of which are predicted to target and modulate genes related to immune response (Zhou et al., 2014). *Chlamys farreri* miRNAs and miRNA expression profiles in response to acute viral necrobiotic virus infection have been obtained by deep sequencing (Chen et al., 2014). These miRNAs may target immune- and stress-related genes for post-transcriptional regulation. Thus, miRNA pathway is completely functional in oysters, but whether it contributes to antiviral defense is still unknown (Green et al., 2015).

In mollusks, the somatic expression of PIWI genes and piRNAs directed against transposons is conserved (Waldron et al., 2018). Metagenomic sequencing has been used to identify a distinct group of RNA virus-derived small ncRNAs in a gastropod mollusk, that is, the dog whelk. However, different from the vsiRNAs of nematodes and arthropods, these vncRNAs have a piRNA-like length distribution broadly at 26–30 nt, which is normal for piRNAs, but they lack the 5′ U and ping-pong signature of piRNA biogenesis. The primary piRNAs derived from an endogenous RNA virus have also been identified in dog whelks. However, whether these vpiRNAs play a role in antiviral defense is still unknown (Waldron et al., 2018).

2.7 Antiviral RNA Silencing Pathways in Other Invertebrates

vncRNAs are absent in RNA virus-infected animals, such as earthworms, sponges, or the sea anemones. This finding suggests that nematode- or arthropod-like antiviral RNAi is absent from these organisms and may challenge the widely held assumption that most invertebrates possess an antiviral RNAi pathway that is similar to that in arthropods and nematodes (Waldron et al., 2018).

3. NON-SPECIFIC ANTIVIRAL IMMUNITY THROUGH PPRS IN INVERTEBRATES

3.1 Nucleic Acid-Induced Antiviral Immunity in Invertebrates

RNAi is the primary antiviral response in invertebrates such as arthropods and nematodes, whereas a protein-based IFN system mediates the major antiviral defense in vertebrates (Ding, 2010; Guo et al., 2018). Although invertebrates lack a canonical IFN response, various nucleic acids can activate non-specific antiviral responses and induce a set of immune-related genes in some invertebrates similar to IFN response in vertebrates (Wang et al., 2013e, 2014, 2015b). The transfection of unrelated ssRNA or dsRNA into sandfly LL5 cells confers considerable resistance to viral infection and reduces viral replication (Wang et al., 2013e). dsRNA delivery into honeybees activates an inducible antiviral response and highly induces a set of novel uncharacterized genes, which are atypical immune responsive genes of insects (Wang et al., 2013e). The dsRNA-induced antiviral defense in honeybees may regulate unique genes via unknown immune pathways. Supplying dsRNA into the syrup can increase the resistance of honeybee hives against dicistrovirus infections (Brutscher et al., 2015; McMenamin et al., 2018).

In shrimp, various nucleic acids, including dsRNA mimics poly(I:C) and poly(C:G), ssRNA mimics CL097 and poly C, and DNA mimic CpG-DNA, activate strong antiviral responses and protects shrimp from WSSV or TSV infection (Wang et al., 2013e). Poly(I:C) injection upregulates a series of genes, 45% of which are completely novel and different from any known genes. Meanwhile, some genes are homologous to IFN system-defined genes, such as IRFs, PKR, ADAR, and other ISGs (Wang et al., 2013e). Shrimp also has a protein-based antiviral immunity possibly mediated by Vago similar to vertebrate IFN responses. However, whether Vago upregulates dsRNA-induced genes remains unknown (Chen et al., 2011; Li et al., 2015). Vago, which was first identified in insects, is a novel antiviral effector induced after viral RNA sensing by Dicer-2 that activates the JAK/STAT pathway to establish a protein-based non-specific antiviral response (Deddouche et al., 2008). Whether dsRNA or ssRNA mimics induce *Vago* via Dicer-2 is still unknown (Paradkar et al., 2012, 2014). Although Vago-dependent antiviral response shows high similarity to IFN response, Vago does not exhibit any sequence homology to IFNs (Deddouche et al., 2008; Paradkar et al., 2012). In mollusk, the *C. gigas* genome encodes the orthologs of RLRs, MAVS, TRAF6, TBK1, IRF family proteins, IFN-like protein (IFNLP), JAK/STAT, and ISGs, which represent almost all the core elements of IFN response in vertebrates (Huang et al., 2017a; Zhang et al., 2015b). The injection of poly(I:C) into *C. gigas* activates IFN-like response; upregulates the expression of ISGs, such as ADAR, viperin, PKR, and IFI44; and limits OsHV-1 replication and distribution. Poly(I:C)-injected *C. gigas* has improved survival rate after OsHV-1 infection later in life. This resistance can even be passed onto the next generations (Green and Speck, 2018). Therefore, nucleic acid sensing or viral infection can trigger protein-based antiviral immunity in invertebrates.

3.2 Toll, IMD, and JAK/STAT Pathways in Antiviral Immunity

In insects, Toll, immune deficiency (IMD), and JAK/STAT are the major innate immune pathways controlling most immune responsive genes induced by microbial pathogens (Fig. 2) (Buchon et al., 2014; Merkling and van Rij, 2013; Myllymaki et al., 2014). These pathways regulate immune factors that modulate systemic immune responses, directly kill microbial pathogens, or participate in phagocytosis, reactive oxygen species production, and melanization cascades (Merkling and van Rij, 2013). Although these pathways mainly function in the defense against bacterial and fungal infections, they also play a role in antiviral immunity. However, the initiation and activation mechanisms during viral infection and effector gene function remain poorly characterized.

3.2.1 Toll Pathway in Invertebrates

The TLR family members are greatly diverse in numbers and types. For example, the nematode *C. elegans* has only one Toll receptor, whereas oysters have at least 83 TLRs. *C. elegans* has no canonical TLR pathway

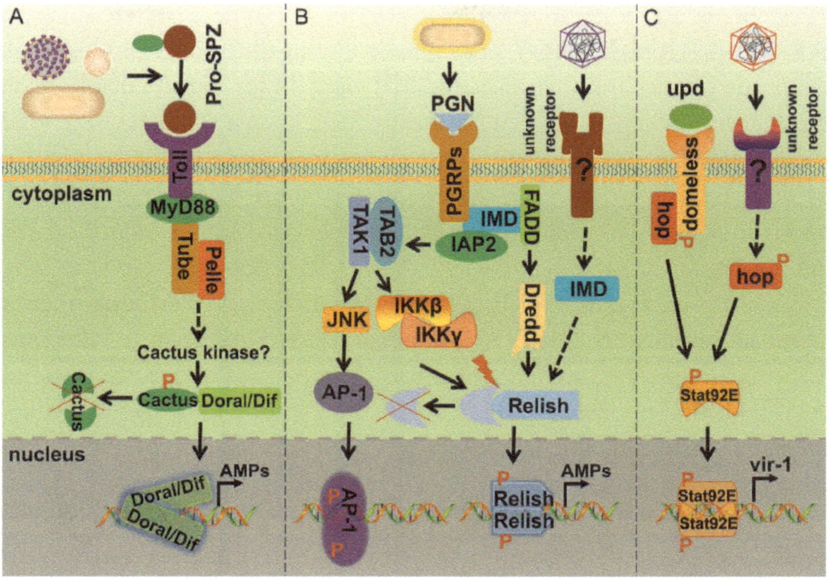

Fig. 2 Immune signaling pathways in insects that share the core elements of mammalian NF-κB, IRF3, and JAK/STAT pathways. (A) Gram-positive bacteria and fungi infections trigger the pro-Spätzle cleavage into active Spätzle, which binds to Toll and activates the intracellular signaling cascade involving MyD88, Tube, Pell, Cactus, and Dorsal/Dif, thereby leading to AMP expression. Certain viruses also activate the Toll pathway and similarly induce AMP expression. During this process, *Spätzle* is required for Toll activation, thereby suggesting that the putative virus sensor should be upstream Spätzle. (B) PGN derived from Gram-negative bacteria activates the IMD pathway and upregulates AMP expression via PGN-recognition protein (PGRP)-LC and -LE receptors involved in IMD, dFADD, IAP2, TAB2/TAK1 complex, IKKβ/IKKγ complex, Dredd, and Relish. IMD activation also induces the JNK pathway through the TAB2/TAK1 complex. Viruses activate the IMD-Relish pathway via unknown receptors and unclear signal transduction mechanisms. (C) The JAK/STAT pathway has an evolutionarily conserved antiviral function from arthropods to mammals. In *Drosophila*, the putative ligands of the receptor domeless, such as upd, upd2, and upd3, activate the JAK kinase hopscotch and induce Stat92E phosphorylation, thereby upregulating antiviral genes, such as *virus-induced RNA-1* (*vir-1*).

components, such as MyD88, IKK, and NF-κB. Thus, TLR/NF-κB signals are absent in this organism. TLR family members are classified into the following categories according to the number of cysteine clusters at the C-terminal ends of leucine-rich repeats (LRRCT): (1) protostome type (P-type) or mccTLR, which has multiple cysteine clusters at LRRCT and (2) vertebrate type (V-type) or sccTLR, which has a single cysteine cluster at LRRCT (Brennan and Gilmore, 2018). V-type TLRs are distributed in vertebrates and some invertebrates. A typical V-type TLR consists of an LRRNT–(LRR)n–LRRCT ectodomain for ligand recognition, a transmembrane region, and a cytoplasmic TIR domain for intracellular signaling transduction (Wang et al., 2012). Meanwhile, P-type TLRs are identified in invertebrates, and they represent an ancient form of TLRs; they do not directly bind to PAMPs; the binding of these TLRs is facilitated by endogenous ligands, similar to the binding process of *Drosophila* Toll (Nie et al., 2018).

Toll, also known as Toll-1, is first characterized as a key component of antifungal immune response in *D. melanogaster.* Subsequently, numerous TLRs have been reported in mammals. V-type TLRs directly recognize PAMPs; however, some invertebrate Tolls are P-type TLRs, such as *Drosophila* Toll-1, which functions as a receptor for the secreted cytokine Spätzle but does not directly recognize pathogens (Wang et al., 2014). Infection by Gram-positive bacteria, fungi, and certain viruses triggers an extracellular proteolytic cascade, resulting in the cleavage of pro-Spätzle into active Spätzle, which then binds to the transmembrane receptor Toll. Subsequently, intracellular signaling involved in MyD88, Tube, Pelle, TRAF6, Cactus, and other factors leads to the activation of NF-κB-like transcription factors Dorsal/Dif. The activated Dorsal translocates to the nucleus, binds to NF-κB responsive elements, and initiates the transcription of a set of immune-related genes, including antimicrobial peptides (AMPs), such as *drosomycin* and *defensin* (Fig. 2). Nine Toll proteins are found in *D. melanogaster.* Among them, Toll-9 is a V-type TLR, whereas the others belong to P-type TLRs. However, Toll-7 directly recognizes viral glycoproteins as a PRR similar to mammalian TLRs and induces antiviral autophagy (Merkling and van Rij, 2013; Wang et al., 2014).

No component of the *Drosophila* Toll pathway acts as the direct sensor of viruses, but viruses, such as FHV, VSV, and DXV, activate this pathway and induce AMP expression. Flies carrying mutations of Toll pathway components, such as *Spätzle*, *Toll*, *tube*, and *pelle*, show increased susceptibility to oral infection but not to the systemic infection of DCV, CrPV, Nora virus, or FHV (Ferreira et al., 2014). MelSV infection induces the upregulation of AMPs (Ferreira et al., 2014). Flies carrying the *Dif* mutant exhibit high viral titers and increased susceptibility and succumb fast after DXV infection. However, the overexpression of *Drosophila* AMPs in transgenic flies cannot confer protection against DXV, suggesting that AMPs may indirectly provide protection against viral infection. In *A. aegypti*, DENV infection induces the upregulation of *Spätzle*, *Toll*, and *Rel1A* and Toll-dependent AMPs, such as *cecropin* and *defensin*. DENV viral load is negatively controlled by Cactus, the ortholog of mammalian IκB that functions as the inhibitor of the NF-κB-like transcription factor of the Toll pathway (Merkling and van Rij, 2013). The knockdown of *MyD88* increases viral titers after DENV infection. The ectopic expression of *cecropin* and *defensin* in transgenic *A. aegypti* inhibits DENV replication. Chemically synthesized cecropin protein inhibits the replication of DENV and CHIKV. AMPs seem to participate in antiviral responses in mosquitoes, but the action mechanism remains unclear (Merkling and van Rij, 2013).

The Spätzle-Toll-MyD88-Pelle-TRAF6-Cactus-Dorsal-AMP cascades are comprehensively studied in crustaceans (Wang et al., 2011, 2013b, c). *LvToll2* is upregulated after WSSV or TSV infection; when transfected into cells, LvToll2 activates AMP promoters (Wang et al., 2012). The knockdown of Toll pathway components reduces the induction of shrimp AMPs, including *penaeidins*, *crustins*, *anti-lipopolysaccharide factors* (*ALFs*), and *lysozymes* in vivo (Wang et al., 2014). *LvSpätzle* is upregulated after WSSV infection. The injection of the recombinant shrimp Spätzle into crayfish upregulates AMP expression, suggesting a functional Spätzle-Toll signaling exists in shrimp (Wang et al., 2014). In *P. monodon*, *PmMyD88*, *PmTRAF6*, and *PmToll* are upregulated by WSSV infection and poly(I:C) stimulation (Deepika et al., 2014). Shrimp *Toll7*, which is grouped with *Drosophila* Toll7, is induced by WSSV infection and is involved in the regulation of AMP expression (Feng et al., 2016). *Procambarus clarkia Toll5* (*PcToll5*) and *Toll6* (*PcToll6*) are remarkably upregulated after WSSV challenge. The knockdown of *PcToll5* and *PcToll6* results in *ALF* downregulation. In *Drosophila* S2, PcToll5 and PcToll6 activate the promoters of both shrimp and insect AMPs (Huang et al., 2018a). WSSV has developed strategies to subvert the shrimp Toll pathway and to facilitate its replication. WSSV infection or WSSV449 overexpression activates LvDorsal and induces its translocation from the cytoplasm to the nucleus, subsequently promoting its replication. LvDorsal strongly activates the promoters of *WSSV069* (*ie1*) and *WSSV303*. In *LvDorsal*-knockdown shrimp, WSSV genomic copies are remarkably reduced (Qiu et al., 2014). Although many viruses activate the shrimp Toll pathway, the direct sensing of viruses or viral nucleic acids via this pathway has not been observed.

CpG oligodeoxynucleotides (ODNs), a kind of PAMP consisted of short synthetic ssDNAs, directly bind to TLR9 and chTLR21 in mammals and chickens, respectively, and induce the production of IFN-α and TNF-α (Sun et al., 2014). LvToll1 and LvToll3 exhibit affinities to CpG ODNs in a dose-dependent manner. The luciferase activities of ISRE reporter induced by LvToll1 and LvToll3 are considerably enhanced by CpG ODNs. TNF-α activities in shrimp hemocytes are significantly enhanced by CpG ODN stimulation, which may depend on the activation of MyD88-dependent Toll pathway (Sun et al., 2014). In shrimp hemocytes, CpG ODNs can be uptaken, internalized, and distributed in the cytoplasm with aggregated signals around the nuclei. The binding between CpG ODNs and LvToll1–3 as well as mature endosomes in hemocytes is necessary for the triggering of immune responses by CpG ODNs (Sun et al., 2014). Moreover, four types of CpG ODNs (CpG ODN-A-D) displayed different binding affinities to LvToll1–3, and the highest binding affinity is observed between CpG ODN-A and LvToll1 (Xu et al., 2016b). CpG ODN-A exhibits the highest upregulation of *LvToll1–3* expression among the known CpG ODN-A-D types. After CpG-A stimulation, the probable IFN level in shrimp plasma is upregulated, and the expression levels of IFN system-like antiviral genes, such as *LvIRF*, *LvVago-4*, and *LvSTAT*, are significantly induced (Xu et al., 2016b). CpG ODNs are the first kind of nucleic acids shown to directly bind to shrimp Toll and activate innate immune responses, suggesting that different activation mechanisms exist between shrimp and *Drosophila* Toll. In *M. japonicus*, MjDorsal translocates into the nucleus after *Staphylococcus aureus* or *Vibrio anguillarum* infection. The knockdown of *MjToll1–3* reduces the nuclear localization and phosphorylation of MjDorsal. Recombinant MjToll1–3 proteins directly bind to Gram-positive bacteria, including *S. aureus* and *B. subtilis*, and Gram-negative bacteria *V. anguillarum* and *Escherichia coli*. ELISA and far Western assays have confirmed that recombinant MjToll1–3 proteins bind to PGN and LPS in a concentration-dependent manner. The knockdown of *MjToll1–3* significantly decreases shrimp AMP induction (Sun et al., 2017). Thus, some shrimp Tolls may exhibit an activation mechanism similar to mammalian TLRs because they can directly bind to PAMPs and activate immune responses.

The genome of the Pacific oyster *Crassostrea gigas* encodes an expanded set of 83 TLRs, 80 of them contain the TIR domain (Zhang et al., 2015a). *CgTLRs* are responsive to diverse aspects of pathogens. Some *CgTLRs* are only induced by a single microbe or by few types of microbes. TLR 03466R is upregulated in response to OsHV-1 infection. The other components of the TLR pathways, including MyD88, IRAKs, TRAFs, TAK1, SARM, TRAM, IKKs, IκB, and NF-κB-like transcription factors, have been identified in oysters. Co-expressed with CgMyD88, five CgTLRs activate

NF-κB signaling, suggesting that canonical MyD88-dependent TLR pathways exist in *C. gigas*. CgTLR1–4 activates NF-κB reporter in a dose-dependent manner, but they cannot directly recognize specific PAMPs (Zhang et al., 2013). *CgTLR-6* is significantly induced in hemocytes after LPS, PGN, or *V. splendidus* stimulation. CgTLR-6 has a broad recognition spectrum for bacteria, including Gram-negative *V. anguillarum* and *V. splendidus* and Gram-positive *S. aureus* and *Micrococcus luteus*, and fungi *Pichia pastoris* but not for *Yarrowia lipolytica*. CgTLR-6 exhibits affinity to LPS and PGN but not mannan (Nie et al., 2018). No TRIF-like molecular is found in *C. gigas* genome, suggesting that mollusks, similar to insects and crustaceans, may lack a TRIF-dependent TLR pathway. In *C. sinensis*, the orthologs of TLR4, TLR13, and MyD88 have been identified. *C. sinensis* TLR13-MyD88-NF-κB signaling is responsive to the challenges of *M. luteus* or *V. anguillarum* in hemocytes. The knockdown of *CsTLR13* abolishes the expression of immune-related genes, including *MyD88*, NF-κB-like transcription factors, and other adaptors involved in the TLR network (Nie et al., 2018). Studies of *C. farreri* have shown that a canonical MyD88-dependent TLR pathway exists. TIR domains from CfTLR and CfMyD88 strongly interact with one another. The knockdown of CfTLR inhibits immune-related gene expression. The extracellular domain of CfTLR is highly homologous to the corresponding parts of mammalian TLRs. CfTLR activates the NF-κB pathway in response to multiple mammalian TLR ligands, including Pam3CSK4, glucan, PGN, poly(I:C), imiquimod, and three CpG types (Wang et al., 2015a). The identification of *C. farreri* TRAM suggests that oyster may use the MyD88-independent TLR pathway, which is absent in insects (Wang et al., 2018a). In the oyster genome, only one sequence shows low similarity to Spätzle but provides multiple stop coding, suggesting that no Spätzle protein is found in the oysters. Thus, in contrast to insect TLRs, oyster TLRs may directly sense and provide protection against microbes. Although oyster TLR pathways are responsive to nucleic acid stimulation, whether oyster TLRs directly recognize viral nucleic acids and initiate antiviral immunity remains unknown (Nie et al., 2018).

3.2.2 IMD Pathway in Invertebrates

In *Drosophila*, Gram-negative bacteria activate the IMD pathway sharing similarities with vertebrate TNFR pathway via PGN-recognition protein (PGRP)-LC and PGRP-LE receptors, as well as intracellular members, including IMD, dFadd, caspase-8 ortholog Dredd, inhibitor of apoptosis 2 (IAP2), TAB2/TAK1 complex, and IKK complex (Kingsolver et al., 2013; Merkling and van Rij, 2013). Relish, another NF-κB-like factor, is activated after its phosphorylation by the IKK complex, and the inhibitory IκB domain is cleaved by Dredd. Activated Relish translocates into the nucleus and induces the expression of AMPs, such as *diptericin* and *cecropin* (Fig. 2). The IMD pathway bifurcates into the JNK pathway at the level of the TAB2/TAK1 complex, which is similar to that in mammalian TNFR pathway (Kingsolver et al., 2013; Merkling and van Rij, 2013).

Flies carrying mutations of *Relish* or *PGRP-LC* show high viral loads and exhibit hypersensitivity to CrPV infection. However, *Relish*-mutated flies do not manifest any phenotype during DCV or DXV infection, suggesting that the IMD pathway-mediated antiviral response is virus specific. IMD pathway components, such as *IMD, dFADD, Dredd, dIKKβ, dIKKγ*, or *Relish*, mutate flies harboring a SINV replicon exhibit increased genomic viral RNA accumulation, whereas normal viral RNA loads are observed in the mutants of Toll pathway members and mutants of *PGRP-LC* and *PGRP-LE* (Avadhanula et al., 2009). PGRP-LC and PGRP-LE serve as receptors that bind to PGNs and initiate the IMD pathway after a bacterial infection occurs. The receptor that detects intracellular viral RNA during IMD pathway activation remains undefined (Merkling and van Rij, 2013). SINV infection induces the expression of IMD pathway-specific AMP *diptericin*. The knockdown of *diptericin* results in developmental defects and mortality at the early pupation stage of flies harboring SINV replicon. When *diptericin* mutant flies are injected with SINV, they show an enhanced viral load. In MelSV-infected flies, the IMD pathway components *PGRP-SB1* and *PGRP-SD*, and downstream AMPs are upregulated (Merkling and van Rij, 2013). DmSTING activates a novel antiviral signaling mechanism with the help of DmIKKβ and Relish, which are the main members of the IMD pathway (Goto et al., 2018).

In *A. gambiae*, ONNV infection downregulates *IKK-1* but activates IMD pathway-controlled AMPs *diptericin* and *cecropin* (Merkling and van Rij, 2013). In U4.4 mosquito cells, the activation of the IMD pathway prior to SFV infection reduces the accumulation of viral genome and subgenome, and thus activated IMD pathway possibly inhibits SFV infection (Kingsolver et al., 2013). *Drosophila* persistently infected with Nora virus demonstrates significant alteration in its genes, particularly those involved in JNK signaling, an alternative branch of the IMD pathway (Kingsolver et al., 2013).

In crustaceans, a similar IMD pathway is identified. Shrimp IMD pathway components, such as IMD, IKKs, IAP2, and Relish, are characterized (Wang et al., 2013d, 2014). Gain-of-function and loss-of-function studies demonstrated that shrimp IMD pathway regulates the expression of AMPs, including *Penaeidins*, *crustins*, *ALFs*, and *lysozymes*. In *P. monodon*, *PmRelish* is upregulated upon the challenges of *V. harveyi*, YHV, and WSSV. The silencing of *PmRelish* decreases the expression levels of shrimp AMPs, such as *Penaeidin-3* and *Penaeidin-5* (Visetnan et al., 2015a, b). *PmRelish*-knocked down shrimps are more susceptible to YHV infection. The accumulation of activated PmRelish in the nucleus is detected in YHV-infected shrimp hemocytes (Visetnan et al., 2015a), and WSSV infection activates the IMD pathway and induces AMP expression. WSSV genome accumulation and vp28 expression levels markedly decrease in *LvRelish*-knocked down shrimp. LvRelish activates the promoter of *WSSV069* (*ie1*) and *WSSV303* through NF-κB-binding sites (Qiu et al., 2014). Thus, WSSV has adopted a mechanism for hijacking the shrimp Toll/IMD-NF-κB pathways as part of their life cycle. This mechanism diverts NF-κB immune regulatory functions to facilitate WSSV replication. In silencing or inhibiting the components of shrimp Toll/IMD-NF-κB pathways, such *LvIKKβ*, *LvIKKε*, and *LvJNK*, WSSV replication is reduced, the outbreak time of WSSV infection is delayed, and shrimp mortality is decreased (Wang et al., 2014).

3.2.3 JAK/STAT Pathway in Invertebrates

The JAK/STAT pathway, which has an evolutionarily conserved antiviral function from arthropods to mammals, is originally identified as an antiviral effector-inducing pathway in mammals (Merkling and van Rij, 2013). This pathway plays a major role in responses to a wide range of growth factors and cytokines. Type I IFNs, including IFN-α and IFN-β, bind to IFN-α receptors composed of IFNAR1 and IFNAR2 subunits and activate Janus kinase 1 (JAK1) and tyrosine kinase 2 (TYK2), resulting in the recruitment of STAT proteins. After phosphorylation, dimerization, and nuclear translocalization, STAT complexes are formed in response to stimulation, inducing numerous IFN-stimulated response element (ISRE)-driven genes, such as ISGs, which can establish a cellular antiviral state (Ivashkiv and Donlin, 2014).

Similar to the mammalian JAK/STAT pathway, the insect JAK/STAT pathway is involved in antiviral immunity (Kingsolver et al., 2013; Merkling and van Rij, 2013). In *Drosophila*, unpaired family cytokines (e.g., upd, upd2, or upd3) bind to the receptor domeless and activate the JAK kinase hopscotch, which then phosphorylates Stat92E. After the dimerization of

Stat92E complexes and their translocalization into the nucleus, Stat92E complexes induce the expression of genes with STAT-binding sites in their promoters, such as virus-induced *RNA-1* (*vir-1*) and *attC* (Fig. 2). *vir-1* transcription is substantially low in flies carrying the loss-of-function mutation of *hopscotch* when infected with DCV. In *Drosophila*, injecting DCV or CrPV induces a set of JAK/STAT-dependent genes. Flies carrying the *hopscotch* mutant have decreased *vir-1* expression levels, exhibit high viral titers, and yield decreased survival rates after DCV, CrPV, DXV, or IIV-6 infection but manifest normal phenotypes after they are infected with FHV, SINV, or VSV. By contrast, flies carrying the mutants of the siRNA pathway components are hypersusceptible to infection of all these viruses (Kemp et al., 2013). *vir-1* overexpression or knockdown has no effect on host resistance to DCV infection, suggesting that *vir-1* is not a direct antiviral effector. However, flies carrying the *attC* mutant are sensitive to viral infection and have increased viral load (Huang et al., 2013). Thus, the *Drosophila* JAK/STAT pathway confers protection against certain viruses but does not have broad antiviral activities.

In *A. aegypti*, the knockdown of *PIAS*, a negative regulator of the JAK/STAT pathway, results in the upregulation of AMPs, such as *cecropin* and *defensin*, suggesting that AMP regulation is a partial proxy of the mosquito JAK/STAT pathway (Samuel et al., 2018). DENV infection induces the upregulation of *domeless* and other JAK/STAT pathway target genes, such as *DENV restriction factor 1* and *2* (*DVRF1* and *2*). The silencing of *domeless* or *hopscotch* causes an increased DENV load, whereas the knockdown of *PIAS* exhibits an enhanced resistance to DENV infection (Cheng et al., 2016). The knockdown of JAK/STAT-dependent DENV-induced *DVRF1* and *2* promotes DENV replication. By contrast, SFV cannot activate the JAK/STAT pathway in *A. albopictus* cell line U4.4. In mosquitoes, a novel antiviral cytokine named *Vago* and induced by Dicer-2-mediated viral RNA sensing also activates JAK/STAT signaling and initiates antiviral immunity through an unknown receptor independent of *domeless* (Paradkar et al., 2012).

In crustaceans, the core JAK/STAT pathway components, including domeless, JAK, SOCSs, and STAT, are involved in antiviral responses (Huang et al., 2016; Song et al., 2015). Shrimp STAT regulates the expression of genes with STAT-binding sites in their promoters. Some of these genes include *the tumor necrosis factor superfamily*, *gamma-IFN-inducible lysosomal thiol reductase*, and *thioester-containing protein 1* (*TEP1*) and *TEP2* (Huang et al., 2016). In shrimps, the knockdown of *TEP1* or *TEP2* promotes the replication of WSSV, indicating that the JAK/STAT pathway effectors TEP1 and TEP2 have a direct antiviral ability. *JAK* or *STAT* silencing induces WSSV replication, and then the JAK/STAT pathway plays positive roles in shrimp antiviral responses. In *L. vannamei* stimulated by poly(I:C) or WSSV, *LvJAK* expression is highly induced. LvJAK itself and LvSTAT activate the *LvJAK* promoter containing multiple STAT-binding sites. *LvJAK* silencing increases mortality rate and viral load suggesting that LvJAK participates in defense against WSSV infection (Song et al., 2015). LvSOCS2, a negative regulator of the JAK/STAT pathway, responds to poly(I:C) and WSSV challenges. LvSTAT activates the promoter of *LvSOCS2* through STAT-binding sites. LvSOCS2 inhibits the expression of *DmVir-1*, an effector gene of the JAK/STAT pathway in *Drosophila*, whereas LvJAK promotes this expression, indicating that LvSOCS2 functions as a negative regulator in this pathway. *LvSOCS2* knockdown increases the resistance to WSSV infection and lessens the viral load (Wang et al., 2016). Therefore, the JAK/STAT pathway plays a positive role in shrimp antiviral immunity. WSSV seems to disrupt the shrimp JAK/STAT pathway to facilitate viral replication at the early stage of infection. At the early stage of WSSV infection, shrimp *STAT* is induced and phosphorylated STAT level increases, resulting in the translocalization of the activated STAT into the nucleus, where it induces *ie1* expression (Wang et al., 2014). *L. vannamei domeless* (*Lvdomeless*), similar to most vertebrate IL-6 receptors, is upregulated after WSSV infection. *Lvdomeless* overexpression activates the WSSV *ie1* promoter through the STAT-binding sites. *Lvdomeless* silencing decreases mortality and the number of WSSV genomic copies suggesting that WSSV highjacks *Lvdomeless* to promote viral replication (Yan et al., 2015). The contradicting roles of the shrimp JAK/STAT pathway in suppressing and increasing viral replication may be attributed to different viral infection stages selected in various experiments, different branches of the STAT pathway or alternative signaling that possibly involves in.

3.2.4 Viral Inhibition of Toll, IMD, and JAK/STAT Pathways

The virus-induced activation of Toll, IMD, and JAK/STAT pathways contribute to antiviral immunity; however, viral sensing and the detailed activation mechanism remain unclear (Wang et al., 2014). The viral suppression of these pathways may further highlight their importance in antiviral defense and the value of future investigations.

Insect viruses have developed strategies to suppress Toll and IMD pathways at the level of NF-κB-like transcription factors, including Relish, Dorsal, and Dif (Bitra et al., 2012). Polydnaviruses, such as *Microplitis demolitor bracovirus* (MdBV), infect parasitoid wasps, which are the main mortality agents of insects, and encode Ank proteins homologous to NF-κB inhibitor IκB. Ank-H4 and Ank-N5 from MdBV bind to Dif, Dorsal, and Relish homodimers with higher affinities than their endogenous inhibitors Cactus and IκB domain of Relish. Ank-H4 and Ank-N5 inhibit the processing and activation of Relish and reduce the downstream expression of AMP (Bitra et al., 2012). Thus, Ank-H4 and Ank-N5 are NF-κB antagonists that prevent Toll and IMD pathway activation and AMP induction.

Diedel, an immunomodulatory cytokine, acts as an IMD pathway inhibitor. *Drosophila Diedel* is strongly induced by viral infection (Lamiable et al., 2016). *Diedel*-mutant flies express elevated levels of IMD pathway target genes. During SINV infection, Diedel is necessary to the promotion of fly survival because it modulate IMD pathway activation. Viral Diedel homologs are identified in several classes of insect DNA viruses belonging to *Entomopoxvirinae*, *Ascoviridae*, and *Baculoviridae*. Viral Diedel homolog encoded by *Spodoptera frugiperda ascovirus 1a* can functionally substitute the cellular Diedel protein, promote host survival, and modulate the activation of the IMD pathway (Lamiable et al., 2016). Although IMD pathway activation is required for defense against some viruses, its antiviral effectors and mechanism remain unknown. The discovery of viral Diedel homologs emphasizes the importance of the IMD pathway in antiviral response.

In *A. aegypti* cells, DENV infection suppresses the induction of Toll- and IMD-dependent AMPs by bacterial challenges. In *A. albopictus* cells, SFV inhibits the immune activation by a constitutively active Toll construct or heat-inactivated *E. coli*, which activate IMD and JAK/STAT pathways. Arboviruses, such as JEV, WNV, and DENV, have developed strategies to interfere with mammalian JAK/STAT pathway activation; these strategies repress JAK phosphorylation and prevent STAT1 and 2 translocations into the nucleus, thereby inhibiting ISG induction (Samuel et al., 2018). In C6/36 mosquito cells, STAT phosphorylation and DNA-binding activity are inhibited by JEV infection (Lin et al., 2004). In U4.4 cells, SFV infection suppresses the activation of the JAK/STAT pathway induced by heat-inactivated bacteria. However, future studies should determine whether arboviruses encode the specific antagonists of JAK/STAT pathways in insects, particularly in mosquitoes. In crustaceans, WSSV-miR-22 targets STAT, downregulates *TEP1* and *TEP2* induction, and promotes WSSV infection (Ren et al., 2015). WSV181 inhibits JAK/STAT signaling and promotes viral replication by controlling Stat92E expression (Wang et al., 2018d). Viruses typically suppress the Toll, IMD, and JAK/STAT pathways as means of circumventing host antiviral immunity and facilitating their infection (Merkling and van Rij, 2013). Nevertheless, whether the suppression of inducible antiviral immunity is a general strategy of arboviruses requires further investigation. The detailed inhibitory mechanism and the effect of such inhibitory action on arbovirus transmission in mosquitoes should also be explored as well.

3.3 RNA Sensing-Induced Antiviral Immunity by Dicer-2

Although the siRNA-mediated gene silencing of viral RNA plays a major role in insect antiviral defense, which represents a specific antiviral immunity, immune-related genes induced by viruses similar to those in mammals contribute to a non-specific antiviral response (Wang et al., 2014). Two main issues include the identification of viral sensors and the determination of related factors responsible for antiviral responses. *D. melanogaster Vago* (*Dm Vago*), which encodes a secreted cysteine-rich polypeptide with 160 residues, is one of the most upregulated genes by DCV and implicated in inducible antiviral response (Deddouche et al., 2008). *Dm Vago* is mainly expressed in the fat body, which is the target organ of DCV. This gene is strongly induced by DCV and SINV infection but weakly upregulated by bacterial challenge. Notably, FHV, which targets and proliferates successfully in the *Drosophila* fat body, cannot modulate *Dm Vago* expression. DCV replication is dramatically promoted in the fat body of *Vago*-mutant *Drosophila*, whereas *Dm Vago* overexpression in these animals restores impaired virus clearance. Thus, *Dm Vago* has a direct role in antiviral response (Deddouche et al., 2008). Although insect Toll, IMD, and JAK/STAT pathways play important roles in antiviral response and respond to viral infection by regulating a set of downstream genes, none is involved in the regulation of *Dm Vago*. Dicer-2, which is known to sense and process viral dsRNAs into siRNAs in the initiation of specific antiviral immunity, is a DCV sensor that triggers a novel signaling pathway to regulate *Dm Vago* expression (Fig. 3). Flies lacking *Dicer-2* have impaired *Dm Vago* induction. Nevertheless, other siRNA pathway components, such as Ago-2 and R2D2, are not required for *Dm Vago* upregulation, indicating that the siRNA pathway is not implicated in *Dm Vago* induction (Deddouche et al., 2008). Uncoupled from its function in siRNA biogenesis, viral RNA sensing by Dicer-2 triggers a novel signaling mechanism to regulate *Dm Vago* expression. Thus, Dicer-2 has dual functions, that is, it not only initiates an antiviral RNAi response but also activates inducible antiviral immunity upon sensing viral RNAs.

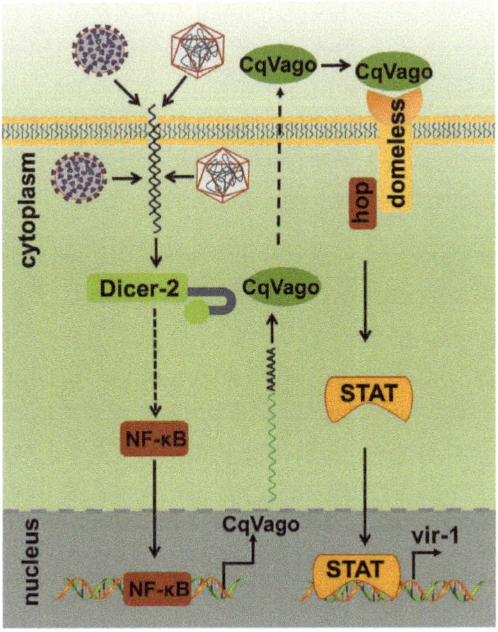

Fig. 3 Insect Dicer-2 is a viral RNA sensor that induces antiviral immunity. In mosquitoes, Dicer-2 senses viral RNA and upregulates the novel antiviral cytokine Vago by activating NF-κB-like transcription factors. Vago activates JAK/STAT signaling and triggers systemic antiviral immunity via an unknown receptor. Cq, *Culex quinquefasciatus*.

In *C. quinquefasciatus*, viral RNA sensing by Dicer-2 activates TRAF but not TRAF-3 and induces the cleavage and activation of the NF-κB protein family Rel2 (Fig. 3) (Paradkar et al., 2014). In mosquitoes, conserved NF-κB-like binding sites are found in *Vago* promoters, which are responsible for its viral-inducible ability. Activated Rel2 binds to NF-κB-binding sites and induces *CqVago* expression in a manner similar to that in mammalian IFN production and antiviral mechanism conducted by RLRs-NF-κB cascades. The identification of this novel and evolutionarily conserved pathway downstream of Dicer-2 provide new insights into the antiviral immunity of mosquitoes and other invertebrates. CqVago functions as a cytokine by activating the JAK/STAT pathway and triggering systemic antiviral immunity in various mosquito tissues, demonstrating high similarity to the action of IFNs in vertebrates (Fig. 3) (Paradkar et al., 2012). WNV induces *CqVago* expression in Hsu cells, and subsequent *CqVago* depletion increases WNV titers. CqVago activates the JAK/STAT pathway and upregulates downstream genes, such as *vir-1*, to achieve its antiviral ability

(Prasad et al., 2013). Silencing *Cqhopscotch*, not *Cqdomeless*, abolishes CqVago-induced WNV suppression (Paradkar et al., 2012). The antiviral ability of CqVago is determined by the upregulation of JAK/STAT downstream genes, such as *vir-1*, via an unknown receptor, which is dependent on hopscotch- and STAT-mediated signaling, but this phenomenon is not attributed to the canonical domeless receptor (Prasad et al., 2013). Thus, CqVago has a function similar to that of mammalian IFNs, that is, it activates the JAK/STAT pathway in neighboring cells, and this activation initiates systemic antiviral response. Vago appears to function as an IFN-like antiviral cytokine in mosquitoes.

Five Vago orthologs (LvVago1–5) are identified in *L. vannamei* (Chen et al., 2011). LvVago1–5 are predicted as secreted proteins containing approximately 102–190 residues and a single von Willebrand factor type C domain with eight conserved cysteine residues similar to those in insect Vagos (Paradkar et al., 2012). *LvDicer-2*, *LvVago-1*, *LvVago-4*, and *LvVago-5* are significantly upregulated by WSSV infection. When challenged by dsRNA mimics poly(C:G), *LvDicer-2* and *LvVago5* are dramatically upregulated. LvDicer-2 and DmDicer-2 activate the promoters of *LvVago-1*, *LvVago-4*, and *LvVago-5* as well as the activation of *DmVago* promoter by DmDicer-2 (Chen et al., 2011). *L. vannamei* IRF-like protein (LvIRF) activates the promoters of *LvVago-4* and *-5* but not *LvVago-1*, *-2*, and *-3* (Li et al., 2015). Stimulated by poly(I:C) and WSSV, the promoter activity of *LvVago-4* is induced. LvIRF overexpression significantly increases the WSSV- and poly(I:C)-induced promoter activity of *LvVago-4*. Recombinant LvIRF binds to *LvVago-4* promoter through ISRE. However, *LvVago-4* promoter having a mutant ISRE cannot bind to LvIRF again. *LvIRF* knockdown represses WSSV-induced *LvVago-4* and *-5* upregulation. The suppression of *LvIRF*, *LvVago-4*, or *LvVago-5* increases mortality after WSSV infection. Therefore, *LvVago-4* and *-5* are activated by LvIRF, an IRF family protein, in a manner similar to those of IFNs in vertebrates. Nevertheless, further studies should determine whether *LvVago-1*, *-2*, and *-3*, which are not regulated by LvIRF, can be activated by NF-κB signaling as a similar mechanism of *CqVago* regulated by TRAF-Rel2 cascade or by other regulatory mechanisms (Paradkar et al., 2014). *Vago* orthologs in *Bombus terrestris* are downregulated after viral infection, while *Dicer-2* is upregulated (Niu et al., 2016). As such, *Vago* may have differentiated regulatory mechanisms in different invertebrates. Thus far, no IRF ortholog has been found in insect genomes, and only one *Vago* has been identified in *Drosophila* and *Culex*, suggesting that the regulatory mechanisms of *Vagos* in insects are different from those of crustaceans. In terms of protein homology, inducible expression, antiviral ability, and functional and regulatory mechanisms, LvVago may function as an IFN-like cytokine in shrimps. Clarifying whether LvVago can activate the JAK/STAT pathway may help further confirm that shrimps possess an IRF-Vago-JAK/STAT regulatory axis, which is similar to the IRF-IFN-JAK/STAT axis of vertebrates (Li et al., 2015). Overall, these findings show that the inducible antiviral mechanism similar to that in vertebrates is present in invertebrates and is required for antiviral immunity.

DmDicer-2 contains an N-terminal DExD/H-box helicase domain and a C-terminal RNase III domain (Fig. 4). The DExD/H-box helicase domain is required for dsRNA sensing, processing, and *Vago* induction, whereas the RNase III domain is essential for siRNA biogenesis (Deddouche et al., 2008). Phylogenetic analysis has revealed that the DExD/H-box helicase domains of insect Dicer-2 is highly homologous to the helicase domains of vertebrate RLRs, including RIG-I and MDA-5, which sense viral RNAs, activate IFNs, and initiate systemic antiviral responses (Fig. 4) (Deddouche et al., 2008). The siRNA-independent antiviral function of insect Dicer-2, which initiates a novel signaling to regulate *Vago*, establishes a parallel between the functions of vertebrate RLRs and insect Dicer-2. The DExD/H helicase domain of DmDicer-2 also senses defective viral genomes (DVGs) as PAMPs and employs them as templates to synthesize viral DNA (vDNA) and circular vDNA (cvDNA), leading to the amplification of vDNA-mediated antiviral RNAi response (Poirier et al., 2018). DVGs are typical PAMPs preferentially sensed by RLRs during viral infection and activate IFN. This finding establishes the conserved roles of DVGs as PAMPs and DExD/H-box helicase containing proteins Dicer-2 and RLRs as their sensors in invertebrates and vertebrates (Poirier et al., 2018). Overall, DExD/H-box helicase containing proteins may represent an evolutionarily conserved set of viral nucleic acid sensors that direct antiviral responses (Fig. 5). DmDicer-2 is also required for the Toll pathway independent of siRNA biogenesis. DmDicer-2 directly binds to the 3′ UTR of Toll, regulates protein translation, and stimulates the proper expression of Toll protein, but DmDicer-2 does not contribute to the maintenance of protein stability or to mRNA transcription (Wang et al., 2015c). *Dicer-2* mutant flies are susceptible to fungal infections. The knockdown of *DmDicer-2* inhibits Toll pathway activation by downregulating Toll pathway-specific AMP *drosomycin* after FHV and VSV infection. Thus, *Dicer-2* may act as a novel class of post-transcriptional regulators of gene expression in immune signaling pathways.

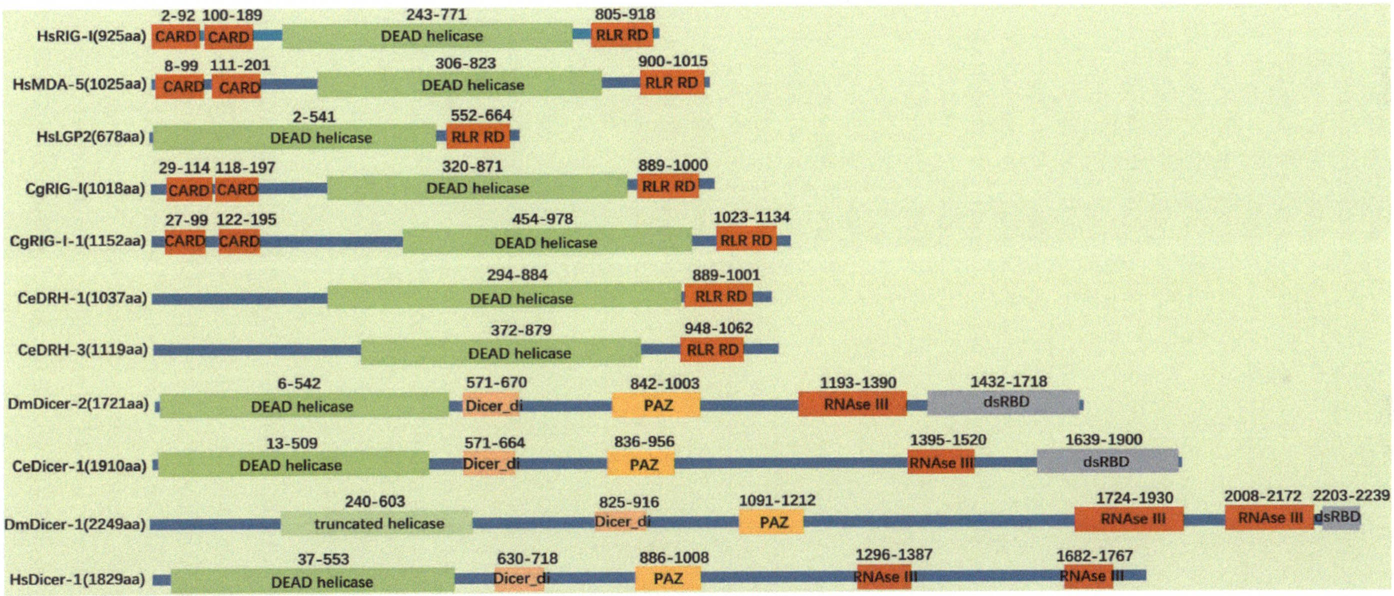

Fig. 4 Domain organization of the DExD/H helicase domain family proteins. CARD, caspase activation and recruitment domain; DEAD helicase, DEAD-box related helicases; RLR RD, RIG-I-like receptor regulatory domain; Dicer_di, Dicer dimerization domain; PAZ, Piwi-Argonaut-Zwille; dsRBD, double-stranded RNA-binding domain. Hs, *Homo sapiens*; Cg, *Crassostrea gigas*; Dm, *Drosophila melanogaster*; Ce, *Caenorhabditis elegans*.

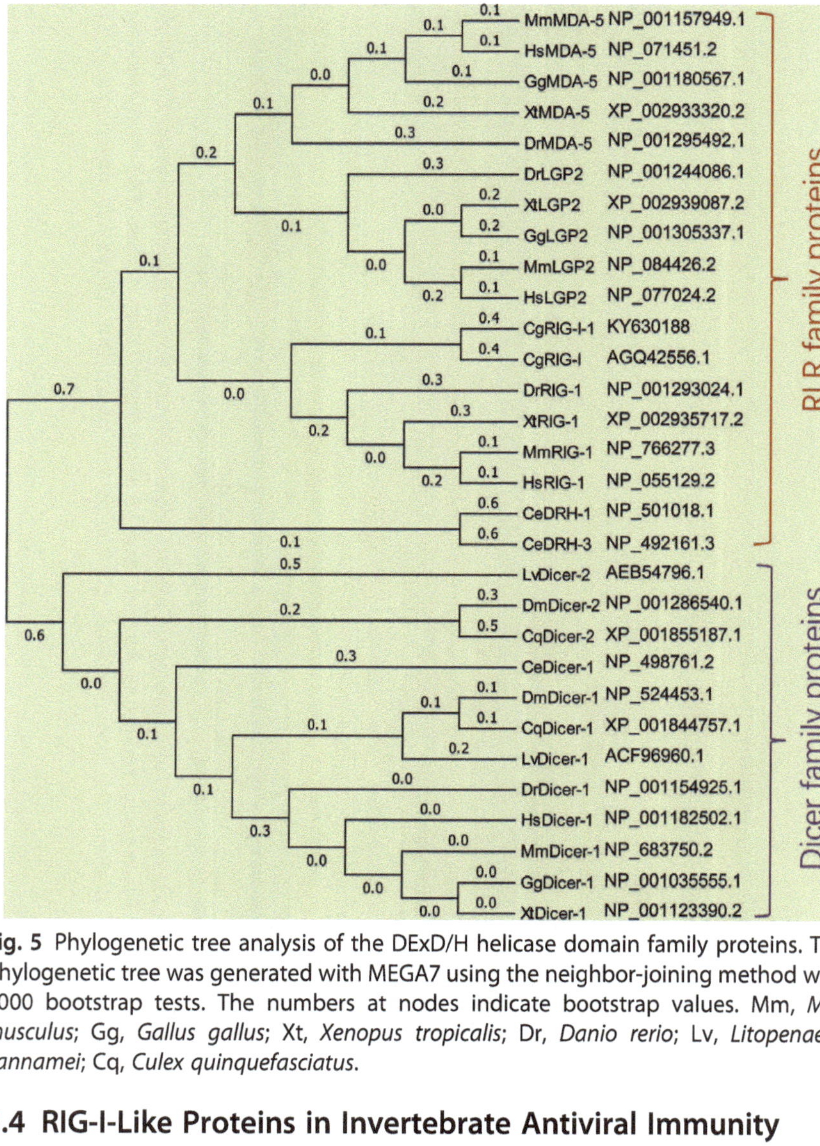

Fig. 5 Phylogenetic tree analysis of the DExD/H helicase domain family proteins. The phylogenetic tree was generated with MEGA7 using the neighbor-joining method with 1000 bootstrap tests. The numbers at nodes indicate bootstrap values. Mm, *Mus musculus*; Gg, *Gallus gallus*; Xt, *Xenopus tropicalis*; Dr, *Danio rerio*; Lv, *Litopenaeus vannamei*; Cq, *Culex quinquefasciatus*.

3.4 RIG-I-Like Proteins in Invertebrate Antiviral Immunity

The *C. gigas* genome encodes the orthologs of RLRs, MAVS, TRAF6, TBK1, IRF family proteins, and IFN-like protein (IFNLP) (Fig. 6) (Huang et al., 2017a; Zhang et al., 2015b). Most of them are responsive to immune challenges. *C. gigas* has 12 RLRs, and 7 of them possess vertebrate type N-terminal CARD structures. CgRIG-I-1 has a typical RLR protein domain structure: the N-terminal CARDs, central RNA helicase domain, and C-terminal regulatory domain (RD) (Fig. 4) (Huang et al., 2017a). Phylogenetic analysis has shown that

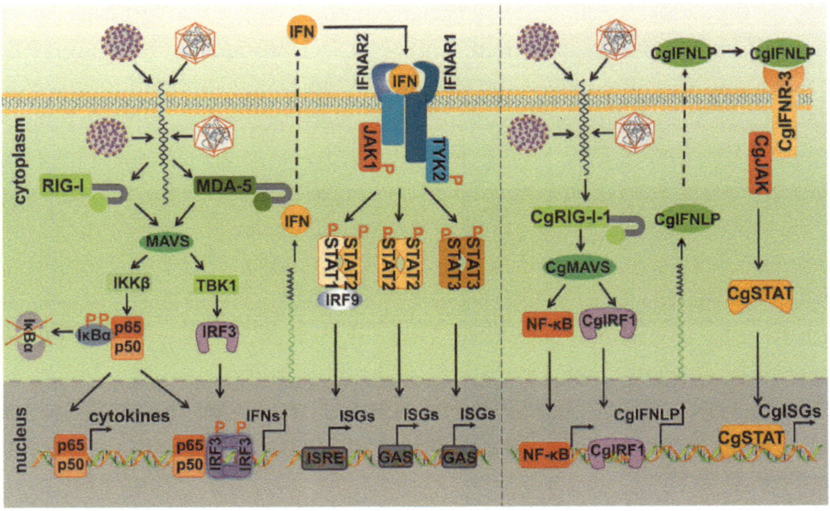

Fig. 6 Proposed working model of the RIG-I pathway in the Pacific oyster *Crassostrea gigas*. In mammals, RIG-I and MDA-5 are critical cytoplasmic dsRNA sensors in the activation of IFN response (*left*). Likewise, in oysters, cytoplasmic dsRNA sensing by CgRIG-I-1 activates transcription factors CgNF-κB and CgIRF1 through CgMAVS, thereby upregulating *CgIFNLP*. CgIFNLP binds with the putative CgIFNLP receptor CgIFNR-3 and activates the ISRE, STAT3, and GAS promoters through the JAK/STAT pathway (*right*). Cg, *Crassostrea gigas*.

CgRIG-I-1 is grouped into the clade of vertebrate RLRs (Fig. 5). CgRIG-I-1 directly binds to poly(I:C) and interacts with CgMAVS via its CARDs. Subsequently, CgMAVS is activated and CgTRAF6 is recruited, and these processes activate the NF-κB pathway. In mammalian cells, CgIRFs trigger the IFN-β promoter and ISRE reporter, which may function downstream of CgMAVS and stimulate an antiviral response. CgIRF-1 binds to ISRE and regulates the expression of *C. gigas IFNLP (CgIFNLP)* as a transcription factor and plays a role in the antiviral immunity of oysters (Fig. 6) (Lu et al., 2018). *CgIRF-1* is induced by poly(I:C) in hemocytes. CgIRF-1 has both cytoplasmic and nuclear localization. Recombinant CgIRF-1 binds to ISRE in vitro, and CgIRF-1 activates the promoter of *CgIFNLP* (Lu et al., 2018). CgIFNLP possesses an IFN domain that is similar to the vertebrate IFNs, and its predicted three-dimensional structure is highly similar to class II helical cytokines. *CgIFNLP* dramatically increases in hemocytes after poly(I:C) stimulation. Recombinant CgIFNLP treatment increases the apoptosis and phagocytosis rates of oyster hemocytes (Zhang et al., 2015b). The proliferation rate of A549 cells substantially decreases when they are treated with recombinant CgIFNLP, which activates the promoters of ISRE, STAT3, and GAS and possibly the JAK/STAT pathway (Zhang et al., 2016). The CgIFNLP receptor CgIFNR-3 is also identified. CgIFNR-3 is associated with CgIFNLP in vitro, and its expression is upregulated by poly(I:C) stimulation in hemocytes. CgIFNR-3 not only functions as a receptor of CgIFNLP but also activates the oyster JAK/STAT pathway (Zhang et al., 2016). Therefore, MAVS-dependent RLR pathways in oysters are useful in the identification of the antiviral mechanisms of invertebrates and understanding of the origin, evolution, and development of vertebrate RLR-mediated antiviral immunity.

In *C. elegans*, three Dicer-related helicases (DRH1–3), which are orthologous to mammalian RLRs (Fig. 5), are encoded, whereas DRH1 and DRH3 are associated with Dicer and required for siRNA biogenesis (Coffman et al., 2017). However, whether *C. elegans* has the same induced antiviral response and whether DRH1–3 plays a role in this response remain unknown.

3.5 STING Signaling in Invertebrates

In vertebrates, cGAS-STING pathway plays an essential role in cytoplasmic DNA sensing and antiviral defense against viral infection (Wang et al., 2018b). Cytoplasmic cGAS senses dsDNA and produces a CDN called 2′3′-cGAMP, which directly binds to STING and induces the phosphorylation of TBK1 and IRF3, resulting in the activation and nuclear translocation of transcription factor IRF3, which in turn activates IFNs and downstream ISGs (Tan et al., 2018; Wu and Chen, 2014). Bacterium-derived CDNs, such as *c*-di-GMP, *c*-di-AMP, and 3′3′-cGAMP, also bind to STING, activate IRF3, and induce IFN production (Fig. 7) (Wang et al., 2018b). cGAS and STING family proteins originated from a common metazoan ancestor and co-evolved in unison. The invertebrate cGAS-STING pathway is unlikely involved in invertebrate antiviral response but is probably implicated in antibacterial immunity given that invertebrate cGAS orthologs lack a zinc ribbon domain participating in dsDNA sensing and invertebrate STING orthologs lack a C-terminal tail (CTT) domain responsive to downstream TBK1 recruitment (Margolis et al., 2017).

Mammalian cGAS and STING originated from metazoans. During metazoan evolution, both cGAS and STING were lost in nematodes and flatworms. NF-κB family proteins, which are important transcription factors of immune pathways, are widely distributed in all metazoans except nematodes (Wu et al., 2014). STING is activated by CDN binding via a CDN-binding domain conserved in different phyla, including Cnidaria,

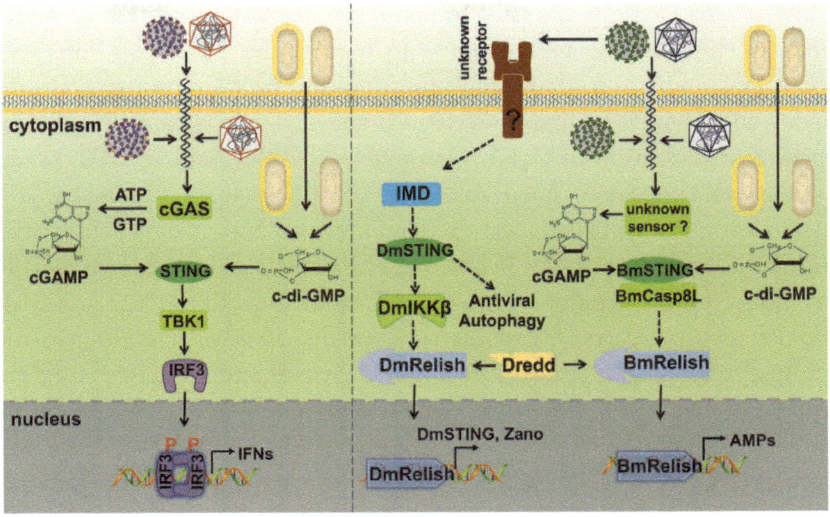

Fig. 7 STING-mediated innate immune response. A proposed working model of the STING pathway in mammals (*left*) and insects (*right*). The function of invertebrate cGAS ortholog has been unidentified to date. Invertebrate STING, including oyster and insect STINGs can directly bind with CDNs and activate immune responses. Up to now, no function of the cGAS ortholog is identified. Picorna-like virus activates the non-canonical IMD pathway involving DmSTING, DmIKKβ, and Relish, thereby upregulating downstream genes, such as antiviral effector Nazo, via an unknown viral sensor (*right*). In fly brain, ZIKV infection induces *DmSTING* expression via the IMD pathway, thereby activating STING-dependent autophagy. In *B. mori* cells, cGAMP is generated via an unknown mechanism after BmNPV infection. cGAMP activates BmSTING and promotes BmRelish cleavage and activation, leading to increased AMP expressions and strong antiviral responses. Bm, *Bombyx mori*.

Mollusca, and Annelida. However, a previous study showed that insect STINGs, including *D. melanogaster* STING (DmSTING), cannot bind to any CDN, particularly *c*-di-AMP, *c*-di-GMP, 3′,3′-cGAMP, and 2′,3′-cGAMP. These negative results do not confirm that insect STING proteins already lost their ability to bind to CDNs because recombinant STINGs contain insoluble hydrophobic N-terminal transmembrane domains, which may cause an unnatural state of recombinant STINGs and inhibit the binding between STINGs and CDNs (Kranzusch et al., 2015). Another study revealed that DmSTING directly binds to *c*-di-GMP and induces immune response partly through IMD-Relish axis (Fig. 7) (Martin et al., 2018). DmSTING binds to cGAMP and *c*-di-AMP under mild conditions, whereas DmSTING binds to *c*-di-GMP only under stringent salt conditions. A STING mutant without an insoluble N-terminal transmembrane domain binds to *c*-di-GMP, whereas a mutant lacking a CDN-binding region abolishes this binding. Compared with mammalian STING, DmSTING lacks CTT, which contains multiple sites for protein post-translational modifications, such as phosphorylation, and for monitoring downstream signaling transduction to IRF3. Mammalian STING lacking CTT cannot activate IRF3 but can maintain its NF-κB activation ability. Consistent with these findings, DmSTING can activate NF-κB but not IRF3. During *Listeria* infection, *c*-di-GMP is released as PAMPs, which activate STING and induce an IFN response in mammals (Martin et al., 2018). *c*-di-GMP stimulation and *Listeria* infection activate the *Drosophila* IMD pathway and upregulate IMD pathway-specific AMPs *AttA* and *CecA2* but not Toll pathway-specific AMP *drosomycin*. The depletion of *DmSTING* reduces the cleavage and activation of Relish, implying that DmSTING functions through Relish, a transcription factor of the canonical IMD pathway. DmSTING-overexpressing flies show a decreased bacterial load, low mortality, and decreased susceptibility to *Listeria* infection (Martin et al., 2018). These flies also exhibit high induction levels of *AttA* and *CecA2* but not *drosomycin*. In DmSTING-overexpressing flies, the silencing of *IMD* or *Relish* reduces AMP induction. In *Drosophila*, the putative cGAS ortholog CG7194 lacks a zinc ribbon domain and a positively charged N-terminus, which are functionally important for DNA binding (Wu et al., 2014). The knockdown of *CG7194* does not affect the mortality rate of flies infected with either *Listeria* or IIV6, suggesting that CG7194 may not be involved in immune defense against these pathogens (Martin et al., 2018). Although *IMD*-mutant flies are increasingly susceptible to CrPV infection, the knockdown of *DmSTING* does not affect AMP induction and mortality. By contrast, DmDicer-2, containing a RIG-I-like helicase domain, senses viral RNAs and induces an effective antiviral response to RNA and DNA viruses, such as CrPV and IIV6, through siRNA-mediated viral RNA degradation (Martin et al., 2018). Thus, Dicer-2-mediated siRNA pathway, but not IMD-Relish signaling, plays a major role in antiviral immunity. Overall, DmSTING directly senses bacterial CDNs, especially *c*-di-GMP, and initiates antibacterial immunity via IMD-Relish signaling in the absence of a functional cGAS ortholog.

A synchronous study has shown that the siRNA pathway is not protective during the neurotropic infection of ZIKV in fly brain (Liu et al., 2018). *DmDicer-2*- or *DmAgo-2*-mutant flies have the same ZIKV RNA levels in their brains as wild-type controls. By contrast, *DmAgo-2* mutant flies infected with SINV have a high viral load in their brains; thus, RNAi pathway is intact and functional but may be suppressed or may lose its antiviral function during ZIKV infection (Liu et al., 2018). Upon ZIKV infection, the IMD pathway-specific AMP *diptericin*, unlike the Toll pathway-specific AMP *drosomycin*, is significantly upregulated in a Relish-dependent manner. In *Relish*-mutant flies, ZIKV infection cannot induce *diptericin* expression; however, increased viral load and animal lethality are observed, suggesting that IMD-Relish signaling plays a key role in ZIKV clearance. Thus, IMD-Relish signaling is activated, which is required for virus control in fly brains following ZIKV infection. The expression of *DmSTING* is significantly induced through IMD-Relish signaling in fly heads by ZIKV infection, and the depletion of *DmSTING* increases the rate of ZIKV replication in flies and neurons (Liu et al., 2018). Autophagy is an ancient immune pathway that targets and kills intracellular bacteria and viruses conserved from flies to humans. ZIKV infection activates autophagy in fly neurons in a DmSTING-dependent manner (Fig. 7) (Liu et al., 2018; Qiu and Zhou, 2018). The blocking of DmSTING-dependent autophagy leads to uncontrolled ZIKV proliferation and animal mortality. The pharmacological activation of autophagy by rapamycin protects fly brains from ZIKV infection. *ATG5*-depleted flies cannot effectively restrict ZIKV infection with increased viral titers (Liu et al., 2018). Upon ZIKV infection, DCV infection induces *DmSTING* expression. *DmSTING*-depleted flies are more susceptible to DCV infection than wide-type controls, but *Relish*- or *Atg5*-mutant flies are susceptibility to DCV infection like wild-type animals (Liu et al., 2018; Qiu and Zhou, 2018). Thus, another DmSTING-dependent antiviral mechanism against DCV exists but is independent of Relish and autophagy. Overall, immune sensing of ZIKV upregulates *DmSTING* expression through IMD-Relish signaling, which activates STING-dependent autophagy, resulting in ZIKV inhibition (Liu et al., 2018). However, the sensing mechanism of ZIKV remains elusive. The sensor should be upstream of or at least parallel to PGRP-LC and PGRP-LE because flies carrying both *PGRP-LC* and *-LE* mutants still show higher susceptibility to ZIKV infection than wild-type flies.

DmSTING activates the DmIKKβ-Relish module of the IMD pathway, induces a set of antiviral effectors, and participates in antiviral response during picorna-like viral infections. The IMD pathway components DmIKKβ and Relish, not DmIKKγ, are required for *Drosophila* defense against the infection of DCV and CrPV (Goto et al., 2018). The infection of Picorna-like virus activates the IMD pathway through DmIKKβ and Relish and induces a set of genes, including *DmSTING*, and a novel evolutionarily conserved antiviral factor *Nazo* (Fig. 7) (Goto et al., 2018). The knockdown of *DmIKKβ* causes a remarkable increase in viral RNA loads after DCV and CrPV infection, but a slight increase in viral RNA loads is observed after VSV infection. However, no effect is observed in the replication of FHV, SINV, and IIV6 (Goto et al., 2018). The overexpression of *DmSTING* results in the strong inhibition of DCV and CrPV replication, but FHV remains unaffected. Flies carrying *DmIKKβ* or *DmSTING* mutant are hypersensitive to DCV and CrPV infections but not to FHV and SINV infections. *DmSTING*-mutant flies are resistant to bacterial and fungal infections and have normal AMP inductions; thus, DmSTING-mediated antiviral response is largely essential for antibacterial and antifungal immunity (Goto et al., 2018). The overexpression of *DmSTING* strongly upregulates *Nazo* expression; however, the silencing of endogenous *DmIKKβ* abolishes *Nazo* induction. Hence, DmSTING functions upstream of DmIKKβ and Relish in this antiviral signaling mode. *DmSTING* is a downstream target gene of the IMD pathway, suggesting that a positive feedback loop exists between DmSTING and DmIKKβ-Relish signaling for the regulation of the antiviral factor expression, including *Nazo* and *DmSTING*, but not AMPs (Goto et al., 2018). Gain-of-function and loss-of-function studies showed that DmSTING and DmIKKβ regulate a similar subset of genes, including a novel antiviral factor *Nazo*. DmSTING overexpression shows a potent antiviral phenotype. When *DmIKKβ*, *Relish*, or *Nazo* is silenced, the antiviral activities of DmSTING are abolished. When key residues responsible for CDN binding of DmSTING are mutated, the in vitro antiviral properties are compromised, suggesting an important role of CDN binding in DmSTING antiviral ability (Zakovic and Levashina, 2018). Nevertheless, the kind of CDNs recognized by DmSTING during viral infection remains unknown. Other canonical CDNs except *c*-di-GMP are poorly recognized by DmSTING (Zakovic and Levashina, 2018). The overexpression of *Nazo* is sufficient to inhibit DCV or CrPV replication but has no effect on FHV. In *Nazo* mutant flies, DCV replication is increased (Goto et al., 2018). Thus, Nazo is the antiviral effector regulated by the DmSTING-DmIKKβ-Relish axis. DmSTING induces a novel antiviral signaling via IMD and Relish upstream of DmIKKβ, which regulates a set of antiviral factors, such as *Nazo*, but remains independent of the canonical PGRPs-IMD-Relish pathway, which has a canonical role in Gram-negative bacterium sensing, AMP induction, and antibacterial immunity. These findings suggest the evolutionarily ancient role of STING as a regulator of antiviral response.

cGAMP is produced in *B. mori* cells after infection of BmNPV, a DNA virus (Hua et al., 2018). In cells and animals, synthesized cGAMP activates immune response by upregulating the expression of *B. mori* AMP *CecA* and *CecB*. *BmSTING* is ubiquitously expressed in various tissues, particularly in hemocytes with prominent expression. After cGAMP stimulation, BmSTINGs assemble in the perinucleus from the cytoplasm, and this phenomenon is similar to the localization of mammalian STINGs upon viral infection or interferon stimulatory DNA (ISD) transfection (Hua et al., 2018). BmSTING overexpression induces *BmCecB* expression and effectively inhibits BmNPV replication, whereas *BmSTING* knockdown reduces BmCecB induction, resulting in a high viral load and extreme vulnerability to viral infection. This observation implies that BmSTING plays a critical role in antiviral immunity. BmSTING promotes BmRelish processing and activates antiviral immunity (Fig. 7). BmNPV infection, cGAMP stimulation, or BmSTING overexpression as well as stimulations of *c*-di-GMP, ISD, poly(dA:dT), or poly(I:C) all induce the cleavage and activation of BmRelish. cGAMP stimulation and BmSTING overexpression promote the translocation of BmRelish from the cytoplasm to the nucleus. BmRelish processing induced by BmSTING dramatically increases AMP expression and accelerates viral clearance (Fig. 7) (Hua et al., 2018). *B. mori* caspase-8-like protein (BmCasp8L) directly interacts with BmSTING and prevents BmSTING-induced BmRelish cleavage and activation. The overexpression of BmCasp8L abolishes the cleavage, nuclear translocation, and activation of BmRelish induced by BmSTING and the stimulation of cGAMP or BmNPV, although these effects are not observed when treatments involving LPS or *E. coli*, are employed. Thus, BmCasp8L may act as a suppressor of the BmSTING-mediated antiviral pathway. Interestingly, cGAMP stimulation decreases BmCasp8L–BmSTING interaction and potent BmRelish activation. The interaction between death-related ced-3/Nedd2-like caspase (BmDredd) and BmSTING promotes BmRelish cleavage, efficiently activates antiviral signaling, and protects insect cells from viral infection. BmCasp8L serves as a switch of BmSTING-mediated BmRelish activation; thus, targeting BmCasp8L may provide a novel approach for enhancing insect antiviral immunity (Hua et al., 2018). Overall, STING-mediated antiviral immunity is conserved in the model insect *B. mori*. cGAMP-activated BmSTING and overexpressed BmSTING induce BmRelish cleavage and activation, resulting in a strong antiviral defense against BmNPV (Hua et al., 2018). However, the antiviral mechanism of BmSTING still requires further investigation. One possibility is that BmSTING upregulates BmRelish-controlled immune effectors, such as AMPs, which directly target and inhibit

BmNPV. Insect cells expressing the activated BmRelish form, which strongly triggers the IMD pathway and induces downstream effectors, are resistant to BmNPV infection. However, other mechanisms are possible. For example, BmNPV may activate not only the STING-dependent antiviral autophagy through a mechanism similar to that of ZIKV in flies but also other alternative immune signaling pathways (Liu et al., 2018). cGAMP is produced in insect cells after BmNPV infection and directly binds to and activates BmSTING, thereby initiating antiviral immune activation (Hua et al., 2018). However, *Drosophila* DNA virus IIV6 cannot induce DmSTING-dependent Relish activation (Goto et al., 2018). Given that insect STING binds to CDNs and induces antiviral response and the putative cGAS orthologs of insects, at least in flies, are not involved in antiviral defense, insect genomes possibly encode non-cGAS DNA sensor that senses viral DNA and produces cGAMP (Martin et al., 2018). *B. mori*, which produces cGAMP after BmNPV infection, also has a cGAS ortholog. Nevertheless, whether this cGAS ortholog or other putative DNA sensors, such as DDX41, contribute to cytoplasmic DNA sensing, cGAMP production and sensing, and antiviral immune activation still requires further investigations (Wu and Chen, 2014).

In crustaceans, a STING ortholog is identified and functionally characterized (Li et al., 2017). *LvSTING* is mainly expressed in the hepatopancreas and the intestine, which are two major immune-related tissues for AMP induction. *LvSTING* is significantly induced in organisms challenged with Gram-negative *V. parahaemolyticus*, a common bacterial pathogen of shrimps. The knockdown of *LvSTING* significantly reduces the induction of *LvPenaeidin-4* by *V. parahaemolyticus* challenge, an important shrimp AMP downstream of the Toll and IMD pathways (Li et al., 2017). The injection of recombinant LvSTING into *LvSTING*-knockdown shrimp restores *LvPenaeidin-4* induction. Thus, the LvSTING regulation of *LvPenaeidin-4* expression may occur via the Toll or IMD pathway or both and via the canonical AMP regulation pathway. The knockdown of *LvSTING* causes high mortality after *V. parahaemolyticus* challenge relative to the mortality rates of the control group. *LvSTING*-knocked down shrimps injected with recombinant LvSTING exhibits increased survival rates and restored resistance to *V. parahaemolyticus* infection (Li et al., 2017). LvSTING may be involved in innate immune regulation through the Toll or IMD pathway or both and is required for shrimp defense against bacterial infection. Our further study also demonstrated that CDNs, including *c*-di-GMP and cGAMP, can directly bind to and activate LvSTING. However, whether LvSTING participates in the signaling transduction of the Toll and IMD pathways and whether LvSTING plays a role in antiviral defense against various shrimp pathogenic viruses, such as WSSV, TSV, and YHV, must be determined. A shrimp DDX41, which is highly similar to vertebrate DDX41 and composed of the conserved domains of DEADc, HELICc, and zinc finger, is identified in *P. monodon* (Soponpong et al., 2018). PmDDX41 is constitutively expressed in various tissues. It is highly expressed in immune-related tissues, including hepatopancreas, gills, and hemocytes, which are implicated in immune responses in crustaceans (He et al., 2015b). *PmDDX41* is upregulated, downregulated, and unchanged after WSSV, YHV, and *V. harveyi* challenge, respectively. The knockdown of *PmDDX41* reduces the expression of shrimp AMPs and some immune signaling-related genes, including *PmIKKβ* and *PmIKKε*. *PmDDX41*-knocked down shrimps quickly die after WSSV infection (Soponpong et al., 2018). Thus, PmDDX41 possibly participates in shrimp antiviral response and regulates the expression of immune-related genes through a DNA-sensing pathway. However, whether PmDDX41 binds to CDNs and interacts with STING as a vertebrate counterpart to induce the production of antiviral effectors should be further explored.

In mollusks, one *C. gigas* STING ortholog (CgSTING) with three N-terminal transmembrane domains followed by a STING domain is identified (Green et al., 2015). CgSTING directly binds to different kinds of CDNs, including *c*-di-AMP, *c*-di-GMP, 2′3′-cGAMP, and 3′3′-cGAMP (Kranzusch et al., 2015; Tang et al., 2016). CgSTING also interacts with and activates CgTBK1, suggesting that CgSTING plays a role in immune response through CgTBK1 and activates downstream IRF family proteins (Tang et al., 2016). Similar to mammalian *STING*, the expression *CgSTING* is upregulated by *Ostreid herpesvirus-1* (OsHV-1) inoculation and poly(I:C) stimulation (Tang et al., 2016), indicating the potential role of CgSTING in immune defense against DNA and RNA viruses. Although *C. gigas* encodes cGAS orthologs, whether CgcGAS senses dsDNA, produces cGAMP, subsequently activates STING to initiate cGAMP-STING-TBK1-IRFs/NF-κB signaling, and upregulates the expression of immune-related genes, such ISGs, still requires further examination (Tang et al., 2016). These investigations may be useful in determining whether STING is initially involved in antibacterial or antiviral immune responses. In *Mytilus galloprovincialis*, one STING ortholog (MgSTING) lacks N-terminal transmembrane domains but has duplicated STING and TIR domains (Gerdol and Venier, 2015). Dimerization is required for mammalian STING activation, and the presence of two STING domains within a single MgSTING protein may provide a functional analogy of STING dimer, which may not need dimerization for its activation. The presence of a TIR domain is likely relevant to signal transduction, which is administered by CTT in mammals but is absent in invertebrates. TIR domain-containing proteins are usually involved in TLR pathways, such as MyD88, TRIF, and TRAM, but whether MgSTING participates in TLR pathway activation needs clarification.

4. CONCLUSIONS

Nucleic acids presented in the viral genome or generated during replication are ideal PAMPs as non-self for PRR sensing across diverse taxa, including in nematodes, mollusks, arthropods (e.g., insects and crustaceans), and vertebrates (Wu and Chen, 2014). Nucleic acid sensing by PRRs activates distinct antiviral responses in different hosts. In nematodes and arthropods, the siRNA pathway is the most efficient and widely used antiviral strategy via the precise degradation of viral RNAs, thereby representing a highly specific antiviral mechanism (Kingsolver et al., 2013; Merkling and van Rij, 2013). Viral infection also induces large-scale immune effects by activating the evolutionarily conserved immune responsive pathways through NF-κB and/or STAT transcription factors. Unlike the RNAi pathway, these pathways do not have a broad antiviral spectrum but exhibit virus-specific effect and protection only against certain viruses in insects. Although both virus and bacteria/fungi activate the NF-κB and STAT pathways, in addition to some genes commonly induced, a set of distinctive effectors induced by viruses is also present (Kingsolver et al., 2013; Merkling and van Rij, 2013). Genes specifically upregulated by bacteria/fungi do not exhibit direct antiviral activity, whereas some virus-induced effectors (e.g., Vago and Nazo) do (Goto et al., 2018; Paradkar et al., 2012). Viruses activate these pathways independent of the canonical receptors, such as Tolls in the Toll pathway, PGRPs in the IMD pathway, and domeless in the JAK/STAT pathway, but via an identified Dicer-2 or other unknown receptors. Thus, non-canonical Toll/IMD-NF-κB and JAK/STAT pathways, such as the Dicer-2-NF-κB-Vago-JAK/STAT and DmSTING-DmIKKβ-Relish-Nazo cascades, are involved in novel receptors and intercellular signaling transduction members during viral infection. Vago orthologs are present in mosquitoes and shrimp but not in nematodes and mollusks, thereby suggesting that Dicer-2-mediated *Vago* induction is evolutionarily conserved within arthropods (Green and Speck, 2018). *Nazo* encodes a predicted transmembrane protein with 140 residues that are highly conserved from flies to humans, but its human ortholog *C19orf12* is poorly characterized. The conservation of the two proteins above may suggest their indispensable roles in antiviral immunity. However, their antiviral mechanism needs further investigations. The nucleic acid- and/or virus infection-stimulated immune pathways may act as an alternative antiviral response when RNAi is absent or suppressed by VSR. Insects are still resistant to viral infection without RNAi. Flies with depleted siRNA pathway show unchanged clearance of the low-level Nora virus infections and stability of persistent infections compared with wild-type controls (Swevers et al., 2018). *Wolbachia*-mediated protection against viral infection is still functional in RNAi mutant flies (Swevers et al., 2018). Nucleic acid sensors and/or virus sensors, subsequently intercellular players involved in signaling transduction, and downstream effectors are still poorly characterized. Further investigations addressing these issues may elucidate the molecular mechanism underlying the inducible non-specific antiviral immunity triggered by nucleic acid or virus sensing and may provide novel strategies to control viral infections in arthropod vectors, such as mosquitoes, and economic important invertebrates, such as oysters, silkworms, honeybees, and shrimps.

The IFN system is the main antiviral strategy of vertebrates, while siRNA-mediated viral RNA degradation is the primary antiviral response in invertebrates and even in plants and fungi (Ding, 2010; Guo et al., 2018). However, siRNA pathway is insignificant in the defense against viruses in some invertebrates, such mollusks, and even insects under particular conditions. *C. gigas* has a clear RLR-mediated antiviral pathway by inducing *IFNLP*, thereby upregulating ISGs that inhibit OsHV-1 replication. The result indicated the significant similarities to vertebrate IFN response. The key elements, namely, TLRs/RLRs-IRFs, IFNLP, JAK/STAT pathway, and ISGs, for IFN response are conserved in oysters and highly upregulated in response to OsHV-1 infection (He et al., 2015a). Thus, whether oysters have expanded nucleic acid sensors, such as Dicer-2 in insects, and novel virus recognition receptors co-operated with RLRs to orchestrate antiviral responses is a topic of interest. In invertebrates, only oysters have functional RLR pathway-mediated antiviral responses by upregulating *IFNLP*. In mammals, IFN and RNAi antiviral systems are incompatible with each other (Green and Speck, 2018). dsRNA triggers a functional RNAi antiviral response in stem cells that lack IFN response. However, in somatic cells, dsRNA only induces IFN response but not siRNA biogenesis. In somatic cells, IFN response turns off RISC, whereas RNAi response restoration inhibits IFN production and competes with the IFN response. Numerous RNA and DNA viruses can naturally infect *C. gigas*. vsiRNA detection and characterization by RNA-seq will determine whether oysters utilize the siRNA pathway to combat viruses. vncRNA libraries from gastropod mollusks infected with RNA viruses have been sequenced. vncRNAs are not siRNAs but most probably piRNAs, thereby suggesting that siRNA-mediated antiviral response may be relatively unimportant for this animal (Waldron et al., 2018). However, whether siRNA and piRNA pathways participate in the antiviral defense of mollusks still require further investigations. Findings from these studies also address whether mollusks have compatible IFN-like response and RNAi antiviral pathway and may aid in understanding why the IFN response supplanted siRNA antiviral pathway as the dominant antiviral immunity in vertebrates (Green and Speck, 2018).

siRNA pathway is involved in the antiviral defense of invertebrates. However, new evidence may challenge this traditional expectation. vncRNAs are absent in earthworms, sponges, or sea anemones infected with RNA viruses (Waldron et al., 2018). This finding suggests that canonical antiviral RNAi is absent from these lineages and challenges the widely held assumption that most invertebrates possess an antiviral RNAi pathway (Waldron et al., 2018). Abundant somatic piRNAs are detected across anciently divergent animals, such as in sponge, sea anemone, starfish, earthworm, dog whelk, and brown alga. Primary vpiRNAs are detected from the parvo/denso-like viruses in the starfish and sea anemone, as well as from rhabdo-like virus 2 in dog whelk. This result suggests that piRNA pathway-mediated virus processing is used in these animals, thereby possibly represent an additional ancient antiviral mechanism (Waldron et al., 2018).

Therefore, although nucleic acid sensing and the related antiviral immunity of invertebrates have evolutionarily conserved genes and pathways, they may not be simply homogenous to each other and to those of vertebrates in certain aspects, thereby displaying high diversities in virus sensing, immune activation, and pathways utilized for defense. The following suggestions will not only advance our understanding of the fundamental knowledge about antiviral immunity but also provide novel ideas in controlling disease vectors and harmful pests and protecting economically or ecologically important invertebrate organisms from infections: obtaining more comprehensive views of these antiviral responses; understanding the mechanisms underlying their activation; identifying novel sensors, new players, and downstream effectors; and evaluating the distinctive contributions of different pathways to antiviral defense.

REFERENCES

Asgari, S., 2015. Regulatory role of cellular and viral microRNAs in insect–virus interactions. Curr. Opin. Insect Sci. 8, 104–110.

Attasart, P., Kaewkhaw, R., Chimwai, C., Kongphom, U., Panyim, S., 2011. Clearance of Penaeus monodon densovirus in naturally pre-infected shrimp by combined ns1 and vp dsRNAs. Virus Res. 159, 79–82.

Avadhanula, V., Weasner, B.P., Hardy, G.G., Kumar, J.P., Hardy, R.W., 2009. A novel system for the launch of alphavirus RNA synthesis reveals a role for the Imd pathway in arthropod antiviral response. PLoS Pathog. 5, e1000582.

Bernier, A., Sagan, S.M., 2018. The diverse roles of microRNAs at the host(−)virus interface. Viruses 10 (8), pii: E440.

Bitra, K., Suderman, R.J., Strand, M.R., 2012. Polydnavirus Ank proteins bind NF-kappaB homodimers and inhibit processing of relish. PLoS Pathog. 8, e1002722.

Blair, C.D., Olson, K.E., 2015. The role of RNA interference (RNAi) in arbovirus-vector interactions. Viruses 7, 820–843.

Brackney, D.E., Beane, J.E., Ebel, G.D., 2009. RNAi targeting of West Nile virus in mosquito midguts promotes virus diversification. PLoS Pathog. 5, e1000502.

Brackney, D.E., Scott, J.C., Sagawa, F., Woodward, J.E., Miller, N.A., Schilkey, F.D., Mudge, J., Wilusz, J., Olson, K.E., Blair, C.D., et al., 2010. C6/36 Aedes albopictus cells have a dysfunctional antiviral RNA interference response. PLoS Negl. Trop. Dis. 4, e856.

Brennan, J.J., Gilmore, T.D., 2018. Evolutionary origins of Toll-like receptor signaling. Mol. Biol. Evol. 35, 1576–1587.

Bronkhorst, A.W., van Rij, R.P., 2014. The long and short of antiviral defense: small RNA-based immunity in insects. Curr. Opin. Virol. 7, 19–28.

Bronkhorst, A.W., van Cleef, K.W., Vodovar, N., Ince, I.A., Blanc, H., Vlak, J.M., Saleh, M.C., van Rij, R.P., 2012. The DNA virus invertebrate iridescent virus 6 is a target of the Drosophila RNAi machinery. Proc. Natl. Acad. Sci. U. S. A. 109, E3604–E3613.

Bronkhorst, A.W., van Cleef, K.W., Venselaar, H., van Rij, R.P., 2014. A dsRNA-binding protein of a complex invertebrate DNA virus suppresses the Drosophila RNAi response. Nucleic Acids Res. 42, 12237–12248.

Brutscher, L.M., Flenniken, M.L., 2015. RNAi and antiviral defense in the honey bee. J. Immunol. Res. 2015, 941897.

Brutscher, L.M., Daughenbaugh, K.F., Flenniken, M.L., 2015. Antiviral defense mechanisms in honey bees. Curr. Opin. Insect Sci. 10, 71–82.

Brutscher, L.M., Daughenbaugh, K.F., Flenniken, M.L., 2017. Virus and dsRNA-triggered transcriptional responses reveal key components of honey bee antiviral defense. Sci. Rep. 7, 6448.

Buchon, N., Silverman, N., Cherry, S., 2014. Immunity in Drosophila melanogaster—from microbial recognition to whole-organism physiology. Nat. Rev. Immunol. 14, 796–810.

Chen, Y.H., Jia, X.T., Zhao, L., Li, C.Z., Zhang, S., Chen, Y.G., Weng, S.P., He, J.G., 2011. Identification and functional characterization of Dicer2 and five single VWC domain proteins of Litopenaeus vannamei. Dev. Comp. Immunol. 35, 661–671.

Chen, G., Zhang, C., Jiang, F., Wang, Y., Xu, Z., Wang, C., 2014. Bioinformatics analysis of hemocyte miRNAs of scallop Chlamys farreri against acute viral necrobiotic virus (AVNV). Fish Shellfish Immunol. 37, 75–86.

Cheng, G., Liu, Y., Wang, P., Xiao, X., 2016. Mosquito defense strategies against viral infection. Trends Parasitol. 32, 177–186.

Chimwai, C., Tongboonsong, P., Namramoon, O., Panyim, S., Attasart, P., 2016. A formulated double-stranded RNA diet for reducing Penaeus monodon densovirus infection in black tiger shrimp. J. Invertebr. Pathol. 134, 23–26.

Chotkowski, H.L., Ciota, A.T., Jia, Y., Puig-Basagoiti, F., Kramer, L.D., Shi, P.Y., Glaser, R.L., 2008. West Nile virus infection of Drosophila melanogaster induces a protective RNAi response. Virology 377, 197–206.

Coffman, S.R., Lu, J., Guo, X., Zhong, J., Jiang, H., Broitman-Maduro, G., Li, W.X., Lu, R., Maduro, M., Ding, S.W., 2017. Caenorhabditis elegans RIG-I homolog mediates antiviral RNA interference downstream of Dicer-dependent biogenesis of viral small interfering RNAs. MBio 8 (2), pii: e00264-17.

Cui, Y., Huang, T., Zhang, X., 2015. RNA editing of microRNA prevents RNA-induced silencing complex recognition of target mRNA. Open Biol. 5, 150126.

Cui, Y., Yang, X., Zhang, X., 2017. Shrimp miR-34 from shrimp stress response to virus infection suppresses tumorigenesis of breast cancer. Mol. Ther. Nucleic Acids 9, 387–398.

Czech, B., Malone, C.D., Zhou, R., Stark, A., Schlingeheyde, C., Dus, M., Perrimon, N., Kellis, M., Wohlschlegel, J.A., Sachidanandam, R., et al., 2008. An endogenous small interfering RNA pathway in Drosophila. Nature 453, 798–802.

Danilchanka, O., Mekalanos, J.J., 2013. Cyclic dinucleotides and the innate immune response. Cell 154, 962–970.

de Faria, I.J., Olmo, R.P., Silva, E.G., Marques, J.T., 2013. dsRNA sensing during viral infection: lessons from plants, worms, insects, and mammals. J. Interferon Cytokine Res. 33, 239–253.

Deddouche, S., Matt, N., Budd, A., Mueller, S., Kemp, C., Galiana-Arnoux, D., Dostert, C., Antoniewski, C., Hoffmann, J.A., Imler, J.-L., 2008. The DExD/H-box helicase Dicer-2 mediates the induction of antiviral activity in Drosophila. Nat. Immunol. 9, 1425.

Deepika, A., Sreedharan, K., Paria, A., Makesh, M., Rajendran, K.V., 2014. Toll-pathway in tiger shrimp (Penaeus monodon) responds to white spot syndrome virus infection: evidence through molecular characterisation and expression profiles of MyD88, TRAF6 and TLR genes. Fish Shellfish Immunol. 41, 441–454.

Ding, S.W., 2010. RNA-based antiviral immunity. Nat. Rev. Immunol. 10, 632–644.

Fabbri, M., Paone, A., Calore, F., Galli, R., Gaudio, E., Santhanam, R., Lovat, F., Fadda, P., Mao, C., Nuovo, G.J., et al., 2012. MicroRNAs bind to Toll-like receptors to induce prometastatic inflammatory response. Proc. Natl. Acad. Sci. U. S. A. 109, E2110–E2116.

Felix, M.A., Ashe, A., Piffaretti, J., Wu, G., Nuez, I., Belicard, T., Jiang, Y., Zhao, G., Franz, C.J., Goldstein, L.D., et al., 2011. Natural and experimental infection of Caenorhabditis nematodes by novel viruses related to nodaviruses. PLoS Biol. 9, e1000586.

Feng, J., Zhao, L., Jin, M., Li, T., Wu, L., Chen, Y., Ren, Q., 2016. Toll receptor response to white spot syndrome virus challenge in giant freshwater prawns (Macrobrachium rosenbergii). Fish Shellfish Immunol. 57, 148–159.

Ferreira, Á.G., Naylor, H., Esteves, S.S., Pais, I.S., Martins, N.E., Teixeira, L., 2014. The Toll-dorsal pathway is required for resistance to viral oral infection in Drosophila. PLoS Pathog. 10, e1004507.

Fire, A., Xu, S., Montgomery, M.K., Kostas, S.A., Driver, S.E., Mello, C.C., 1998. Potent and specific genetic interference by double-stranded RNA in Caenorhabditis elegans. Nature 391, 806–811.

Fung, E., Hill, K., Hogendoorn, K., Glatz, R.V., Napier, K.R., Bellgard, M.I., Barrero, R.A., 2018. De novo assembly of honey bee RNA viral genomes by tapping into the innate insect antiviral response pathway. J. Invertebr. Pathol. 152, 38–47.

Galbraith, D.A., Yang, X., Nino, E.L., Yi, S., Grozinger, C., 2015. Parallel epigenomic and transcriptomic responses to viral infection in honey bees (Apis mellifera). PLoS Pathog. 11, e1004713.

Galiana-Arnoux, D., Dostert, C., Schneemann, A., Hoffmann, J.A., Imler, J.L., 2006. Essential function in vivo for Dicer-2 in host defense against RNA viruses in Drosophila. Nat. Immunol. 7, 590–597.

Gammon, D.B., Mello, C.C., 2015. RNA interference-mediated antiviral defense in insects. Curr. Opin. Insect Sci. 8, 111–120.

Gerdol, M., Venier, P., 2015. An updated molecular basis for mussel immunity. Fish Shellfish Immunol. 46, 17–38.

Goertz, G.P., Fros, J.J., Miesen, P., Vogels, C.B., van der Bent, M.L., Geertsema, C., Koenraadt, C.J., van Rij, R.P., van Oers, M.M., Pijlman, G.P., 2016. Noncoding subgenomic flavivirus RNA is processed by the mosquito RNA interference machinery and determines West Nile virus transmission by Culex pipiens mosquitoes. J. Virol. 90, 10145–10159.

Goic, B., Vodovar, N., Mondotte, J.A., Monot, C., Frangeul, L., Blanc, H., Gausson, V., Vera-Otarola, J., Cristofari, G., Saleh, M.C., 2013. RNA-mediated interference and reverse transcription control the persistence of RNA viruses in the insect model Drosophila. Nat. Immunol. 14, 396–403.

Goto, A., Okado, K., Martins, N., Cai, H., Barbier, V., Lamiable, O., Troxler, L., Santiago, E., Kuhn, L., Paik, D., et al., 2018. The kinase IKKbeta regulates a STING- and NF-kappaB-dependent antiviral response pathway in Drosophila. Immunity 49 (2), 225.e4–234.e4.

Gottwein, E., Cullen, B.R., 2008. Viral and cellular microRNAs as determinants of viral pathogenesis and immunity. Cell Host Microbe 3, 375–387.

Goubau, D., Deddouche, S., Reis e Sousa, C., 2013. Cytosolic sensing of viruses. Immunity 38, 855–869.

Green, T.J., Speck, P., 2018. Antiviral defense and innate immune memory in the oyster. Viruses 10 (3), pii: E133.

Green, T.J., Raftos, D., Speck, P., Montagnani, C., 2015. Antiviral immunity in marine molluscs. J. Gen. Virol. 96, 2471–2482.

Guo, X., Zhang, R., Wang, J., Ding, S.W., Lu, R., 2013. Homologous RIG-I-like helicase proteins direct RNAi-mediated antiviral immunity in C. elegans by distinct mechanisms. Proc. Natl. Acad. Sci. U. S. A. 110, 16085–16090.

Guo, Z., Li, Y., Ding, S.W., 2018. Small RNA-based antimicrobial immunity. Nat. Rev. Immunol. https://doi.org/10.1038/s41577-018-0071-x (Epub ahead of print).

He, Y., Zhang, X., 2012. Comprehensive characterization of viral miRNAs involved in white spot syndrome virus (WSSV) infection. RNA Biol. 9, 1019–1029.

He, Y., Yang, K., Zhang, X., 2014. Viral microRNAs targeting virus genes promote virus infection in shrimp in vivo. J. Virol. 88, 1104–1112.

He, Y., Jouaux, A., Ford, S.E., Lelong, C., Sourdaine, P., Mathieu, M., Guo, X., 2015a. Transcriptome analysis reveals strong and complex antiviral response in a mollusc. Fish Shellfish Immunol. 46, 131–144.

He, Y., Ju, C., Zhang, X., 2015b. Roles of small RNAs in the immune defense mechanisms of crustaceans. Mol. Immunol. 68, 399–403.

He, Y., Ma, T., Zhang, X., 2017a. The mechanism of synchronous precise regulation of two shrimp white spot syndrome virus targets by a viral microRNA. Front. Immunol. 8, 1546.

He, Y., Sun, Y., Zhang, X., 2017b. Noncoding miRNAs bridge virus infection and host autophagy in shrimp in vivo. FASEB J. 31, 2854–2868.

Hess, A.M., Prasad, A.N., Ptitsyn, A., Ebel, G.D., Olson, K.E., Barbacioru, C., Monighetti, C., Campbell, C.L., 2011. Small RNA profiling of Dengue virus-mosquito interactions implicates the PIWI RNA pathway in anti-viral defense. BMC Microbiol. 11, 45.

Ho, T., Yasri, P., Panyim, S., Udomkit, A., 2011. Double-stranded RNA confers both preventive and therapeutic effects against Penaeus stylirostris densovirus (PstDNV) in Litopenaeus vannamei. Virus Res. 155, 131–136.

Hua, X., Li, B., Song, L., Hu, C., Li, X., Wang, D., Xiong, Y., Zhao, P., He, H., Xia, Q., et al., 2018. Stimulator of interferon genes (STING) provides insect antiviral immunity by promoting Dredd caspase-mediated NF-kappaB activation. J. Biol. Chem. 293, 11878–11890.
Huang, T., Zhang, X., 2012. Functional analysis of a crustacean microRNA in host-virus interactions. J. Virol. 86, 12997–13004.
Huang, T., Zhang, X., 2013. Host defense against DNA virus infection in shrimp is mediated by the siRNA pathway. Eur. J. Immunol. 43, 137–146.
Huang, T., Xu, D., Zhang, X., 2012. Characterization of host microRNAs that respond to DNA virus infection in a crustacean. BMC Genomics 13, 159.
Huang, Z., Kingsolver, M.B., Avadhanula, V., Hardy, R.W., 2013. An antiviral role for antimicrobial peptides during the arthropod response to alphavirus replication. J. Virol. 87, 4272–4280.
Huang, T., Cui, Y., Zhang, X., 2014. Involvement of viral microRNA in the regulation of antiviral apoptosis in shrimp. J. Virol. 88, 2544–2554.
Huang, Y., Wang, W., Ren, Q., 2016. Two host microRNAs influence WSSV replication via STAT gene regulation. Sci. Rep. 6, 23643.
Huang, B., Zhang, L., Du, Y., Xu, F., Li, L., Zhang, G., 2017a. Characterization of the mollusc RIG-I/MAVS pathway reveals an archaic antiviral signalling framework in invertebrates. Sci. Rep. 7, 8217.
Huang, J.Y., Kang, S.T., Chen, I.T., Chang, L.K., Lin, S.S., Kou, G.H., Chu, C.Y., Lo, C.F., 2017b. Shrimp miR-10a is co-opted by white spot syndrome virus to increase viral gene expression and viral replication. Front. Immunol. 8, 1084.
Huang, X., Fejes Toth, K., Aravin, A.A., 2017c. piRNA biogenesis in Drosophila melanogaster. Trends Genet. 33, 882–894.
Huang, Y., Han, K., Wang, W., Ren, Q., 2017d. Host microRNA-217 promotes white spot syndrome virus infection by targeting tube in the Chinese mitten crab (Eriocheir sinensis). Front. Cell. Infect. Microbiol. 7, 164.
Huang, Y., Chen, Y., Hui, K., Ren, Q., 2018a. Cloning and characterization of two Toll receptors (PcToll5 and PcToll6) in response to white spot syndrome virus in the red swamp crayfish Procambarus clarkii. Front. Physiol. 9, 936.
Huang, Y., Wang, W., Xu, Z., Pan, J., Zhao, Z., Ren, Q., 2018b. Eriocheir sinensis microRNA-7 targets crab Myd88 to enhance white spot syndrome virus replication. Fish Shellfish Immunol. 79, 274–283.
Hussain, M., Asgari, S., 2014. MicroRNA-like viral small RNA from Dengue virus 2 autoregulates its replication in mosquito cells. Proc. Natl. Acad. Sci. U. S. A. 111, 2746–2751.
Hussain, M., Taft, R.J., Asgari, S., 2008. An insect virus-encoded microRNA regulates viral replication. J. Virol. 82, 9164–9170.
Hussain, M., Abraham, A.M., Asgari, S., 2010. An ascovirus-encoded RNase III autoregulates its expression and suppresses RNA interference-mediated gene silencing. J. Virol. 84, 3624–3630.
Hussain, M., Torres, S., Schnettler, E., Funk, A., Grundhoff, A., Pijlman, G.P., Khromykh, A.A., Asgari, S., 2012. West Nile virus encodes a microRNA-like small RNA in the 3′ untranslated region which up-regulates GATA4 mRNA and facilitates virus replication in mosquito cells. Nucleic Acids Res. 40, 2210–2223.
Itsathitphaisarn, O., Thitamadee, S., Weerachatyanukul, W., Sritunyalucksana, K., 2017. Potential of RNAi applications to control viral diseases of farmed shrimp. J. Invertebr. Pathol. 147, 76–85.
Ivashkiv, L.B., Donlin, L.T., 2014. Regulation of type I interferon responses. Nat. Rev. Immunol. 14, 36–49.
Iwasaki, Y.W., Siomi, M.C., Siomi, H., 2015. PIWI-interacting RNA: its biogenesis and functions. Annu. Rev. Biochem. 84, 405–433.

Jayachandran, B., Hussain, M., Asgari, S., 2012. RNA interference as a cellular defense mechanism against the DNA virus baculovirus. J. Virol. 86, 13729–13734.
Kaewkascholkul, N., Somboonviwat, K., Asakawa, S., Hirono, I., Tassanakajon, A., Somboonwiwat, K., 2016. Shrimp miRNAs regulate innate immune response against white spot syndrome virus infection. Dev. Comp. Immunol. 60, 191–201.
Kakumani, P.K., Ponia, S.S., S, R.K., Sood, V., Chinnappan, M., Banerjea, A.C., Medigeshi, G.R., Malhotra, P., Mukherjee, S.K., Bhatnagar, R.K., 2013. Role of RNA interference (RNAi) in Dengue virus replication and identification of NS4B as an RNAi suppressor. J. Virol. 87, 8870–8883.
Kang, L., Wang, M., Cao, X., Tang, S., Xia, D., Shen, X., Zhao, Q., 2018. Inhibition of expression of BmNPV cg30 by bmo-miRNA-390 is a host response to baculovirus invasion. Arch. Virol. 163, 2719–2725.
Kawai, T., Akira, S., 2011. Toll-like receptors and their crosstalk with other innate receptors in infection and immunity. Immunity 34, 637–650.
Keene, K.M., Foy, B.D., Sanchez-Vargas, I., Beaty, B.J., Blair, C.D., Olson, K.E., 2004. RNA interference acts as a natural antiviral response to O'nyong-nyong virus (Alphavirus; Togaviridae) infection of Anopheles gambiae. Proc. Natl. Acad. Sci. U. S. A. 101, 17240–17245.
Kemp, C., Imler, J.L., 2009. Antiviral immunity in Drosophila. Curr. Opin. Immunol. 21, 3–9.
Kemp, C., Mueller, S., Goto, A., Barbier, V., Paro, S., Bonnay, F., Dostert, C., Troxler, L., Hetru, C., Meignin, C., et al., 2013. Broad RNA interference-mediated antiviral immunity and virus-specific inducible responses in Drosophila. J. Immunol. 190, 650–658.
Kingsolver, M.B., Huang, Z., Hardy, R.W., 2013. Insect antiviral innate immunity: pathways, effectors, and connections. J. Mol. Biol. 425, 4921–4936.
Kranzusch, P.J., Wilson, S.C., Lee, A.S., Berger, J.M., Doudna, J.A., Vance, R.E., 2015. Ancient origin of cGAS-STING reveals mechanism of universal 2′,3' cGAMP signaling. Mol. Cell 59, 891–903.
Labreuche, Y., Veloso, A., de la Vega, E., Gross, P.S., Chapman, R.W., Browdy, C.L., Warr, G.W., 2010. Non-specific activation of antiviral immunity and induction of RNA interference may engage the same pathway in the Pacific white leg shrimp Litopenaeus vannamei. Dev. Comp. Immunol. 34, 1209–1218.
Lamiable, O., Kellenberger, C., Kemp, C., Troxler, L., Pelte, N., Boutros, M., Marques, J.T., Daeffler, L., Hoffmann, J.A., Roussel, A., et al., 2016. Cytokine Diedel and a viral homologue suppress the IMD pathway in Drosophila. Proc. Natl. Acad. Sci. U. S. A. 113, 698–703.
Lee, R.C., Feinbaum, R.L., Ambros, V., 1993. The C. elegans heterochronic gene lin-4 encodes small RNAs with antisense complementarity to lin-14. Cell 75, 843–854.
Leger, P., Lara, E., Jagla, B., Sismeiro, O., Mansuroglu, Z., Coppee, J.Y., Bonnefoy, E., Bouloy, M., 2013. Dicer-2- and Piwi-mediated RNA interference in Rift Valley fever virus-infected mosquito cells. J. Virol. 87, 1631–1648.
Leggewie, M., Schnettler, E., 2018. RNAi-mediated antiviral immunity in insects and their possible application. Curr. Opin. Virol. 32, 108–114.
Lehmann, S.M., Kruger, C., Park, B., Derkow, K., Rosenberger, K., Baumgart, J., Trimbuch, T., Eom, G., Hinz, M., Kaul, D., et al., 2012. An unconventional role for miRNA: let-7 activates Toll-like receptor 7 and causes neurodegeneration. Nat. Neurosci. 15, 827–835.
Lewis, S.H., Quarles, K.A., Yang, Y., Tanguy, M., Frezal, L., Smith, S.A., Sharma, P.P., Cordaux, R., Gilbert, C., Giraud, I., et al., 2018. Pan-arthropod analysis reveals somatic piRNAs as an ancestral defence against transposable elements. Nat. Ecol. Evol. 2, 174–181.
Li, H., Li, W.X., Ding, S.W., 2002. Induction and suppression of RNA silencing by an animal virus. Science 296, 1319–1321.

Li, C., Li, H., Chen, Y., Chen, Y., Wang, S., Weng, S.P., Xu, X., He, J., 2015. Activation of Vago by interferon regulatory factor (IRF) suggests an interferon system-like antiviral mechanism in shrimp. Sci. Rep. 5, 15078.

Li, H., Wang, S., Lŭ, K., Yin, B., Xiao, B., Li, S., He, J., Li, C., 2017. An invertebrate STING from shrimp activates an innate immune defense against bacterial infection. FEBS Lett. 591, 1010–1017.

Lin, C.C., Chou, C.M., Hsu, Y.L., Lien, J.C., Wang, Y.M., Chen, S.T., Tsai, S.C., Hsiao, P.W., Huang, C.J., 2004. Characterization of two mosquito STATs, AaSTAT and CtSTAT. Differential regulation of tyrosine phosphorylation and DNA binding activity by lipopolysaccharide treatment and by Japanese encephalitis virus infection. J. Biol. Chem. 279, 3308–3317.

Liu, H., Soderhall, K., Jiravanichpaisal, P., 2009. Antiviral immunity in crustaceans. Fish Shellfish Immunol. 27, 79–88.

Liu, C., Li, F., Sun, Y., Zhang, X., Yuan, J., Yang, H., Xiang, J., 2016. Virus-derived small RNAs in the penaeid shrimp Fenneropenaeus chinensis during acute infection of the DNA virus WSSV. Sci. Rep. 6, 28678.

Liu, Y., Gordesky-Gold, B., Leney-Greene, M., Weinbren, N.L., Tudor, M., Cherry, S., 2018. Inflammation-induced, STING-dependent autophagy restricts Zika virus infection in the Drosophila brain. Cell Host Microbe 24 (57–68), e53.

Loy, J.D., Mogler, M.A., Loy, D.S., Janke, B., Kamrud, K., Scura, E.D., Harris, D.L., Bartholomay, L.C., 2012. dsRNA provides sequence-dependent protection against infectious myonecrosis virus in Litopenaeus vannamei. J. Gen. Virol. 93, 880–888.

Lu, R., Maduro, M., Li, F., Li, H.W., Broitman-Maduro, G., Li, W.X., Ding, S.W., 2005. Animal virus replication and RNAi-mediated antiviral silencing in Caenorhabditis elegans. Nature 436, 1040–1043.

Lu, M., Yang, C., Li, M., Yi, Q., Lu, G., Wu, Y., Qu, C., Wang, L., Song, L., 2018. A conserved interferon regulation factor 1 (IRF-1) from Pacific oyster Crassostrea gigas functioned as an activator of IFN pathway. Fish Shellfish Immunol. 76, 68–77.

Margolis, S.R., Wilson, S.C., Vance, R.E., 2017. Evolutionary origins of cGAS-STING signaling. Trends Immunol. 38, 733–743.

Marques, J.T., Imler, J.L., 2016. The diversity of insect antiviral immunity: insights from viruses. Curr. Opin. Microbiol. 32, 71–76.

Martin, M., Hiroyasu, A., Guzman, R.M., Roberts, S.A., Goodman, A.G., 2018. Analysis of Drosophila STING reveals an evolutionarily conserved antimicrobial function. Cell Reports 23, 3537–3550.e3536.

Maruekawong, K., Tirasophon, W., Panyim, S., Attasart, P., 2018. Involvement of LvSID-1 in dsRNA uptake in Litopenaeus vannamei. Aquaculture 482, 65–72.

McMenamin, A.J., Daughenbaugh, K.F., Parekh, F., Pizzorno, M.C., Flenniken, M.L., 2018. Honey bee and bumble bee antiviral defense. Viruses 10 (8), pii: E395.

Mehrabadi, M., Hussain, M., Matindoost, L., Asgari, S., 2015. The baculovirus antiapoptotic p35 protein functions as an inhibitor of the host RNA interference antiviral response. J. Virol. 89, 8182–8192.

Merkling, S.H., van Rij, R.P., 2013. Beyond RNAi: antiviral defense strategies in Drosophila and mosquito. J. Insect Physiol. 59, 159–170.

Miesen, P., Girardi, E., van Rij, R.P., 2015. Distinct sets of PIWI proteins produce arbovirus and transposon-derived piRNAs in Aedes aegypti mosquito cells. Nucleic Acids Res. 43, 6545–6556.

Miesen, P., Ivens, A., Buck, A.H., van Rij, R.P., 2016a. Small RNA profiling in dengue virus 2-infected Aedes mosquito cells reveals viral piRNAs and novel host miRNAs. PLoS Negl. Trop. Dis. 10, e0004452.

Miesen, P., Joosten, J., van Rij, R.P., 2016b. PIWIs go viral: arbovirus-derived piRNAs in vector mosquitoes. PLoS Pathog. 12, e1006017.

Monsanto-Hearne, V., Tham, A.L.Y., Wong, Z.S., Asgari, S., Johnson, K.N., 2017. Drosophila miR-956 suppression modulates ectoderm-expressed 4 and inhibits viral replication. Virology 502, 20–27.
Morazzani, E.M., Wiley, M.R., Murreddu, M.G., Adelman, Z.N., Myles, K.M., 2012. Production of virus-derived ping-pong-dependent piRNA-like small RNAs in the mosquito soma. PLoS Pathog. 8, e1002470.
Mueller, S., Gausson, V., Vodovar, N., Deddouche, S., Troxler, L., Perot, J., Pfeffer, S., Hoffmann, J.A., Saleh, M.C., Imler, J.L., 2010. RNAi-mediated immunity provides strong protection against the negative-strand RNA vesicular stomatitis virus in Drosophila. Proc. Natl. Acad. Sci. U. S. A. 107, 19390–19395.
Mussabekova, A., Daeffler, L., Imler, J.L., 2017. Innate and intrinsic antiviral immunity in Drosophila. Cell. Mol. Life Sci. 74, 2039–2054.
Myles, K.M., Wiley, M.R., Morazzani, E.M., Adelman, Z.N., 2008. Alphavirus-derived small RNAs modulate pathogenesis in disease vector mosquitoes. Proc. Natl. Acad. Sci. U. S. A. 105, 19938–19943.
Myllymaki, H., Valanne, S., Ramet, M., 2014. The Drosophila imd signaling pathway. J. Immunol. 192, 3455–3462.
Nayak, A., Tassetto, M., Kunitomi, M., Andino, R., 2013. RNA interference-mediated intrinsic antiviral immunity in invertebrates. Curr. Top. Microbiol. Immunol. 371, 183–200.
Nie, L., Cai, S.Y., Shao, J.Z., Chen, J., 2018. Toll-like receptors, associated biological roles, and signaling networks in non-mammals. Front. Immunol. 9, 1523.
Nilsen, P., Karlsen, M., Sritunyalucksana, K., Thitamadee, S., 2017. White spot syndrome virus VP28 specific double-stranded RNA provides protection through a highly focused siRNA population. Sci. Rep. 7, 1028.
Niu, J., Meeus, I., Smagghe, G., 2016. Differential expression pattern of Vago in bumblebee (Bombus terrestris), induced by virulent and avirulent virus infections. Sci. Rep. 6, 34200.
Obbard, D.J., Jiggins, F.M., Halligan, D.L., Little, T.J., 2006. Natural selection drives extremely rapid evolution in antiviral RNAi genes. Curr. Biol. 16, 580–585.
Paradkar, P.N., Trinidad, L., Voysey, R., Duchemin, J.-B., Walker, P.J., 2012. Secreted Vago restricts West Nile virus infection in *Culex* mosquito cells by activating the Jak-STAT pathway. Proc. Natl. Acad. Sci. U. S. A. 109, 18915–18920.
Paradkar, P.N., Duchemin, J.-B., Voysey, R., Walker, P.J., 2014. Dicer-2-dependent activation of Culex Vago occurs via the TRAF-Rel2 signaling pathway. PLoS Negl. Trop. Dis. 8, e2823.
Petit, M., Mongelli, V., Frangeul, L., Blanc, H., Jiggins, F., Saleh, M.C., 2016. piRNA pathway is not required for antiviral defense in Drosophila melanogaster. Proc. Natl. Acad. Sci. U. S. A. 113, E4218–E4227.
Poirier, E.Z., Goic, B., Tome-Poderti, L., Frangeul, L., Boussier, J., Gausson, V., Blanc, H., Vallet, T., Loyd, H., Levi, L.I., et al., 2018. Dicer-2-dependent generation of viral DNA from defective genomes of RNA viruses modulates antiviral immunity in insects. Cell Host Microbe 23 (353–365), e358.
Pompon, J., Manuel, M., Ng, G.K., Wong, B., Shan, C., Manokaran, G., Soto-Acosta, R., Bradrick, S.S., Ooi, E.E., Misse, D., et al., 2017. Dengue subgenomic flaviviral RNA disrupts immunity in mosquito salivary glands to increase virus transmission. PLoS Pathog. 13, e1006535.
Prasad, A.N., Brackney, D.E., Ebel, G.D., 2013. The role of innate immunity in conditioning mosquito susceptibility to West Nile virus. Viruses 5, 3142–3170.
Qiu, Y., Zhou, X., 2018. STING: from mammals to insects. Cell Host Microbe 24, 5–7.
Qiu, W., Zhang, S., Chen, Y.-G., Wang, P.-H., Xu, X.-P., Li, C.-Z., Chen, Y.-H., Fan, W.-Z., Yan, H., Weng, S.-P., et al., 2014. Litopenaeus vannamei NF-κB is required for WSSV replication. Dev. Comp. Immunol. 45, 156–162.

Ren, Q., Huang, Y., He, Y., Wang, W., Zhang, X., 2015. A white spot syndrome virus microRNA promotes the virus infection by targeting the host STAT. Sci. Rep. 5, 18384.

Ren, Q., Huang, X., Cui, Y., Sun, J., Wang, W., Zhang, X., 2017. Two white spot syndrome virus microRNAs target the dorsal gene to promote virus infection in Marsupenaeus japonicus shrimp. J. Virol. 91 (8), pii: e02261-16.

Robalino, J., Bartlett, T., Shepard, E., Prior, S., Jaramillo, G., Scura, E., Chapman, R.W., Gross, P.S., Browdy, C.L., Warr, G.W., 2005. Double-stranded RNA induces sequence-specific antiviral silencing in addition to nonspecific immunity in a marine shrimp: convergence of RNA interference and innate immunity in the invertebrate antiviral response? J. Virol. 79, 13561–13571.

Sabin, L.R., Cherry, S., 2013. Small creatures use small RNAs to direct antiviral defenses. Eur. J. Immunol. 43, 27–33.

Sabin, L.R., Zhou, R., Gruber, J.J., Lukinova, N., Bambina, S., Berman, A., Lau, C.K., Thompson, C.B., Cherry, S., 2009. Ars2 regulates both miRNA- and siRNA-dependent silencing and suppresses RNA virus infection in Drosophila. Cell 138, 340–351.

Saleh, M.C., van Rij, R.P., Hekele, A., Gillis, A., Foley, E., O'Farrell, P.H., Andino, R., 2006. The endocytic pathway mediates cell entry of dsRNA to induce RNAi silencing. Nat. Cell Biol. 8, 793–802.

Samuel, G.H., Wiley, M.R., Badawi, A., Adelman, Z.N., Myles, K.M., 2016. Yellow fever virus capsid protein is a potent suppressor of RNA silencing that binds double-stranded RNA. Proc. Natl. Acad. Sci. U. S. A. 113, 13863–13868.

Samuel, G.H., Adelman, Z.N., Myles, K.M., 2018. Antiviral immunity and virus-mediated antagonism in disease vector mosquitoes. Trends Microbiol. 26, 447–461.

Sanchez-Vargas, I., Scott, J.C., Poole-Smith, B.K., Franz, A.W., Barbosa-Solomieu, V., Wilusz, J., Olson, K.E., Blair, C.D., 2009. Dengue virus type 2 infections of Aedes aegypti are modulated by the mosquito's RNA interference pathway. PLoS Pathog. 5, e1000299.

Santos, D., Wynant, N., Van den Brande, S., Verdonckt, T.W., Mingels, L., Peeters, P., Kolliopoulou, A., Swevers, L., Vanden Broeck, J., 2018. Insights into RNAi-based antiviral immunity in Lepidoptera: acute and persistent infections in Bombyx mori and Trichoplusia ni cell lines. Sci. Rep. 8, 2423.

Schnettler, E., Sterken, M.G., Leung, J.Y., Metz, S.W., Geertsema, C., Goldbach, R.W., Vlak, J.M., Kohl, A., Khromykh, A.A., Pijlman, G.P., 2012. Noncoding flavivirus RNA displays RNA interference suppressor activity in insect and mammalian cells. J. Virol. 86, 13486–13500.

Schott, D.H., Cureton, D.K., Whelan, S.P., Hunter, C.P., 2005. An antiviral role for the RNA interference machinery in Caenorhabditis elegans. Proc. Natl. Acad. Sci. U. S. A. 102, 18420–18424.

Schuster, S., Zirkel, F., Kurth, A., van Cleef, K.W., Drosten, C., van Rij, R.P., Junglen, S., 2014. A unique nodavirus with novel features: mosinovirus expresses two subgenomic RNAs, a capsid gene of unknown origin, and a suppressor of the antiviral RNA interference pathway. J. Virol. 88, 13447–13459.

Shu, L., Zhang, X., 2017. Shrimp miR-12 suppresses white spot syndrome virus infection by synchronously triggering antiviral phagocytosis and apoptosis pathways. Front. Immunol. 8, 855.

Shu, L., Li, C., Zhang, X., 2016. The role of shrimp miR-965 in virus infection. Fish Shellfish Immunol. 54, 427–434.

Singh, C.P., Singh, J., Nagaraju, J., 2012. A baculovirus-encoded microRNA (miRNA) suppresses its host miRNA biogenesis by regulating the exportin-5 cofactor ran. J. Virol. 86, 7867–7879.

Singh, C.P., Singh, J., Nagaraju, J., 2014. Bmnpv-miR-3 facilitates BmNPV infection by modulating the expression of viral P6.9 and other late genes in Bombyx mori. Insect Biochem. Mol. Biol. 49, 59–69.
Skalsky, R.L., Cullen, B.R., 2010. Viruses, microRNAs, and host interactions. Annu. Rev. Microbiol. 64, 123–141.
Slonchak, A., Hussain, M., Torres, S., Asgari, S., Khromykh, A.A., 2014. Expression of mosquito microRNA Aae-miR-2940-5p is downregulated in response to West Nile virus infection to restrict viral replication. J. Virol. 88, 8457–8467.
Song, X., Zhang, Z., Wang, S., Li, H., Zuo, H., Xu, X., Weng, S., He, J., Li, C., 2015. A Janus kinase in the JAK/STAT signaling pathway from Litopenaeus vannamei is involved in antiviral immune response. Fish Shellfish Immunol. 44, 662–673.
Soponpong, S., Amparyup, P., Tassanakajon, A., 2018. A cytosolic sensor, PmDDX41, mediates antiviral immune response in black tiger shrimp Penaeus monodon. Dev. Comp. Immunol. 81, 291–302.
Sun, R., Wang, M., Wang, L., Yue, F., Yi, Q., Huang, M., Liu, R., Qiu, L., Song, L., 2014. The immune responses triggered by CpG ODNs in shrimp Litopenaeus vannamei are associated with LvTolls. Dev. Comp. Immunol. 43, 15–22.
Sun, J.-J., Xu, S., He, Z.-H., Shi, X.-Z., Zhao, X.-F., Wang, J.-X., 2017. Activation of Toll pathway is different between Kuruma shrimp and Drosophila. Front. Immunol. 8, 1151.
Swevers, L., Liu, J., Smagghe, G., 2018. Defense mechanisms against viral infection in Drosophila: RNAi and non-RNAi. Viruses 10 (5), pii: E230.
Takeuchi, O., Akira, S., 2010. Pattern recognition receptors and inflammation. Cell 140, 805–820.
Tan, X., Sun, L., Chen, J., Chen, Z.J., 2018. Detection of microbial infections through innate immune sensing of nucleic acids. Annu. Rev. Microbiol. 72, 447–478.
Tang, X., Huang, B., Zhang, L., Li, L., Zhang, G., 2016. TANK-binding kinase-1 broadly affects oyster immune response to bacteria and viruses. Fish Shellfish Immunol. 56, 330–335.
Tassetto, M., Kunitomi, M., Andino, R., 2017. Circulating immune cells mediate a systemic RNAi-based adaptive antiviral response in Drosophila. Cell 169 (314–325), e313.
Tirasophon, W., Yodmuang, S., Chinnirunvong, W., Plongthongkum, N., Panyim, S., 2007. Therapeutic inhibition of yellow head virus multiplication in infected shrimps by YHV-protease dsRNA. Antiviral Res. 74, 150–155.
Trobaugh, D.W., Klimstra, W.B., 2017. MicroRNA regulation of RNA virus replication and pathogenesis. Trends Mol. Med. 23, 80–93.
Ulvila, J., Parikka, M., Kleino, A., Sormunen, R., Ezekowitz, R.A., Kocks, C., Ramet, M., 2006. Double-stranded RNA is internalized by scavenger receptor-mediated endocytosis in Drosophila S2 cells. J. Biol. Chem. 281, 14370–14375.
van Cleef, K.W., van Mierlo, J.T., Miesen, P., Overheul, G.J., Fros, J.J., Schuster, S., Marklewitz, M., Pijlman, G.P., Junglen, S., van Rij, R.P., 2014. Mosquito and Drosophila entomobirnaviruses suppress dsRNA- and siRNA-induced RNAi. Nucleic Acids Res. 42, 8732–8744.
van Mierlo, J.T., Bronkhorst, A.W., Overheul, G.J., Sadanandan, S.A., Ekstrom, J.O., Heestermans, M., Hultmark, D., Antoniewski, C., van Rij, R.P., 2012. Convergent evolution of argonaute-2 slicer antagonism in two distinct insect RNA viruses. PLoS Pathog. 8, e1002872.
van Mierlo, J.T., Overheul, G.J., Obadia, B., van Cleef, K.W., Webster, C.L., Saleh, M.C., Obbard, D.J., van Rij, R.P., 2014. Novel Drosophila viruses encode host-specific suppressors of RNAi. PLoS Pathog. 10, e1004256.
van Rij, R.P., Saleh, M.C., Berry, B., Foo, C., Houk, A., Antoniewski, C., Andino, R., 2006. The RNA silencing endonuclease Argonaute 2 mediates specific antiviral immunity in Drosophila melanogaster. Genes Dev. 20, 2985–2995.

Varjak, M., Donald, C.L., Mottram, T.J., Sreenu, V.B., Merits, A., Maringer, K., Schnettler, E., Kohl, A., 2017. Characterization of the Zika virus induced small RNA response in Aedes aegypti cells. PLoS Negl. Trop. Dis. 11, e0006010.

Verbruggen, B., Bickley, L.K., van Aerle, R., Bateman, K.S., Stentiford, G.D., Santos, E.M., Tyler, C.R., 2016. Molecular mechanisms of white spot syndrome virus infection and perspectives on treatments. Viruses 8 (1), pii: E23.

Visetnan, S., Supungul, P., Hirono, I., Tassanakajon, A., Rimphanitchayakit, V., 2015a. Activation of PmRelish from Penaeus monodon by yellow head virus. Fish Shellfish Immunol. 42, 335–344.

Visetnan, S., Supungul, P., Tang, S., Hirono, I., Tassanakajon, A., Rimphanitchayakit, V., 2015b. YHV-responsive gene expression under the influence of PmRelish regulation. Fish Shellfish Immunol. 47, 572–581.

Waldron, F.M., Stone, G.N., Obbard, D.J., 2018. Metagenomic sequencing suggests a diversity of RNA interference-like responses to viruses across multicellular eukaryotes. PLoS Genet. 14, e1007533.

Wang, Z., Zhu, F., 2017. MicroRNA-100 is involved in shrimp immune response to white spot syndrome virus (WSSV) and Vibrio alginolyticus infection. Sci. Rep. 7, 42334.

Wang, Z., Zhu, F., 2018. Different roles of a novel shrimp microRNA in white spot syndrome virus (WSSV) and Vibrio alginolyticus infection. Dev. Comp. Immunol. 79, 21–30.

Wang, X.H., Aliyari, R., Li, W.X., Li, H.W., Kim, K., Carthew, R., Atkinson, P., Ding, S.W., 2006. RNA interference directs innate immunity against viruses in adult Drosophila. Science 312, 452–454.

Wang, P.H., Gu, Z.H., Wan, D.H., Zhang, M.Y., Weng, S.P., Yu, X.Q., He, J.G., 2011. The shrimp NF-kappaB pathway is activated by white spot syndrome virus (WSSV) 449 to facilitate the expression of WSSV069 (ie1), WSSV303 and WSSV371. PLoS One 6, e24773.

Wang, P.H., Liang, J.P., Gu, Z.H., Wan, D.H., Weng, S.P., Yu, X.Q., He, J.G., 2012. Molecular cloning, characterization and expression analysis of two novel tolls (LvToll2 and LvToll3) and three putative Spatzle-like Toll ligands (LvSpz1-3) from Litopenaeus vannamei. Dev. Comp. Immunol. 36, 359–371.

Wang, P.H., Gu, Z.H., Wan, D.H., Liu, B.D., Huang, X.D., Weng, S.P., Yu, X.Q., He, J.G., 2013a. The shrimp IKK-NF-kappaB signaling pathway regulates antimicrobial peptide expression and may be subverted by white spot syndrome virus to facilitate viral gene expression. Cell. Mol. Immunol. 10, 423–436.

Wang, P.H., Gu, Z.H., Wan, D.H., Zhu, W.B., Qiu, W., Chen, Y.G., Weng, S.P., Yu, X.Q., He, J.G., 2013b. Litopenaeus vannamei Toll-interacting protein (LvTollip) is a potential negative regulator of the shrimp Toll pathway involved in the regulation of the shrimp antimicrobial peptide gene penaeidin-4 (PEN4). Dev. Comp. Immunol. 40, 266–277.

Wang, P.H., Gu, Z.H., Wan, D.H., Zhu, W.B., Qiu, W., Weng, S.P., Yu, X.Q., He, J.G., 2013c. Litopenaeus vannamei sterile-alpha and armadillo motif containing protein (LvSARM) is involved in regulation of Penaeidins and antilipopolysaccharide factors. PLoS One 8, e52088.

Wang, P.H., Wan, D.H., Gu, Z.H., Qiu, W., Chen, Y.G., Weng, S.P., Yu, X.Q., He, J.G., 2013d. Analysis of expression, cellular localization, and function of three inhibitors of apoptosis (IAPs) from Litopenaeus vannamei during WSSV infection and in regulation of antimicrobial peptide genes (AMPs). PLoS One 8, e72592.

Wang, P.H., Yang, L.S., Gu, Z.H., Weng, S.P., Yu, X.Q., He, J.G., 2013e. Nucleic acid-induced antiviral immunity in shrimp. Antiviral Res. 99, 270–280.

Wang, P.H., Huang, T., Zhang, X., He, J.G., 2014. Antiviral defense in shrimp: from innate immunity to viral infection. Antiviral Res. 108, 129–141.

Wang, M., Wang, L., Guo, Y., Sun, R., Yue, F., Yi, Q., Song, L., 2015a. The broad pattern recognition spectrum of the Toll-like receptor in mollusk Zhikong scallop Chlamys farreri. Dev. Comp. Immunol. 52, 192–201.

Wang, P.H., Weng, S.P., He, J.G., 2015b. Nucleic acid-induced antiviral immunity in invertebrates: an evolutionary perspective. Dev. Comp. Immunol. 48, 291–296.

Wang, Z., Wu, D., Liu, Y., Xia, X., Gong, W., Qiu, Y., Yang, J., Zheng, Y., Li, J., Wang, Y.-F., et al., 2015c. *Drosophila* Dicer-2 has an RNA interference–independent function that modulates Toll immune signaling. Sci. Adv. 1.

Wang, S., Song, X., Zhang, Z., Li, H., Lu, K., Yin, B., He, J., Li, C., 2016. Shrimp with knockdown of LvSOCS2, a negative feedback loop regulator of JAK/STAT pathway in Litopenaeus vannamei, exhibit enhanced resistance against WSSV. Dev. Comp. Immunol. 65, 289–298.

Wang, M., Wang, L., Jia, Z., Yi, Q., Song, L., 2018a. The various components implied the diversified Toll-like receptor (TLR) signaling pathway in mollusk Chlamys farreri. Fish Shellfish Immunol. 74, 205–212.

Wang, P.-H., Fung, S.-Y., Gao, W.-W., Deng, J.-J., Cheng, Y., Chaudhary, V., Yuen, K.-S., Ho, T.-H., Chan, C.-P., Zhang, Y., et al., 2018b. A novel transcript isoform of STING that sequesters cGAMP and dominantly inhibits innate nucleic acid sensing. Nucleic Acids Res. 46, 4054–4071.

Wang, P.H., Ye, Z.W., Deng, J.J., Siu, K.L., Gao, W.W., Chaudhary, V., Cheng, Y., Fung, S.Y., Yuen, K.S., Ho, T.H., et al., 2018c. Inhibition of AIM2 inflammasome activation by a novel transcript isoform of IFI16. EMBO Rep. 19 (10), pii: e45737.

Wang, W., Pan, C., Huang, Z., Yuan, H., Chen, J., 2018d. WSV181 inhibits JAK/STAT signaling and promotes viral replication in Drosophila. Dev. Comp. Immunol. 92, 20–28.

Wang, Y., Jin, B., Liu, P., Li, J., Chen, X., Gu, J., 2018e. piRNA profiling of dengue virus type 2-infected Asian Tiger mosquito and midgut tissues. Viruses 10 (4), pii: E213.

Whisnant, A.W., Bogerd, H.P., Flores, O., Ho, P., Powers, J.G., Sharova, N., Stevenson, M., Chen, C.H., Cullen, B.R., 2013. In-depth analysis of the interaction of HIV-1 with cellular microRNA biogenesis and effector mechanisms. mBio 4, e000193.

Wilkins, C., Dishongh, R., Moore, S.C., Whitt, M.A., Chow, M., Machaca, K., 2005. RNA interference is an antiviral defence mechanism in Caenorhabditis elegans. Nature 436, 1044–1047.

Wu, J., Chen, Z.J., 2014. Innate immune sensing and signaling of cytosolic nucleic acids. Annu. Rev. Immunol. 32, 461–488.

Wu, Y., Lü, L., Yang, L.-S., Weng, S.-P., Chan, S.-M., He, J.-G., 2007. Inhibition of white spot syndrome virus in Litopenaeus vannamei shrimp by sequence-specific siRNA. Aquaculture 271, 21–30.

Wu, Q., Luo, Y., Lu, R., Lau, N., Lai, E.C., Li, W.X., Ding, S.W., 2010. Virus discovery by deep sequencing and assembly of virus-derived small silencing RNAs. Proc. Natl. Acad. Sci. U. S. A. 107, 1606–1611.

Wu, Y.L., Wu, C.P., Liu, C.Y., Hsu, P.W., Wu, E.C., Chao, Y.C., 2011. A non-coding RNA of insect HzNV-1 virus establishes latent viral infection through microRNA. Sci. Rep. 1, 60.

Wu, J., Sun, L., Chen, X., Du, F., Shi, H., Chen, C., Chen, Z.J., 2013. Cyclic GMP-AMP is an endogenous second messenger in innate immune signaling by cytosolic DNA. Science 339, 826–830.

Wu, X., Wu, F.H., Wang, X., Wang, L., Siedow, J.N., Zhang, W., Pei, Z.M., 2014. Molecular evolutionary and structural analysis of the cytosolic DNA sensor cGAS and STING. Nucleic Acids Res. 42, 8243–8257.

Wu, P., Qin, G., Qian, H., Chen, T., Guo, X., 2016. Roles of miR-278-3p in IBP2 regulation and Bombyx mori cytoplasmic polyhedrosis virus replication. Gene 575, 264–269.

Xu, J., Han, F., Zhang, X., 2007. Silencing shrimp white spot syndrome virus (WSSV) genes by siRNA. Antiviral Res. 73, 126–131.
Xu, J., Wang, X.F., Chen, P., Liu, F.T., Zheng, S.C., Ye, H., Mo, M.H., 2016a. RNA interference in moths: mechanisms, applications, and progress. Genes (Basel) 7 (10), pii: E88.
Xu, J., Zhao, D., Sun, M., Wang, L., Jia, Z., Liu, R., Wang, L., Song, L., 2016b. A-type CpG ODN with higher binding affinity to LvToll1 could probably activate downstream IFN system-like antiviral response in shrimp Litopenaeus vannamei. Invertebr. Surviv. J. 13, 397–410.
Xu, X., Yuan, J., Yang, L., Weng, S., He, J., Zuo, H., 2016c. The dorsal/miR-1959/Cactus feedback loop facilitates the infection of WSSV in Litopenaeus vannamei. Fish Shellfish Immunol. 56, 397–401.
Yan, H., Zhou, Y., Liu, Y., Deng, Y., Chen, X., 2014. miR-252 of the Asian tiger mosquito Aedes albopictus regulates dengue virus replication by suppressing the expression of the dengue virus envelope protein. J. Med. Virol. 86, 1428–1436.
Yan, M., Li, C., Su, Z., Liang, Q., Li, H., Liang, S., Weng, S., He, J., Xu, X., 2015. Identification of a JAK/STAT pathway receptor domeless from Pacific white shrimp Litopenaeus vannamei. Fish Shellfish Immunol. 44, 26–32.
Yang, G., Yang, L., Zhao, Z., Wang, J., Zhang, X., 2012. Signature miRNAs involved in the innate immunity of invertebrates. PLoS One 7, e39015.
Yang, L., Yang, G., Zhang, X., 2014. The miR-100-mediated pathway regulates apoptosis against virus infection in shrimp. Fish Shellfish Immunol. 40, 146–153.
Zakovic, S., Levashina, E.A., 2018. Insects go on a STING operation to tackle intracellular invaders. Immunity 49, 195–197.
Zambon, R.A., Vakharia, V.N., Wu, L.P., 2006. RNAi is an antiviral immune response against a dsRNA virus in Drosophila melanogaster. Cell. Microbiol. 8, 880–889.
Zhang, Y., He, X., Yu, F., Xiang, Z., Li, J., Thorpe, K.L., Yu, Z., 2013. Characteristic and functional analysis of toll-like receptors (TLRs) in the lophotrocozoan, Crassostrea gigas, reveals ancient origin of TLR-mediated innate immunity. PLoS One 8, e76464.
Zhang, L., Li, L., Guo, X., Litman, G.W., Dishaw, L.J., Zhang, G., 2015a. Massive expansion and functional divergence of innate immune genes in a protostome. Sci. Rep. 5, 8693.
Zhang, R., Liu, R., Wang, W., Xin, L., Wang, L., Li, C., Song, L., 2015b. Identification and functional analysis of a novel IFN-like protein (CgIFNLP) in Crassostrea gigas. Fish Shellfish Immunol. 44, 547–554.
Zhang, R., Liu, R., Xin, L., Chen, H., Li, C., Wang, L., Song, L., 2016. A CgIFNLP receptor from Crassostrea gigas and its activation of the related genes in human JAK/STAT signaling pathway. Dev. Comp. Immunol. 65, 98–106.
Zhou, Z., Wang, L., Song, L., Liu, R., Zhang, H., Huang, M., Chen, H., 2014. The identification and characteristics of immune-related microRNAs in haemocytes of oyster Crassostrea gigas. PLoS One 9, e88397.
Zhu, M., Wang, J., Deng, R., Wang, X., 2016. Functional regulation of an Autographa californica nucleopolyhedrovirus-encoded microRNA, AcMNPV-miR-1, in baculovirus replication. J. Virol. 90, 6526–6537.
Zografidis, A., Van Nieuwerburgh, F., Kolliopoulou, A., Apostolou-Karampelis, K., Head, S.R., Deforce, D., Smagghe, G., Swevers, L., 2015. Viral small-RNA analysis of Bombyx mori larval midgut during persistent and pathogenic cytoplasmic Polyhedrosis virus infection. J. Virol. 89, 11473–11486.
Zuo, H., Yuan, J., Chen, Y., Li, S., Su, Z., Wei, E., Li, C., Weng, S., Xu, X., He, J., 2016. A microRNA-mediated positive feedback regulatory loop of the NF-kappaB pathway in Litopenaeus vannamei. J. Immunol. 196, 3842–3853.

FSC
www.fsc.org
MIX
Papier | Fördert
gute Waldnutzung
FSC® C083411

Kolibri 360 GmbH
Ferdinand-Jühlke-Straße 7
99095 Erfurt, Deutschland
produktsicherheit@kolibri360.de